Teacher Edition

Eureka Math Precalculus Module 1

Special thanks go to the Gordan A. Cain Center and to the Department of Mathematics at Louisiana State University for their support in the development of *Eureka Math*.

Published by the non-profit Great Minds.

Printed in the U.S.A.
This book may be purchased from the publisher at eureka-math.org
10 9 8 7 6 5 4 3 2 1

ISBN 978-1-63255-644-8

Table of Contents[1]

Complex Numbers and Transformations

[1]Each lesson is ONE day, and ONE day is considered a 45-minute period.

Precalculus and Advanced Topics • Module 1

Complex Numbers and Transformations

OVERVIEW

Module 1 sets the stage for expanding students' understanding of transformations by first exploring the notion of linearity in an algebraic context ("Which familiar algebraic functions are linear?"). This quickly leads to a return to the study of complex numbers and a study of linear transformations in the complex plane. Thus, Module 1 builds on standards **N-CN.A.1** and **N-CN.A.2** introduced in the Algebra II course and standards **G-CO.A.2**, **G-CO.A.4**, and **G-CO.A.5** introduced in the Geometry course.

Topic A opens with a study of common misconceptions by asking questions such as "For which numbers a and b does $(a+b)^2 = a^2 + b^2$ happen to hold?"; "Are there numbers a and b for which $\frac{1}{a+b} = \frac{1}{a} + \frac{1}{b}$?"; and so on. This second equation has only complex solutions, which launches a study of quotients of complex numbers and the use of conjugates to find moduli and quotients (**N-CN.A.3**). The topic ends by classifying real and complex functions that satisfy linearity conditions. (A function L is linear if, and only if, there is a real or complex value w such that $L(z) = wz$ for all real or complex z.) Complex number multiplication is emphasized in the last lesson.

In Topic B, students develop an understanding that when complex numbers are considered points in the Cartesian plane, complex number multiplication has the geometric effect of a rotation followed by a dilation in the complex plane. This is a concept that has been developed since Algebra II and builds upon standards **N-CN.A.1** and **N-CN.A.2**, which, when introduced, were accompanied with the observation that multiplication by i has the geometric effect of rotating a given complex number $90°$ about the origin in a counterclockwise direction. The algebraic inverse of a complex number (its reciprocal) provides the inverse geometric operation. Analysis of the angle of rotation and the scale of the dilation brings a return to topics in trigonometry first introduced in Geometry (**G-SRT.C.6**, **G-SRT.C.7**, **G-SRT.C.8**) and expanded on in Algebra II (**F-TF.A.1**, **F-TF.A.2**, **F-TF.C.8**). It also reinforces the geometric interpretation of the modulus of a complex number and introduces the notion of the argument of a complex number.

The theme of Topic C is to highlight the effectiveness of changing notations and the power provided by certain notations such as matrices. By exploiting the connection to trigonometry, students see how much complex arithmetic is simplified. By regarding complex numbers as points in the Cartesian plane, students can begin to write analytic formulas for translations, rotations, and dilations in the plane and revisit the ideas of high school Geometry (**G-CO.A.2**, **G-CO.A.4**, **G-CO.A.5**) in this light. Taking this work one step further, students develop the 2×2 matrix notation for planar transformations represented by complex number arithmetic. This work sheds light on how geometry software and video games efficiently perform rigid motion calculations. Finally, the flexibility implied by 2×2 matrix notation allows students to study additional matrix transformations (shears, for example) that do not necessarily arise from our original complex number thinking context.

In Topic C, the study of vectors and matrices is introduced through a coherent connection to transformations and complex numbers. Students learn to see matrices as representing transformations in the plane and develop understanding of multiplication of a matrix by a vector as a transformation acting on a point in the plane (**N-VM.C.11**, **N-VM.C.12**). While more formal study of multiplication of matrices occurs in Module 2, in Topic C, students are exposed to initial ideas of multiplying 2×2 matrices including a geometric interpretation of matrix invertibility and the meaning of the zero and identity matrices (**N-VM.C.8**, **N-VM.C.10**). **N-VM.C.8** is introduced in a strictly geometric context and is expanded upon more formally in Module 2. **N-VM.C.8** is assessed secondarily, in the context of other standards but not directly, in the Mid- and End-of-Module Assessments until Module 2.

The Mid-Module Assessment follows Topic B. The End-of-Module Assessment follows Topic C.

Focus Standards

Perform arithmetic operations with complex numbers.

N-CN.A.3 (+) Find the conjugate of a complex number; use conjugates to find moduli and quotients of complex numbers.

Represent complex numbers and their operations on the complex plane.

N-CN.B.4 (+) Represent complex numbers on the complex plane in rectangular and polar form (including real and imaginary numbers), and explain why the rectangular and polar forms of a given complex number represent the same number.

N-CN.B.5 (+) Represent addition, subtraction, multiplication, and conjugation of complex numbers geometrically on the complex plane; use properties of this representation for computation. *For example,* $\left(-1+\sqrt{3}i\right)^3 = 8$ *because* $\left(-1+\sqrt{3}i\right)$ *has modulus* 2 *and argument* $120°$.

N-CN.B.6 (+) Calculate the distance between numbers in the complex plane as the modulus of the difference, and the midpoint of a segment as the average of the numbers at its endpoints.

Perform operations on matrices and use matrices in applications.

N-VM.C.8 (+) Add, subtract, and multiply matrices of appropriate dimensions.

N-VM.C.10[2] (+) Understand that the zero and identity matrices play a role in matrix addition and multiplication similar to the role of 0 and 1 in the real numbers. The determinant of a square matrix is nonzero if and only if the matrix has a multiplicative inverse.

N-VM.C.11 (+) Multiply a vector (regarded as a matrix with one column) by a matrix of suitable dimensions to produce another vector. Work with matrices as transformations of vectors.

N-VM.C.12 (+) Work with 2×2 matrices as transformations of the plane, and interpret the absolute value of the determinant in terms of area.

---

[2]N.VM and G.CO standards are included in the context of defining transformations of the plane rigorously using complex numbers and 2×2 matrices and linking rotations and reflections to multiplication by complex number and/or by 2×2 matrices to show how geometry software and video games work.

Foundational Standards

Reason quantitatively and use units to solve problems.

N-Q.A.2 Define appropriate quantities for the purpose of descriptive modeling.★

Perform arithmetic operations with complex numbers.

N-CN.A.1 Know there is a complex number i such that $i^2 = -1$, and every complex number has the form $a + bi$ with a and b real.

N-CN.A.2 Use the relation $i^2 = -1$ and the commutative, associative, and distributive properties to add, subtract, and multiply complex numbers.

Use complex numbers in polynomial identities and equations.

N-CN.C.7 Solve quadratic equations with real coefficients that have complex solutions.

N-CN.C.8 (+) Extend polynomial identities to the complex numbers. *For example, rewrite $x^2 + 4$ as $(x + 2i)(x - 2i)$.*

Interpret the structure of expressions.

A-SSE.A.1 Interpret expressions that represent a quantity in terms of its context.★

a. Interpret parts of an expression, such as terms, factors, and coefficients.

b. Interpret complicated expressions by viewing one or more of their parts as a single entity. *For example, interpret $P(1 + r)^n$ as the product of P and a factor not depending on P.*

Write expressions in equivalent forms to solve problems.

A-SSE.B.3 Choose and produce an equivalent form of an expression to reveal and explain properties of the quantity represented by the expression.★

a. Factor a quadratic expression to reveal the zeros of the function it defines.

b. Complete the square in a quadratic expression to reveal the maximum or minimum value of the function it defines.

c. Use the properties of exponents to transform expressions for exponential functions. *For example the expression 1.15^t can be rewritten as $\left(1.15^{1/12}\right)^{12t} \approx 1.012^{12t}$ to reveal the approximate equivalent monthly interest rate if the annual rate is 15%.*

Create equations that describe numbers or relationships.★

A-CED.A.1 Create equations and inequalities in one variable and use them to solve problems. *Include equations arising from linear and quadratic functions, and simple rational and exponential functions.*

A-CED.A.2 Create equations in two or more variables to represent relationships between quantities; graph equations on coordinate axes with labels and scales.

A-CED.A.3 Represent constraints by equations or inequalities, and by systems of equations and/or inequalities, and interpret solutions as viable or non-viable options in a modeling context. *For example, represent inequalities describing nutritional and cost constraints on combinations of different foods.*

A-CED.A.4 Rearrange formulas to highlight a quantity of interest, using the same reasoning as in solving equations. *For example, rearrange Ohm's law* $V = IR$ *to highlight resistance* R.

Understand solving equations as a process of reasoning and explain the reasoning.

A-REI.A.1 Explain each step in solving a simple equation as following from the equality of numbers asserted at the previous step, starting from the assumption that the original equation has a solution. Construct a viable argument to justify a solution method.

Solve equations and inequalities in one variable.

A-REI.B.3 Solve linear equations and inequalities in one variable, including equations with coefficients represented by letters.

Solve systems of equations.

A-REI.C.6 Solve systems of linear equations exactly and approximately (e.g., with graphs), focusing on pairs of linear equations in two variables.

Experiment with transformations in the plane.

G-CO.A.2 Represent transformations in the plane using, e.g., transparencies and geometry software; describe transformations as functions that take points in the plane as inputs and give other points as outputs. Compare transformations that preserve distance and angle to those that do not (e.g., translation versus horizontal stretch).

G-CO.A.4 Develop definitions of rotations, reflections, and translations in terms of angles, circles, perpendicular lines, parallel lines, and line segments.

G-CO.A.5 Given a geometric figure and a rotation, reflection, or translation, draw the transformed figure using, e.g., graph paper, tracing paper, or geometry software. Specify a sequence of transformations that will carry a given figure onto another.

Extend the domain of trigonometric functions using the unit circle.

F-TF.A.1 Understand radian measure of an angle as the length of the arc on the unit circle subtended by the angle.

F-TF.A.2 Explain how the unit circle in the coordinate plane enables the extension of trigonometric functions to all real numbers, interpreted as radian measures of angles traversed counterclockwise around the unit circle.

F-TF.A.3 (+) Use special triangles to determine geometrically the values of sine, cosine, tangent for $\pi/3$, $\pi/4$, and $\pi/6$, and use the unit circle to express the values of sine, cosine, and tangent for $\pi - x$, $\pi + x$, and $2\pi - x$ in terms of their values for x, where x is any real number.

Prove and apply trigonometric identities.

F-TF.C.8 Prove the Pythagorean identity $\sin^2(\theta) + \cos^2(\theta) = 1$ and use it to find $\sin(\theta)$, $\cos(\theta)$, or $\tan(\theta)$ given $\sin(\theta)$, $\cos(\theta)$, or $\tan(\theta)$ and the quadrant of the angle.

Focus Standards for Mathematical Practice

MP.2 **Reason abstractly and quantitatively.** Students come to recognize that multiplication by a complex number corresponds to the geometric action of a rotation and dilation from the origin in the complex plane. Students apply this knowledge to understand that multiplication by the reciprocal provides the inverse geometric operation to a rotation and dilation. Much of the module is dedicated to helping students quantify the rotations and dilations in increasingly abstract ways so they do not depend on the ability to visualize the transformation. That is, they reach a point where they do not need a specific geometric model in mind to think about a rotation or dilation. Instead, they can make generalizations about the rotation or dilation based on the problems they have previously solved.

MP.3 **Construct viable arguments and critique the reasoning of others.** Throughout the module, students study examples of work by algebra students. This work includes a number of common mistakes that algebra students make, but it is up to students to decide about the validity of the argument. Deciding on the validity of the argument focuses students on justification and argumentation as they work to decide when purported algebraic identities do or do not hold. In cases where they decide that the given student work is incorrect, students work to develop the correct general algebraic results and justify them by reflecting on what they perceived as incorrect about the original student solution.

MP.4 **Model with mathematics.** As students work through the module, they become attuned to the geometric effect that occurs in the context of complex multiplication. However, initially it is unclear to them why multiplication by complex numbers entails specific geometric effects. In the module, students create a model of computer animation in the plane. The focus of the mathematics in the computer animation is such that students come to see rotating and translating as dependent on matrix operations and the addition of 2×1 vectors. Thus, their understanding becomes more formal with the notion of complex numbers.

Terminology

New or Recently Introduced Terms

- **Argument** (The *argument* of the complex number z is the radian (or degree) measure of the counterclockwise rotation of the complex plane about the origin that maps the initial ray (i.e., the ray corresponding to the positive real axis) to the ray from the origin through the complex number z in the complex plane. The argument of z is denoted $\arg(z)$.)
- **Bound Vector** (A *bound vector* is a directed line segment (an *arrow*). For example, the directed line segment $\overrightarrow{AB}$ is a bound vector whose initial point (or *tail*) is A and terminal point (or *tip*) is B.

  Bound vectors are *bound* to a particular location in space. A bound vector $\overrightarrow{AB}$ has a magnitude given by the length of $\overline{AB}$ and direction given by the ray $\overrightarrow{AB}$. Many times, only the magnitude and direction of a bound vector matters, not its position in space. In that case, any translation of that bound vector is considered to represent the same free vector.)
- **Complex Number** (A *complex number* is a number that can be represented by a point in the complex plane. A complex number can be expressed in two forms:
  1. The *rectangular form* of a complex number z is $a + bi$ where z corresponds to the point (a, b) in the complex plane, and i is the imaginary unit. The number a is called the *real part* of $a + bi$, and the number b is called the *imaginary part* of $a + bi$. Note that both the real and imaginary parts of a complex number are themselves real numbers.
  2. For $z \neq 0$, the *polar form* of a complex number z is $r(\cos(\theta) + i\sin(\theta))$ where $r = |z|$ and $\theta = \arg(z)$, and i is the imaginary unit.)
- **Complex Plane** (The *complex plane* is a Cartesian plane equipped with addition and multiplication operators defined on ordered pairs by the following:
  - Addition: $(a, b) + (c, d) = (a + c, b + d)$.

    When expressed in rectangular form, if $z = a + bi$ and $w = c + di$, then $z + w = (a + c) + (b + d)i$.
  - Multiplication: $(a, b) \cdot (c, d) = (ac - bd, ad + bc)$.

    When expressed in rectangular form, if $z = a + bi$ and $w = c + di$, then $z \cdot w = (ac - bd) + (ad + bc)i$. The horizontal axis corresponding to points of the form $(x, 0)$ is called the *real axis,* and a vertical axis corresponding to points of the form $(0, y)$ is called the *imaginary axis*.)
- **Conjugate** (The *conjugate* of a complex number of the form $a + bi$ is $a - bi$. The conjugate of z is denoted $\overline{z}$.)
- **Determinant of 2×2 Matrix** (The *determinant of the* 2×2 *matrix* $\begin{bmatrix} a & b \\ c & d \end{bmatrix}$ is the number computed by evaluating $ad - bc$ and is denoted by $\det\left(\begin{bmatrix} a & b \\ c & d \end{bmatrix}\right)$.)

- **Determinant of 3×3 Matrix** (The *determinant of the* 3×3 *matrix* $\begin{bmatrix} a_{11} & a_{12} & a_{13} \\ a_{21} & a_{22} & a_{23} \\ a_{31} & a_{32} & a_{33} \end{bmatrix}$ is the number computed by evaluating the expression,

$$a_{11}\det\left(\begin{bmatrix} a_{22} & a_{23} \\ a_{32} & a_{33} \end{bmatrix}\right) - a_{12}\det\left(\begin{bmatrix} a_{21} & a_{23} \\ a_{31} & a_{33} \end{bmatrix}\right) + a_{13}\det\left(\begin{bmatrix} a_{21} & a_{22} \\ a_{31} & a_{32} \end{bmatrix}\right),$$

  and is denoted by $\det\left(\begin{bmatrix} a_{11} & a_{12} & a_{13} \\ a_{21} & a_{22} & a_{23} \\ a_{31} & a_{32} & a_{33} \end{bmatrix}\right)$.)
- **Directed Graph** (A *directed graph* is an ordered pair $D(V, E)$ with
  - V a set whose elements are called *vertices* or *nodes*, and
  - E a set of ordered pairs of vertices, called *arcs* or *directed edges*.)
- **Directed Segment** (A *directed segment* $\overrightarrow{AB}$ is the line segment AB together with a direction given by connecting an initial point A to a terminal point B.)
- **Free Vector** (A *free vector* is the equivalence class of all directed line segments (*arrows*) that are equivalent to each other by translation. For example, scientists often use free vectors to describe physical quantities that have magnitude and direction only, *freely* placing an arrow with the given magnitude and direction anywhere in a diagram where it is needed. For any directed line segment in the equivalence class defining a free vector, the directed line segment is said to be a *representation* of the free vector or is said to *represent* the free vector.)
- **Identity Matrix** (The $n \times n$ *identity matrix* is the matrix whose entry in row i and column i for $1 \le i \le n$ is 1 and whose entries in row i and column j for $1 \le i, j \le n$, and $i \ne j$ are all zero. The identity matrix is denoted by I.)
- **Imaginary Axis** (See *complex plane*.)
- **Imaginary Number** (An *imaginary number* is a complex number that can be expressed in the form bi where b is a real number.)
- **Imaginary Part** (See *complex number*.)
- **Imaginary Unit** (The *imaginary unit*, denoted by i, is the number corresponding to the point $(0,1)$ in the complex plane.)
- **Incidence Matrix** (The *incidence matrix of a network diagram* is the $n \times n$ matrix such that the entry in row i and column j is the number of edges that start at node i and end at node j.)
- **Inverse Matrix** (An $n \times n$ matrix A is *invertible* if there exists an $n \times n$ matrix B so that $AB = BA = I$, where I is the $n \times n$ identity matrix. The matrix B, when it exists, is unique and is called the *inverse of* A and is denoted by A^{-1}.)
- **Linear Function** (A function $f: \mathbb{R} \to \mathbb{R}$ is called a *linear function* if it is a polynomial function of degree one, that is, a function with real number domain and range that can be put into the form $f(x) = mx + b$ for real numbers m and b. A linear function of the form $f(x) = mx + b$ is a linear transformation only if $b = 0$.)

- **Linear Transformation** (A function $L: \mathbb{R}^n \to \mathbb{R}^n$ for a positive integer n is a *linear transformation* if the following two properties hold:
  - $L(\mathbf{x} + \mathbf{y}) = L(\mathbf{x}) + L(\mathbf{y})$ for all $\mathbf{x}, \mathbf{y} \in \mathbb{R}^n$, and
  - $L(k\mathbf{x}) = k \cdot L(\mathbf{x})$ for all $\mathbf{x} \in \mathbb{R}^n$ and $k \in \mathbb{R}$,

  where $\mathbf{x} \in \mathbb{R}^n$ means that $\mathbf{x}$ is a point in $\mathbb{R}^n$.)
- **Linear Transformation Induced by Matrix A** (Given a 2×2 matrix A, the *linear transformation induced by matrix A* is the linear transformation L given by the formula $L\left(\begin{bmatrix} x \\ y \end{bmatrix}\right) = A \cdot \begin{bmatrix} x \\ y \end{bmatrix}$. Given a 3×3 matrix A, the *linear transformation induced by matrix A* is the linear transformation L given by the formula $L\left(\begin{bmatrix} x \\ y \\ z \end{bmatrix}\right) = A \cdot \begin{bmatrix} x \\ y \\ z \end{bmatrix}$.)
- **Matrix** (An $m \times n$ *matrix* is an ordered list of nm real numbers, $a_{11}, a_{12}, \ldots, a_{1n}, a_{21}, a_{22}, \ldots, a_{2n}, \ldots, a_{m1}, a_{m2}, \ldots, a_{mn}$, organized in a rectangular array of m rows and n columns:
  $\begin{bmatrix} a_{11} & a_{12} & \cdots & a_{1n} \\ a_{21} & a_{22} & \cdots & a_{2n} \\ \vdots & \vdots & \ddots & \vdots \\ a_{m1} & a_{m2} & \cdots & a_{mn} \end{bmatrix}$. The number a_{ij} is called the *entry in row i and column j.)*
- **Matrix Difference** (Let A be an $m \times n$ matrix whose entry in row i and column j is a_{ij}, and let B be an $m \times n$ matrix whose entry in row i and column j is b_{ij}. Then, the *matrix difference $A - B$* is the $m \times n$ matrix whose entry in row i and column j is $a_{ij} - b_{ij}$.)
- **Matrix Product** (Let A be an $m \times n$ matrix whose entry in row i and column j is a_{ij}, and let B be an $n \times p$ matrix whose entry in row i and column j is b_{ij}. Then, the *matrix product AB* is the $m \times p$ matrix whose entry in row i and column j is $a_{i1}b_{1j} + a_{i2}b_{2j} + \cdots + a_{in}b_{nj}$.)
- **Matrix Scalar Multiplication** (Let k be a real number, and let A be an $m \times n$ matrix whose entry in row i and column j is a_{ij}. Then, the *scalar product $k \cdot A$* is the $m \times n$ matrix whose entry in row i and column j is $k \cdot a_{ij}$.)
- **Matrix Sum** (Let A be an $m \times n$ matrix whose entry in row i and column j is a_{ij}, and let B be an $m \times n$ matrix whose entry in row i and column j is b_{ij}. Then, the *matrix sum $A + B$* is the $m \times n$ matrix whose entry in row i and column j is $a_{ij} + b_{ij}$.)
- **Modulus** (The *modulus* of a complex number z, denoted $|z|$, is the distance from the origin to the point corresponding to z in the complex plane. If $z = a + bi$, then $|z| = \sqrt{a^2 + b^2}$.)
- **Network Diagram** (A *network diagram* is a graphical representation of a directed graph where the n vertices are drawn as circles with each circle labeled by a number 1 through n and the directed edges are drawn as segments or arcs with the arrow pointing from the tail vertex to the head vertex.)

- **Opposite Vector** (For a vector $\vec{v}$ represented by the directed line segment $\overrightarrow{AB}$, the *opposite vector*, denoted $-\vec{v}$, is the vector represented by the directed line segment $\overrightarrow{BA}$. If $\vec{v} = \begin{bmatrix} v_1 \\ v_2 \\ \vdots \\ v_n \end{bmatrix}$ in $\mathbb{R}^n$, then $-\vec{v} = \begin{bmatrix} -v_1 \\ -v_2 \\ \vdots \\ -v_n \end{bmatrix}$.)
- **Polar Form of a Complex Number** (The *polar form of a complex number* z is $r(\cos(\theta) + i\sin(\theta))$ where $r = |z|$ and $\theta = \arg(z)$.)
- **Position Vector** (For a point $P(v_1, v_2, \dots, v_n)$ in $\mathbb{R}^n$, the *position vector* $\vec{v}$, denoted by $\begin{bmatrix} v_1 \\ v_2 \\ \vdots \\ v_n \end{bmatrix}$ or $\langle v_1, v_2, \dots, v_n \rangle$, is a free vector $\vec{v}$ that is represented by the directed line segment $\overrightarrow{OP}$ from the origin $O(0,0,0,\dots,0)$ to the point P. The real number v_i is called the ith *component* of the vector $\vec{v}$.)
- **Real Coordinate Space** (For a positive integer n, the n-dimensional *real coordinate space*, denoted $\mathbb{R}^n$, is the set of all n-tuple of real numbers equipped with a distance function d that satisfies
$$d[(x_1, x_2, \dots, x_n), (y_1, y_2, \dots, y_n)] = \sqrt{(y_1 - x_1)^2 + (y_2 - x_1)^2 + \cdots + (y_n - x_n)^2}$$
for any two points in the space. One-dimensional real coordinate space is called a *number line,* and the two-dimensional real coordinate space is called the *Cartesian plane.)*
- **Rectangular Form of a Complex Number** (The *rectangular form of a complex number* z is $a + bi$ where z corresponds to the point (a, b) in the complex plane and i is the imaginary unit. The number a is called the *real part* of $a + bi$, and the number b is called the *imaginary part* of $a + bi$.)
- **Translation by a Vector in Real Coordinate Space** (A *translation by a vector* $\vec{v}$ *in* $\mathbb{R}^n$ is the translation transformation $T_{\vec{v}}: \mathbb{R}^n \to \mathbb{R}^n$ given by the map that takes $\vec{x} \mapsto \vec{x} + \vec{v}$ for all $\vec{x}$ in $\mathbb{R}^n$. If $\vec{v} = \begin{bmatrix} v_1 \\ v_2 \\ \vdots \\ v_n \end{bmatrix}$ in $\mathbb{R}^n$, then $T_{\vec{v}}\left(\begin{bmatrix} x_1 \\ x_2 \\ \vdots \\ x_n \end{bmatrix}\right) = \begin{bmatrix} x_1 + v_1 \\ x_2 + v_2 \\ \vdots \\ x_n + v_n \end{bmatrix}$ for all $\vec{x}$ in $\mathbb{R}^n$.)

- **Vector Addition** (For vectors $\vec{v}$ and $\vec{w}$ in $\mathbb{R}^n$, the sum $\vec{v}+\vec{w}$ is the vector whose i^{th} component is the sum of the ith components of $\vec{v}$ and $\vec{w}$ for $1 \le i \le n$. If $\vec{v} = \begin{bmatrix} v_1 \\ v_2 \\ \vdots \\ v_n \end{bmatrix}$ and $\vec{w} = \begin{bmatrix} w_1 \\ w_2 \\ \vdots \\ w_n \end{bmatrix}$ in $\mathbb{R}^n$, then $\vec{v}+\vec{w} = \begin{bmatrix} v_1 + w_1 \\ v_2 + w_2 \\ \vdots \\ v_n + w_n \end{bmatrix}$.)
- **Vector Magnitude** (The *magnitude* or *length* of a vector $\vec{v}$, denoted $|\vec{v}|$ or $\|\vec{v}\|$, is the length of any directed line segment that represents the vector. If $\vec{v} = \begin{bmatrix} v_1 \\ v_2 \\ \vdots \\ v_n \end{bmatrix}$ in $\mathbb{R}^n$, then $|\vec{v}| = \sqrt{v_1^2 + v_2^2 + \cdots + v_n^2}$, which is the distance from the origin to the associated point $P(v_1, v_2, \ldots, v_n)$.)
- **Vector Representation of a Complex Number** (The *vector representation of a complex number* z is the position vector $\vec{z}$ associated to the point z in the complex plane. If $z = a + bi$ for two real numbers a and b, then $\vec{z} = \begin{bmatrix} a \\ b \end{bmatrix}$.)
- **Vector Scalar Multiplication** (For a vector $\vec{v}$ in $\mathbb{R}^n$ and a real number k, the scalar product $k \cdot \vec{v}$ is the vector whose i^{th} component is the product of k and the ith component of $\vec{v}$ for $1 \le i \le n$. If k is a real number and $\vec{v} = \begin{bmatrix} v_1 \\ v_2 \\ \vdots \\ v_n \end{bmatrix}$ in $\mathbb{R}^n$, then $k \cdot \vec{v} = \begin{bmatrix} kv_1 \\ kv_2 \\ \vdots \\ kv_n \end{bmatrix}$.)
- **Vector Subtraction** (For vectors $\vec{v}$ and $\vec{w}$, the difference $\vec{v} - \vec{w}$ is the sum of $\vec{v}$ and the opposite of $\vec{w}$; that is, $\vec{v} - \vec{w} = \vec{v} + (-\vec{w})$. If $\vec{v} = \begin{bmatrix} v_1 \\ v_2 \\ \vdots \\ v_n \end{bmatrix}$ and $\vec{w} = \begin{bmatrix} w_1 \\ w_2 \\ \vdots \\ w_n \end{bmatrix}$ in $\mathbb{R}^n$, then $\vec{v} - \vec{w} = \begin{bmatrix} v_1 - w_1 \\ v_2 - w_2 \\ \vdots \\ v_n - w_n \end{bmatrix}$.)
- **Zero Matrix** (The $m \times n$ *zero matrix* is the $m \times n$ matrix in which all entries are equal to zero. For example, the 2×2 zero matrix is $\begin{bmatrix} 0 & 0 \\ 0 & 0 \end{bmatrix}$, and the 3×3 zero matrix is $\begin{bmatrix} 0 & 0 & 0 \\ 0 & 0 & 0 \\ 0 & 0 & 0 \end{bmatrix}$.)
- **Zero Vector** (The *zero vector* in $\mathbb{R}^n$ is the vector in which each component is equal to zero. For example, the zero vector in $\mathbb{R}^2$ is $\begin{bmatrix} 0 \\ 0 \end{bmatrix}$, and the zero vector in $\mathbb{R}^3$ is $\begin{bmatrix} 0 \\ 0 \\ 0 \end{bmatrix}$.)

Familiar Terms and Symbols[3]

- Dilation
- Rectangular Form
- Rotation
- Translation

Suggested Tools and Representations

- Geometer's Sketchpad software
- Graphing calculator
- Wolfram Alpha software

Preparing to Teach a Module

Preparation of lessons will be more effective and efficient if there has been an adequate analysis of the module first. Each module in *A Story of Functions* can be compared to a chapter in a book. How is the module moving the plot, the mathematics, forward? What new learning is taking place? How are the topics and objectives building on one another? The following is a suggested process for preparing to teach a module.

Step 1: Get a preview of the plot.

A: Read the Table of Contents. At a high level, what is the plot of the module? How does the story develop across the topics?

B: Preview the module's Exit Tickets to see the trajectory of the module's mathematics and the nature of the work students are expected to be able to do.

[3]These are terms and symbols students have seen previously.

Note: When studying a PDF file, enter "Exit Ticket" into the search feature to navigate from one Exit Ticket to the next.

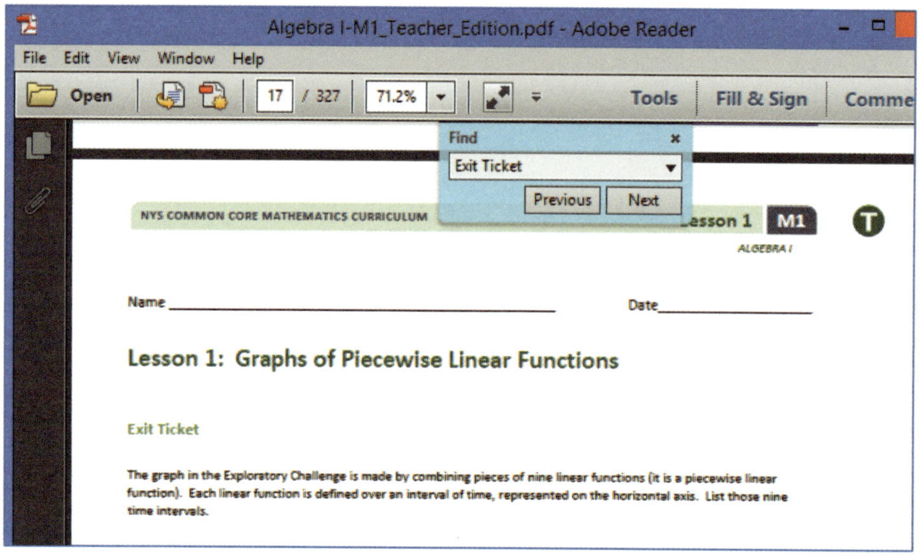

Step 2: Dig into the details.

A: Dig into a careful reading of the Module Overview. While reading the narrative, liberally reference the lessons and Topic Overviews to clarify the meaning of the text—the lessons demonstrate the strategies, show how to use the models, clarify vocabulary, and build understanding of concepts.

B: Having thoroughly investigated the Module Overview, read through the Student Outcomes of each lesson (in order) to further discern the plot of the module. How do the topics flow and tell a coherent story? How do the outcomes move students to new understandings?

Step 3: Summarize the story.

Complete the Mid- and End-of-Module Assessments. Use the strategies and models presented in the module to explain the thinking involved. Again, liberally reference the lessons to anticipate how students who are learning with the curriculum might respond.

Preparing to Teach a Lesson

A three-step process is suggested to prepare a lesson. It is understood that at times teachers may need to make adjustments (customizations) to lessons to fit the time constraints and unique needs of their students. The recommended planning process is outlined below. Note: The ladder of Step 2 is a metaphor for the teaching sequence. The sequence can be seen not only at the macro level in the role that this lesson plays in the overall story, but also at the lesson level, where each rung in the ladder represents the next step in understanding or the next skill needed to reach the objective. To reach the objective, or the top of the ladder, all students must be able to access the first rung and each successive rung.

Step 1: Discern the plot.

A: Briefly review the module's Table of Contents, recalling the overall story of the module and analyzing the role of this lesson in the module.

B: Read the Topic Overview related to the lesson, and then review the Student Outcome(s) and Exit Ticket of each lesson in the topic.

C: Review the assessment following the topic, keeping in mind that assessments can be found midway through the module and at the end of the module.

Step 2: Find the ladder.

A: Work through the lesson, answering and completing each question, example, exercise, and challenge.

B: Analyze and write notes on the new complexities or new concepts introduced with each question or problem posed; these notes on the sequence of new complexities and concepts are the rungs of the ladder.

C: Anticipate where students might struggle, and write a note about the potential cause of the struggle.

D: Answer the Closing questions, always anticipating how students will respond.

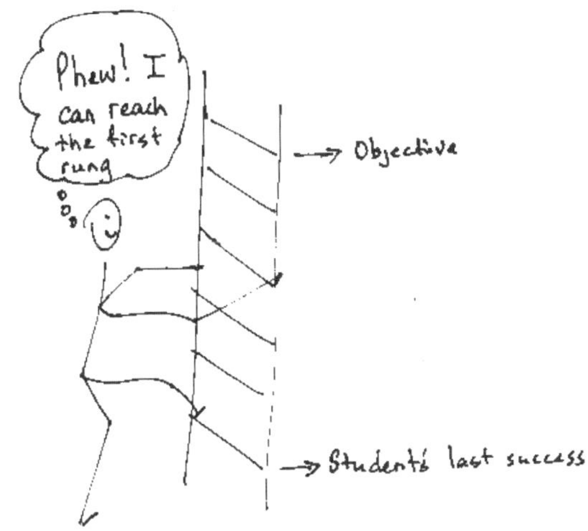

Step 3: Hone the lesson.

Lessons may need to be customized if the class period is not long enough to do all of what is presented and/or if students lack prerequisite skills and understanding to move through the entire lesson in the time allotted. A suggestion for customizing the lesson is to first decide upon and designate each question, example, exercise, or challenge as either "Must Do" or "Could Do."

A: Select "Must Do" dialogue, questions, and problems that meet the Student Outcome(s) while still providing a coherent experience for students; reference the ladder. The expectation should be that the majority of the class will be able to complete the "Must Do" portions of the lesson within the allocated time. While choosing the "Must Do" portions of the lesson, keep in mind the need for a balance of dialogue and conceptual questioning, application problems, and abstract problems, and a balance between students using pictorial/graphical representations and abstract representations. Highlight dialogue to be included in the delivery of instruction so that students have a chance to articulate and consolidate understanding as they move through the lesson.

B: "Must Do" portions might also include remedial work as necessary for the whole class, a small group, or individual students. Depending on the anticipated difficulties, the remedial work might take on different forms as suggested in the chart below.

Anticipated Difficulty	"Must Do" Remedial Problem Suggestion
The first problem of the lesson is too challenging.	Write a short sequence of problems on the board that provides a ladder to Problem 1. Direct students to complete those first problems to empower them to begin the lesson.
There is too big of a jump in complexity between two problems.	Provide a problem or set of problems that bridge student understanding from one problem to the next.
Students lack fluency or foundational skills necessary for the lesson.	Before beginning the lesson, do a quick, engaging fluency exercise[4]. Before beginning any fluency activity for the first time, assess that students have conceptual understanding of the problems in the set and that they are poised for success with the easiest problem in the set.
More work is needed at the concrete or pictorial level.	Provide manipulatives or the opportunity to draw solution strategies.
More work is needed at the abstract level.	Add a set of abstract problems to be completed toward the end of the lesson.

C: "Could Do" problems are for students who work with greater fluency and understanding and can, therefore, complete more work within a given time frame.

D: At times, a particularly complex problem might be designated as a "Challenge!" problem to provide to advanced students. Consider creating the opportunity for students to share their "Challenge!" solutions with the class at a weekly session or on video.

E: If the lesson is customized, be sure to carefully select Closing questions that reflect such decisions, and adjust the Exit Ticket if necessary.

Assessment Summary

Assessment Type	Administered	Format	Standards Addressed
Mid-Module Assessment Task	After Topic B	Constructed response with rubric	N-CN.A.3, N-CN.B.4, N-CN.B.5, N-CN.B.6
End-of-Module Assessment Task	After Topic C	Constructed response with rubric	N-CN.B.4, N-CN.B.5, N-VM.C.8, N-VM.C.10, N-VM.C.11, N-VM.C.12

[4]Look for fluency suggestions at www.eureka-math.org.

Topic A

A Question of Linearity

N-CN.A.3, N-CN.B.4

Focus Standards:	N-CN.A.3	(+) Find the conjugate of a complex number; use conjugates to find moduli and quotients of complex numbers.
	N-CN.B.4	(+) Represent complex numbers on the complex plane in rectangular and polar form (including real and imaginary numbers), and explain why the rectangular and polar forms of a given complex number represent the same number.
Instructional Days:	8	
Lessons 1–2:	Wishful Thinking—Does Linearity Hold? (E, E)[1]	
Lesson 3:	Which Real Number Functions Define a Linear Transformation? (S)	
Lessons 4–5:	An Appearance of Complex Numbers (P, P)	
Lesson 6:	Complex Numbers as Vectors (P)	
Lessons 7–8:	Complex Number Division (P, P)	

Linear transformations are a unifying theme of Module 1 Topic A. In Lesson 1, students are introduced to the term *linear transformation* and its definition. A function is a linear transformation if it satisfies the conditions $f(x+y) = f(x) + f(y)$ and $kf(x) = f(kx)$. Students contrast this to their previous understanding of a linear transformation, which was likely a function whose graph is a straight line. This idea of linearity is revisited as students study complex numbers and their transformations in Topic B and matrices in Topic C. Lesson 1 begins as students look at common mistakes made in algebra and asks questions such as "For which numbers a and b does $(a+b)^2 = a^2 + b^2$ happen to hold?" Students discover that these statements are usually false by substituting real number values for the variables and then exploring values that make the statements true. Lesson 2 continues this exploration asking, "Are there numbers a and b for which $\frac{1}{a+b} = \frac{1}{a} + \frac{1}{b}$?" and so on. This exercise has only complex solutions, which launches a study of complex numbers. Lesson 3 concludes this study of misconceptions by defining a linear function (a function whose graph is a line) and explaining the difference between a linear function and a linear transformation. The concept of a linear transformation is developed in the first three lessons and is revisited throughout the module. Linear transformations are important because they help students link complex numbers and their

[1]Lesson Structure Key: **P**-Problem Set Lesson, **M**-Modeling Cycle Lesson, **E**-Exploration Lesson, **S**-Socratic Lesson

transformations to matrices as this module progresses. Linear transformations are also essential in college mathematics as they are a foundational concept in linear algebra. Lessons 4 and 5 begin the study of complex numbers defining i geometrically by rotating the number line 90° and thus giving a "number" i with the property $i^2 = -1$. Students then add, subtract, and multiply complex numbers (**N-CN.A.2**). Lesson 6 explores complex numbers as vectors. Lessons 7 and 8 conclude Topic A with the study of quotients of complex numbers and the use of conjugates to find moduli and quotients (**N-CN.A.3**). Linearity is revisited when students classify real and complex functions that satisfy linearity conditions. (A function L is linear if and only if there is a real or complex value w such that $L(z) = wz$ for all real or complex z.) Complex number multiplication is again emphasized in Lesson 8. This topic focuses on MP.3 as students study common mistakes that algebra students make and determine the validity of the statements.

Lesson 1: Wishful Thinking—Does Linearity Hold?

Student Outcomes

- Students learn when ideal linearity properties do and do not hold for classes of functions studied in previous years.
- Students develop familiarity with linearity conditions.

Lesson Notes

This is a two-day lesson that introduces a new definition of a linear transformation and looks at common mistakes that students make when assuming that all linear functions meet the requirements for this new definition. A linear transformation is not equivalent to a linear function, which is a function whose graph is a line and can be written as $y = mx + b$. In this sequence of lessons, a linear transformation is defined as it is in linear algebra courses, which is that a function is linear if it satisfies two conditions: $f(x + y) = f(x) + f(y)$ and $f(kx) = kf(x)$. This definition leads to surprising results when students study the function $f(x) = 3x + 1$. Students apply this new definition of linear transformation to classes of functions learned in previous years and explore why the conditions for linearity sometimes produce false statements. Students then solve to find specific solutions when the conditions for linearity produce true statements, giving the appearance that a linear function is a linear transformation when it is not. In Lesson 1, students explore polynomials and radical equations. Lesson 2 extends this exploration to trigonometric, rational, and logarithmic functions. Lessons 1 and 2 focus on linearity for real-numbered inputs but lead to the discovery of complex solutions and launch the study of complex numbers. This study includes operations on complex numbers as well as the use of conjugates to find moduli and quotients.

Classwork

Exploratory Challenge (13 minutes)

In this Exploratory Challenge, students work individually while discussing the steps as a class. Students complete the exercises in pairs with the class coming together at the end to present their findings and to watch a video.

- Wouldn't it be great if functions were sensible and behaved the way we expected them to do?
- Let $f(x) = 2x$ and $g(x) = 3x + 1$.
- Write down three facts that you know about $f(x)$ and $g(x)$.
  - *Answers will vary. Both graphs are straight lines. $f(x)$ has a y-intercept of 0. $g(x)$ has a y-intercept of 1. The slope of $f(x)$ is 2. The slope of $g(x)$ is 3.*
- Which of these functions is linear?
  - *Students will probably say both because they are applying a prior definition of a linear function: $y = mx + b$.*
- Introduce the following definition: A function is a linear transformation if $f(x + y) = f(x) + f(y)$ and $f(kx) = kf(x)$.

MP.3

- Based on this definition, which function is a linear transformation? Explain how you know.
  - *$f(x) = 2x$ is a linear transformation because $2(x + y) = 2x + 2y$ and $2(kx) = k(2x)$.*
  - *$g(x) = 3x + 1$ is not a linear transformation because $3(x + y) + 1 \neq (3x + 1) + (3y + 1)$ and $3(kx) + 1 \neq k(3x + 1)$.*
- Is $h(x) = 2x - 3$ a linear transformation? Explain.
  - *$h(x) = 2x - 3$ is not a linear transformation because $2(x + y) - 3 \neq (2x - 3) + (2y - 3)$ and $2(kx) - 3 \neq k(2x - 3)$.*
- Is $p(x) = \frac{1}{2}x$ a linear transformation? Explain.
  - *$p(x) = \frac{1}{2}x$ is a linear transformation because $\frac{1}{2}(x + y) = \frac{1}{2}x + \frac{1}{2}y$ and $\frac{1}{2}(kx) = k\left(\frac{1}{2}x\right)$.*
- Let $g(x) = x^2$.
- Is $g(x)$ a linear transformation?
  - *No. $(x + y)^2 \neq x^2 + y^2$, and $(ax)^2 \neq a(x)^2$.*
- A common mistake made by many math students is saying $(a + b)^2 = a^2 + b^2$. How many of you have made this mistake before?

Scaffolding:

- Remind students that $(a + b)^2$ means $(a + b)(a + b)$.
- Have students complete the following chart for different values of a and b and the expressions.

a	b	$(a+b)^2$	a^2	b^2	a^2+b^2
1	2	9	1	4	5
2	3	25	4	9	13

MP.3

- Does $(a + b)^2 = a^2 + b^2$? Justify your claim.
- Substitute some values of a and b into this equation to show that this statement is not generally true.
  - *Answers will vary, but students could choose $a = 1$ and $b = 1$. In this case, $(1 + 1)^2 = 1^2 + 1^2$ leads to $4 = 2$, which we know is not true. There are many other choices.*
- Did anyone find values of a and b that made this statement true?
  - *Answers will vary but could include $a = 0$, $b = 0$ or $a = 1$, $b = 0$ or $a = 0$, $b = 1$.*
- We can find all values of a and b for which this statement is true by solving for one of the variables. I want half the class to solve this equation for a and the other half to solve for b.
  - *Expanding the left side and then combining like terms gives* $a^2 + 2ab + b^2 = a^2 + b^2 2ab = 0$. This leads to $a = 0$ if students are solving for a and $b = 0$ if students are solving for b.
- We have solutions for two different variables. Can you explain this to your neighbor?
  - *If $a = 0$ and/or $b = 0$, the statement $(a + b)^2 = a^2 + b^2$ is true.*
- Take a moment and discuss with your neighbor what we have just shown. What statement is true for all real values of a and b?
  - *$(a + b)^2 = a^2 + b^2$ is true for only certain values of a and b, namely, if either or both variables equal 0. The statement that is true for all real numbers is $(a + b)^2 = a^2 + 2ab + b^2$.*
- A function is a linear transformation when the following are true: $f(kx) = kf(x)$ and $f(x + y) = f(x) + f(y)$. We call this function a *linear transformation*.
- Repeat what I have just said to your neighbor.
  - *Students repeat.*

- Look at the functions $f(x) = 2x$ and $g(x) = x^2$ listed above. Which is a linear transformation? Explain.
  - *$f(x) = 2x$ is a linear transformation because $f(ax) = af(x)$ and $f(x + y) = f(x) + f(y)$.*
    $g(x) = x^2$ is not a linear transformation because $g(ax) \neq ag(x)$ and $g(x + y) \neq g(x) + g(y)$.

Linear transformations are introduced in Lessons 1 and 2. This leads to the discussion in Lesson 3 on when functions are linear transformations. In Lesson 3, students discover that a function whose graph is a line may or may not be a linear transformation.

Exercises (10 minutes)

In the exercises below, instruct students to work in pairs and to go through the same steps that they went through in the Exploratory Challenge. Call the class back together, and have groups present their results. All groups can be assigned both examples, or half the class can be assigned Exercise 1 and the other half Exercise 2. Exercise 2 is slightly more difficult than Exercise 1.

Scaffolding:

- For advanced learners, assign Exercises 1 and 2 with no leading question.
- Monitor group work, and target some groups with more specific questions to help them with the algebra needed. For example, remind them that

$$\left(\sqrt{a} + \sqrt{b}\right)^2 = \left(\sqrt{a} + \sqrt{b}\right)\left(\sqrt{a} + \sqrt{b}\right).$$

- Also, remind them how to multiply binomials.

MP.3 & MP.8

Exercises

Look at these common mistakes that students make, and answer the questions that follow.

1. **If $f(x) = \sqrt{x}$, does $f(a + b) = f(a) + f(b)$, when a and b are not negative?**

   a. **Can we find a counterexample to refute the claim that $f(a + b) = f(a) + f(b)$ for all nonnegative values of a and b?**

   Answers will vary, but students could choose $a = 1$ and $b = 1$. In this case, $\sqrt{1+1} = \sqrt{1} + \sqrt{1}$, or $\sqrt{2} = 2$, which we know is not true. There are many other choices.

   b. **Find some nonnegative values for a and b for which the statement, by coincidence, happens to be true.**

   Answers will vary but could include $a = 0$, $b = 0$ or $a = 4$, $b = 0$ or $a = 0$, $b = 16$.

   c. **Find all values of a and b for which the statement is true. Explain your work and the results.**

$$\sqrt{a+b} = \sqrt{a} + \sqrt{b}$$
$$(\sqrt{a+b})^2 = \left(\sqrt{a} + \sqrt{b}\right)^2$$
$$a + b = a + 2\sqrt{ab} + b$$
$$2\sqrt{ab} = 0$$
$$ab = 0,$$

   which leads to $a = 0$ if students are solving for a and $b = 0$ if students are solving for b.

   Anytime $a = 0$ and/or $b = 0$, then $\sqrt{a+b} = \sqrt{a} + \sqrt{b}$, and the equation is true.

   d. **Why was it necessary for us to consider only nonnegative values of a and b?**

   If either variable is negative, then we would be taking the square root of a negative number, which is not a real number, and we are only addressing real-numbered inputs and outputs here.

e. Does $f(x) = \sqrt{x}$ display ideal linear properties? Explain.

No, because $\sqrt{a+b} \neq \sqrt{a} + \sqrt{b}$ for all real values of the variables.

2. If $f(x) = x^3$, does $f(a+b) = f(a) + f(b)$?

a. Substitute in some values of a and b to show this statement is not true in general.

Answers will vary, but students could choose $a = 1$ and $b = 1$. In this case, $(1+1)^3 = 1^3 + 1^3$ or $8 = 2$, which we know is not true. There are many other choices.

b. Find some values for a and b for which the statement, by coincidence, happens to work.

Answers will vary but could include $a = 0, b = 0$ or $a = 2, b = 0$ or $a = 0, b = 3$.

MP.3 & MP.8

c. Find all values of a and b for which the statement is true. Explain your work and the results.

$$(a+b)^3 = a^3 + b^3$$
$$a^3 + 3a^2b + 3ab^2 + b^3 = a^3 + b^3$$
$$3a^2b + 3ab^2 = 0$$
$$3ab(a+b) = 0,$$

which leads to $a = 0$, $b = 0$, and $a = -b$.

Anytime $a = 0$ and/or $b = 0$ or $a = -b$, then $(a+b)^3 = a^3 + b^3$, and the equation is true.

d. Is this true for all positive and negative values of a and b? Explain and prove by choosing positive and negative values for the variables.

Yes, since $= -b$, if a is positive, the equation would be true if b was negative. Likewise, if a is negative, the equation would be true if b was positive. Answers will vary. If $a = 2$ and $b = -2$, $(2 + (-2))^3 = (2)^3 + (-2)^3$, meaning $0^3 = 8 + (-8)$ or $0 = 0$. If $a = -2$ and $b = 2$, $((-2) + 2)^3 = (-2)^3 + (2)^3$, meaning $0^3 = (-8) + 8$, or $0 = 0$. Therefore, this statement is true for all positive and negative values of a and b.

e. Does $f(x) = x^3$ display ideal linear properties? Explain.

No, because $(a+b)^3 \neq a^3 + b^3$ for all real values of the variables.

Extension Discussion (14 minutes, optional)

As a class, watch this video (7 minutes) that shows another way to justify Exercise 1 (http://www.jamestanton.com/?p=677). Discuss what the groups discovered in the exercises and what was shown in the video. If time allows, let the groups present their findings and discuss similarities and differences.

Closing (3 minutes)

Ask students to perform a 30-second Quick Write explaining what was learned today using these questions as a guide.

- When does $(a+b)^2 = a^2 + b^2$? How do you know?
  - *When $a = 0$ and/or $b = 0$*
- When does $\sqrt{a+b} = \sqrt{a} + \sqrt{b}$? How do you know?
  - *When $a = 0$ and/or $b = 0$*
- Are $a = 0$ and/or $b = 0$ always the values when functions display ideal linear properties?
  - *No. It depends on the function. Sometimes these values work, and other times they do not. Sometimes there are additional values that work such as with the function $f(x) = x^3$, when $a = -b$ also works.*
- When does a function display ideal linear properties?
  - *When $f(x+y) = f(x) + f(y)$ and $f(kx) = kf(x)$*

Exit Ticket (5 minutes)

Name ______________________________ Date ____________________

Lesson 1: Wishful Thinking—Does Linearity Hold?

Exit Ticket

1. Xavier says that $(a+b)^2 \neq a^2 + b^2$ but that $(a+b)^3 = a^3 + b^3$. He says that he can prove it by using the values $a = 2$ and $b = -2$. Shaundra says that both $(a+b)^2 = a^2 + b^2$ and $(a+b)^3 = a^3 + b^3$ are true and that she can prove it by using the values of $a = 7$ and $b = 0$ and also $a = 0$ and $b = 3$. Who is correct? Explain.

2. Does $f(x) = 3x + 1$ display ideal linear properties? Explain.

Exit Ticket Sample Solutions

1. Xavier says that $(a+b)^2 \neq a^2+b^2$ but that $(a+b)^3 = a^3+b^3$. He says that he can prove it by using the values $a=2$ and $b=-2$. Shaundra says that both $(a+b)^2 = a^2+b^2$ and $(a+b)^3 = a^3+b^3$ are true and that she can prove it by using the values of $a=7$ and $b=0$ and also $a=0$ and $b=3$. Explain.

   Neither is correct. Both have chosen values that just happen to work in one or both of the equations. In the first equation, anytime $a=0$ and/or $b=0$, the statement is true. In the second equation, anytime $a=0$ and/or $b=0$ and also when $a=-b$, the statement is true. If they tried other values such as $a=1$ and $b=1$, neither statement would be true.

2. Does $f(x)=3x+1$ display ideal linear properties? Explain.

   No. $f(ax)=3ax+1$, but $af(x)=3ax+a$ These are not equivalent.
   Also, $f(x+y)=3(x+y)+1=3x+3y+1$, but $f(x)+f(y)=3x+1+3y+1=3x+3y+2$.
   They are not equivalent, so the function does not display ideal linear properties.

Problem Set Sample Solutions

Assign students some or all of the functions to investigate. All students should attempt Problem 4 to set up the next lesson. It is hoped that students may give some examples that are studied in Lesson 2.

Study the statements given in Problems 1–3. Prove that each statement is false, and then find all values of a and b for which the statement is true. Explain your work and the results.

1. If $f(x)=x^2$, does $f(a+b)=f(a)+f(b)$?

   Answers that prove the statement false will vary but could include $a=2$ and $b=-2$.

   This statement is true when $a=0$ and/or $b=0$.

2. If $f(x)=x^{\frac{1}{3}}$, does $f(a+b)=f(a)+f(b)$?

   Answers that prove the statement false will vary but could include $a=1$ and $b=1$.

   This statement is true when $a=0$ and/or $b=0$ and when $a=-b$.

3. If $f(x)=\sqrt{4x}$, does $f(a+b)=f(a)+f(b)$?

   Answers that prove the statement false will vary but could include $a=-1$ and $b=1$.

   This statement is true when $a=0$ and/or $b=0$.

4. Think back to some mistakes that you have made in the past simplifying or expanding functions. Write the statement that you assumed was correct that was not, and find numbers that prove your assumption was false.

   Answers will vary but could include $\sin(x+y)=\sin(x)+\sin(y)$, which is false when x and y equal $45°$, $\log(2a)=2\log(a)$, which is false for $a=1$.

   $10^{a+b}=10^a+10^b$, which is false for $a,b=1$, $\frac{1}{a+b}=\frac{1}{a}+\frac{1}{b}$, which is false for $a,b=1$.

Lesson 2: Wishful Thinking—Does Linearity Hold?

Student Outcomes

- Students learn when ideal linearity properties do and do not hold for classes of functions studied in previous years.
- Students develop familiarity with linearity conditions.

Lesson Notes

This is the second day of a two-day lesson that looks at common mistakes that students make, all based on assuming linearity holds for all functions. In Lesson 1, students were introduced to a new definition of linearity. $f(x)$ is a linear transformation if $f(x+y) = f(x) + f(y)$ and $f(kx) = kf(x)$. Students continue to explore linearity by looking at common student mistakes. In Lesson 1, students explored polynomials and radical equations. Lesson 2 extends this exploration to trigonometric, rational, and logarithmic functions. The last exercise in this lesson, Exercise 4, has no real solutions, leading to the discovery of complex solutions, and this launches the study of complex numbers. This study includes operations on complex numbers as well as using the conjugates to find moduli and quotients.

Classwork

In this Exploratory Challenge, students work individually while discussing the steps as a class. The exercises are completed in pairs with the class coming together at the end to present their findings and to watch a video.

Opening Exercise (8 minutes)

In the last problem of the Problem Set from Lesson 1, students were asked to use what they had learned in Lesson 1 and then to think back to some mistakes that they had made in the past simplifying or expanding functions and to show that the mistakes were based on false assumptions. In this Opening Exercise, students give examples of some of their misconceptions. Have students put the examples on the board. Some of these are studied directly in Lesson 2, and others can be assigned as part of classwork, homework, or as extensions.

- In the last problem of the Problem Set from Lesson 1, you were asked to think back to some mistakes that you had made in the past simplifying or expanding functions. Show me some examples that you wrote down, and I will ask some of you to put your work on the board.
  - *Answers will vary but could include mistakes such as* $\sin(x+y) = \sin(x) + \sin(y)$, $\log(2a) = 2\log(a)$, $10^{a+b} = 10^a + 10^b$, $\frac{1}{a+b} = \frac{1}{a} + \frac{1}{b}$, *and many others.*

Note: Emphasize that these examples are errors, not true mathematical statements.

Pick a couple of the simpler examples of mistakes that are not covered in class, and talk about those, going through the steps of Lesson 1. For each example of a mistake, have students verify with numbers that the equation is not true for all real numbers and then find a solution that is true for all real numbers. Indicate to students the statements that are to be reviewed in class. List other statements that may be reviewed later.

Exploratory Challenge (14 minutes)

> *Scaffolding:*
> Students may need a reminder of how to convert between radians and degrees and critical trigonometric function values. Create a chart for students to complete that lists degrees, radian measure, $\sin(x)$, and $\cos(x)$. A copy of a table follows this lesson in the student materials.

- In Lesson 1, we discovered that not all functions are linear transformations. Today, we will study some different functions.

MP.3

- Let's start by looking at a trigonometric function. Is $f(x) = \sin(x)$ a linear transformation? Explain why or why not.
  - *No,* $\sin(x + y) \neq \sin(x) + \sin(y)$ *and* $\sin(ax) \neq a\sin(x)$.
- Does $\sin(x + y) = \sin(x) + \sin(y)$ for all real values of x and y?
  - *Answers may vary.*
- Substitute some values of x and y into this equation.

Some students may use degrees and others radians. Allow students to choose. Alternatively, assign half of students to use degrees and the other half to use radians. Compare answers.

- Did anyone find values of x and y that produced a true statement?
  - *Answers will vary but could include* $x = 0°, y = 0°$ *or* $x = 180°, y = 180°$ *or* $x = 0°, y = 90°$ *or the equivalent in radians* $x = 0, y = 0$ *or* $x = \pi, y = \pi$ *or* $x = 0, y = \frac{\pi}{2}$.
- If you used degrees, compare your answers to the answers of a neighbor who used radians. What do you notice?
  - *The answers will be the same but a different measure. For example,* $x = 180°, y = 180°$ *is the same as* $x = \pi, y = \pi$ *because* $180° = \pi$ rad. *For example,* $\sin(\pi + \pi) = \sin(2\pi) = 0$, *and* $\sin(\pi) + \sin(\pi) = 0 + 0 = 0$, *so the statement is true.*
- Did anyone find values of x and y that produced a false statement? Explain.
  - *Answers will vary but could include* $x = 45°, y = 45°$ *or* $x = 30°, y = 30°$ *or* $x = 30°, y = 60°$ *or the equivalent in radians* $x = \frac{\pi}{4}, y = \frac{\pi}{4}$ *or* $x = \frac{\pi}{6}, y = \frac{\pi}{6}$ *or* $x = \frac{\pi}{6}, y = \frac{\pi}{3}$. *For example,* $\sin\left(\frac{\pi}{4} + \frac{\pi}{4}\right) = \sin\left(\frac{\pi}{2}\right) = 1$, *but* $\sin\left(\frac{\pi}{4}\right) + \sin\left(\frac{\pi}{4}\right) = \frac{\sqrt{2}}{2} + \frac{\sqrt{2}}{2} = \sqrt{2}$, *so the statement is false.*
- Is this function a linear transformation? Explain this to your neighbor.
  - *This function is not a linear transformation because* $\sin(x + y) \neq \sin(x) + \sin(y)$ *for all real numbers.*

Exercises (15 minutes)

In the exercises below, allow students to work through the problems in pairs. Circulate and give students help as needed. Call the class back together, and have groups present their results. All groups can be assigned Exercises 1–4. For advanced groups, ask students to find the imaginary solutions to Exercise 4, and/or assign some of the more complicated examples that students brought to class from the Lesson 1 Problem Set and presented in the Opening Exercise.

Exercises

1. **Let $f(x) = \sin(x)$. Does $f(2x) = 2f(x)$ for all values of x? Is it true for any values of x? Show work to justify your answer.**

   No. If $x = \frac{\pi}{2}$, $\sin\left(2\left(\frac{\pi}{2}\right)\right) = \sin(\pi) = 0$, but $2\sin\left(\frac{\pi}{2}\right) = 2(1) = 2$, so the statement does not hold for every value of x. It is true anytime $\sin(x) = 0$, so for $x = 0$, $x = \pm\pi, x = \pm 2\pi$.

2. **Let $f(x) = \log(x)$. Find a value for a such that $f(2a) = 2f(a)$. Is there one? Show work to justify your answer.**

$$\log(2a) = 2\log(a)$$
$$\log(2a) = \log(a^2)$$
$$2a = a^2$$
$$a^2 - 2a = 0$$
$$a(a-2) = 0$$

   Thus, $a = 2$ or $a = 0$. Because 0 is not in the domain of the logarithmic function, the only solution is $a = 2$.

MP.3

3. **Let $f(x) = 10^x$. Show that $f(a+b) = f(a) + f(b)$ is true for $a = b = \log(2)$ and that it is not true for $a = b = 2$.**

   For $a = b = \log(2)$

   $f(a+b) = 10^{(\log(2)+\log(2))} = 10^{2\log(2)} = 10^{\log(2^2)} = 2^2 = 4$

   $f(a) + f(b) = 10^{\log 2} + 10^{\log 2} = 2 + 2 = 4$

   Therefore, $f(a+b) = f(a) + f(b)$.

   For $a = b = 2$

   $f(a+b) = 10^{2+2} = 10^4 = 10000$

   $f(a) + f(b) = 10^2 + 10^2 = 100 + 100 - 200$

   Therefore, $f(a+b) \neq f(a) + f(b)$.

4. **Let $f(x) = \frac{1}{x}$. Are there any real numbers a and b so that $f(a+b) = f(a) + f(b)$? Explain.**

   Neither a nor b can equal zero since they are in the denominator of the rational expressions.

$$\frac{1}{a+b} = \frac{1}{a} + \frac{1}{b}$$
$$\frac{1}{a+b}ab(a+b) = \frac{1}{a}ab(a+b) + \frac{1}{b}ab(a+b)$$
$$ab = a(a+b) + b(a+b)$$
$$ab = a^2 + ab + ab + b^2$$
$$ab = a^2 + 2ab + b^2$$
$$ab = (a+b)^2$$

   This means that ab must be a positive number. Simplifying further, we get $0 = a^2 + ab + b^2$. The sum of three positive numbers will never equal zero, so there are no real solutions for a and b.

5. **What do your findings from these exercises illustrate about the linearity of these functions? Explain.**

   Answers will vary but should address that in each case, the function is not a linear transformation because it does not hold to the conditions $f(a+b) = f(a) + f(b)$ and $f(cx) = c(f(x))$ for all real-numbered inputs.

Scaffolding:

- For advanced learners, have students determine the general solution that works for all real numbers.
- Monitor group work, and target some groups with more specific questions to help them with the algebra needed. Students may need a reminder of the properties of logarithms such as $a\log(x) = \log(x^a)$.
- Some groups may need to complete the trigonometry value table before starting the exercises.

Closing (3 minutes)

As a class, have a discussion using the following questions.

- What did you notice about the solutions of trigonometric functions? Why?
  - *There are more solutions that work for trigonometric functions because they are cyclical.*
- Which functions were hardest to find solutions that worked? Why?
  - *Answers will vary, but many students may say logarithmic or exponential functions.*
- Are $a = 0$ and/or $b = 0$ always solutions? Explain.
  - *No. It depends on the function.*
  - *For example,* $\cos(0+0) \neq \cos(0) + \cos(0)$ *and* $10^{(0+0)} \neq 10^0 + 10^0$.
- Are trigonometric, exponential, and logarithmic functions linear transformations? Explain.
  - *No. They do not meet the conditions required for linearity:*
    $f(a+b) = f(a) + f(b)$ *and* $f(cx) = c\big(f(x)\big)$ *for all real-numbered inputs*

Exit Ticket (5 minutes)

Name ______________________________ Date ______________

Lesson 2: Wishful Thinking—Does Linearity Hold?

Exit Ticket

1. Koshi says that he knows that $\sin(x+y) = \sin(x) + \sin(y)$ because he has substituted in multiple values for x and y, and they all work. He has tried $x = 0°$ and $y = 0°$, but he says that usually works, so he also tried $x = 45°$ and $y = 180°$, $x = 90°$ and $y = 270°$, and several others. Is Koshi correct? Explain your answer.

2. Is $f(x) = \sin(x)$ a linear transformation? Why or why not?

Exit Ticket Sample Solutions

1. Koshi says that he knows that $\sin(x+y) = \sin(x) + \sin(y)$ because he has substituted in multiple values for x and y, and they all work. He has tried $x = 0°$ and $y = 0°$, but he says that usually works, so he also tried $x = 45°$ and $y = 180°$, $x = 90°$ and $y = 270°$, and several others. Is Koshi correct? Explain your answer.

   Koshi is not correct. He happened to pick values that worked, most giving at least one value of $\sin(x) = 0$. *If he had chosen other values such as* $x = 30°$ *and* $y = 60°$, $\sin(30° + 60°) = \sin(90°) = 1$, *but* $\sin(30°) + \sin(60°) = \frac{1}{2} + \frac{\sqrt{3}}{2}$, *so the statement that* $\sin(30° + 60°) = \sin(30°) + \sin(60°)$ *is false.*

2. Is $f(x) = \sin(x)$ a linear transformation? Why or why not?

   No. $\sin(x+y) \neq \sin(x) + \sin(y)$ *and* $\sin(ax) \neq a\sin(x)$

Problem Set Sample Solutions

Assign students some or all of the functions to investigate. Problems 1–4 are all trigonometric functions, Problem 5 is a rational function, and Problems 6 and 7 are logarithmic functions. These can be divided up. Problem 8 sets up Lesson 3 but is quite challenging.

Examine the equations given in Problems 1–4, and show that the functions $f(x) = \cos(x)$ and $g(x) = \tan(x)$ are not linear transformations by demonstrating that they do not satisfy the conditions indicated for all real numbers. Then, find values of x and/or y for which the statement holds true.

1. $\cos(x+y) = \cos(x) + \cos(y)$

   Answers that prove the statement false will vary but could include $x = 0$ *and* $y = 0$.

   This statement is true when $x = 1.9455$, *or* $111.47°$, *and* $y = 1.9455$, *or* $111.47°$. *This will be difficult for students to find without technology.*

2. $\cos(2x) = 2\cos(x)$

   Answers that prove the statement false will vary but could include $x = 0$ *or* $x = \frac{\pi}{2}$.

   This statement is true when $x = 1.9455$, *or* $111.47°$. *This will be difficult for students to find without technology.*

3. $\tan(x+y) = \tan(x) + \tan(y)$

   Answers that prove the statement false will vary but could include $x = \frac{\pi}{4}$ *and* $y = \frac{\pi}{4}$.

   This statement is true when $x = 0$ *and* $y = 0$.

4. $\tan(2x) = 2\tan(x)$

   Answers that prove the statement false will vary but could include $x = \frac{\pi}{4}$ *and* $y = \frac{\pi}{4}$.

   This statement is true when $x = 0$ *and* $y = 0$.

5. Let $f(x) = \frac{1}{x^2}$. Are there any real numbers a and b so that $f(a+b) = f(a) + f(b)$? Explain.

Neither a nor b can equal zero since they are in the denominator of the fractions.

If $f(a+b) = f(a) + f(b)$, then $\frac{1}{(a+b)^2} = \frac{1}{a^2} + \frac{1}{b^2}$.

Multiplying each term by $a^2b^2(a+b)^2$, we get

$$a^2b^2(a+b)^2\frac{1}{(a+b)^2} = a^2b^2(a+b)^2\frac{1}{a^2} + a^2b^2(a+b)^2\frac{1}{b^2}$$
$$a^2b^2 = b^2(a+b)^2 + a^2(a+b)^2$$
$$a^2b^2 = (a^2+b^2)(a+b)^2$$
$$a^2b^2 = a^4 + 2a^3b + 2a^2b^2 + 2ab^3 + b^4$$
$$a^2b^2 = a^4 + 2ab(a^2+b^2) + 2a^2b^2 + b^4$$
$$0 = a^4 + 2ab(a^2+b^2) + a^2b^2 + b^4$$

The terms a^4, a^2b^2, and b^4 are positive because they are even-numbered powers of nonzero numbers. We established in the lesson that $ab = (a+b)^2$ and, therefore, is also positive.

The product $2ab(a^2+b^2)$ must then also be positive.

The sum of four positive numbers will never equal zero, so there are no real solutions for a and b.

6. Let $f(x) = \log(x)$. Find values of a such that $f(3a) = 3f(a)$.

$$\log(3a) = 3\log(a)$$
$$\log(3a) = \log(a)^3$$
$$3a = a^3$$
$$3 = a^2$$
$$a = \sqrt{3}$$

This is true for the value of a when $3a = a^3$ that is in the domain, which is $a = \sqrt{3}$.

7. Let $f(x) = \log(x)$. Find values of a such that $f(ka) = kf(a)$.

This is true for the values of a when $ka = a^k$ that are in the domain of the function.

8. Based on your results from the previous two problems, form a conjecture about whether $f(x) = \log(x)$ represents a linear transformation.

The function is not an example of a linear transformation. The condition $f(ka) = kf(a)$ does not hold for all values of a, for example, nonzero values of c and $a = 1$.

9. Let $f(x) = ax^2 + bx + c$.

a. Describe the set of all values for a, b, and c that make $f(x+y) = f(x) + f(y)$ valid for all real numbers x and y.

This is challenging for students, but the goal is for them to realize that $a = 0$, $c = 0$, and any real number b. They may understand that $a = 0$, but $c = 0$ could be more challenging. The point is that it is unusual for functions to satisfy this condition for all real values of x and y. This is discussed in detail in Lesson 3.

$$f(x+y) = a(x+y)^2 + b(x+y) + c = ax^2 + 2axy + ay^2 + bx + by + c$$
$$f(x) + f(y) = ax^2 + bx + c + ay^2 + by + c$$

$$f(x+y) = f(x) + f(y)$$
$$ax^2 + 2axy + ay^2 + bx + by + c = ax^2 + bx + c + ay^2 + by + c$$
$$2axy + c = c + c$$
$$2axy = c$$

Therefore, the set of values that satisfies this equation for all real numbers x and y is $a = 0$, any real number b, and $c = 0$.

b. What does your result indicate about the linearity of quadratic functions?

Answers will vary but should address that quadratic functions are not linear transformations, since they only meet the condition $f(x+y) = f(x) + f(y)$ when $a = 0$.

Trigonometry Table

Angle Measure (x Degrees)	Angle Measure (x Radians)	$\sin(x)$	$\cos(x)$
0	0	0	1
30	$\frac{\pi}{6}$	$\frac{1}{2}$	$\frac{\sqrt{3}}{2}$
45	$\frac{\pi}{4}$	$\frac{\sqrt{2}}{2}$	$\frac{\sqrt{2}}{2}$
60	$\frac{\pi}{3}$	$\frac{\sqrt{3}}{2}$	$\frac{1}{2}$
90	$\frac{\pi}{2}$	1	0

Lesson 3: Which Real Number Functions Define a Linear Transformation?

Student Outcomes

- Students develop facility with the properties that characterize linear transformations.
- Students learn that a mapping $L: \mathbb{R} \to \mathbb{R}$ is a linear transformation if and only if $L(x) = ax$ for some real number a.

Lesson Notes

This lesson begins with two examples of functions that were explored in Lessons 1–2, neither of which is a linear transformation. Next, students explore the function $f(x) = 5x$, followed by the more general $f(x) = ax$, proving that these functions satisfy the requirements for linear transformations. The rest of the lesson is devoted to proving that functions of the form $f(x) = ax$ are, in fact, the *only* linear transformations from $\mathbb{R}$ to $\mathbb{R}$.

Classwork

Opening Exercise (4 minutes)

Opening Exercise

Recall from the previous two lessons that a linear transformation is a function f that satisfies two conditions: (1) $f(x + y) = f(x) + f(y)$ and (2) $f(kx) = kf(x)$. Here, k refers to any real number, and x and y represent arbitrary elements in the domain of f.

a. **Let $f(x) = x^2$. Is f a linear transformation? Explain why or why not.**

Let x be 2 and y be 3. $f(2 + 3) = f(5) = 5^2 = 25$, but $f(2) + f(3) = 2^2 + 3^2 = 4 + 9 = 13$. Since these two values are different, we can conclude that f is not a linear transformation.

b. **Let $g(x) = \sqrt{x}$. Is g a linear transformation? Explain why or why not.**

Let x be 2 and y be 3. $g(2 + 3) = g(5) = \sqrt{5}$, but $g(2) + g(3) = \sqrt{2} + \sqrt{3}$, which is not equal to $\sqrt{5}$. This means that g is not a linear transformation.

Discussion (9 minutes): A Linear Transformation

- The exercises you just did show that neither $f(x) = x^2$ nor $g(x) = \sqrt{x}$ is a linear transformation. Let's look at a third function together.

MP.3

- Let $h(x) = 5x$. Does h satisfy the requirements for a linear transformation? Take a minute to explore this question on your own, and then explain your thinking with a partner.
  - *First, we need to check the addition requirement:* $h(x + y) = 5(x + y) = 5x + 5y$.
    $h(x) + h(y) = 5x + 5y$. *Thus, we do indeed have* $h(x + y) = h(x) + h(y)$. *So far, so good.*
    Now, we need to check the multiplication requirement: $h(kx) = 5(kx) = 5kx$.
    $kh(x) = k \cdot 5x = 5kx$. *Thus, we also have* $h(kx) = kh(x)$.
    Therefore, h *satisfies both of the requirements for a linear transformation.*
- So, now we know that $h(x) = 5x$ is a linear transformation. Can you generate your own example of a linear transformation? Write down a conjecture, and share it with another student.
  - *Answers will vary.*
- Do you think that every function of the form $h(x) = ax$ is a linear transformation? Let's check to make sure that the requirements are satisfied.
  - $h(x + y) = a(x + y) = ax + ay$
    $h(x) + h(y) = ax + ay$
    Thus, $h(x + y) = h(x) + h(y)$, *as required.*
    $h(kx) = a(kx) = akx$
    $kh(x) = k \cdot ax = akx$
    Thus, $h(kx) = kh(x)$, *as required.*
    This proves that $h(x) = ax$, *with* a *any real number, is indeed a linear transformation.*

MP.3

- What about $f(x) = 5x + 3$? Since the graph of this equation is a straight line, we know that it represents a linear function. Does that mean that it automatically meets the technical requirements for a linear transformation? Write down a conjecture, and then take a minute to see if you are correct.
  - *If* f *is a linear transformation, then it must have the addition property.*
    $f(x + y) = 5(x + y) + 3 = 5x + 5y + 3$
    $f(x) + f(y) = (5x + 3) + (5y + 3) = 5x + 5y + 6$
    Clearly, these two expressions are not the same for all values of x *and* y, *so* f *fails the requirements for a linear transformation.*
- A bit surprising, isn't it? The graph is a straight line, and it is 100% correct to say that f is a linear function. But at the same time, it does not meet the technical requirements for a linear transformation. It looks as though *some* linear functions are considered linear transformations, but not all of them are. Let's try to understand what is going on here.
- Does anything strike you about the graph of $f(x) = 5x$ as compared to the graph of $f(x) = 5x + 3$?
  - *The first graph passes through the origin; the second one does not.*
- Do you think it is necessary for a graph to pass through the origin in order to be considered a linear transformation? Let's explore this question together. We have shown that every function of the form $f(x) = ax$ is a linear transformation. Are there other functions that map real numbers to real numbers that are linear in this sense, or are these the only kind that do? Let's see what we can learn about these questions.

Discussion (6 minutes): The Addition Property

- Suppose we have a linear transformation L that takes a real number as an input and produces a real number as an output. We can write $L: \mathbb{R} \to \mathbb{R}$ to denote this.
- Now, suppose that L takes 2 to 8 and 3 to 12; that is, $L(2) = 8$ and $L(3) = 12$.
- Where does L take 5? Can you calculate the value of $L(5)$?
  - *Since $5 = 2 + 3$, we know that $L(5) = L(2 + 3)$. Since L is a linear transformation, this must be the same as $L(2) + L(3)$, which is $8 + 12 = 20$. So, $L(5)$ must be 20.*
- For practice, find out where L takes 7 and 8. That is, find $L(7)$ and $L(8)$.
  - $L(7) = L(5 + 2) = L(5) + L(2) = 20 + 8 = 28$
  - $L(8) = L(5 + 3) = L(5) + L(3) = 20 + 12 = 32$
- What can we learn about linear transformations through these examples? Let's dig a little deeper.
- We used the facts that $L(2) = 8$ and $L(3) = 12$ to figure out that $L(5) = 20$. What is the relationship between the three inputs here? What is the relationship among the three outputs?
  - *$5 = 2 + 3$, so the third input is the sum of the first two inputs.*
  - *$20 = 8 + 12$, so the third output is the sum of the first two outputs.*
- Do these examples give you a better understanding of the property $L(x + y) = L(x) + L(y)$? This statement is saying that if you know what a linear transformation does to any two inputs x and y, then you know for sure what it does to their sum $x + y$. In particular, to get the output for $x + y$, you just have to add the outputs $L(x)$ and $L(y)$, just as we did in the example above, where we figured out that $L(5)$ must be 20.
- What do you suppose all of this means in terms of the graph of L? Let's plot each of the input-output pairs we have generated so far and then see what we can learn.

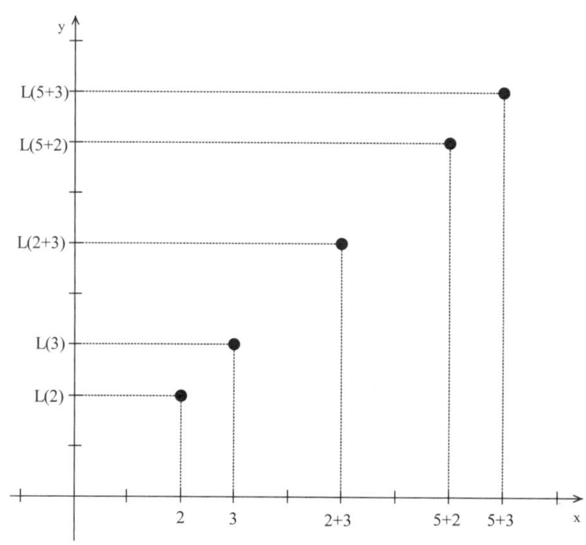

- What do you notice about this graph?
  - *It looks as though the points lie on a line through the origin.*
- Can we be absolutely sure of this? Let's keep exploring to find out if this is really true.

Discussion (4 minutes): The Multiplication Property

- Let's again suppose that $L(2) = 8$.
- Can you figure out where L takes 6?
  - *Since $6 = 3 \cdot 2$, we know that $L(6) = L(3 \cdot 2)$. Since L is a linear transformation, this must be the same as $3 \cdot L(2)$, which is $3 \cdot 8 = 24$. So, $L(6)$ must be 24.*
- For practice, find out where L takes 4 and 8. That is, find $L(4)$ and $L(8)$.
  - $L(4) = L(2 \cdot 2) = 2 \cdot L(2) = 2 \cdot 8 = 16$
  - $L(8) = L(4 \cdot 2) = 4 \cdot L(2) = 4 \cdot 8 = 32$
- We computed $L(8)$ earlier using the addition property, and now we have computed it again using the multiplication property. Are the results the same?
  - *Yes. In both cases, we have $L(8) = 32$.*
- Does this work give you a feel for what the multiplication property is all about? Let's summarize our work in the last few examples. Suppose you know that, for a certain input x, L produces output y, so that $L(x) = y$. The multiplication property is saying that if you triple the input from x to $3x$, you will also triple the output from y to $3y$. This is the meaning of the statement $L(3x) = 3 \cdot L(x)$, or more generally, $L(kx) = kL(x)$.
- Once again, let's see what all of this means in terms of the graph of L. We will plot the input-output pairs we generated.

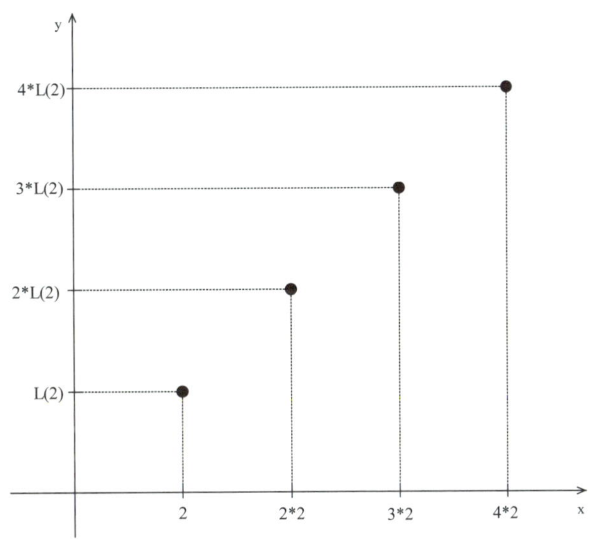

- Does this graph look like you expected it to?
  - *Yes. It is a straight line through the origin, just like before.*

Discussion (4 minutes): Opposites

> *Scaffolding:*
> If students are struggling to compute $L(-2)$, point out that $-2 = -1 \cdot 2$, and then ask them to apply the multiplication property for linear transformations.

- So, we used the fact that $L(2) = 8$ to figure out that $L(4) = 16$, $L(6) = 24$, and $L(8) = 32$.
- What about the negative multiples of 2? Can you figure out $L(-2)$?
  - $L(-2) = L(-1 \cdot 2) = -1 \cdot L(2) = -1 \cdot 8 = -8$
- For practice, find $L(-4)$, $L(-6)$, and $L(-8)$.
  - $L(-4) = L(-1 \cdot 4) = -1 \cdot L(4) = -1 \cdot 16 = -16$
  - $L(-6) = L(-1 \cdot 6) = -1 \cdot L(6) = -1 \cdot 24 = -24$
  - $L(-8) = L(-1 \cdot 8) = -1 \cdot L(8) = -1 \cdot 32 = -32$
- Look carefully at what L does to a number and its opposite. For instance, compare the outputs for 2 and -2, for 4 and -4, etc. What do you notice?
  - *We see that* $L(2) = 8$ *and* $L(-2) = -8$. *We also see that* $L(4) = 16$ *and* $L(-4) = -16$.
- Can you take your observation and formulate a general conjecture?
  - *It looks as though* $L(-x) = -L(x)$.
- This says that if you know what L does to a particular input x, then you know for sure that L takes the opposite input, $-x$, to the opposite output, $-L(x)$.
- Now, prove that your conjecture is true in all cases.
  - $L(-x) = L(-1 \cdot x) = -1 \cdot L(x) = -L(x)$
- Once again, let's collect all of this information in graphical form.

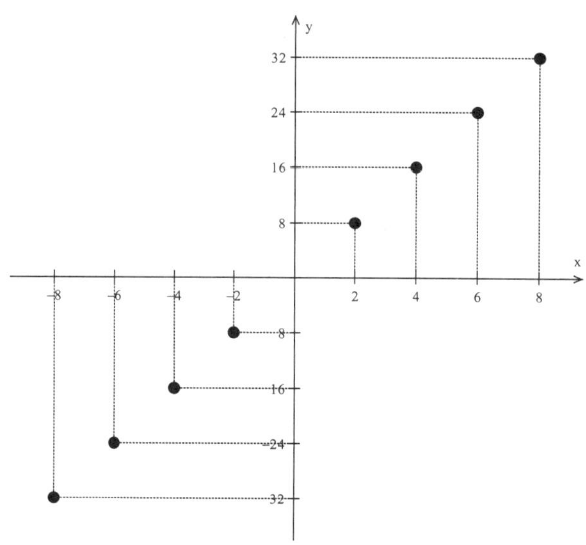

- All signs point to a straight-line graph that passes through the origin. But we have not yet shown that the graph actually contains the origin. Let's turn our attention to that question now.

Discussion (5 minutes): Zero

- If the graph of L contains the origin, then L must take 0 to 0. Does this really have to be the case?
- How can we use the addition property to our advantage here? Can you form the number 0 from the inputs we already have information about?
  - *We know that* $2 + -2 = 0$*, so maybe that can help. Since* $L(2) = 8$ *and* $L(-2) = -8$*, we can now figure out* $L(0)$*.* $L(0) = L(2 + -2) = L(2) + L(-2) = 8 + -8 = 0$.
- So, it really is true that $L(0) = 0$. What does this tell us about the graph of L?
  - *The graph contains the point* $(0,0)$*, which is the origin.*
- In summary, if you give the number 0 as an input to a linear transformation $L(x)$, then the output is sure to be 0.
- Quickly: Is $f(x) = x + 1$ a linear transformation? Why or why not?
  - *No. It cannot be a linear transformation because* $f(0) = 0 + 1 = 1$*, and a linear transformation cannot transform* 0 *into* 1.
- For practice, use the fact that $L(6) = 24$ to show that $L(0) = 0$.
  - *We already showed that* $L(-6) = -24$*, so* $L(0) = L(6 + -6) = L(6) + L(-6) = 24 - 24 = 0$.
- We have used the addition property to show that $L(0) = 0$. Do you think it is possible to use the multiplication property to reach the same conclusion?
  - *Yes.* 0 *is a multiple of* 2*, so we can write* $L(0) = L(0 \cdot 2) = 0 \cdot L(2) = 0 \cdot 8 = 0$.
- So now, we have two pieces of evidence that corroborate our hypothesis that the graph of L passes through the origin. We can now officially add $(0,0)$ to our graph.

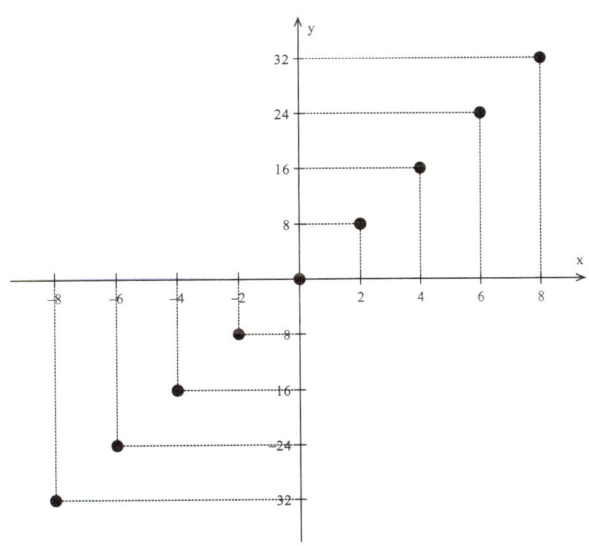

- We originally said that the graph looks like a line through the origin. What is the equation of that line?
  - *The equation of the line that contains all of these points is* $y = 4x$.

Discussion (3 minutes): The Complete Graph of L

How can we be sure that the graph of $L(x)$ is identical to the graph of $y = 4x$? In theory, there could be points on $L(x)$ that are not on $y = 4x$, or vice versa. Are these graphs in fact identical? Perhaps we can show once and for all that $L(x) = 4x$.

> *Scaffolding:*
>
> Advanced students could be challenged to pursue this question without specific guidance. In other words, ask students if the graph of $y = 4x$ is identical to the graph of $L(x)$, and then let them investigate on their own and justify their responses.

- We know that $L(2) = 8$. What is $L(1)$?
  - $L(1) = L\left(\frac{1}{2} \cdot 2\right) = \frac{1}{2} \cdot L(2) = \frac{1}{2} \cdot 8 = 4$
- How might we use the multiplication property to compute $L(x)$ for an arbitrary input x?
  - *Since* $x = x \cdot 1$, *we have* $L(x) = L(x \cdot 1) = x \cdot L(1) = x \cdot 4$.
- We have shown that, for any input x, $L(x) = 4x$ is a formula that gives the output under the linear transformation L. So, the complete graph of L looks like this:

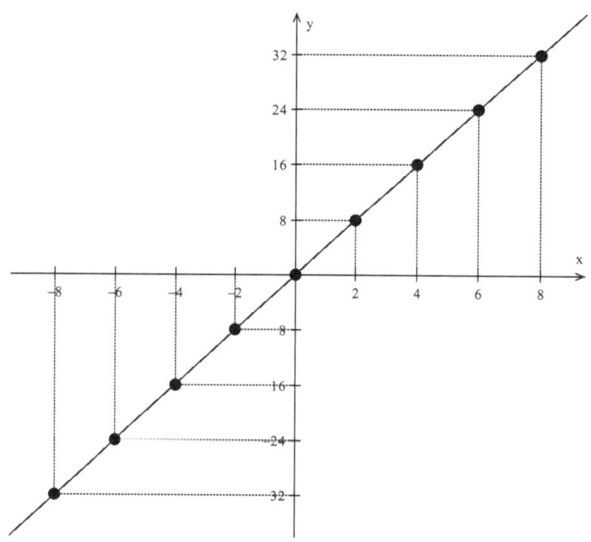

Discussion (2 minutes): General Linear Transformations $\mathbb{R} \rightarrow \mathbb{R}$

- All of the work we did to reach the conclusion that $L(x) = 4x$ was based on just one assumption: We took $L(2) = 8$ as a given, and the rest of our conclusions were worked out from the properties of linear transformations. Now, let's show that *every* linear transformation $L: \mathbb{R} \rightarrow \mathbb{R}$ has the form $L(x) = ax$.
- Show that $L(x) = x \cdot L(1)$.
  - *Since* $x = x \cdot 1$, *we have* $L(x) = L(x \cdot 1) = x \cdot L(1)$.
- Since L produces real numbers as outputs, there is some number a corresponding to $L(1)$. So, let's define $a = L(1)$. Now, we have that $L(x) = x \cdot L(1) = x \cdot a$, which means that every linear transformation looks like $L(x) = ax$. There are no other functions $\mathbb{R} \rightarrow \mathbb{R}$ that can possibly satisfy the requirements for a linear transformation.

Closing (2 minutes)

- Write down a summary of what you learned in the lesson today, and then share your summary with a partner.
  - *Every function of the form* $L(x) = ax$ *is a linear transformation.*
  - *Every linear transformation* $L: \mathbb{R} \to \mathbb{R}$ *corresponds to a formula* $L(x) = ax$.
  - *Linear transformations take the origin to the origin; that is,* $L(0) = 0$.
  - *Linear transformations are odd functions; that is,* $L(-x) = -L(x)$.
  - *The graph of a linear transformation* $L: \mathbb{R} \to \mathbb{R}$ *is a straight line.*

Exit Ticket (6 minutes)

Name ______________________________ Date ____________________

Lesson 3: Which Real Number Functions Define a Linear Transformation?

Exit Ticket

Suppose you have a linear transformation $f: \mathbb{R} \to \mathbb{R}$, where $f(3) = 9$ and $f(5) = 15$.

1. Use the addition property to compute $f(8)$ and $f(13)$.

2. Find $f(12)$ and $f(10)$. Show your work.

3. Find $f(-3)$ and $f(-5)$. Show your work.

4. Find $f(0)$. Show your work.

5. Find a formula for $f(x)$.

6. Draw the graph of the function $y = f(x)$.

Exit Ticket Sample Solutions

Suppose you have a linear transformation $f: \mathbb{R} \to \mathbb{R}$, where $f(3) = 9$ and $f(5) = 15$.

1. **Use the addition property to compute $f(8)$ and $f(13)$.**

   $f(8) = f(3+5) = f(3) + f(5) = 9 + 15 = 24$

   $f(13) = f(8+5) = f(8) + f(5) = 24 + 15 = 39$

2. **Find $f(12)$ and $f(10)$. Show your work.**

   $f(12) = f(4 \cdot 3) = 4 \cdot f(3) = 4 \cdot 9 = 36$

   $f(10) = f(2 \cdot 5) = 2 \cdot f(5) = 2 \cdot 15 = 30$

3. **Find $f(-3)$ and $f(-5)$. Show your work.**

   $f(-3) = -f(3) = -9$

   $f(-5) = -f(5) = -15$

4. **Find $f(0)$. Show your work.**

   $f(0) = f(3 + -3) = f(3) + f(-3) = 9 + -9 = 0$

5. **Find a formula for $f(x)$.**

   We know that there is some number a such that $f(x) = ax$, and since $f(3) = 9$, the value of $a = 3$. In other words, $f(x) = 3x$. We can also check to see if $f(5) = 15$ is consistent with $a = 3$, which it is.

6. **Draw the graph of the function $y = f(x)$.**

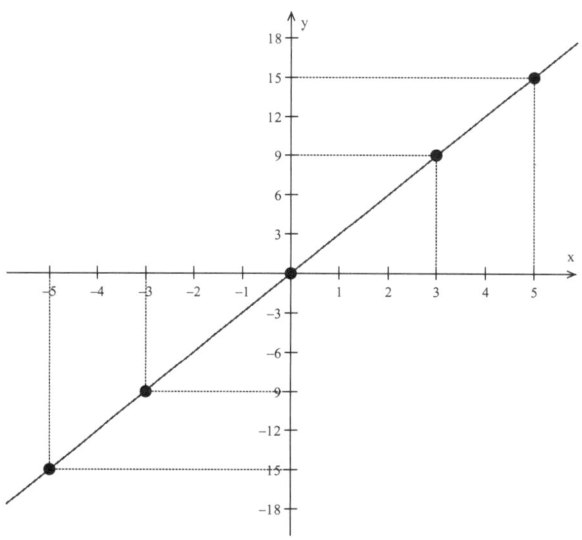

Problem Set Sample Solutions

The first problem provides students with practice in the core skills for this lesson. The second problem is a series of exercises in which students explore concepts of linearity in the context of integer-valued functions as opposed to real-valued functions. The third problem plays with the relation $f(x+y)=f(x)+f(y)$, exchanging addition for multiplication in one or both expressions.

1. Suppose you have a linear transformation $f: \mathbb{R} \to \mathbb{R}$, where $f(2)=1$ and $f(4)=2$.

a. Use the addition property to compute $f(6)$, $f(8)$, $f(10)$, and $f(12)$.

$f(6)=f(2+4)=f(2)+f(4)=1+2=3$

$f(8)=f(2+6)=f(2)+f(6)=1+3=4$

$f(10)=f(4+6)=f(4)+f(6)=2+3=5$

$f(12)=f(10+2)=f(10)+f(2)=5+1=6$

b. Find $f(20)$, $f(24)$, and $f(30)$. Show your work.

$f(20)=f(10\cdot 2)=10\cdot f(2)=10\cdot 1=10$

$f(24)=f(6\cdot 4)=6\cdot f(4)=6\cdot 2=12$

$f(30)=f(15\cdot 2)=15\cdot f(2)=15\cdot 1=15$

c. Find $f(-2)$, $f(-4)$, and $f(-8)$. Show your work.

$f(-2)=f(-1\cdot 2)=-1\cdot f(2)=-1\cdot 1=-1$

$f(-4)=f(-1\cdot 4)=-1\cdot f(4)=-1\cdot 2=-2$

$f(-8)=f(-2\cdot 4)=-2\cdot f(4)=-2\cdot 2=-4$

d. Find a formula for $f(x)$.

We know there is some number a such that $f(x)=ax$, and since $f(2)=1$, then the value of $a=\frac{1}{2}$.

In other words, $f(x)=\frac{x}{2}$. We can also check to see if $f(4)=2$ is consistent with $a=\frac{1}{2}$, which it is.

e. Draw the graph of the function $f(x)$.

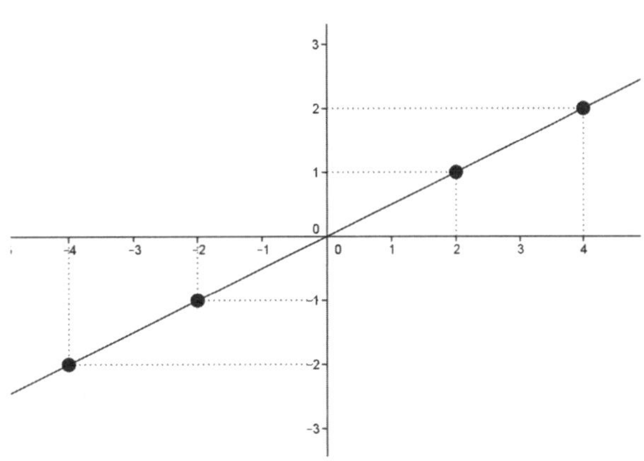

2. The symbol $\mathbb{Z}$ represents the set of integers, and so $g: \mathbb{Z} \to \mathbb{Z}$ represents a function that takes integers as inputs and produces integers as outputs. Suppose that a function $g: \mathbb{Z} \to \mathbb{Z}$ satisfies $g(a+b) = g(a) + g(b)$ for all integers a and b. Is there necessarily an integer k such that $g(n) = kn$ for all integer inputs n?

   a. Let $k = g(1)$. Compute $g(2)$ and $g(3)$.

   $g(2) = g(1+1) = g(1) + g(1) = k + k = 2k$

   $g(3) = g(1+1+1) = g(1) + g(1) + g(1) = k + k + k = 3k$

   b. Let n be any positive integer. Compute $g(n)$.

   $g(n) = g(1 + \cdots + 1) = g(1) + \cdots + g(1) = k + \cdots + k = nk$

   c. Now, consider $g(0)$. Since $g(0) = g(0+0)$, what can you conclude about $g(0)$?

   $g(0) = g(0+0) = g(0) + g(0)$. *By subtracting* $g(0)$ *from both sides of the equation, we get* $g(0) = 0$.

   d. Lastly, use the fact that $g(n + -n) = g(0)$ to learn something about $g(-n)$, where n is any positive integer.

   $g(0) = g(n + -n) = g(n) + g(-n)$. *Since we know that* $g(0) = 0$*, we have* $g(n) + g(-n) = 0$. *This tells us that* $g(-n) = -g(n)$.

   e. Use your work above to prove that $g(n) = kn$ for every integer n. Be sure to consider the fact that n could be positive, negative, or 0.

   We showed that if n *is a positive integer, then* $g(n) = kn$*, where* $k = g(1)$*. Also, since* $k \cdot 0 = 0$ *and we showed that* $g(0) = 0$*, we have* $g(0) = k \cdot 0$*. Finally, if* n *is a negative integer, then* $-n$ *is positive, which means* $g(-n) = k(-n) = -kn$*. But, since* $g(-n) = -g(n)$*, we have* $g(n) = -g(-n) = -(-kn) = kn$*. Thus, in all cases,* $g(n) = kn$*.*

3. In the following problems, be sure to consider all kinds of functions: polynomial, rational, trigonometric, exponential, logarithmic, etc.

   a. Give an example of a function $f: \mathbb{R} \to \mathbb{R}$ that satisfies $f(x \cdot y) = f(x) + f(y)$.

   Any logarithmic function works; for instance, $f(x) = \log(x)$.

   b. Give an example of a function $g: \mathbb{R} \to \mathbb{R}$ that satisfies $g(x+y) = g(x) \cdot g(y)$.

   Any exponential function works; for instance, $g(x) = 2^x$.

   c. Give an example of a function $h: \mathbb{R} \to \mathbb{R}$ that satisfies $h(x \cdot y) = h(x) \cdot h(y)$.

   Any power of x *works; for instance,* $h(x) = x^3$.

EUREKA MATH™

Lesson 4: An Appearance of Complex Numbers

Student Outcomes

- Students solve quadratic equations with complex solutions.
- Students understand the geometric origins of the imaginary unit i in terms of 90-degree rotations. Students use this understanding to see why $i^2 = -1$.

Lesson Notes

This lesson begins with an exploration of an equation that arose in Lesson 1 in the context of studying linear transformations. To check the solutions to this equation, students need a variety of skills involving the arithmetic of complex numbers. The purpose of this phase of the lesson is to point to the need to review and extend students' knowledge of complex number arithmetic. This phase of the lesson continues with a second example of a quadratic equation with complex solutions, which is solved by completing the square.

The second phase of the lesson involves a review of the theory surrounding complex numbers. In particular, i is introduced as a multiplier that induces a 90-degree rotation of the coordinate plane and which satisfies the equation $i^2 = -1$.

Classwork

Opening Exercise (2 minutes)

Scaffolding:

If students need help answering the question in the Opening Exercise, ask them, "What are the properties of a linear transformation?" If necessary, cue them to check whether or not $R(a+b) = R(a) + R(b)$.

MP.3

Opening Exercise

Is $R(x) = \frac{1}{x}$ a linear transformation? Explain how you know.

$R(2+3) = R(5) = \frac{1}{5}$, but $R(2) + R(3) = \frac{1}{2} + \frac{1}{3} = \frac{5}{6}$, which is not the same as $\frac{1}{5}$. This means that the reciprocal function does not preserve addition, and so it is not a linear transformation.

Example 1 (8 minutes)

- Apparently, it is not true in general that $R(2+x) = R(2) + R(x)$, since this statement is false when $x = 3$. But this does not mean that there are *no* values of x that make the equation true. Let's see if we can produce at least one solution.
- Solve the equation $\frac{1}{2+x} = \frac{1}{2} + \frac{1}{x}$.
- What is the first step in solving this equation?
  - *We can multiply both sides by $2x(2+x)$:*

$$2x = x(2+x) + 2(2+x)$$

- How might we continue?
  - *We can apply the distributive property:*

$$2x = 2x + x^2 + 4 + 2x$$

- Evidently, we are dealing with a quadratic equation. What are some techniques you know for solving quadratic equations?
  - *We could try factoring, we could **complete the square**, or we could use the quadratic formula.*
- Let's solve this equation by completing the square. Work on this problem until you have a perfect square equal to a number, and then stop.

Circulate throughout the class, monitoring students' work and providing assistance as needed.

$$-4 = x^2 + 2x$$
$$1 - 4 = x^2 + 2x + 1$$
$$-3 = (x + 1)^2$$

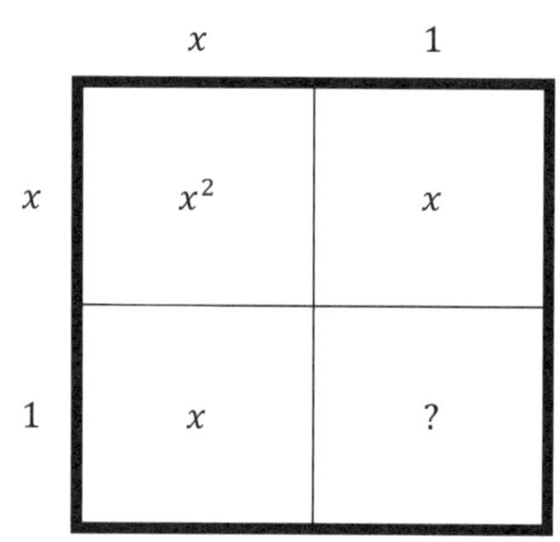

> *Scaffolding:*
> An area diagram can be used to help students understand why it was necessary to add 1 to both sides of the equation to create a perfect square.

- Do you notice anything unusual about the equation at this point?
  - *We have a quantity whose square is equal to a negative number.*
- What does that tell you about the solutions to the equation?
  - *No real number can satisfy the equation, so the solutions must be **complex numbers**.*
- Go ahead and find the solutions.

$$\sqrt{(x+1)^2} = \sqrt{-3}$$
$$x + 1 = \pm i\sqrt{3}$$
$$x = -1 \pm i\sqrt{3}$$

- Do these solutions satisfy the original equation $R(2 + x) = R(2) + R(x)$? How can we tell?
  - *We need to check to see whether or not* $\frac{1}{2+(-1+i\sqrt{3})} = \frac{1}{2} + \frac{1}{-1+i\sqrt{3}}$
- In order to ascertain whether or not these two expressions are equal, we need to review and extend the things we learned about complex numbers in Algebra II. But first, let's do some additional work with quadratics that have complex solutions.

Example 2 (13 minutes)

- Solve the equation $3x^2 + 5x + 7 = 1$.
- Recall that we can use an area diagram to help us visualize the process of completing the square. A first attempt might look something like this:

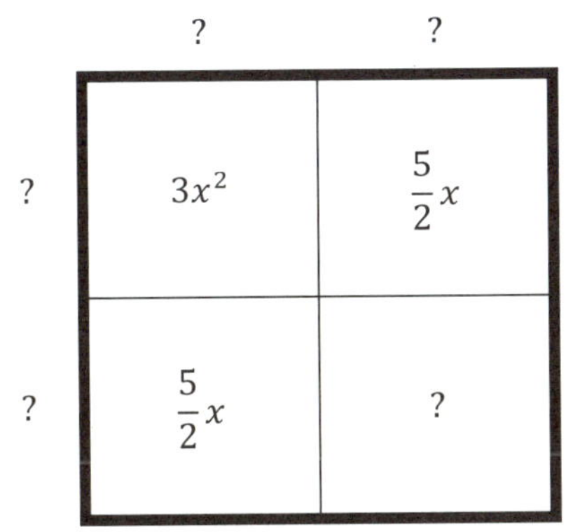

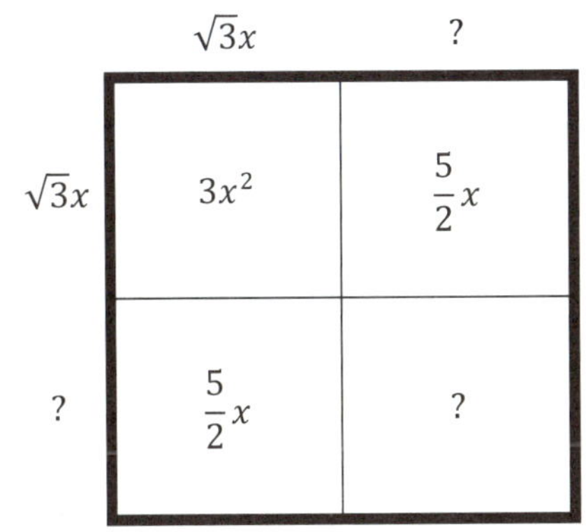

- Do you notice anything awkward about this initial diagram?
  - *The square roots and the fractions are a bit awkward to work with.*
- How can we get around these awkward points? What would make this problem easier to handle?
  - *If the coefficient of x^2 were a perfect square, then we would not have a radical to contend with. If the coefficient of x were even, then we would not have a fraction to contend with.*
- We can multiply both sides of the equation by any number we choose. Let's be strategic about this. What multiplier could we choose that would create a perfect square for the x^2-term?
  - *If we multiply both sides of the equation by 3, the leading coefficient becomes 9, a perfect square.*
- Go ahead and multiply by 3, and see what you get.

$$3x^2 + 5x + 7 = 1$$
$$3(3x^2 + 5x + 7) = 3(1)$$
$$9x^2 + 15x + 21 = 3$$

- Now, let's deal with the x-term. What multiplier could we choose that would make the x-term even, without disturbing the requirement about having a perfect square in the leading term?
  - *We could multiply both sides of the equation by 4, which is both even and a perfect square.*
- Go ahead and multiply by 4, and see what you get.

$$9x^2 + 15x + 21 = 3$$
$$4(9x^2 + 15x + 21) = 4(3)$$
$$36^2 + 60x + 84 = 12$$
$$36^2 + 60x + 84 = 12$$

- Because we took the simple steps of multiplying by 3 and then by 4, the algebra will now be much easier to handle.

- Go ahead and complete the square now. Use an area diagram to help you do this. Then, solve the equation completely.

	?	?
?	$36x^2$	$30x$
?	$30x$	?

	$6x$	5
$6x$	$36x^2$	$30x$
5	$30x$	?

$$36x^2 + 60x + 84 = 12$$
$$36x^2 + 60x = 12 - 84 = -72$$
$$36x^2 + 60x + 25 = -72 + 25$$
$$(6x + 5)^2 = -47$$
$$6x + 5 = \pm i\sqrt{47}$$
$$x = \frac{-5 \pm i\sqrt{47}}{6}$$
$$x = -\frac{5}{6} \pm i\frac{\sqrt{47}}{6}$$

- Quickly answer the following question in your notebook:

  When you want to complete the square of a quadratic expression $ax^2 + bx + c$, what conditions on a and b make the process go smoothly?
  - *It is desirable to convert a into a perfect square and to convert b into an even number.*
- Let's generalize the work we did with the example above.
- Take the expression $ax^2 + bx + c$, and multiply each term by $4a$. What do you get?
  - $4a(ax^2 + bx + c) = 4a^2x^2 + 4abx + 4ac$
- How does this connect to the summary point you wrote in your notebook?
  - *$4a^2 = (2a)^2$, so it is a perfect square. Also, $4ab = 2(2ab)$, so it is even.*
- You may also recognize that the expression $4ac$ is a component from the general quadratic formula. Using $4a$ as a multiplier is useful indeed.

Exercise 1 (4 minutes)

Exercises

1. Solve $5x^2 - 3x + 17 = 9$.

$$5x^2 - 3x + 17 = 9$$
$$5x^2 - 3x = 9 - 17 = -8$$
$$20(5x^2 - 3x) = 20(-8)$$
$$100x^2 - 60x = -160$$
$$100x^2 - 60x + 9 = -160 + 9$$
$$(10x - 3)^2 = -151$$
$$x = \frac{3 \pm i\sqrt{151}}{10} = \frac{3}{10} \pm i\frac{\sqrt{151}}{10}$$

	$10x$	-3
$10x$	$100x^2$	$-30x$
-3	$-30x$	?

- Take about 30 seconds to write down what you have learned so far today, and then share what you wrote with another student.
- Now that we have practiced solving a few equations with complex solutions, we are going to conduct a general review of things we know about complex numbers, starting with the definition of i.

Discussion (5 minutes): The Geometry of Multiplication by i

Recall that multiplying by -1 rotates the number line about the origin through 180 degrees.

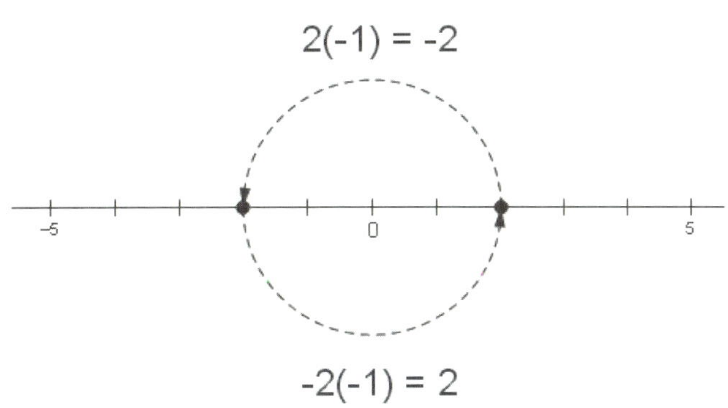

- You may remember that the number i is the multiplier that rotates the number line through 90 degrees.

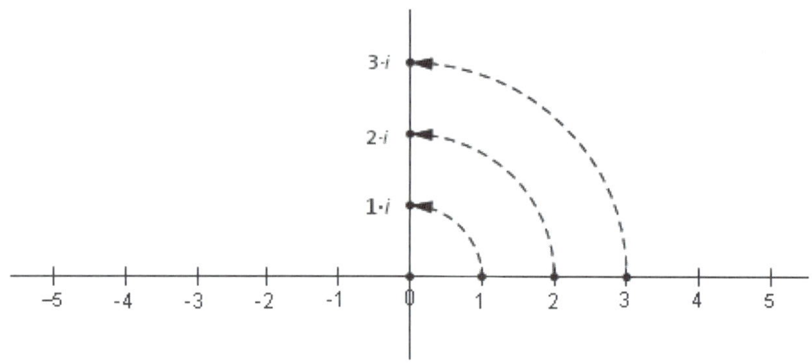

- If we take a point on the vertical axis and multiply it by i, what would you expect to see geometrically?
  - *This should produce another 90-degree rotation.*

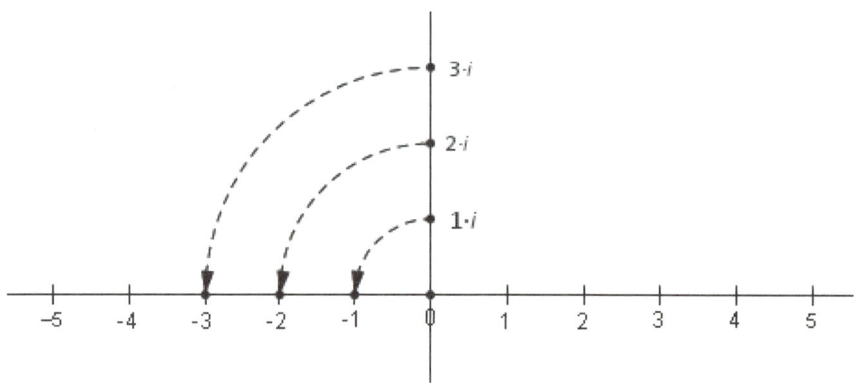

- Now we have performed two 90-degree rotations, which is the same as a 180-degree rotation. This means that multiplying a number by i twice is the same as multiplying the number by -1.
- Knowing that $i \cdot ix = i^2 \cdot x$, what do the above observations suggest must be true about the number i?
  - *If x is any real number, we have $i^2 \cdot x = -1 \cdot x$, which means that $i^2 = -1$.*

Example 3 (2 minutes)

- We know that multiplying by i rotates a point through 90 degrees, and multiplying by i^2 rotates a point through 180 degrees. What do you suppose multiplying by i^3 does? What about i^4?
  - *It would seem as though this should produce three 90-degree rotations, which is 270 degrees. If multiplying by i^4 is the same as doing four 90-degree rotations, then that would make 360 degrees.*
- So, i^4 takes a point back to where it started. In light of the fact that $i^2 = -1$, does this make sense?
  - *Yes, because $i^4 = i^2 \cdot i^2$, which is $(-1)(-1)$, which is just 1. Multiplying by 1 takes a point to itself, so the 360-degree rotation does indeed make sense.*

Exercises 2–4 (3 minutes)

Ask students to work independently on these problems, and then discuss them as a whole class.

2. **Use the fact that $i^2 = -1$ to show that $i^3 = -i$. Interpret this statement geometrically.**

   We have $i^3 = i^2 \cdot i = (-1) \cdot i = -i$. Multiplying by i rotates a point through 90 degrees, and multiplying by -1 rotates it 180 degrees farther. This makes sense with our earlier conjecture that multiplying by i^3 would induce a 270-degree rotation.

3. **Calculate i^6**

   $i^6 = i^2 \cdot i^2 \cdot i^2 = (-1)(-1)(-1) = (1)(-1) = -1$

4. **Calculate i^5**

   $i^5 = i^2 \cdot i^2 \cdot i = (-1)(-1)(i) = (1)(i) = i$

Closing (3 minutes)

Ask students to write responses to the following questions, and then have them share their responses in pairs. Then, briefly discuss the responses as a whole class.

- What is important to know about i from a geometric point of view?
  - *Multiplication by i rotates a point in the plane counterclockwise about the origin through 90 degrees.*
- What is important to know about i from an algebraic point of view?
  - *The number i satisfies the equation $i^2 = -1$.*

Exit Ticket (5 minutes)

Name ______________________________ Date ____________

Lesson 4: An Appearance of Complex Numbers

Exit Ticket

1. Solve the equation below.

$$2x^2 - 3x + 9 = 4$$

2. What is the geometric effect of multiplying a number by i^4? Explain your answer using words or pictures, and then confirm your answer algebraically.

Exit Ticket Sample Solutions

1. Solve the equation below.

$$2x^2 - 3x + 9 = 4$$

$$2x^2 - 3x = -5$$
$$16x^2 - 24x = -40$$
$$16x^2 - 24x + 9 = -40 + 9$$
$$(4x - 3)^2 = -31$$
$$4x - 3 = \pm i\sqrt{31}$$
$$x = \frac{3 \pm i\sqrt{31}}{4} = \frac{3}{4} \pm i\frac{\sqrt{31}}{4}$$

2. What is the geometric effect of multiplying a number by i^4? Explain your answer using words or pictures, and then confirm your answer algebraically.

If you multiply a number by i four times, you would expect to see four 90-degree rotations. This amounts to a 360-degree rotation. In other words, each point is mapped back to itself. This makes sense algebraically as well since the work below shows that $i^4 = 1$.

$i^4 = i^2 \cdot i^2 = -1 \cdot -1 = 1$

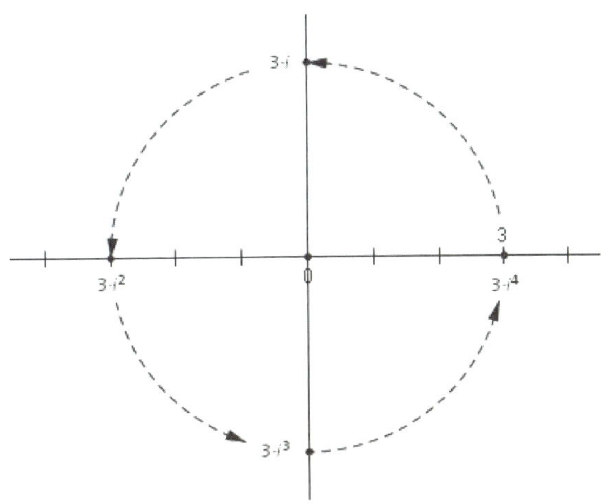

Problem Set Sample Solutions

1. Solve the equation below.

$$5x^2 - 7x + 8 = 2$$

$$5x^2 - 7x = -6$$
$$100x^2 - 140x = -120$$
$$100x^2 - 140x + 49 = -120 + 49$$
$$(10x - 7)^2 = -71$$
$$x = \frac{7 \pm i\sqrt{71}}{10} = \frac{7}{10} \pm i\frac{\sqrt{71}}{10}$$

2. **Consider the equation $x^3 = 8$.**

   a. **What is the first solution that comes to mind?**

   It is easy to see that $2^3 = 8$, so 2 is a solution.

   b. **It may not be easy to tell at first, but this equation actually has three solutions. To find all three solutions, it is helpful to consider $x^3 - 8 = 0$, which can be rewritten as $(x-2)(x^2+2x+4) = 0$ (check this for yourself). Find all of the solutions to this equation.**

$$x^2 + 2x + 4 = 0$$
$$x^2 + 2x = -4$$
$$x^2 + 2x + 1 = -4 + 1$$
$$(x+1)^2 = -3$$
$$x = -1 \pm i\sqrt{3}$$

   The solutions to $x^3 - 8 = 0$ are 2, $-1 + i\sqrt{3}$, and $-1 - i\sqrt{3}$.

3. **Make a drawing that shows the first 5 powers of i (i.e., $i^1, i^2, \ldots, i^5$), and then confirm your results algebraically.**

$i^1 = i$
$i^2 = -1$
$i^3 = i^2 \cdot i = -1 \cdot i = -i$
$i^4 = i^2 \cdot i^2 = -1 \cdot -1 = 1$
$i^5 = i^4 \cdot i = 1 \cdot i = i$

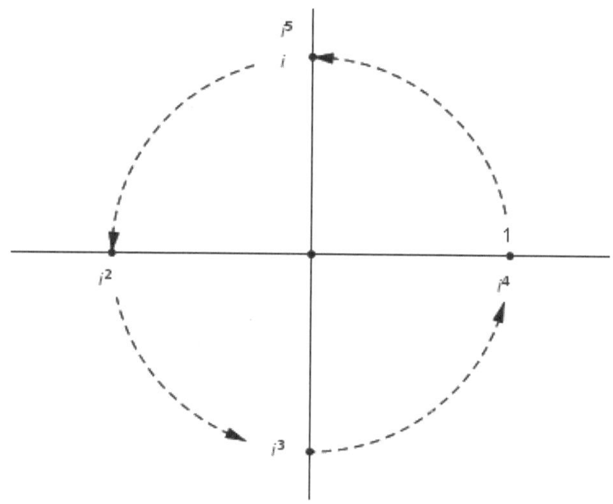

4. **What is the value of i^{99}? Explain your answer using words or drawings.**

   Multiplying by i four times is equivalent to rotating through $4 \cdot 90 = 360$ degrees, which is a complete rotation. Since $99 = 4 \cdot 24 + 3$, multiplying by i for 99 times is equivalent to performing 24 complete rotations, followed by three 90-degree rotations. Thus, $i^{99} = -i$.

5. **What is the geometric effect of multiplying a number by $-i$? Does your answer make sense to you? Give an explanation using words or drawings.**

If we multiply a number by i and then by -1, we get a quarter turn followed by a half turn. This is equivalent to a three-quarters turn in the counterclockwise direction, which is the same as a quarter turn in the clockwise direction. This makes sense because we would expect multiplication by $-i$ to have the opposite effect as multiplication by i, and so it feels right to say that multiplying by $-i$ rotates a point in the opposite direction by the same amount.

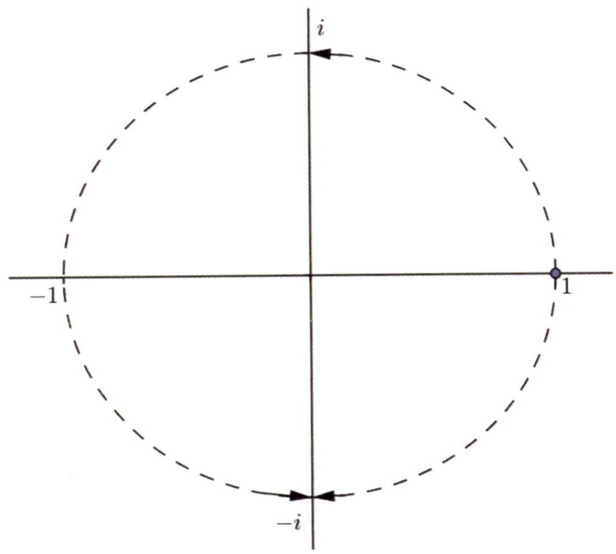

Lesson 5: An Appearance of Complex Numbers

Student Outcomes

- Students describe complex numbers and represent them as points in the complex plane.
- Students perform arithmetic with complex numbers, including addition, subtraction, scalar multiplication, and complex multiplication.

Lesson Notes

MP.7 In this lesson, complex numbers are formally described (**N-CN.A.1**), and students review how to represent complex numbers as points in the complex plane (**N-CN.B.4**). Students look for and make use of structure as they see similarities between plotting ordered pairs of real numbers in the coordinate plane and plotting complex numbers in the complex plane.

MP.7 Next, students review the mechanics involved in adding complex numbers, subtracting complex numbers, multiplying a complex number by a scalar, and multiplying a complex number by a second complex number. Students look for and make use of structure as they see similarities between the process of multiplying two binomials and the process of multiplying two complex numbers.

Classwork

Opening Exercise (2 minutes)

Opening Exercise

Write down two fundamental facts about i that you learned in the previous lesson.

Multiplication by i induces a 90-degree counterclockwise rotation about the origin. Also, i satisfies the equation $i^2 = -1$.

Discussion (5 minutes): Describing Complex Numbers

- What do you recall about the meanings of the following terms? Briefly discuss what you remember with a partner, providing examples of each kind of number as you are able. After you have had a minute to share with one another, we will review each term as a whole class.
  - *Real number*
  - *Imaginary number*
  - *Complex number*

After students talk in pairs, bring the class together, and ask a few individual students to share an example of each kind of number.

- *Examples of real numbers:* $5, \frac{1}{2}, 0.865, -4, 0$
- *Examples of imaginary numbers:* $3i, 5i, -2i$
- *Examples of complex numbers:* $3+4i, 5-6i$

In the previous lesson, we reviewed the definition of the imaginary unit i. We can also form multiples of i, such as $2i, 3i, 4i, -10i$. The multiples of i are called *imaginary numbers*. As you know, the term *real number* refers to numbers like 3, -12, 0, $\frac{3}{5}$, $\sqrt{2}$, and so forth, none of which have an imaginary component. If we combine a real number and an imaginary number, we get expressions like these: $5+2i$, $4-3i$, $-6+10i$. These numbers are called *complex numbers*. In general, a complex number has the form $a+bi$, where a and b are both real numbers. The number a is called the *real component*, and the number b is called the *imaginary component*.

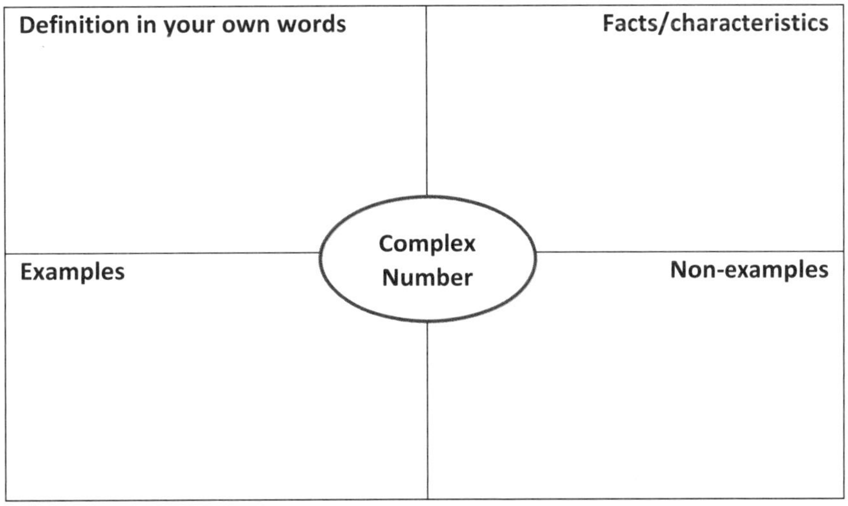

Discussion (5 minutes): Visualizing Complex Numbers

- Visualization is an extremely important tool in mathematics. How do we visually represent real numbers?
  - *Real numbers can be represented as points on a number line.*
- How do you suppose we could visually represent a complex number? Do you think we could use a number line just like the one we use for real numbers?
  - *Since it takes two real numbers a and b to describe a complex number $a+bi$, we cannot just use a single number line.*

In fact, we need two number lines to represent a complex number. The standard way to represent a complex number is to create what mathematicians call the *complex plane*. In the complex plane, the x-axis is used to represent the real component of a complex number, and the y-axis is used to represent its imaginary component. For instance, the complex number $5+2i$ has a real component of 5, so we take the point that is 5 units along the x-axis. The imaginary component is 2, so we take the point that is 2 units along the y-axis. In this way, we can associate the complex number $5+2i$ with the point $(5,2)$ as shown on the next page. This seemingly simple maneuver, associating complex numbers with points in the plane, turns out to have profound implications for our studies in this module.

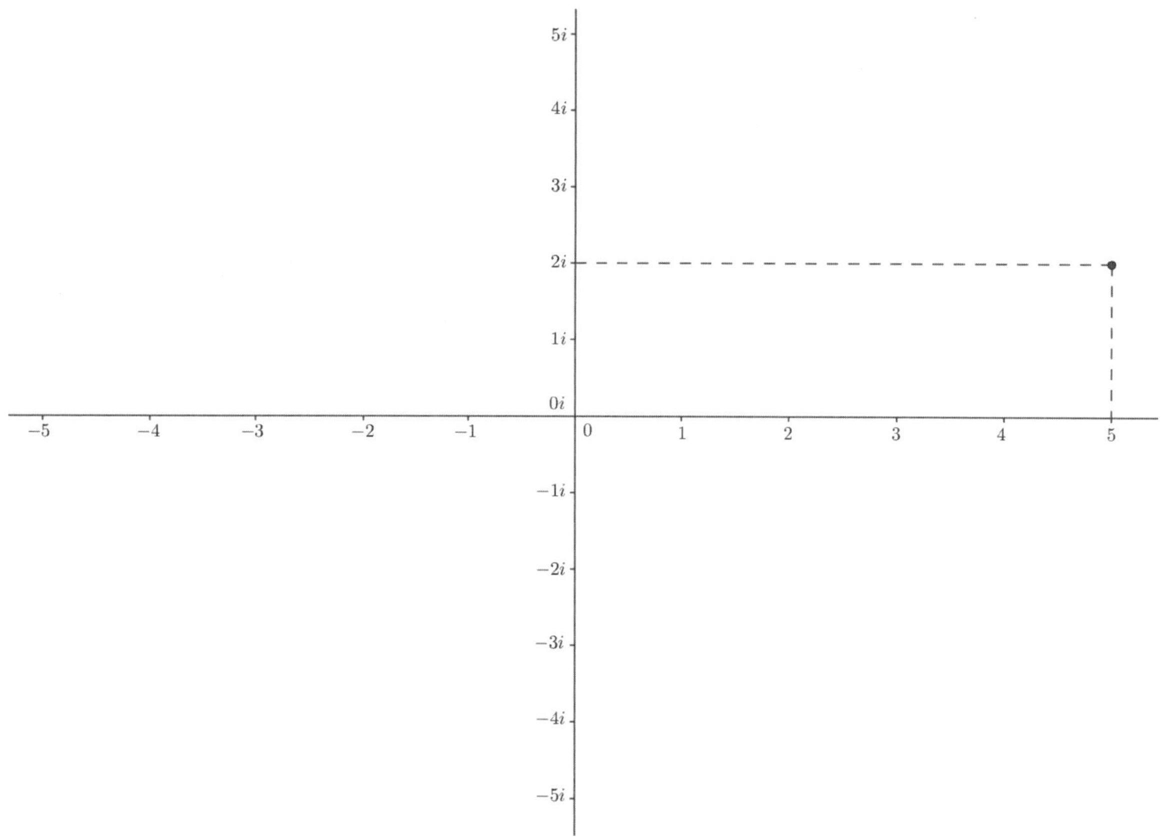

Discussion: Visualizing Complex Numbers

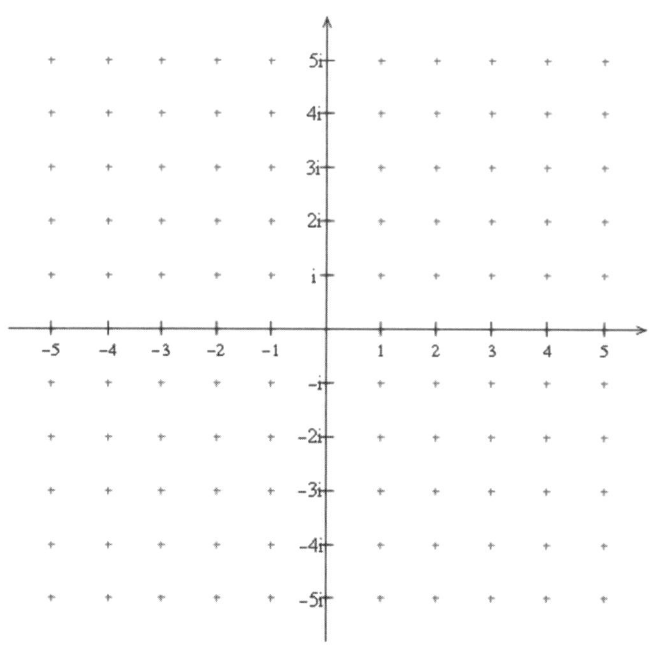

Exercises 1–3 (2 minutes)

Allow students time to respond to the following questions and to discuss their responses with a partner. Then, bring the class together, and allow a few individual students to share their responses with the class.

Exercises 1–7

1. **Give an example of a real number, an imaginary number, and a complex number. Use examples that have not already been discussed in the lesson.**

   Answers will vary.

2. **In the complex plane, what is the horizontal axis used for? What is the vertical axis used for?**

   The horizontal axis is used to represent the real component of a complex number. The vertical axis is used to represent the imaginary component.

3. **How would you represent $-4+3i$ in the complex plane?**

   The complex number $-4+3i$ corresponds to the point $(-4,3)$ in the coordinate plane.

Example 1 (6 minutes): Scalar Multiplication

Let's consider what happens when we multiply a real number by a complex number.

$$9(-8+10i)$$

MP.7

- Does this remind you of a situation that is handled by a property of real numbers?
  - *This expression resembles the form $a(b+c)$, which can be handled using the distributive property.*
- The distributive property tells us that $a(b+c)=ab+ac$, but the ordinary version of this property only applies when a, b, and c are real numbers. In fact, we can extend the use of the distributive property to include cases that involve complex numbers.

$$9(-8+10i)=9(-8)+9(10i)=-72+90i$$

- Let's explore this operation from a geometric point of view.
- If $w=3+i$, what do you suppose $2w$ looks like in the complex plane? Compute $2w$, and then plot both w and $2w$ in the complex plane.
  - $2w=2(3+i)=6+2i$

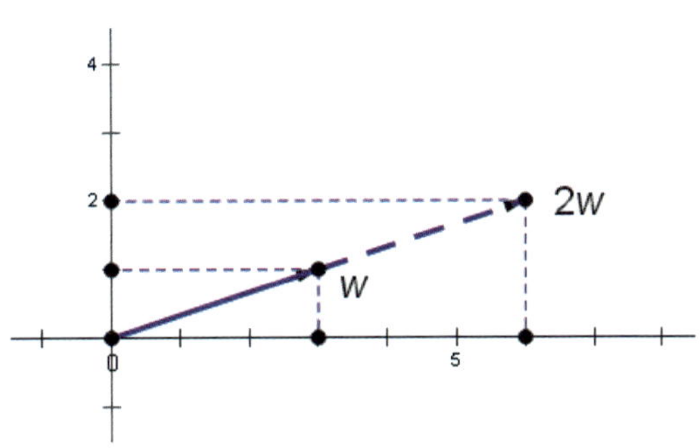

MP.7

- How would you describe the relationship between w and $2w$?
  - *The points representing w and $2w$ are on the same line through the origin. The distance from 0 to $2w$ is twice as long as the distance from 0 to w.*
- Notice that the real component of w was transformed from 3 to $2(3) = 6$, and the imaginary component of w was transformed from 1 to 2. We could say that each component got scaled up by a factor of 2. For this reason, multiplying by a real number is referred to as *scalar multiplication*, and since real numbers have this kind of scaling effect, they are sometimes called *scalars*.

Example 2 (7 minutes): Multiplying Complex Numbers

- Let's look now at an example that involves multiplying a complex number by another complex number.

$$(8 + 7i)(10 - 5i)$$

> *Scaffolding:*
>
> Advanced learners may be challenged to perform the multiplication without any cues from the teacher.

- What situation involving real numbers does this remind you of?
  - *It resembles the situation where you are multiplying two binomials: $(a + b)(c + d)$.*
- Although we are working with complex numbers, the distributive property still applies. Multiply the terms using the distributive property.

$$(8 + 7i)(10 - 5i) = 8(10 - 5i) + 7i(10 - 5i) = 80 - 40i + 70i - 35i^2$$

- What could we do next?
  - *We can combine the two i-terms into a single term:*

$$80 - 40i + 70i - 35i^2 = 80 + 30i - 35i^2$$

- Does anything else occur to you to try here?
  - *We know that $i^2 = -1$, so we can write*

$$80 + 30i - 35i^2 = 80 + 30i - 35(-1)$$
$$80 + 30i - 35(-1) = 80 + 30i + 35 = 115 + 30i$$

> *Scaffolding:*
>
> The diagram below can be used to multiply complex numbers in much the same way that it can be used to multiply two binomials. Some students may find it helpful to organize their work in this way.

- Summing up: We started with two complex numbers $8 + 7i$ and $10 - 5i$, we multiplied them together, and we produced a new complex number $115 + 30i$.

	8	$7i$
10	80	$70i$
$-5i$	$-40i$	$-35i^2$

Example 3 (5 minutes): Addition and Subtraction

- Do you recall the procedures for adding and subtracting complex numbers? Go ahead and give these two problems a try.

$$(-10 - 3i) + (-6 + 6i)$$
$$(9 - 6i) - (-3 - 10i)$$

Give students an opportunity to try these problems. Walk around the room, and monitor students' work, and then call on students to share what they have done.

$$(-10 - 3i) + (-6 + 6i) = (-10 + -6) + (-3i + 6i) = -16 + 3i$$
$$(9 - 6i) - (-3 - 10i) = [9 - (-3)] + [-6i - (-10i)] = 12 + 4i$$

- The key points to understand here are these:
  - To add two complex numbers, add the real components and the imaginary components separately.
  - To subtract two complex numbers, subtract the real components and the imaginary components separately.
- In the lesson today, we saw that when we multiply a complex number by a scalar, we get a new complex number that is simply a scaled version of the original number. Now, suppose we were to take two complex numbers z and w. How do you suppose $z + w$, $z - w$, and $z \cdot w$ are related geometrically? These questions will be explored in the upcoming lessons.

Exercises 4–7 (4 minutes)

Tell students to perform the following exercises for practice and then to compare their answers with a partner. Call on students at random to share their answers.

For Exercises 4–7, let $a = 1 + 3i$ and $b = 2 - i$.

4. **Find $a + b$. Then, plot a, b, and $a + b$ in the complex plane.**

   $a + b = 3 + 2i$

5. **Find $a - b$. Then, plot a, b, and $a - b$ in the complex plane.**

   $a - b = -1 + 4i$

6. **Find $2a$. Then, plot a and $2a$ in the complex plane.**

   $2a = 2 + 6i$

7. **Find $a \cdot b$. Then, plot a, b, and $a \cdot b$ in the complex plane.**

$$\begin{aligned} a \cdot b &= (1 + 3i)(2 - i) \\ &= 2 - i + 6i - 3i^2 \\ &= 2 + 5i + 3 \\ &= 5 + 5i \end{aligned}$$

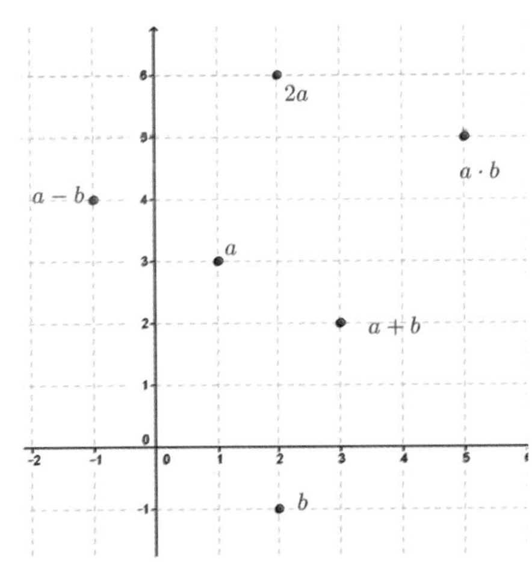

Closing (3 minutes)

Ask students to respond to the following questions in their notebooks, and then give them a minute to share their responses with a partner.

- What is the complex plane used for?
  - *The complex plane is used to represent complex numbers visually.*
- What operations did you learn to perform on complex numbers?
  - *We learned how to add, subtract, and multiply two complex numbers, as well as how to perform scalar multiplication on complex numbers.*
- Which of the four fundamental operations was not discussed in this lesson? This topic will be treated in an upcoming lesson.
  - *We did not discuss how to divide two complex numbers.*

Exit Ticket (6 minutes)

Name ______________________________ Date ________________

Lesson 5: An Appearance of Complex Numbers

Exit Ticket

In Problems 1–4, perform the indicated operations. Write each answer as a complex number $a + bi$.

1. Let $z_1 = -2 + i$, $z_2 = 3 - 2i$, and $w = z_1 + z_2$. Find w, and graph z_1, z_2, and w in the complex plane.

2. Let $z_1 = -1 - i$, $z_2 = 2 + 2i$, and $w = z_1 - z_2$. Find w, and graph z_1, z_2, and w in the complex plane.

3. Let $z = -2 + i$ and $w = -2z$. Find w, and graph z and w in the complex plane.

4. Let $z_1 = 1 + 2i$, $z_2 = 2 - i$, and $w = z_1 \cdot z_2$. Find w, and graph z_1, z_2, and w in the complex plane.

Exit Ticket Sample Solutions

In Problems 1–4, perform the indicated operations. Write each answer as a complex number $a + bi$.

1. **Let $z_1 = -2 + i$, $z_2 = 3 - 2i$, and $w = z_1 + z_2$. Find w, and graph z_1, z_2, and w in the complex plane.**

   $w = 1 - i$

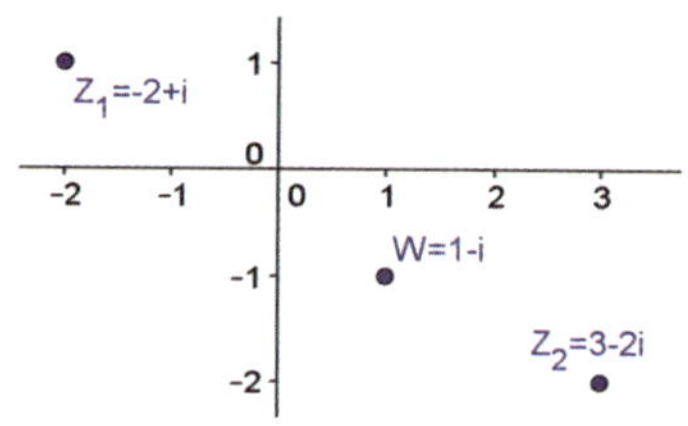

2. **Let $z_1 = -1 - i$, $z_2 = 2 + 2i$, and $w = z_1 - z_2$. Find w, and graph z_1, z_2, and w in the complex plane.**

   $w = -3 - 3i$

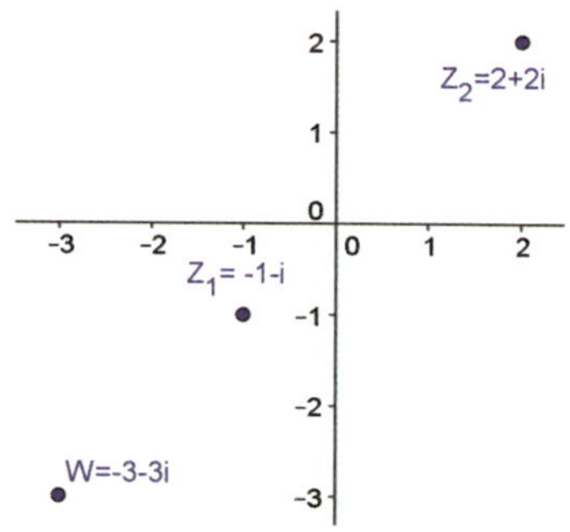

3. **Let $z = -2 + i$ and $w = -2z$. Find w, and graph z and w in the complex plane.**

   $w = 4 - 2i$

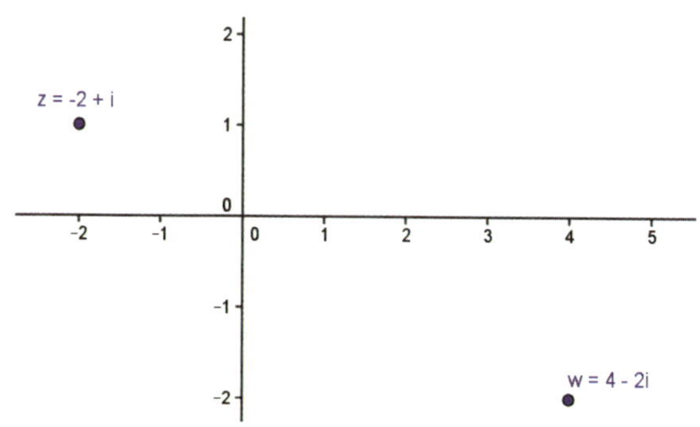

4. **Let $z_1 = 1 + 2i$, $z_2 = 2 - \text{i}$, and $w = z_1 \cdot z_2$. Find w, and graph z_1, z_2, and w in the complex plane.**

   $$\begin{aligned} w &= (1 + 2i)(2 - i) \\ &= 2 - i + 4i + 2 \\ &= 4 + 3i \end{aligned}$$

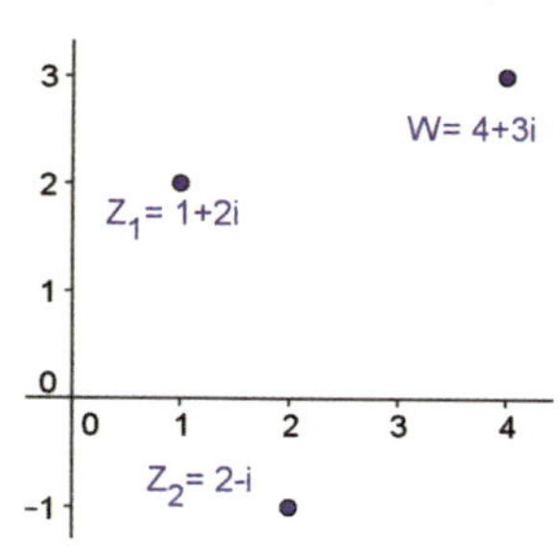

Problem Set Sample Solutions

Problems 1–4 involve the relationships among the set of real numbers, the set of imaginary numbers, and the set of complex numbers. Problems 5–9 involve practice with the core set of arithmetic skills from this lesson. Problems 10–12 involve the complex plane. A reproducible complex plane is provided at the end of the lesson should the teacher choose to hand out copies for the Problem Set.

1. **The number 5 is a real number. Is it also a complex number? Try to find values of a and b so that $5 = a + bi$.**

   Because $5 = 5 + 0i$, 5 is a complex number.

2. **The number $3i$ is an imaginary number and a multiple of i. Is it also a complex number? Try to find values of a and b so that $3i = a + bi$.**

   Because $3i = 0 + 3i$, $3i$ is a complex number.

3. **Daria says that every real number is a complex number. Do you agree with her? Why or why not?**

   For any real number a, $a = a + 0i$, so Daria is correct.

4. **Colby says that every imaginary number is a complex number. Do you agree with him? Why or why not?**

   An imaginary number $bi = 0 + bi$, so Colby is correct.

In Problems 5–9, perform the indicated operations. Report each answer as a complex number $w = a + bi$, and graph it in a complex plane.

5. **Given $z_1 = -9 + 5i$, $z_2 = -10 - 2i$, find $w = z_1 + z_2$, and graph z_1, z_2, and w.**

   $w = -19 + 3i$

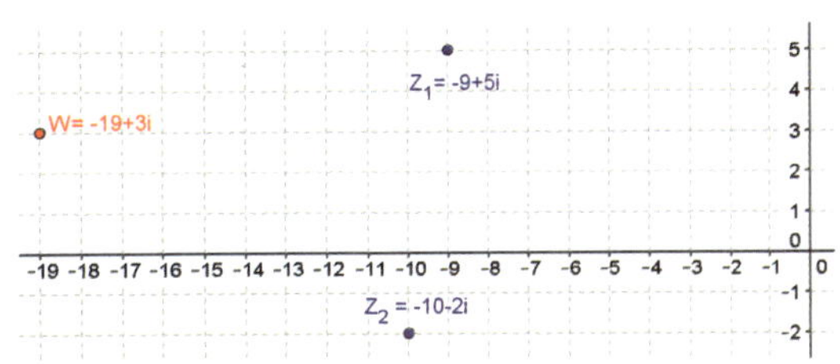

6. **Given $z_1 = -4 + 10i$, $z_2 = -7 - 6i$, find $w = z_1 - z_2$, and graph z_1, z_2, and w.**

   $w = (-4 + 10i) - (-7 - 6i)$
   $= 3 + 16i$

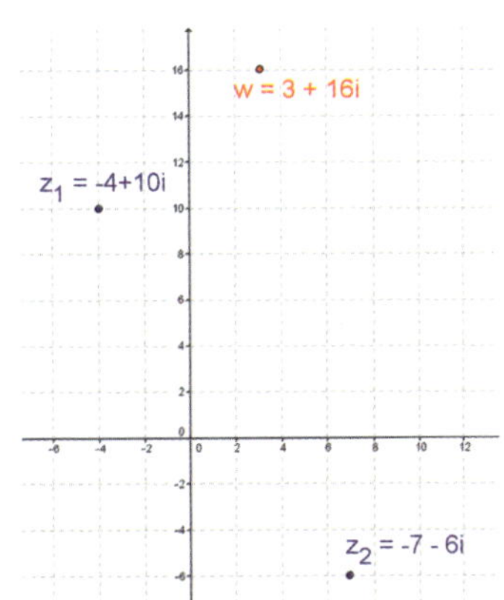

7. Given $z_1 = 3\sqrt{2} + 2i$, $z_2 = \sqrt{2} - i$, find $w = z_1 - z_2$, and graph z_1, z_2, and w.

$w = (3\sqrt{2} + 2i) - (\sqrt{2} - i)$
$= 2\sqrt{2} + 3i$

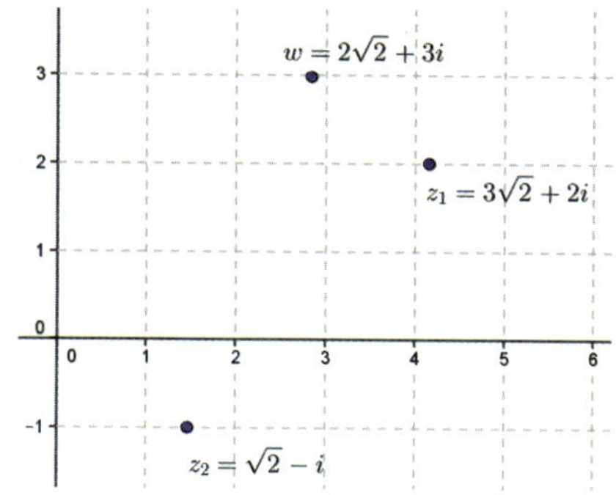

8. Given $z_1 = 3$, $z_2 = -4 + 8i$, find $w = z_1 \cdot z_2$, and graph z_1, z_2, and w.

$w = -12 - 24i$

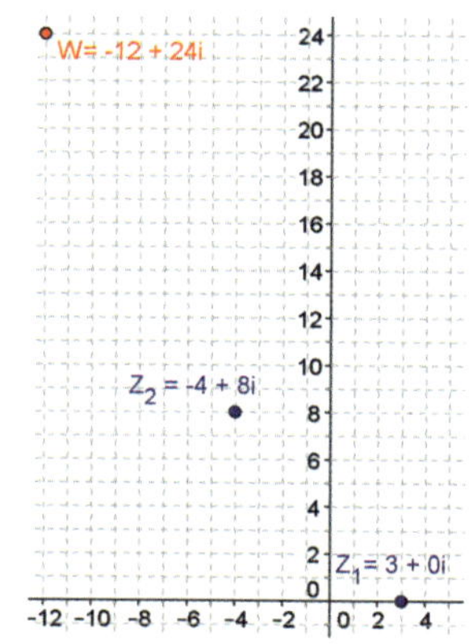

9. Given $z_1 = \frac{1}{4}$, $z_2 = 12 - 4i$, find $w = z_1 \cdot z_2$, and graph z_1, z_2, and w.

$w = 3 - i$

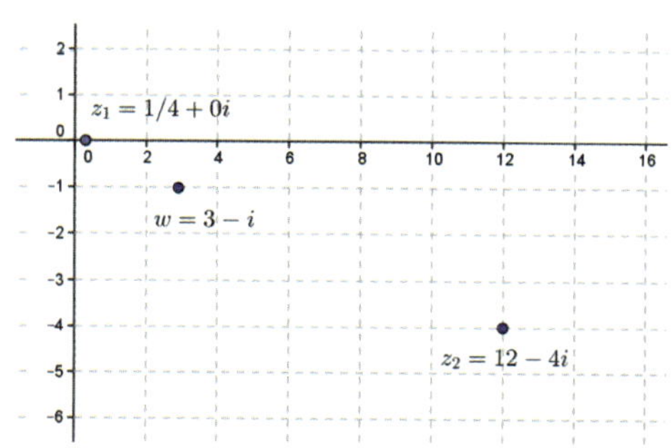

10. Given $z_1 = -1$, $z_2 = 3 + 4i$, find $w = z_1 \cdot z_2$, and graph z_1, z_2, and w.

$w = -3 - 4i$

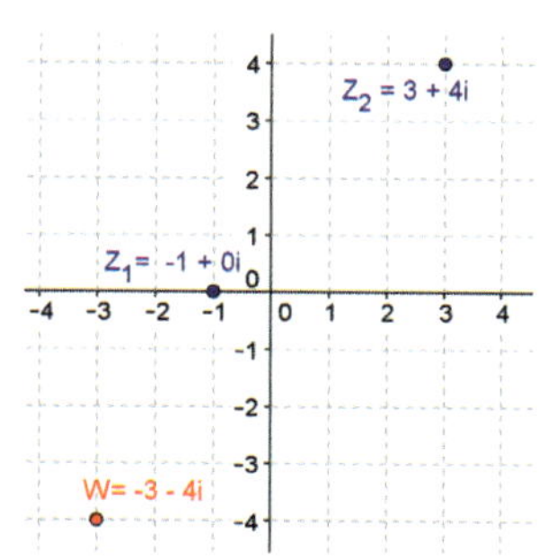

11. Given $z_1 = 5 + 3i$, $z_2 = -4 - 2i$, find $w = z_1 \cdot z_2$, and graph z_1, z_2, and w.

$$w = (5 + 3i)(-4 - 2i)$$
$$= -20 - 10i - 12i - 6i^2$$
$$= -20 - 22i - 6(-1)$$
$$= -14 - 22i$$

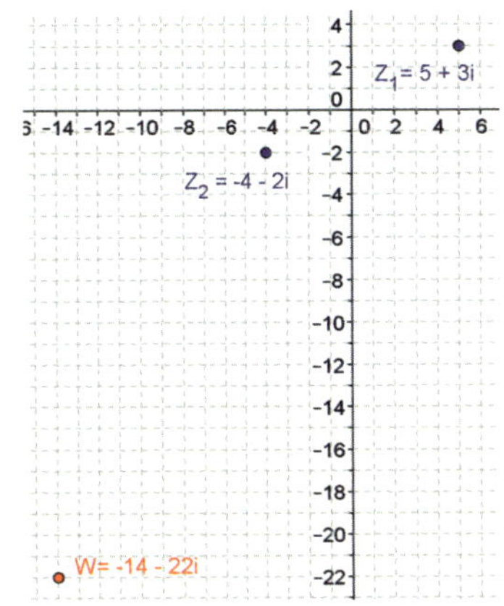

12. Given $z_1 = 1 + i$, $z_2 = 1 + i$, find $w = z_1 \cdot z_2$, and graph z_1, z_2, and w.

$$w = (1 + i)(1 + i)$$
$$= 1 + 2i + i^2$$
$$= 1 + 2i - 1$$
$$= 2i$$

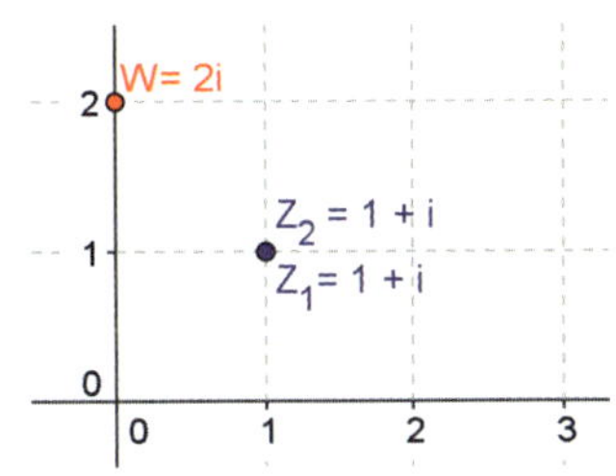

13. Given $z_1 = 3$, $z_2 = i$, find $w = z_1 \cdot z_2$, and graph z_1, z_2, and w.

$$w = i \cdot 3$$
$$= 3i$$

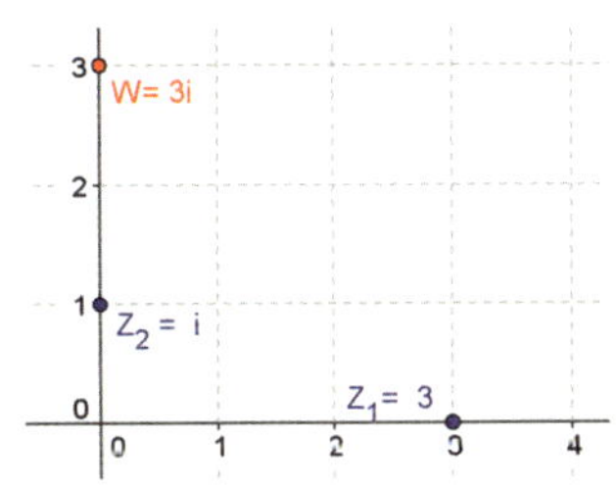

14. Given $z_1 = 4 + 3i$, $z_2 = i$, find $w = z_1 \cdot z_2$, and graph z_1, z_2, and w.

$$w = i(4 + 3i)$$
$$= -3 + 4i$$

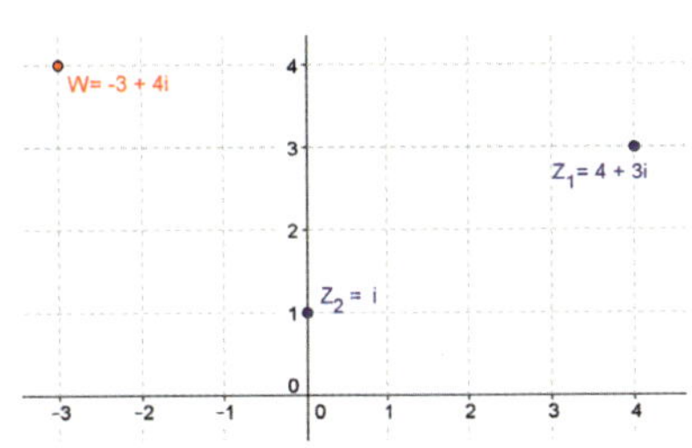

15. Given $z_1 = 2\sqrt{2} + 2\sqrt{2}i$, $z_2 = -\sqrt{2} + \sqrt{2}i$, find $w = z_1 \cdot z_2$, and graph z_1, z_2, and w.

$w = (2\sqrt{2} + 2\sqrt{2}i)(-\sqrt{2} + \sqrt{2}i)$

$= -4 + 4i - 4i - 4$

$= -8$

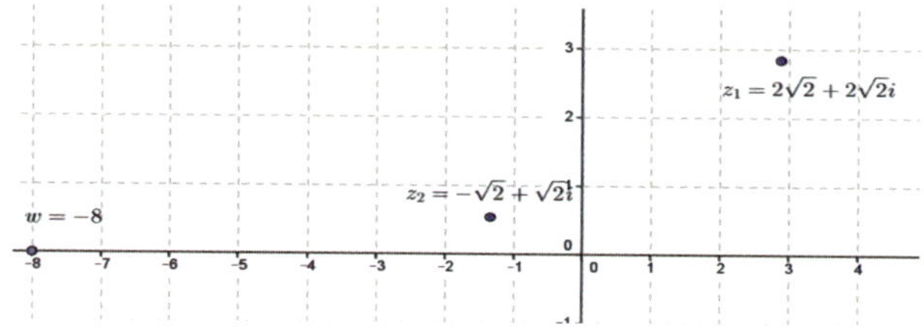

16. Represent $w = -4 + 3i$ as a point in the complex plane.

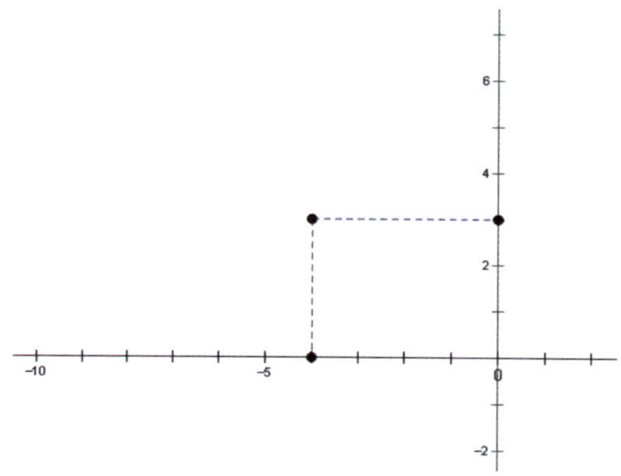

17. Represent $2w$ as a point in the complex plane. $2w = 2(-4 + 3i) = -8 + 6i$

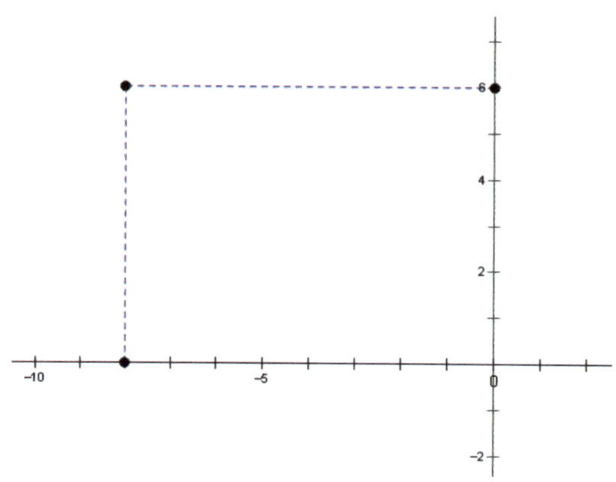

18. **Compare the positions of w and $2w$ from Problems 10 and 11. Describe what you see. (Hint: Draw a segment from the origin to each point.)**

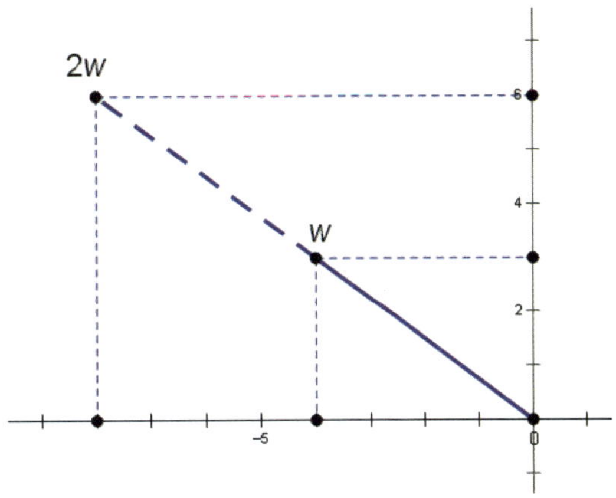

The points 0, w, and $2w$ all lie on the same line. The distance from 0 to $2w$ is twice as great as the distance from 0 to w. The segment to $2w$ is a scaled version of the segment to w, with scale factor 2.

Complex Plane Reproducible

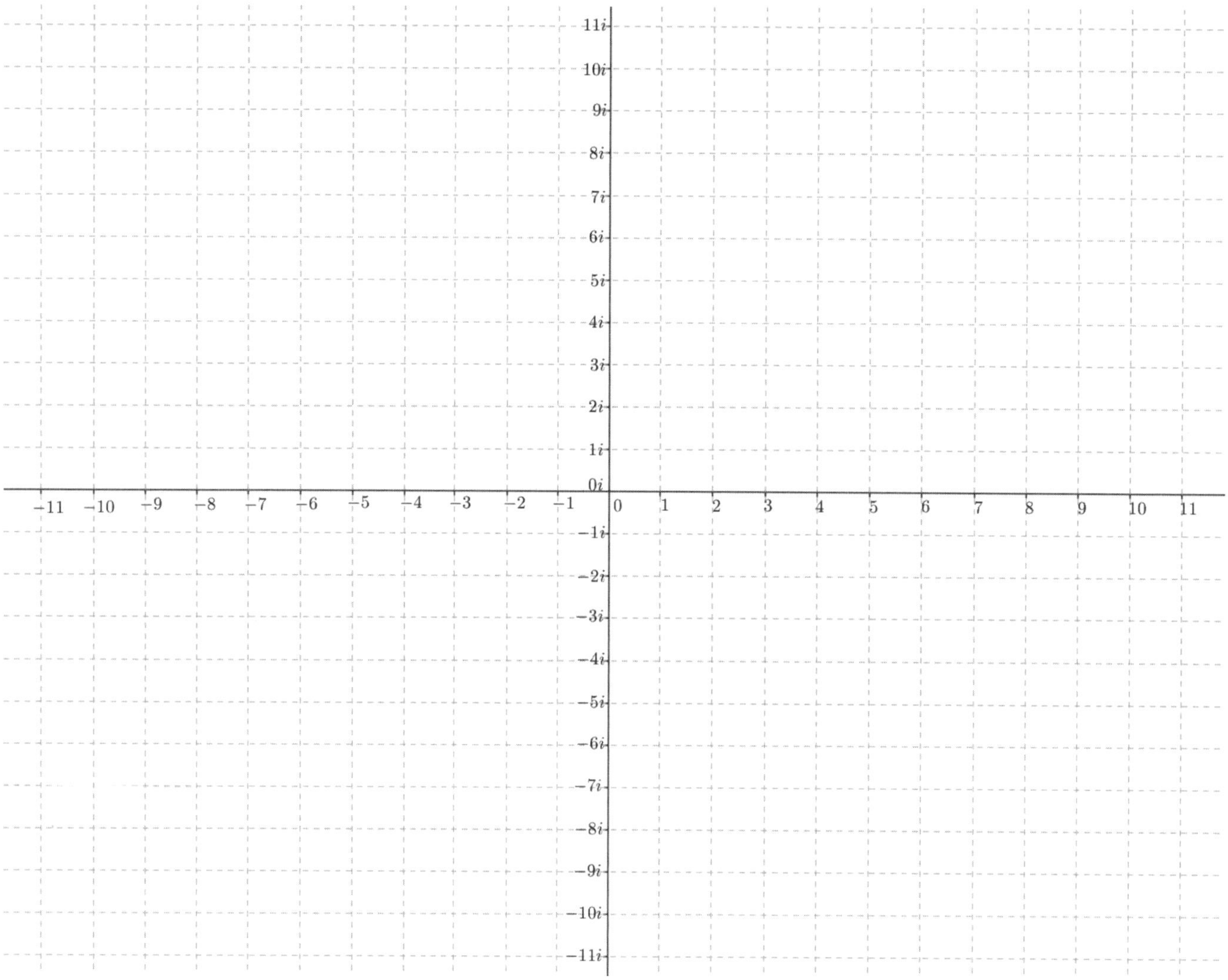

Lesson 6: Complex Numbers as Vectors

Student Outcomes

- Students represent complex numbers as vectors.
- Students represent complex number addition and subtraction geometrically using vectors.

Lesson Notes

Students studied vectors as directed line segments in Grade 8, and in this lesson, vectors are used to represent complex numbers in the coordinate plane. This representation presents a geometric interpretation of addition and subtraction of complex numbers and is needed to make the case in Lesson 15 that when multiplying two complex numbers z and w, the argument of the product is the sum of the arguments: $\arg(zw) = \arg(z) + \arg(w)$.

This lesson aligns with **N-CN.B.5**: Represent addition, subtraction, multiplication, and conjugation of complex numbers geometrically on the complex plane; use properties of this representation for computation.

The following vocabulary terms from Grade 8 are needed in this lesson:

VECTOR: A *vector* associated to the directed line segment $\overrightarrow{AB}$ is any directed segment that is congruent to the directed segment $\overrightarrow{AB}$ using only translations of the plane.

DIRECTED SEGMENT: A *directed segment* $\overrightarrow{AB}$ is the line segment AB together with a direction given by connecting an initial point A to a terminal point B.

The study of vectors forms a vital part of this course; notation for vectors varies across different contexts and curricula. These materials refer to a vector as $\mathbf{v}$ (lowercase, bold, non-italicized) or $\langle 4, 5\rangle$ or in column format, $\begin{pmatrix}4\\5\end{pmatrix}$ or $\begin{bmatrix}4\\5\end{bmatrix}$.

"Let $\mathbf{v} = \langle 4,5\rangle$" is used to establish a name for the vector $\langle 4, 5\rangle$.

This curriculum avoids stating $\mathbf{v} = \langle 4,5\rangle$ without the word *let* preceding the equation when naming a vector unless it is absolutely clear from the context that a vector is being named. However, the "=" continues to be used to describe vector equations, like $\mathbf{v} + \mathbf{w} = \mathbf{w} + \mathbf{v}$, as has been done with equations throughout all other grades.

The vector from A to B is referred to as "vector $\overrightarrow{AB}$"—notice, this is a ray with a full arrow. This notation is consistent with the way vectors are introduced in Grade 8 and is also widely used in postsecondary textbooks to describe both a ray and a vector depending on the context. To avoid confusion in this curriculum, the context is provided or strongly implied, so it is clear whether the full arrow indicates a vector or a ray. For example, when referring to a ray from A passing through B, "ray $\overrightarrow{AB}$" is used, and when referring to a vector from A to B, "vector $\overrightarrow{AB}$" is used. Students should be encouraged to think about the context of the problem and not just rely on a hasty inference based on the symbol.

The magnitude of a vector is signified as $\|\mathbf{v}\|$ (lowercase, bold, non-italicized).

Classwork

Opening Exercise (4 minutes)

The Opening Exercise reviews complex number arithmetic. This example is revisited later in the lesson when the geometric interpretation of complex addition and subtraction using a vector representation of complex numbers is studied. Students should work on these exercises either individually or in pairs.

Opening Exercise

Perform the indicated arithmetic operations for complex numbers $z = -4 + 5i$ and $w = -1 - 2i$.

a. $z + w$

$z + w = -5 + 3i$

b. $z - w$

$z - w = -3 + 7i$

c. $z + 2w$

$z + 2w = -6 + i$

d. $z - z$

$z - z = 0 + 0i$

e. **Explain how you add and subtract complex numbers.**

Add or subtract the real components and the imaginary components separately.

Discussion (6 minutes)

- In Lesson 5, we represented a complex number $a + bi$ as the point (a, b) in the coordinate plane. Another way we can represent a complex number in the coordinate plane is as a vector. Recall the definition of a vector from Grade 8, which is that a vector $\overrightarrow{AB}$ is a directed segment from point A in the plane to point B, which we draw as an arrow from point A to point B. Since we can represent a complex number $z = a + bi$ as the point (a, b) in the plane, and we can use a vector to represent the directed segment from the origin to the point (a, b), we can represent a complex number as a vector in the plane. The vector representing the complex number $z = -3 + 4i$ is shown.

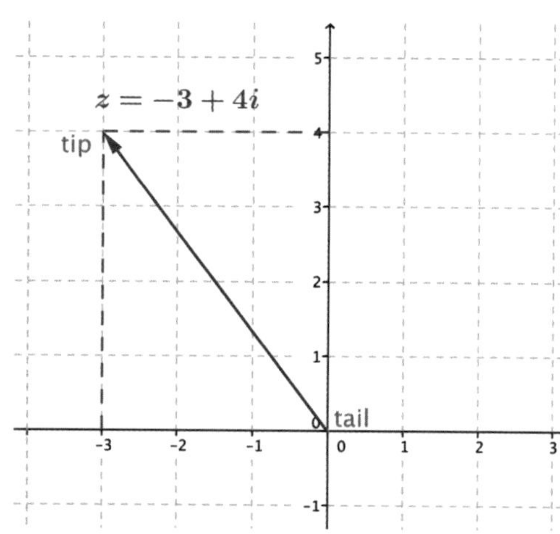

- The length of a vector $\overrightarrow{AB}$ is the distance from the tail A of the vector to the tip B. For our purposes, the tail is the origin, and the tip is the point $z = (a + bi)$ in the coordinate plane.

- A vector consists of a length and a direction. To get from point A to point B, you move a distance AB in the direction of the vector $\overrightarrow{AB}$. So, to move from the origin to point $z = a + bi$, we move the length of $a + bi$ in the direction of the line from the origin to (a, b). (This idea is important later in the lesson when we use vectors to add and subtract complex numbers.)
- What is the length of the vector that represents the complex number $z_1 = -3 + 4i$?
    - *Since $\sqrt{(-3-0)^2 + (4-0)^2} = \sqrt{9+16} = 5$, the length of the vector that represents z_1 is 5.*
- What is the length of the vector that represents the complex number $z_2 = 2 - 7i$?
    - *Since $\sqrt{(2-0)^2 + (-7-0)^2} = \sqrt{4+49} = \sqrt{53}$, the length of the vector that represents z_2 is $\sqrt{53}$.*
- What is the length of the vector that represents the complex number $z_3 = a + bi$?
    - *Since $\sqrt{(a-0)^2 + (b-0)^2} = \sqrt{a^2+b^2}$, the length of the vector that represents z_3 is $\sqrt{a^2+b^2}$.*

Exercise 1 (8 minutes)

Have students work Exercise 1 in pairs or small groups. Circulate to be sure that students are correctly plotting the complex numbers in the plane and correctly computing the lengths of the resulting vectors.

Exercises

1. The length of the vector that represents $z_1 = 6 - 8i$ is 10 because $\sqrt{6^2 + (-8)^2} = \sqrt{100} = 10$.

a. Find at least seven other complex numbers that can be represented as vectors that have length 10.

There are an infinite number of complex numbers that meet this criteria; the most obvious are 10, $6+8i$, $8+6i$, $10i$, $-6+8i$, $-8+6i$, -10, $-8-6i$, $-6-8i$, $-10i$, $8-6i$, and $-8+6i$. The associated vectors for these numbers are shown in the sample response for part (b).

Scaffolding:

Students struggling to find these values may want to work with the more general formula $a^2 + b^2 = 100$. Choose either a or b to create an equation they can solve. This also helps students see the relation to a circle in part (c).

b. Draw the vectors on the coordinate axes provided below.

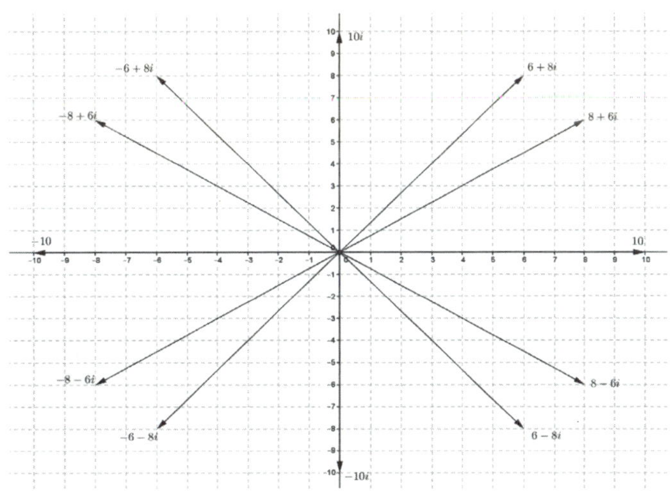

MP.7

c. What do you observe about all of these vectors?

Students should observe that the tips of the vectors lie on the circle of radius 10 centered at the origin.

Discussion (10 minutes)

Proceed slowly through this Discussion, drawing plenty of figures to clarify the process of adding two vectors.

MP.2 & MP.3

- Using vectors, how can we add two complex numbers? We know from the Opening Exercise that if $z = -4 + 5i$ and $w = -1 - 2i$, $z + w = -5 + 3i$. How could we show this is true using vectors?
  - *We know that if we think of vectors as a length and a direction, the sum $z + w$ is the distance we need to move from the origin in the direction of the vector $z + w$ to get to the point $z + w$. We can get from the origin to point $z + w$ by first moving from the origin to point z and then moving from point z to point w.*
- Using coordinates, this means that to find $z + w$, we do the following:
  1. Starting at the origin, move 4 units left and 5 units up to the tip of the vector representing z.
  2. From point z, move 1 unit left and 2 units down.
  3. The resulting point is $z + w$.
- Using vectors, we can locate point $z + w$ by the tip-to-tail method: Translate the vector that represents w so that the tip of z is at the same point as the tail of the new vector that represents w. The tip of this new translated vector is the sum $z + w$. See the sequence of graphs below.

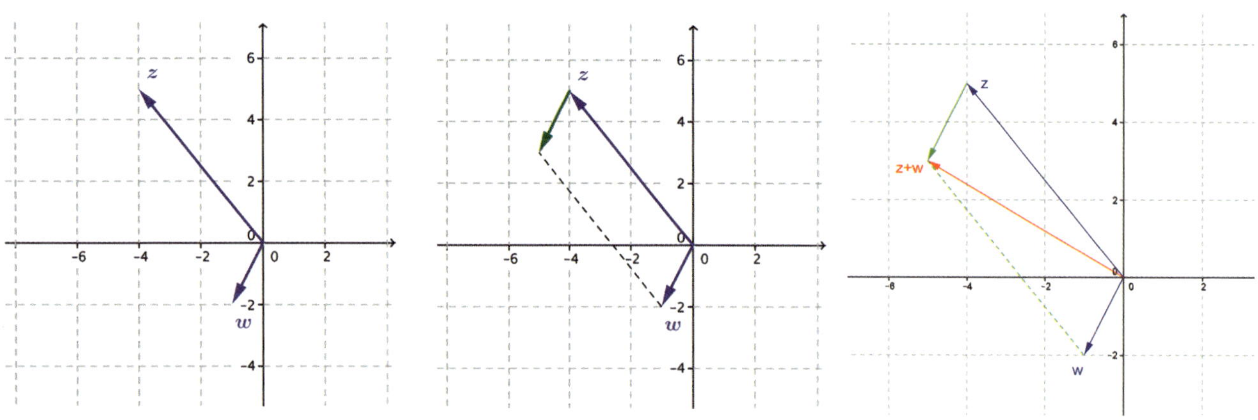

MP.7

- How would we find $z - w$ using vectors?
  - *We could think of $z - w$ as $z + (-w)$ since we already know how to add vectors.*
- Thinking of a vector as a length and a direction, how does $-w$ relate to w?
  - *The vector $-w$ would have the same length as w but the opposite direction.*

Scaffolding:

For advanced students, discuss how to find $z - w$ using the original parallelogram; that is, $z - w$ is the vector from the tip of w to the tip of z and then translated to the origin. Discuss how this is the same as subtraction in one dimension.

- Yes. So, we need to add $-w$ to z. To find $-w$, we reverse its direction. See the sequence of graphs below.

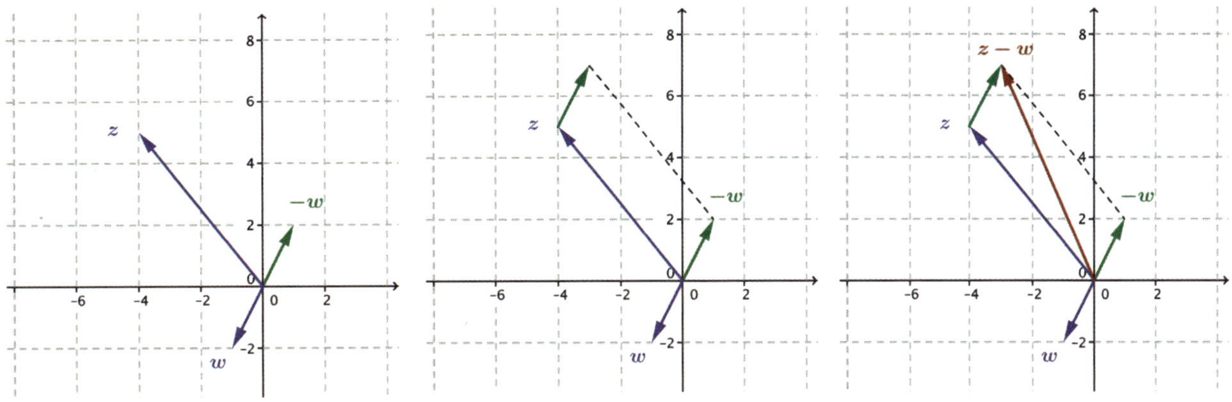

- In the Opening Exercise, we found that $z - w = -3 + 7i$. Does that agree with our calculation using vectors?
  - *Yes*

Exercises 2–6 (8 minutes)

Have students work on these exercises in pairs or small groups.

> *Scaffolding:*
>
> Model an example such as $2z + w$ for struggling students before asking them to work on these exercises.

2. In the Opening Exercise, we computed $z + 2w$. Calculate this sum using vectors.

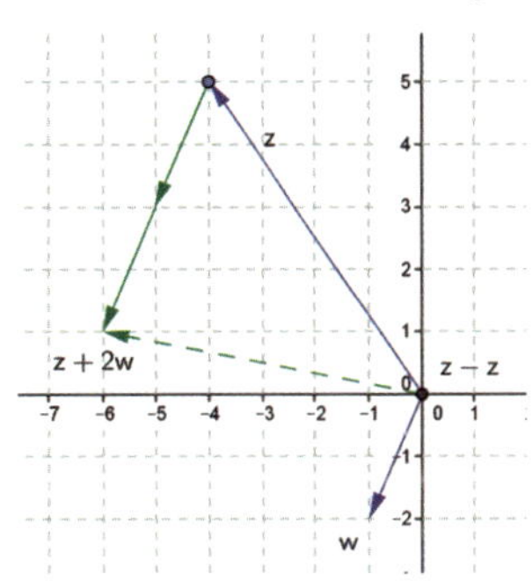

3. In the Opening Exercise, we also computed $z - z$. Calculate this sum using vectors.

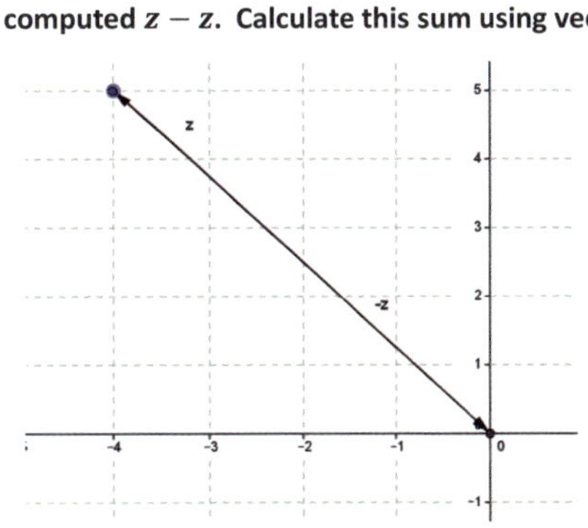

4. **For the vectors u and v pictured below, draw the specified sum or difference on the coordinate axes provided.**

   a. $\mathbf{u} + \mathbf{v}$

   b. $\mathbf{v} - \mathbf{u}$

   c. $2\mathbf{u} - \mathbf{v}$

   d. $-\mathbf{u} - 3\mathbf{v}$

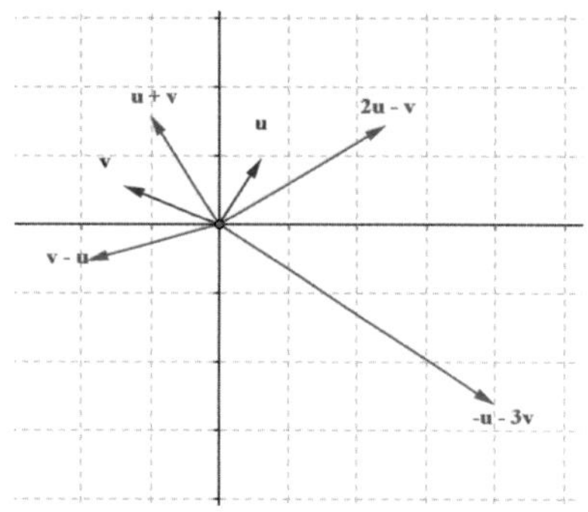

5. **Find the sum of $4 + i$ and $-3 + 2i$ geometrically.**

   $1 + 3i$

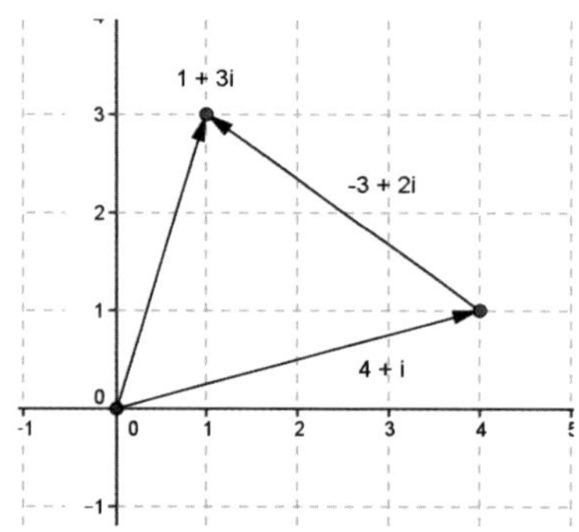

6. **Show that $(7 + 2i) - (4 - i) = 3 + 3i$ by representing the complex numbers as vectors.**

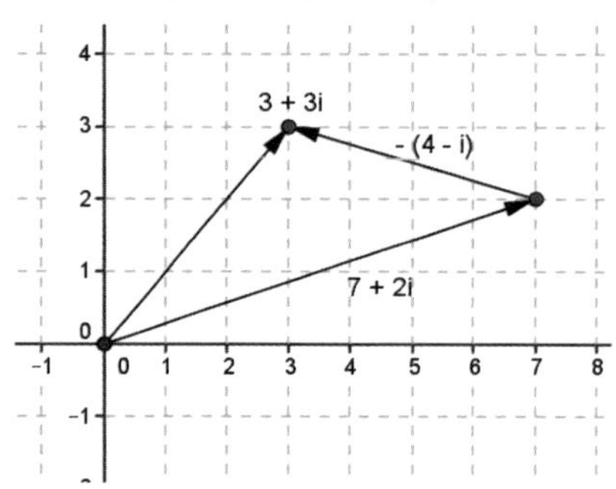

Closing (4 minutes)

Ask students to write in their journals or notebooks to explain the process of representing a complex number by a vector and the processes for adding and subtracting two vectors.

Exit Ticket (5 minutes)

Name ______________________________ Date ______________

Lesson 6: Complex Numbers as Vectors

Exit Ticket

Let $z = -1 + 2i$ and $w = 2 + i$. Find the following, and verify each geometrically by graphing z, w, and each result.

a. $z + w$

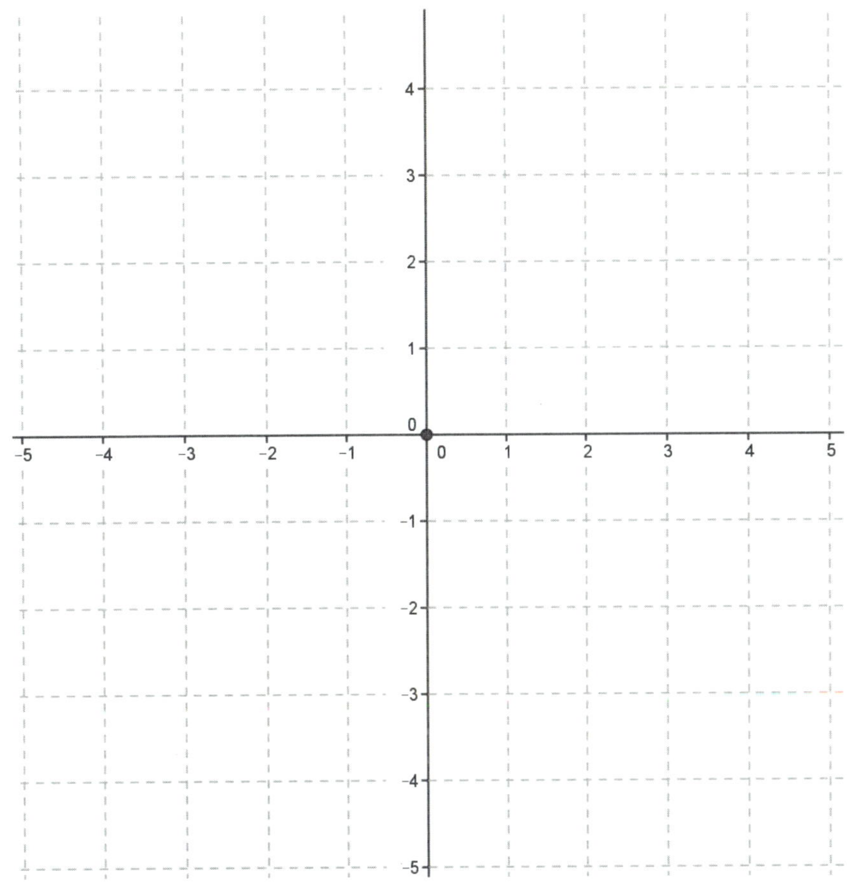

b. $z - w$

c. $2z - w$

d. $w - z$

Exit Ticket Sample Solutions

Let $z = -1 + 2i$ and $w = 2 + i$. Find the following, and verify each geometrically by graphing z, w, and each result.

a. $z + w$

$1 + 3i$

b. $z - w$

$-3 + i$

c. $2z - w$

$-4 + 3i$

d. $w - z$

$3 - i$

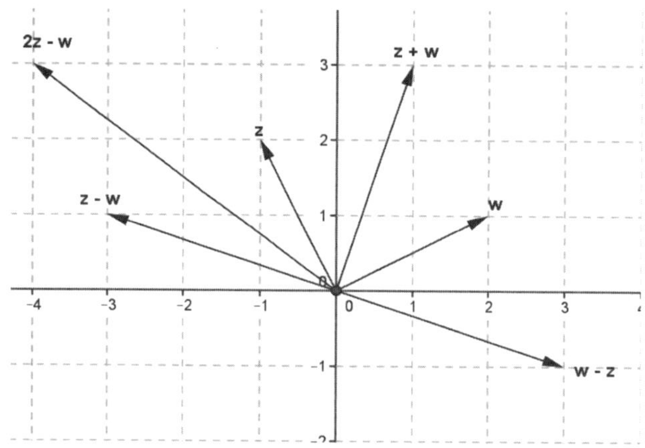

Problem Set Sample Solutions

1. Let $z = 1 + i$ and $w = 1 - 3i$. Find the following. Express your answers in $a + bi$ form.

a. $z + w$

$1 + i + 1 - 3i = 2 - 2i$

b. $z - w$

$$1 + i - (1 - 3i) = 1 + i - 1 + 3i$$
$$= 0 + 4i$$

c. $4w$

$4(1 - 3i) = 4 - 12i$

d. $3z + w$

$$3(1 + i) + 1 - 3i = 3 + 3i + 1 - 3i$$
$$= 4 + 0i$$

e. $-w - 2z$

$$-(1 - 3i) - 2(1 + i) = -1 + 3i - 2 - 2i$$
$$= -3 + i$$

f. What is the length of the vector representing z?

The length of the vector representing z is $\sqrt{1^2 + 1^2} = \sqrt{2}$.

g. What is the length of the vector representing w?

The length of the vector representing w *is* $\sqrt{1^2 + (-3)^2} = \sqrt{10}$.

2. Let $u = 3 + 2i$, $v = 1 + i$, and $w = -2 - i$. Find the following. Express your answer in $a + bi$ form, and represent the result in the plane.

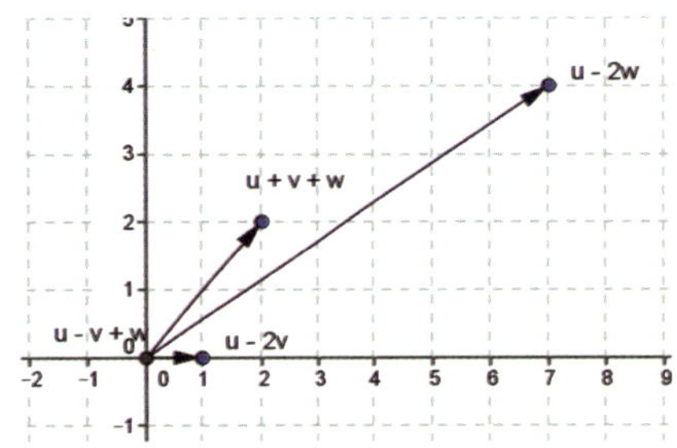

a. $u - 2v$

$$3 + 2i - 2(1 + i) = 3 + 2i - 2 - 2i$$
$$= 1 + 0i$$

b. $u - 2w$

$$3 + 2i - 2(-2 - i) = 3 + 2i + 4 + 2i$$
$$= 7 + 4i$$

c. $u + v + w$

$$3 + 2i + 1 + i - 2 - i = 2 + 2i$$

d. $u - v + w$

$$3 + 2i - (1 + i) - 2 - i = 3 + 2i - 1 - i - 2 - i$$
$$= 0 + 0i$$

e. What is the length of the vector representing u?

The length of the vector representing u *is* $\sqrt{3^2 + 2^2} = \sqrt{13}$.

f. What is the length of the vector representing $u - v + w$?

The length of the vector representing $u - v + w = \sqrt{0^2 + 0^2} = \sqrt{0} = 0$.

3. Find the sum of $-2 - 4i$ and $5 + 3i$ geometrically.

$3 - i$

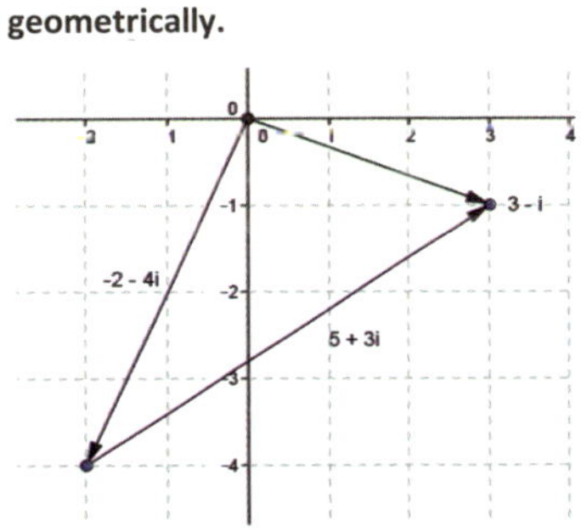

4. Show that $(-5-6i)-(-8-4i)=3-2i$ by representing the complex numbers as vectors.

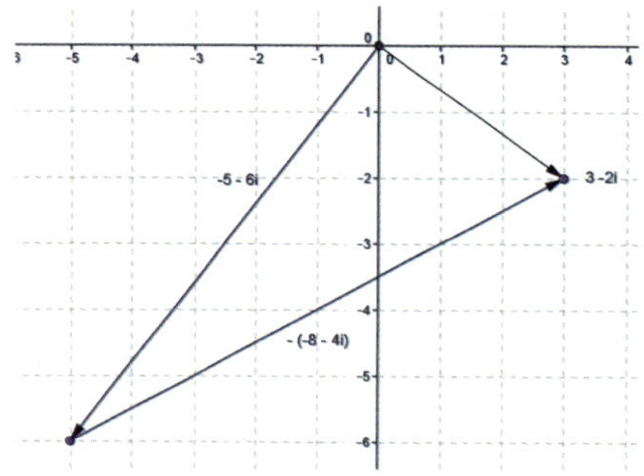

5. Let $z_1=a_1+b_1i$, $z_2=a_2+b_2i$, and $z_3=a_3+b_3i$. Prove the following using algebra or by showing with vectors.

   a. $z_1+z_2=z_2+z_1$

$$\begin{aligned} \boldsymbol{z_1+z_2} &= (a_1+b_1i)+(a_2+b_2i) \\ &= (a_2+b_2i)+(a_1+b_1i) \\ &= z_2+z_1 \end{aligned}$$

   b. $z_1+(z_2+z_3)=(z_1+z_2)+z_3$

$$\begin{aligned} z_1+(z_2+z_3) &= (a_1+b_1i)+\big((a_2+b_2i)+(a_3+b_3i)\big) \\ &= \big((a_1+b_1i)+(a_2+b_2i)\big)+(a_3+b_3i) \\ &= (z_1+z_2)+z_3 \end{aligned}$$

6. Let $z=-3-4i$ and $w=-3+4i$.

   a. Draw vectors representing z and w on the same set of axes.

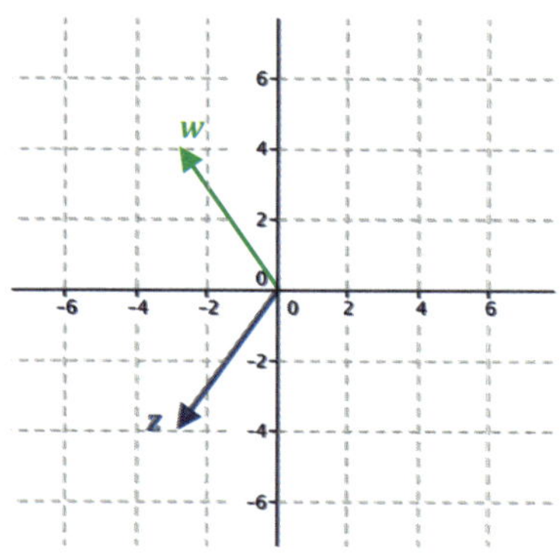

   b. What are the lengths of the vectors representing z and w?

   The length of the vector representing z is $\sqrt{(-3)^2+(-4)^2}=\sqrt{25}=5$.

   The length of the vector representing w is $\sqrt{(-3)^2+4^2}=\sqrt{25}=5$.

   c. Find a new vector, u_z, such that u_z is equal to z divided by the length of the vector representing z.

   $$u_z=\frac{-3-4i}{5}=\frac{-3}{5}-\frac{4}{5}i$$

d. Find u_w, such that u_w is equal to w divided by the length of the vector representing w.

$$u_w = \frac{-3+4i}{5} = \frac{-3}{5} + \frac{4}{5}i$$

e. Draw vectors representing u_z and u_w on the same set of axes as part (a).

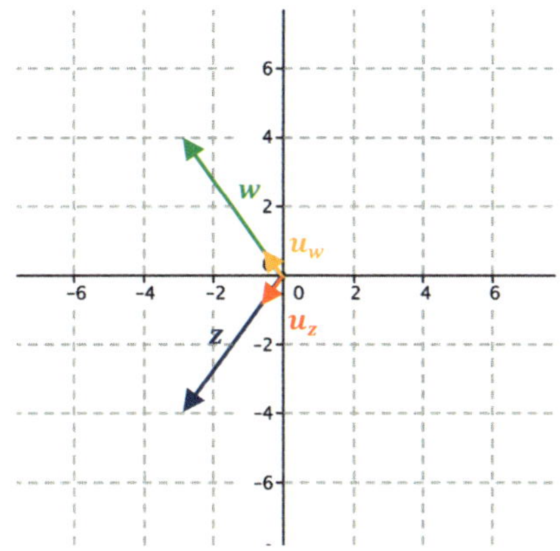

f. What are the lengths of the vectors representing u_z and u_w?

The length of the vector representing u_z *is* $\sqrt{\left(-\frac{3}{5}\right)^2 + \left(-\frac{4}{5}\right)^2} = \sqrt{\frac{9}{25} + \frac{16}{25}} = \sqrt{\frac{25}{25}} = \sqrt{1} = 1.$

The length of the vector representing u_w *is* $\sqrt{\left(-\frac{3}{5}\right)^2 + \left(\frac{4}{5}\right)^2} = \sqrt{\frac{9}{25} + \frac{16}{25}} = \sqrt{\frac{25}{25}} = \sqrt{1} = 1.$

g. Compare the vectors representing u_z to z and u_w to w. What do you notice?

The vectors representing u_z and u_w are in the same direction as z and w, respectively, but their lengths are only 1.

h. What is the value of u_z times u_w?

$$\left(\frac{3}{5} - \frac{4}{5}i\right)\left(\frac{3}{5} + \frac{4}{5}i\right) = \left(\frac{3}{5}\right)^2 - \left(\frac{4}{5}i\right)^2 = \frac{9}{25} + \frac{16}{25} = 1$$

i. What does your answer to part (h) tell you about the relationship between u_z and u_w?

Since their product is 1, we know that u_z and u_w are reciprocals of each other.

7. Let $z = a + bi$.

a. Let u_z be represented by the vector in the direction of z with length 1. How can you find u_z? What is the value of u_z?

Find the length of z, and then divide z by its length.

$$u_z = \frac{a+bi}{\sqrt{a^2+b^2}}$$

b. Let u_w be the complex number that when multiplied by u_z, the product is 1. What is the value of u_w?

From Problem 4, we expect $u_w = \dfrac{a-bi}{\sqrt{a^2+b^2}}$. *Multiplying, we get*

$$\frac{a+bi}{\sqrt{a^2+b^2}} \cdot \frac{a-bi}{\sqrt{a^2+b^2}} = \frac{a^2-(bi)^2}{a^2+b^2}$$
$$= \frac{a^2+b^2}{a^2+b^2}$$
$$= 1$$

c. What number could we multiply z by to get a product of 1?

Since we know that u_z *is equal to* z *divided by the length of* z *and that* $u_z \cdot u_w = 1$, *we get*

$$z \cdot \frac{1}{\sqrt{a^2+b^2}} \cdot \frac{a-bi}{\sqrt{a^2+b^2}} = z \cdot \frac{a-bi}{a^2+b^2} = 1$$

So, multiplying z *by* $\dfrac{a-bi}{a^2+b^2}$ *will result in a product of* 1.

8. Let $z = -3 + 5i$.

a. Draw a picture representing $z + w = 8 + 2i$.

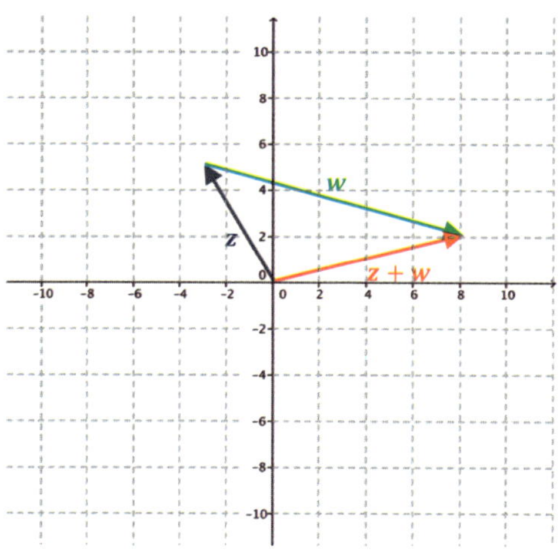

b. What is the value of w?

$w = 11 - 3i$

EUREKA MATH™

Lesson 7: Complex Number Division

Student Outcomes

- Students determine the multiplicative inverse of a complex number.
- Students determine the conjugate of a complex number.

Lesson Notes

This is the first day of a two-day lesson on complex number division and applying this knowledge to further questions about linearity. In this lesson, students find the multiplicative inverse of a complex number. Students see the connection between the conjugate of a complex number and its multiplicative inverse. This sets the stage for our study of complex number division in Lesson 8.

Classwork

Opening Exercise (5 minutes)

To get ready for our work in this lesson, we review complex number operations that students have previously studied in Algebra II, as well as $a + bi$ form. For our work in Lessons 7 and 8, students need to understand the real and imaginary components of complex numbers.

Opening Exercise

Perform the indicated operations. Write your answer in $a + bi$ form. Identify the real part of your answer and the imaginary part of your answer.

a. $(2 + 3i) + (-7 - 4i)$

$-5 - i$, -5 is real, and $-i$ is imaginary.

b. $i^2(-4i)$

$4i$, there is no real component, and $4i$ is imaginary.

c. $3i - (-2 + 5i)$

$2 - 2i$, 2 is real, and $-2i$ is imaginary.

d. $(3 - 2i)(-7 + 4i)$

$-13 + 26i$, -13 is real, and $26i$ is imaginary.

e. $(-4 - 5i)(-4 + 5i)$

41, 41 is real, and there is no imaginary component.

Discussion (5 minutes)

- In real number arithmetic, what is the multiplicative inverse of 5?
  - $\frac{1}{5}$
- How do you know? In other words, what is a multiplicative inverse?
  - $5\left(\frac{1}{5}\right) = 1$; *a number times its multiplicative inverse is always equal to* 1.
- The role of the multiplicative inverse is to get back to the identity.
- Is there a multiplicative inverse of i?

Allow students to really think about this and discuss this among themselves. Then, follow with the questions below.

- Is there a complex number z such that $z \cdot i = 1$?
  - $\frac{1}{i}$
- Can you find another way to say $\frac{1}{i}$? Explain your answer.
  - $-i$ *because* $i \cdot -i = -(i^2) = -(-1) = 1$. *Students could also mention* i^3 *as a possibility.*
- In today's lesson, we look further at the multiplicative inverse of complex numbers.

Scaffolding:

- If students do not see the pattern, have them do a few additional examples. Find the multiplicative inverses of $-1 + 2i$, $-2 - 7i$, $3 + 10i$, and $4 - i$.
- To help students see the pattern of the multiplicative inverse, have them compare the inverses of $2 + 3i$, $2 - 3i$, $-2 + 3i$, and $-2 - 3i$.
- For advanced students, have them work independently in pairs through the examples and exercises without leading questions. Be sure to check to make sure their general formula is correct before they begin the exercises.

Exercise 1 (2 minutes)

Exercises

1. **What is the multiplicative inverse of $2i$?**

   $$\frac{1}{2i} = \frac{1}{2} \cdot \frac{1}{i} = \frac{1}{2} \cdot (-i) = -\frac{1}{2}i$$

Example 1 (8 minutes)

Students were able to reason what the multiplicative inverse of i was in the Discussion, but the multiplicative inverse of a complex number in the form of $p + qi$ is more difficult to find. In this example, students find the multiplicative inverse of a complex number by multiplying by a complex number in general form and solving the resulting system of equations.

- Does $3 + 4i$ have a multiplicative inverse?
  - *Yes,* $\frac{1}{3+4i}$
- Is there a complex number $p + qi$ such that $(3 + 4i)(p + qi) = 1$?

Students will have to think about this answer. Give them a couple of minutes, and then proceed with the example.

- Let's begin by expanding this binomial. What equation do you get?
  - $3p + 3iq + 4ip + 4i^2q = 1$, *so* $3p + 3iq + 4ip - 4q = 1$.
- Group the real terms and the complex terms, and rewrite the equation.
  - $(3p - 4q) + (3q + 4p)i = 1$
- Look at the right side of the equation. What do you notice?
  - *The number* 1 *is real, and there is no imaginary component.*
- What would the real terms have to be equal to? The imaginary terms?
  - *The real terms must equal* 1, *and the imaginary terms must equal* 0.
- Set up that system of equations.
  - $3p - 4q = 1$ *and* $4p + 3q = 0$
- Solve this system of equations for p and q.
  - $p = \frac{3}{25}, q = -\frac{4}{25}$
- This suggests that $\frac{1}{3+4i} = \ldots?$
  - $\frac{3}{25} + \frac{-4}{25}i = \frac{3-4i}{25}$
- Does the product of $3 + 4i$ and $\frac{3-4i}{25}$ equal 1? Confirm that they are multiplicative inverses by performing this calculation. Check your work with a neighbor.

Students should confirm that their result was correct.

Note: Students can use their prior knowledge of conjugates and radicals from Algebra II for a simpler method of finding the inverse for Examples 1 and 2. If students see this connection, allow this, but be sure that students see the connection and understand the math behind this concept.

Exercise 2 (3 minutes)

> **2. Find the multiplicative inverse of $5 + 3i$.**
>
> $\frac{5-3i}{34}$

Example 2 (8 minutes)

In Example 2, students look at patterns between the complex numbers and their multiplicative inverses from Example 1 and Exercise 2 and then find the general formula for the multiplicative inverse of any number.

MP.8

- Without doing any work, can you tell me what the multiplicative inverse of $3 - 4i$ and $5 - 3i$ would be?
  - $\frac{3+4i}{25}$ *and* $\frac{5+3i}{34}$
- Explain to your neighbor in words how to find the multiplicative inverse of a complex number.
  - *Change the sign between the real and imaginary terms, and then divide by the sum of the squares of the coefficients of each term.*

- If $z = a + bi$, do you remember the name of $\bar{z} = a - bi$ from Algebra II?
  - *The conjugate*
- Let's develop a general formula for the multiplicative inverse of any number of the form $z = a + bi$. Using what we did earlier in this example, what might we do?
  - *Multiply by another complex number $(p + qi)$, and set the product equal to 1.*
- Solve $(a + bi)(p + qi) = 1$. Show each step, and explain your work to your neighbor.
  - $ap + aqi + bpi + bqi^2 = 1$ *Expand the binomial.*
  - $ap + aqi + bpi - bq = 1$ *Simplify the equation.*
  - $ap - bq = 1$ and $aq + bp = 0$ *Set the real terms equal to 1 and the imaginary terms equal to 0.*
  - $p = \frac{a}{a^2 + b^2}$ *and* $q = -\frac{b}{a^2 + b^2}$ *Solve the system of equations for p and q.*
- What is the general formula of the multiplicative inverse of $z = a + bi$?
  - $\frac{a}{a^2+b^2} + \frac{-b}{a^2+b^2}i$ *or* $\frac{a - bi}{a^2 + b^2}$
- Does this agree with what you discovered earlier in the example?
  - *Yes*
- Explain how to find the multiplicative inverse of a complex number using the term *conjugate*.
  - *To find the multiplicative inverse of a complex number, $a + bi$, take the conjugate of the number, and divide by $a^2 + b^2$.*

Exercises 3–7 (6 minutes)

In these exercises, students practice using the general formula for finding the multiplicative inverse of a number. The goal is to show students that this formula works for all numbers, real or complex.

State the conjugate of each number, and then using the general formula for the multiplicative inverse of $z = a + bi$, find the multiplicative inverse.

3. $3 + 4i$

$3 - 4i; \frac{3 - 4i}{3^2 + 4^2} = \frac{3 - 4i}{25}$

4. $7 - 2i$

$7 + 2i; \frac{7 - (-2)i}{7^2 + (-2)^2} = \frac{7 + 2i}{53}$

5. i

$-i; \frac{0 - 1i}{0^2 + (1)^2} = \frac{-i}{1} = -i$

6. 2

$2; \frac{2 - 0i}{2^2 + 0^2} = \frac{2}{4} = \frac{1}{2}$

7. Show that $a = -1 + \sqrt{3}i$ and $b = 2$ satisfy $\frac{1}{a+b} = \frac{1}{a} + \frac{1}{b}$.

Finding a common denominator of the right side, and then simplifying:

$$\begin{aligned}
\frac{1}{a} + \frac{1}{b} &= \frac{1}{-1+\sqrt{3}i} + \frac{1}{2} \\
&= \frac{(2)1}{(2)} + \frac{(-1+\sqrt{3}i)\cdot 1}{(-1+\sqrt{3}i)\cdot 2} \\
&= \frac{1+\sqrt{3}i}{2(-1+\sqrt{3}i)} \\
&= \frac{1+\sqrt{3}i}{2(-1+\sqrt{3}i)} \cdot \frac{1-\sqrt{3}i}{1-\sqrt{3}i} \\
&= \frac{1-3i^2}{2(-1+\sqrt{3}i+\sqrt{3}i-3i^2)} \\
&= \frac{1+3}{2(-1+2\sqrt{3}i+3)} \\
&= \frac{4}{2(2+2\sqrt{3}i)} \\
&= \frac{4}{4(1+\sqrt{3}i)} \\
&= \frac{1}{1+\sqrt{3}i}
\end{aligned}$$

$$\begin{aligned}
\frac{1}{a+b} &= \frac{1}{-1+\sqrt{3}i+2} \\
&= \frac{1}{1+\sqrt{3}i}
\end{aligned}$$

The two expressions are equal for the given values of a and b.

Closing (3 minutes)

Allow students to think about the questions below in pairs, and then pull the class together to wrap up the discussion.

- Was it necessary to use the formula for Exercise 6? Explain.
  - *No. The number 2 is a real number, so the multiplicative inverse was its reciprocal.*
- Look at Exercises 3–6. What patterns did you discover in the formats of the real and complex numbers and their multiplicative inverses?
  - *For any complex number $z = a + bi$, the multiplicative inverse has the format $\frac{a-bi}{a^2+b^2}$ when simplified.*
  - *This formula works for real and complex numbers, but for real numbers it is easier just to find the reciprocal.*

Exit Ticket (5 minutes)

Name ______________________________ Date ______________

Lesson 7: Complex Number Division

Exit Ticket

1. Find the multiplicative inverse of $3 - 2i$. Verify that your solution is correct by confirming that the product of $3 - 2i$ and its multiplicative inverse is 1.

2. What is the conjugate of $3 - 2i$?

Exit Ticket Sample Solutions

1. **Find the multiplicative inverse of $3-2i$. Verify that your solution is correct by confirming that the product of $3-2i$ and its multiplicative inverse is 1.**

   If $a+bi$ is the multiplicative inverse of $3-2i$, then

   $$(3-2i)(a+bi) = 1+0i(3-2i)(a+bi)$$
   $$3a+3bi-2ai-2bi^2 = 1+0i$$
   $$3a+3bi-2ai+2b = 1+0i.$$

   $3a+2b=1$ *and* $(3b-2a)i=0i$, *so* $3b-2a=0$.

   $a=\frac{3}{13}, b=\frac{2}{13},$ *so the multiplicative inverse* $a+bi=\frac{3+2i}{13}$.

   Verification: $(3-2i)\left(\frac{3}{13}+\frac{2i}{13}\right)=\frac{9}{13}+\frac{6i}{13}-\frac{6i}{13}-\frac{4i^2}{13}=\frac{9}{13}+\frac{4}{13}=1$

2. **What is the conjugate of $3-2i$?**

   $3+2i$

Problem Set Sample Solutions

Problems 1 and 2 are easy entry problems that allow students to practice operations on complex numbers and the algebra involved in such operations, including solving systems of equations. These problems also reinforce that complex numbers have a real component and an imaginary component. Problem 3 is more difficult. Most students should attempt part (a), but part (b) is optional and sets the stage for the next lesson. All skills practiced in this Problem Set are essential for success in Lesson 8.

1. **State the conjugate of each complex number. Then, find the multiplicative inverse of each number, and verify by multiplying by $a+bi$ and solving a system of equations.**

   a. $-5i$

   Conjugate: $5i$

   $$-5i(a+bi)=1$$
   $$-5ai-5bi^2=1$$
   $$-5ai+5b=1$$
   $$-5a=0, 5b=1$$
   $$a=0, b=\frac{1}{5}$$

   Multiplicative inverse: $0+\frac{1}{5}i=\frac{1}{5}i$

b. $5 - \sqrt{3}i$

Conjugate: $(5 + \sqrt{3}i)$

$$(5 - \sqrt{3}i)(a + bi) = 1$$
$$5a + 5bi - \sqrt{3}ai - \sqrt{3}bi^2 = 1$$
$$5a + 5bi - \sqrt{3}ai + \sqrt{3}b = 1$$
$$5a + \sqrt{3}b = 1, 5b - \sqrt{3}a = 0$$
$$a = \frac{5}{28}, b = \frac{\sqrt{3}}{28}$$

Multiplicative Inverse: $\frac{5+\sqrt{3}i}{28}$

2. **Find the multiplicative inverse of each number, and verify using the general formula to find multiplicative inverses of numbers of the form $z = a + bi$.**

a. i^3

$i^3 = -i = 0 - i$

Multiplicative inverse: $\frac{0-(-1)i}{0^2+(-1)^2} = \frac{i}{1} = i$

b. $\frac{1}{3}$

$\frac{1}{3} = \frac{1}{3} + 0i$

Multiplicative inverse: $\frac{\frac{1}{3} - 0i}{\left(\frac{1}{3}\right)^2 + 0^2} = \frac{\frac{1}{3}}{\frac{1}{9}} = \frac{1}{3} \cdot \frac{9}{1} = 3$

c. $\frac{\sqrt{3}-i}{4}$

$\frac{\sqrt{3} - i}{4} = \frac{\sqrt{3}}{4} + \frac{-1}{4}i$

Multiplicative inverse: $\frac{\frac{\sqrt{3}}{4}-\left(\frac{-1}{4}\right)i}{\left(\frac{\sqrt{3}}{4}\right)^2+\left(\frac{-1}{4}\right)^2} = \frac{\frac{\sqrt{3}}{4}+\frac{1}{4}i}{\frac{3}{16}+\frac{1}{16}} = \frac{\frac{\sqrt{3}+i}{4}}{\frac{4}{16}} = \frac{\frac{\sqrt{3}+i}{4}}{\frac{1}{4}} = \frac{\sqrt{3}+i}{4} \cdot \frac{4}{1} = \sqrt{3} + i$

d. $1 + 2i$

Multiplicative inverse: $\frac{1-2i}{(1)^2+(-2)^2} = \frac{1-2i}{5} = \frac{1}{5} - \frac{2i}{5}$

e. $4 - 3i$

Multiplicative inverse: $\frac{4+3i}{(4)^2+(-3)^2} = \frac{4+3i}{25} = \frac{4}{25} + \frac{3i}{25}$

f. $2+3i$

Multiplicative inverse: $\frac{2-3i}{(2)^2+(-3)^2}=\frac{2-3i}{13}=\frac{2}{13}-\frac{3i}{13}$

g. $-5-4i$

Multiplicative inverse: $\frac{-5+4i}{(-5)^2+(-4)^2}=\frac{-5+4i}{41}=-\frac{5}{41}+\frac{4i}{41}$

h. $-3+2i$

Multiplicative inverse: $\frac{-3-2i}{(-3)^2+(2)^2}=\frac{-3-2i}{13}=-\frac{3}{13}-\frac{2i}{13}$

i. $\sqrt{2}+i$

Multiplicative inverse: $\frac{\sqrt{2}-i}{\left(\sqrt{2}\right)^2+(1)^2}=\frac{\sqrt{2}-i}{3}=\frac{\sqrt{2}}{3}-\frac{i}{3}$

j. $3-\sqrt{2}\cdot i$

Multiplicative inverse: $\frac{3+\sqrt{2}i}{(3)^2+\left(\sqrt{2}\right)^2}=\frac{3+\sqrt{2}i}{11}=\frac{3}{11}+\frac{\sqrt{2}i}{11}$

k. $\sqrt{5}+\sqrt{3}\cdot i$

Multiplicative inverse: $\frac{\sqrt{5}-\sqrt{3}i}{\left(\sqrt{5}\right)^2+\left(-\sqrt{3}\right)^2}=\frac{\sqrt{5}-\sqrt{3}i}{8}=\frac{\sqrt{5}}{8}-\frac{\sqrt{3}i}{8}$

3. **Given $z_1=1+i$ and $z_2=2+3i$.**

a. **Let $w=z_1\cdot z_2$. Find w and the multiplicative inverse of w.**

$w=(1+i)(2+3i)=-1+5i$

Multiplicative inverse: $\frac{-1-5i}{1+25}=-\frac{1}{26}-\frac{5i}{26}$

b. **Show that the multiplicative inverse of w is the same as the product of the multiplicative inverses of z_1 and z_2.**

$z_1=1+i$; *Multiplicative inverse:* $\frac{1-i}{1+1}=\frac{1-i}{2}$

$z_2=2+3i$; *Multiplicative inverse:* $\frac{2-3i}{4+9}=\frac{2-3i}{13}$

$$\begin{aligned}z_1\cdot z_2&=\left(\frac{1-i}{2}\right)\left(\frac{2-3i}{13}\right)\\&=\frac{2-3i-2i-3}{26}\\&=\frac{-1-5i}{26}\\&=-\frac{1}{26}-\frac{5i}{26}\end{aligned}$$

Lesson 8: Complex Number Division

Student Outcomes

- Students determine the modulus and conjugate of a complex number.
- Students use the concept of conjugate to divide complex numbers.

Lesson Notes

This is the second day of a two-day lesson on complex number division and applying this knowledge to further questions about linearity. In Lesson 7, students studied the multiplicative inverse. In this lesson, students study the numerator and denominator of the multiplicative inverse and their relationship to the conjugate and modulus. The lesson culminates with complex number division.

Classwork

Opening Exercise (3 minutes)

Students practice using the formula for the multiplicative inverse derived in Lesson 7 as a lead into this lesson.

> **Opening Exercise**
>
> **Use the general formula to find the multiplicative inverse of each complex number.**
>
> a. $2 + 3i$
>
> $\frac{2 - 3i}{13}$
>
> b. $-7 - 4i$
>
> $\frac{-7 + 4i}{65}$
>
> c. $-4 + 5i$
>
> $\frac{-4 - 5i}{41}$

Discussion (2 minutes)

- Look at the complex numbers given in the Opening Exercise and the numerators of the multiplicative inverses. Do you notice a pattern? Explain.
  - *The real term is the same in both the original complex number and its multiplicative inverse, but the imaginary term in the multiplicative inverse is the opposite of the imaginary term in the original complex number.*

> *Scaffolding:*
>
> Use a Frayer diagram to define **conjugate**. See Lesson 5 for an example.

- If the complex number $z = a + bi$, what is the numerator of its multiplicative inverse?
  - $a - bi$
- Features of the multiplicative inverse formula often reappear in complex number arithmetic, so mathematicians have given these features names. The conjugate of a complex number $a + bi$ is $a - bi$. Repeat that with me.
  - *The conjugate of* $a + bi$ *is* $a - bi$.

Exercises 1–4 (2 minutes)

Have students quickly complete the exercises individually, and then follow up with the questions below. This would be a good exercise to do as a rapid white board exchange.

Exercises 1–11

Find the conjugate, and plot the complex number and its conjugate in the complex plane. Label the conjugate with a prime symbol.

1. A: $3 + 4i$

   A': $3 - 4i$

2. B: $-2 - i$

   B': $-2 + i$

3. C: 7

   C': 7

4. D: $4i$

   D': $-4i$

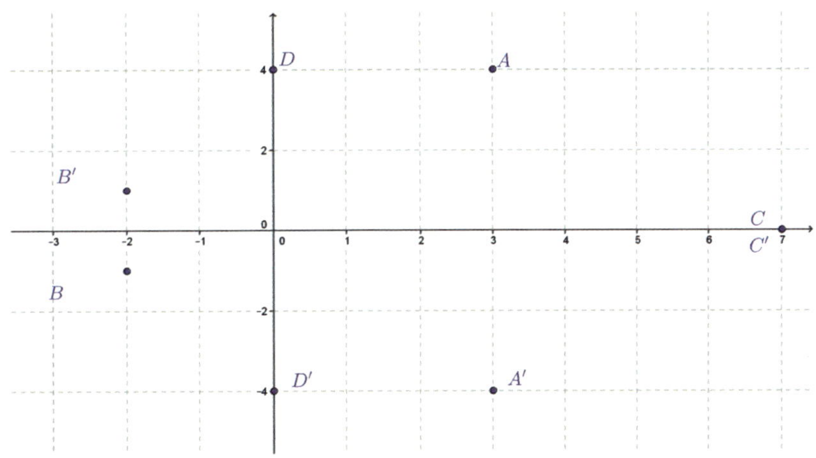

Discussion (8 minutes)

- Does 7 have a complex conjugate? If so, what is it? Explain your answer.
  - *Yes.* $7 = 7 + 0i$, *so the complex conjugate would be* $7 - 0i = 7$.
- What is the complex conjugate of $4i$? Explain.
  - $4i = 0 + 4i$; *the complex conjugate is* $0 - 4i = -4i$.
- If $z = a + bi$, then the conjugate of z is denoted $\bar{z}$. That means $\bar{z} = a - bi$.
- What is the geometric effect of taking the conjugate of a complex number? **MP.7**
  - *The complex conjugate reflects the complex number across the real axis.*
- What can you say about the conjugate of the conjugate of a complex number?
  - *The conjugate of the conjugate is the original number.*

> *Scaffolding:*
> - Describe how $3 + 4i$ is related to $3 - 4i$.
> - Describe how $-2 - i$ is related to $-2 + i$.
> - In the Discussion, instead of using variables, use numbers.

MP.3

- Is $\overline{z+w} = \bar{z} + \bar{w}$ always true? Explain.
  - *Yes. Answers will vary. Students could plug in different complex numbers for z and w and show that they work or use a general formula argument. If $z = a + bi$ and $w = c + di$, $z + w = (a + c) + (b + d)i$, and $\overline{z+w} = (a + c) - (b + d)i$. $\bar{z} = a - bi$, and $\bar{w} = c - di$, so $\bar{z} + \bar{w} = (a + c) - (b + d)i$. Therefore, $\overline{z+w} = \bar{z} + \bar{w}$ is always true.*
- Is $\overline{z \cdot w} = \bar{z} \cdot \bar{w}$ always true? Explain.
  - *Yes. Answers will vary. Using the general formula argument: If $z = a + bi$ and $w = c + di$, then $z \cdot w = ac + adi + bci - bd = (ac - bd) + (ad + bc)i$. $\overline{z \cdot w} = (ac - bd) - (ad + bc)i$. $\bar{z} = a - bi$, and $\bar{w} = c - di$, so $\bar{z} \cdot \bar{w} = ac - adi - bci - bd = (ac - bd) - (ad + bc)i$. Therefore, $\overline{z \cdot w} = \bar{z} \cdot \bar{w}$ is always true.*
- Now, let's look at the denominator of the multiplicative inverse. Remind me how we find the denominator.
  - *$a^2 + b^2$, the sum of the squares of the real term and the coefficient of the imaginary term*
- Does this remind of you something that we have studied?
  - *The Pythagorean theorem, and if we take the square root, the distance formula*
- Mathematicians have given this feature a name, too. The modulus of a complex number $a + bi$ is the real number $\sqrt{a^2 + b^2}$. Repeat that with me.
  - *The modulus of a complex number $a + bi$ is the real number $\sqrt{a^2 + b^2}$.*

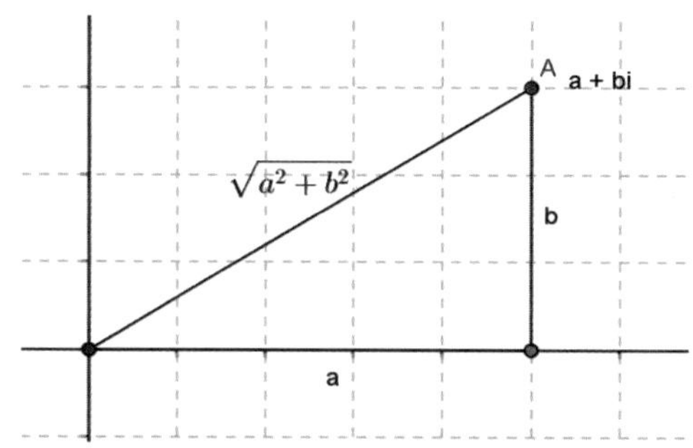

Exercises 5–8 (4 minutes)

Have students quickly complete the exercises individually, and then follow up with the questions below. This would also be a good exercise to do as a rapid white board exchange.

Find the modulus.

5. $3 + 4i$

   $\sqrt{3^2 + 4^2} = \sqrt{25} = 5$

6. $-2 - i$

   $\sqrt{(-2)^2 + (-1)^2} = \sqrt{5}$

7. 7

   $\sqrt{7^2 + 0^2} = \sqrt{49} = 7$

8. $4i$

   $\sqrt{0^2 + (4)^2} = \sqrt{16} = 4$

Discussion (3 minutes)

- If $z = a + bi$, then the modulus of z is denoted $|z|$. This means $|z| = \sqrt{a^2 + b^2}$.
- If $z = a + bi$ is a point in the complex plane, what is the geometric interpretation of $|z|$?
  - *The modulus is the distance of the point from the origin in the complex plane.*
- The notation for the modulus of a complex number matches the notation for the absolute value of a real number. Do you think this is a coincidence? If a complex number is real, what can you say about its modulus?
  - *The modulus is the number.*
- Explain to your neighbor what you have learned about the conjugate and the modulus of a complex number.
  - *The conjugate of a complex number $a + bi$ is $a - bi$; taking the conjugate of a complex number reflects the number over the real axis.*
  - *The modulus of complex number $a + bi$ is $\sqrt{a^2 + b^2}$; the modulus represents the distance from the origin to the point $a + bi$ in the complex plane.*

Exercises 9–11 (6 minutes)

Students should complete Exercises 9–11 in pairs. For advanced learners, assign all problems. Assign only one problem to other groups. Bring the class back together to debrief.

Given $z = a + bi$.

9. Show that for all complex numbers z, $|iz| = |z|$.

$|iz| = |i(a + bi)| = |ai - b| = |-b + ai| = \sqrt{(-b)^2 + a^2} = \sqrt{a^2 + b^2} = |z|$

10. Show that for all complex numbers z, $z \cdot \bar{z} = |z|^2$.

$$z \cdot \bar{z} = (a + bi)(a - bi) = a^2 + b^2$$
$$|z|^2 = \left(\sqrt{a^2 + b^2}\right)^2 = a^2 + b^2$$
$$z \cdot \bar{z} = |z|^2$$

11. Explain the following: Every nonzero complex number z has a multiplicative inverse. It is given by $\frac{1}{z} = \frac{\bar{z}}{|z|}$.

The multiplicative inverse of $a + bi = \frac{a-bi}{a^2+b^2} = \frac{\bar{z}}{|z|}$.

Example (5 minutes)

In this example, students divide complex numbers by multiplying the numerator and denominator by the conjugate. Do this as a whole-class discussion.

Scaffolding:
- For advanced learners, assign this example without leading questions.
- Target some groups for individual instruction.

Example

$$\frac{2-6i}{2+5i}$$

$$\frac{-26-22i}{29}$$

- In this example, we are going to divide these two complex numbers. Complex number division is different from real number division, and the quotient also looks different.
- To divide complex numbers, we want to make the denominator a real number. We need to multiply the denominator by a complex number that makes it a real number. Multiply the denominator by its conjugate. What type of product do you get?
  - $(2+5i)(2-5i) = 4-10i+10i-25i^2 = 4+25 = 29$
    You get a real number.
- The result of multiplying a complex number by its conjugate is always a real number.
- The goal is to rewrite this expression $\frac{2-6i}{2+5i}$ as an equivalent expression with a denominator that is a real number. We now know that we must multiply the denominator by its conjugate. What about the numerator? What must we multiply the numerator by in order to obtain an equivalent expression?
  - *We must multiply the numerator by the same expression,* $2-5i$.
- Perform that operation, and check your answer with a neighbor.
  - $\frac{2-6i}{2+5i}\cdot\frac{2-5i}{2-5i} = \frac{4-10i-12i+30i^2}{4-25i^2} = \frac{4-22i-30}{4+25} = \frac{-26-22i}{29}$
- Tell your neighbor how to divide complex numbers.
  - *Multiply the numerator and denominator by the conjugate of the denominator.*

Exercises 12–13 (5 minutes)

Have students complete the exercises and then check answers and explain their work to a neighbor.

Exercises 12–13

Divide.

12. $\frac{3+2i}{-2-7i}$

$$\frac{3+2i}{-2-7i}\cdot\frac{-2+7i}{-2+7i} = \frac{-6+21i-4i-14}{4+49} = \frac{-20+17i}{53}$$

13. $\frac{3}{3-i}$

$$\frac{3}{3-i}\cdot\frac{3+i}{3+i}=\frac{9+3i}{9+1}=\frac{9+3i}{10}$$

Closing (2 minutes)

Allow students to think about the questions below in pairs, and then pull the class together to wrap up the discussion.

- What is the conjugate of $a+bi$? What is the geometric effect of this conjugate in the complex plane?
  - *$a-bi$; the conjugate is a reflection of the complex number across the real axis.*
- What is the modulus of $a+bi$? What is the geometric effect of the modulus in the complex plane?
  - *$\sqrt{a^2+b^2}$; the modulus is the distance of the point from the origin in the complex plane.*
- How is the conjugate used in complex number division?
  - *Multiply by a ratio in which both the numerator and denominator are the conjugate.*

Exit Ticket (5 minutes)

Name ______________________________ Date ______________

Lesson 8: Complex Number Division

Exit Ticket

1. Given $z = 4 - 3i$
   a. What does $\bar{z}$ mean?

   b. What does $\bar{z}$ do to z geometrically?

   c. What does $|z|$ mean both algebraically and geometrically?

2. Describe how to use the conjugate to divide $2 - i$ by $3 + 2i$, and then find the quotient.

Exit Ticket Sample Solutions

1. Given $z = 4 - 3i$.

   a. What does $\bar{z}$ mean?

   $\bar{z}$ means the conjugate of z, which is $4 + 3i$.

   b. What does $\bar{z}$ do to z geometrically?

   $\bar{z}$ is the reflection of z across the real axis.

   c. What does $|z|$ mean both algebraically and geometrically?

   $|z|$ is a modulus of z, which is a real number.

   $|z|$ is the distance from the point $z = 4 - 3i$ to the origin in the complex plane.

$$\begin{aligned} |z| &= \sqrt{a^2 + b^2} \\ &= \sqrt{(4)^2 + (-3)^2} \\ &= \sqrt{16 + 9} \\ &= \sqrt{25} \\ &= 5 \end{aligned}$$

2. Describe how to use the conjugate to divide $2 - i$ by $3 + 2i$, and then find the quotient.

   When $3 + 2i$ is multiplied by its conjugate of $3 - 2i$, the denominator is a real number, which is necessary.

   Multiply by $\frac{3-2i}{3-2i}$.

$$\frac{2 - i}{3 + 2i} = \frac{(2 - i)(3 - 2i)}{(3 + 2i)(3 - 2i)} = \frac{6 - 4i - 3i - 2}{9 + 4} = \frac{4 - 7i}{13} = \frac{4}{13} - \frac{7}{13}i$$

Problem Set Sample Solutions

Problems 1–3 are easy problems and allow students to practice finding the conjugate and modulus and dividing complex numbers. Problems 4–6 are more difficult. Students can use examples or a geometrical approach to explain their reasoning. Problem 5 is a preview of the effect of adding or subtracting complex numbers in terms of geometrical interpretations. Students need to find and compare the modulus, r_n, and φ_n in order to come to their assumptions.

1. Let $z = 4 - 3i$ and $w = 2 - i$. Show that

   a. $|z| = |\bar{z}|$

   $|z| = \sqrt{(4)^2 + (-3)^3} = \sqrt{16 + 9} = \sqrt{25} = 5$

   $\bar{z} = 4 + 3i, |\bar{z}| = \sqrt{(4)^2 + (3)^2} = \sqrt{16 + 9} = \sqrt{25} = 5$

   Therefore, $|z| = |\bar{z}|$.

b. $\left|\frac{1}{z}\right| = \frac{1}{|\bar{z}|}$

$\frac{1}{z} = \frac{1}{4-3i} = \frac{(4+3i)}{(4-3i)(4+3i)} = \frac{4+3i}{25} = \frac{4}{25} + \frac{3}{25}i$; therefore,

$$\left|\frac{1}{z}\right| = \sqrt{\left(\frac{4}{25}\right)^2 + \left(\frac{3}{25}\right)^2} = \sqrt{\frac{16}{(25)^2} + \frac{9}{(25)^2}} = \sqrt{\frac{25}{(25)^2}} = \frac{1}{5}.$$

Since $|\bar{z}| = 5$; therefore, $\frac{1}{|\bar{z}|} = \frac{1}{5}$, which equals $\left|\frac{1}{z}\right| = \frac{1}{5}$.

c. If $|z| = 0$, must it be that $z = 0$?

Yes. Let $z = a + bi$, and then $|z| = \sqrt{(a)^2 + (b)^2}$. If $|z| = 0$, it indicates that $\sqrt{(a)^2 + (b)^2} = 0$. Since $(a)^2 + (b)^2$ both are positive real numbers, the only values of a and b that will make the equation true is that a and b have to be 0, which means $z = 0 + 0i = 0$.

d. Give a specific example to show that $|z + w|$ usually does not equal $|z| + |w|$.

Answers vary, but $z = 3 + 2i$ and $w = 3 - 2i$ will work.

$$z + w = 6 + 0i$$
$$|z + w| = \sqrt{(6)^2 + (0)^2} = 6$$

$|z| + |w| = \sqrt{(3)^2 + (2)^2} + \sqrt{(3)^2 + (-2)^2} = 2\sqrt{13}$, which is not equal to 6.

2. Divide.

a. $\frac{1-2i}{2i}$

$\frac{(1-2i)(i)}{2i(i)} = \frac{2+i}{-2}$ or $-1 - \frac{1}{2}i$

b. $\frac{5-2i}{5+2i}$

$\frac{(5-2i)(5-2i)}{(5+2i)(5-2i)} = \frac{25-20i-4}{25+4} = \frac{21-20i}{29}$ or $\frac{21}{29} - \frac{20}{29}i$

c. $\frac{\sqrt{3} - 2i}{-2 - \sqrt{3}i}$

$$\frac{(\sqrt{3} - 2i)(-2 + \sqrt{3}i)}{(-2 - \sqrt{3}i)(-2 + \sqrt{3}i)} = \frac{-2\sqrt{3} + 3i + 4i + 2\sqrt{3}}{4 + 3} = \frac{7i}{7} = i$$

3. Prove that $|zw| = |z| \cdot |w|$ for complex numbers z and w.

Since $|z|^2 = z \cdot \bar{z}$; therefore, $|zw|^2 = (zw)(\overline{zw}) = (zw)(\bar{z}\bar{w}) = z\bar{z}w\bar{w} = |z|^2 \cdot |w|^2$.

Now we have $|zw|^2 = |z|^2 \cdot |w|^2$; therefore, $|zw| = |z| \cdot |w|$.

4. Given $z = 3 + i$, $w = 1 + 3i$.

a. Find $z + w$, and graph z, w, and $z + w$ on the same complex plane. Explain what you discover if you draw line segments from the origin to those points z, w, and $z + w$. Then, draw line segments to connect w to $z +$ and $z + w$ to z.

$z + w = 4 + 4i$

Students should discover that the lines form a parallelogram. They then can graphically see that the lengths of the two sides are greater than the diagonal, $|z + w| \leq |z| + |w|$.

b. Find $-w$, and graph z, w, and $z - w$ on the same complex plane. Explain what you discover if you draw line segments from the origin to those points z, w, and $z - w$. Then, draw line segments to connect w to $z - w$ and $z - w$ to z.

$z - w = 2 - 2i$

Students should discover that the lines form a parallelogram. They then can graphically see that the lengths of the two sides are greater than the diagonal, $|z - w| \leq |z| + |w|$.

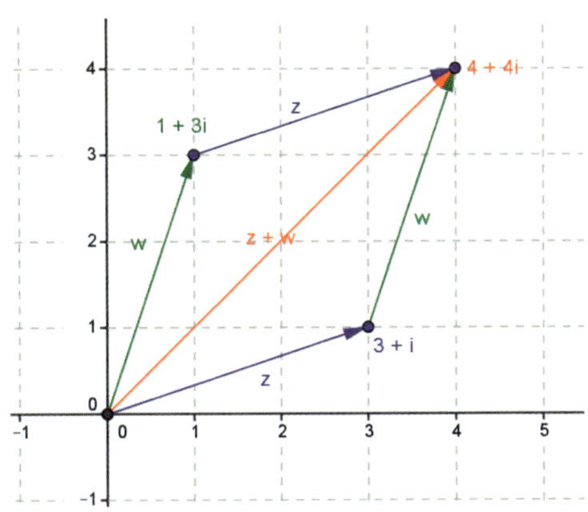

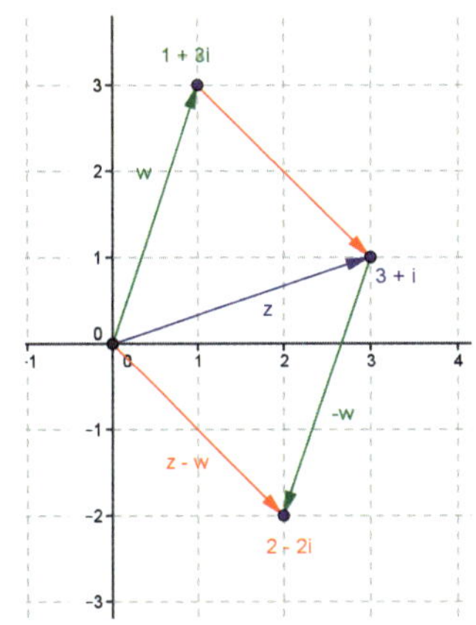

5. Explain why $|z + w| \leq |z| + |w|$ and $|z - w| \leq |z| + |w|$ geometrically. (Hint: Triangle inequality theorem)

By using Example 5, we can apply the triangle inequality theorem into these two formulas.

Topic B

Complex Number Operations and Transformations

N-CN.A.3, N-CN.B.4, N-CN.B.5, N-CN.B.6

Focus Standards:	N-CN.A.3	(+) Find the conjugate of a complex number; use conjugates to find moduli and quotients of complex numbers.
	N-CN.B.4	(+) Represent complex numbers on the complex plane in rectangular and polar form (including real and imaginary numbers), and explain why the rectangular and polar forms of a given complex number represent the same number.
	N-CN.B.5	(+) Represent addition, subtraction, multiplication, and conjugation of complex numbers geometrically on the complex plane; use properties of this representation for computation. *For example,* $\left(-1+\sqrt{3}i\right)^3 = 8$ *because* $\left(-1+\sqrt{3}i\right)$ *has a modulus of* 2 *and an argument of* $120°$.
	N-CN.B.6	(+) Calculate the distance between numbers in the complex plane as the modulus of the difference and the midpoint of a segment as the average of the number at its endpoints.
Instructional Days:	9	
Lessons 9–10:	The Geometric Effect of Some Complex Arithmetic (P, P)[1]	
Lessons 11–12:	Distance and Complex Numbers (P, E)	
Lesson 13:	Trigonometry and Complex Numbers (P)	
Lesson 14:	Discovering the Geometric Effect of Complex Multiplication (E)	
Lesson 15:	Justifying the Geometric Effect of Complex Multiplication (S)	
Lesson 16:	Representing Reflections with Transformations (P)	
Lesson 17:	The Geometric Effect of Multiplying by a Reciprocal (E)	

In Topic B, students develop an understanding of the geometric effect of operations on complex numbers. In Lesson 9, students explore what happens to a point in the complex plane when complex numbers are added and subtracted, leading to Lesson 10's study of the effect of multiplication (**N-CN.B.5**). Students revisit the

[1]Lesson Structure Key: **P**-Problem Set Lesson, **M**-Modeling Cycle Lesson, **E**-Exploration Lesson, **S**-Socratic Lesson

idea of linearity in Lesson 10, determining whether given complex functions are linear transformations. Students discover that some complex functions are linear transformations and others are not. Students understand that when complex numbers are considered points in the Cartesian plane, complex number multiplication has the geometric effect of a rotation followed by a dilation in the complex plane. In Lesson 11, students use the distance and midpoint formulas they studied in Geometry to find the distance of a complex point from the origin and the midpoint of two complex numbers (**N-CN.B.6**). Lesson 12 extends this concept as students play the Leap Frog game, repeatedly finding the midpoints of pairs of complex numbers. They discover that when starting with three fixed midpoints, a series of moves jumping across midpoints leads back to the starting point. Students then explore what happens when they start with only one or two fixed midpoints and find that a series of jumps never returns to the starting point. Students verify these results using the midpoint formula derived earlier in the lesson. Lesson 13 introduces the modulus and argument and polar coordinates as students study the geometric effect of complex number multiplication, leading to writing the complex number in polar form (**N-CN.B.4**). If a complex number z has argument θ and modulus r, then it can be written in polar form, $z = r(\cos(\theta) + i\sin(\theta))$. Students explain why complex numbers can be written in either rectangular or polar form and why the two forms are equivalent. Lessons 14–16 continue the study of multiplication by complex numbers, leading students to the understanding that the geometric effect of multiplying by a complex number, w, is a rotation of the argument of w followed by a dilation with scale factor the modulus of w. Lesson 17 concludes this topic as students discover that the multiplicative inverse of a complex number (i.e., its reciprocal) provides the inverse geometric operation, which leads to the complex conjugate and division of complex numbers (**N-CN.A.3**). Mathematical practices highlighted in the topic include MP.2 as students reason abstractly and quantitatively about complex numbers and the geometric effects of operations involving complex numbers and use previous results to predict rotations and dilations produced.

Lesson 9: The Geometric Effect of Some Complex Arithmetic

Student Outcomes

- Students represent addition, subtraction, and conjugation of complex numbers geometrically on the complex plane.

Lesson Notes

In the last few lessons, students have informally seen the geometric effects of complex conjugates and of multiplying by i. This is the first of a two-day lesson in which students further explore the geometric interpretations of complex arithmetic. This lesson focuses on the geometric effects of adding and subtracting complex numbers.

Classwork

Opening (3 minutes)

Students have had previous exposure to some geometric effects of complex numbers. Ask them to answer the following questions and to share their responses with a neighbor.

- Describe the geometric effect of multiplying a complex number by i.
- Describe the geometric effect of a complex conjugate.

> *Scaffolding:*
> - Use concrete examples such as asking students to describe the geometric effect of multiplying $2 + i$ by i and $3 + 2i$ by $3 - 2i$.
> - Help students see the effect of multiplying by complex numbers by plotting ($2 + i$ and $-1 + 2i$) and ($2 + i$ and $2 - i$) in the complex plane and asking how they are geometrically related.

Discussion (5 minutes)

The points in the complex plane are similar to points in the coordinate plane. The real part of the complex number is represented on the horizontal axis and the imaginary part on the vertical axis. This and the next lesson show that complex arithmetic causes reflections, translations, dilations, and rotations to points in the complex plane.

Begin the lesson by having students share their responses to the opening questions. As responses are shared, provide a visual depiction of each effect on the board. Use this as an opportunity to review notation as well.

- In coordinate geometry, what would happen to a point (x, y) if we rotated it $90°$ counterclockwise?
  - *The point (x, y) would map to $(-y, x)$.*
- Describe the geometric effect of multiplying a complex number by i.
  - *Multiplying a complex number by i induces a $90°$ rotation about the origin.*

- What would happen if we continued to multiply by i?
  - *Each time we multiply by* i *results in another counterclockwise rotation of* $90°$ *about the origin; for example, multiplying by* i *twice results in a* $180°$ *rotation about the origin, and multipliyng by* i *four times results in a full rotation about the origin.*

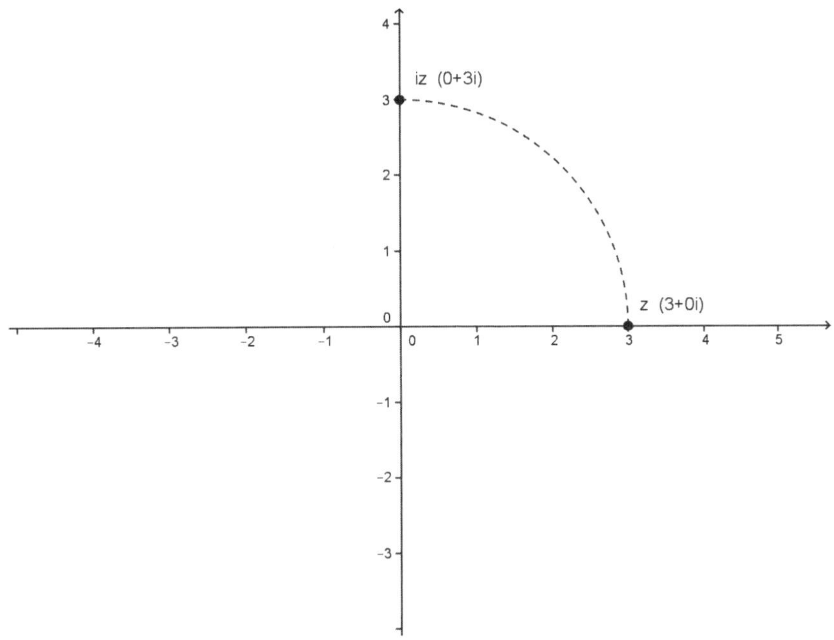

- Describe the geometric effect of taking the complex conjugate.
  - *A complex conjugate reflects the complex number across the real axis.*

Remind students that the notation for the conjugate of z is $\bar{z}$.

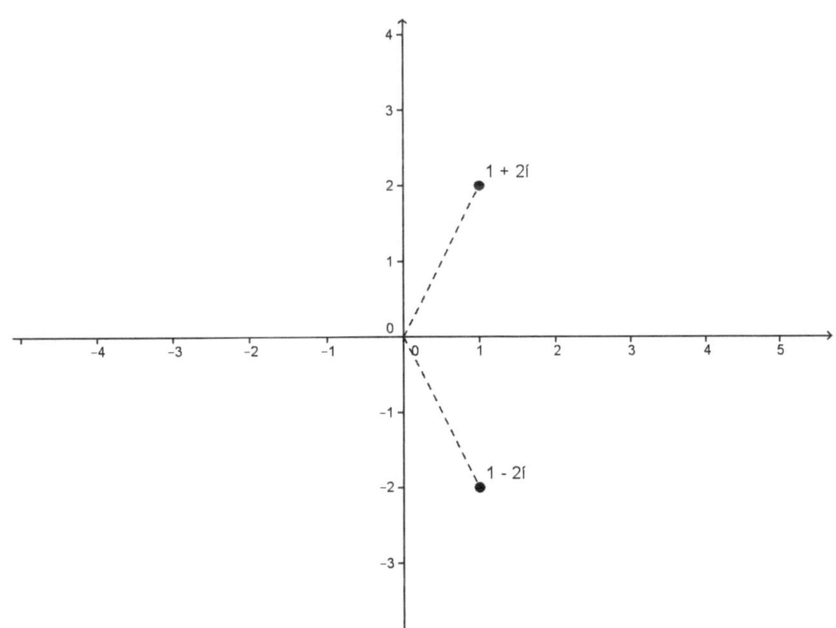

Exercise 1 (5 minutes)

Have students answer this exercise individually and share their responses with a neighbor. Then, continue with the following discussion.

> *Scaffolding:*
>
> - Encourage advanced learners to write the general rule for Exercise 1.
> - If students struggle to answer the question posed in Exercise 1, encourage them to plot a complex number like $-3+4i$ and to use it to find the reflection.

Exercises

1. **Taking the conjugate of a complex number corresponds to reflecting a complex number about the real axis. What operation on a complex number induces a reflection across the imaginary axis?**

   For a complex number $a+bi$, the reflection across the imaginary axis is $-a+bi$. Alternatively, for a complex number z, the reflection across the imaginary axis is $-\bar{z}$.

Students may have answered that the reflection of $a+bi$ across the imaginary axis is $-a+bi$. Discuss as a class how to write this in terms of the *conjugate* of the complex number.

- Is it possible to write $-a+bi$ another way? (Recall that the complex number z can be written as $a+bi$.)
- Begin by factoring out -1: $-1(a-bi)$.
- Replace $a-bi$ with $\bar{z}$: $-\bar{z}$.

Exercises 2–3 (8 minutes)

In this exercise, students explore the geometric effects of addition and subtraction to the points in the complex plane. Let students work in small groups. Before students begin, ask them to write a conjecture about the effect of adding a real number (e.g., 2) to a complex number.

2. **Given the complex numbers $w=-4+3i$ and $z=2-5i$, graph each of the following:**
   - a. w
   - b. z
   - c. $w+2$
   - d. $z+2$
   - e. $w-1$
   - f. $z-1$

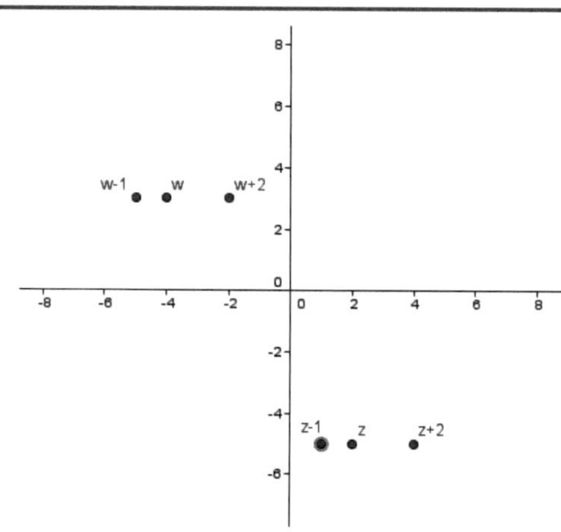

MP.8

3. **Describe in your own words the geometric effect adding or subtracting a real number has on a complex number.**

   Adding a real number to a complex number shifts the point to the right on the real (horizontal) axis, while subtracting a real number shifts the point to the left.

When students have finished the exercise, confirm as a class the answer to Exercise 3.

- Did your conjecture match the answer to Exercise 3?
  - *Answers will vary.*

Some students may no doubt have guessed that adding a positive real value (i.e., $w+2$) to the complex number would shift the point vertically instead of horizontally. They may be confusing the translation of a function, such as $f(x)=x^2$, with that of a complex number. Make clear that even though comparisons are made between the complex and coordinate planes, the geometric effects are different. Use the following discussion points to clarify.

- What is the effect of adding a constant to a function like $f(x)=x^2$? (For example, $f(x)=x^2+2$.)
  - *The graph of the parabola would shift upward 2 units.*
- How does this differ from adding the real number 2 to a complex number?
  - *The point representing the complex number would shift two units to the right, not vertically like the function.*

Exercises 4–5 (5 minutes)

Students continue to explore the geometric effects of addition and subtraction to the points in the complex plane. Let students work in small groups.

4. **Given the complex numbers $w=-4+3i$ and $z=2-5i$, graph each of the following:**
   a. w
   b. z
   c. $w+i$
   d. $z+i$
   e. $w-2i$
   f. $z-2i$

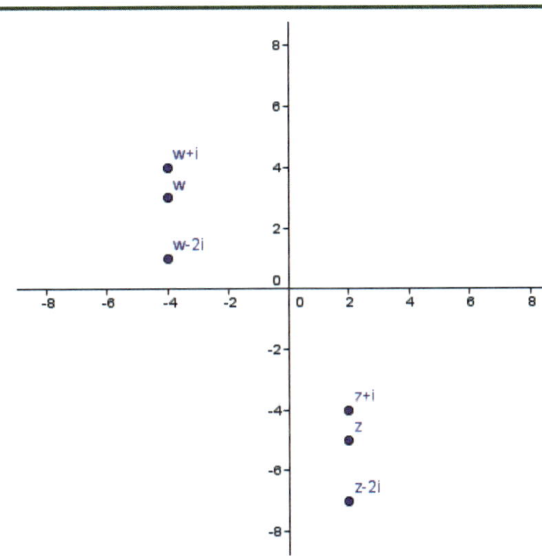

MP.8

5. **Describe in your own words the geometric effect adding or subtracting an imaginary number has on a complex number.**

   Adding an imaginary number to a complex number shifts the point up the imaginary (vertical) axis, while subtracting an imaginary number shifts the point down.

Discussion (5 minutes)

Now that the class has explored the effect of adding and subtracting real and imaginary parts to a complex number, bring both concepts together.

- Given the complex numbers $z = -6 + i$ and $w = 2 + 5i$, how would you describe the translation of the point z compared to $z + w$?
  - *The point z would shift 2 units to the right and 5 units up.*

Represent the translation on the complex plane, and point out that a right triangle is formed. Encourage students to think about how to describe the translation other than simply stating that the point shifts left/right or up/down.

Note: At this time, do not explicitly state to students that the distance between the complex numbers is the modulus of the difference, as that is covered in a later lesson.

- In what other way could we describe or quantify the relationship between z and $z + w$?
  - *The distance between the two points. We could use the Pythagorean theorem to determine the missing side of the right triangle.*

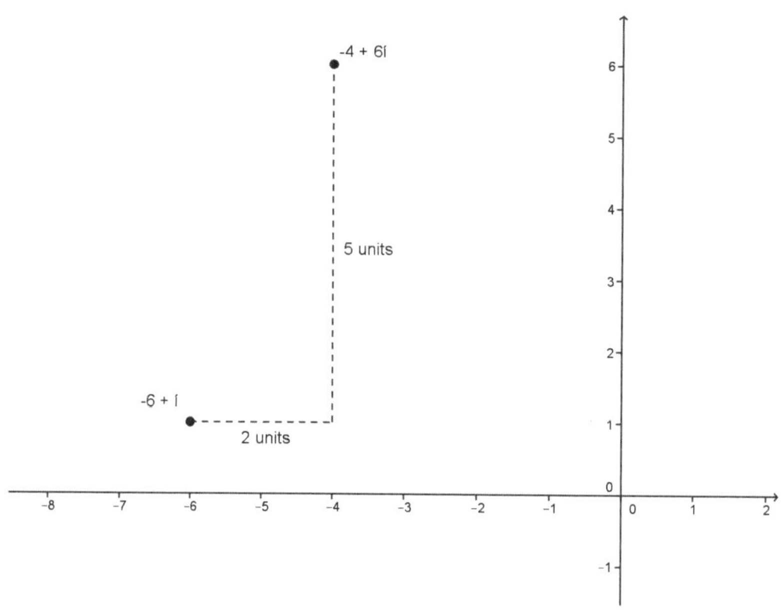

Example 1 (6 minutes)

Work through this example as a whole-class discussion. Encourage advanced learners to attempt the whole problem on their own.

> **Example**
>
> **Given the complex number z, find a complex number w such that $z + w$ is shifted $\sqrt{2}$ units in a southwest direction.**

- Begin by plotting the complex number. What does it mean for the point to be shifted in a southwest direction?
  - *The point shifts to the left and down the same number of units.*
- A right triangle is formed. What are the values of the legs and the hypotenuse?
  - *The legs are both x, and the hypotenuse is $\sqrt{2}$.*

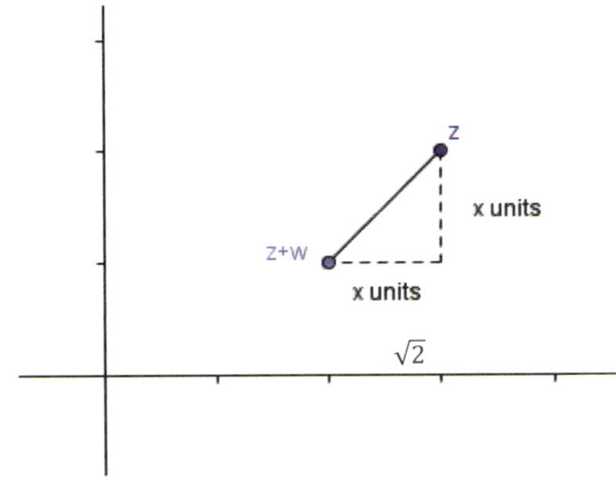

Give students an opportunity to solve for x on their own and use the information to determine the complex number w.

  - $x^2 + x^2 = \left(\sqrt{2}\right)^2$
  - $2x^2 = 2$, *so* $x = 1$.
  - *Since the point was shifted* 1 *unit down and* 1 *unit to the left, the complex number must be* $-1 - i$.

Closing (3 minutes)

Have students summarize the key ideas of the lesson in writing or by talking to a neighbor. Take this opportunity to informally assess student understanding. The Lesson Summary provides some of the key ideas from the lesson.

Lesson Summary

- **The conjugate, $\overline{z}$, of a complex number z reflects the point across the real axis.**
- **The negative conjugate, $-\overline{z}$, of a complex number z reflects the point across the imaginary axis.**
- **Adding or subtracting a real number to a complex number shifts the point left or right on the real (horizontal) axis.**
- **Adding or subtracting an imaginary number to a complex number shifts the point up or down on the imaginary (vertical) axis.**

Exit Ticket (5 minutes)

Name ______________________________ Date ______________

Lesson 9: The Geometric Effect of Some Complex Arithmetic

Exit Ticket

1. Given $z = 3 + 2i$ and $w = -2 - i$, plot the following in the complex plane:
   a. z
   b. w
   c. $z - 2$
   d. $w + 3i$
   e. $w + z$

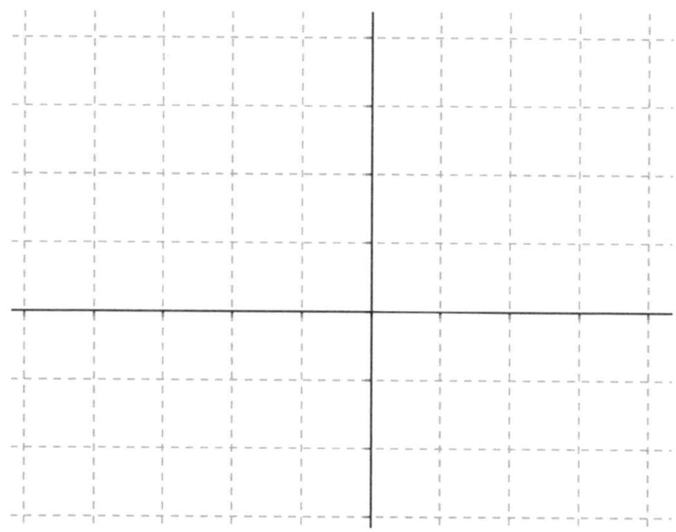

2. Given $z = a + bi$, what complex number represents the reflection of z about the imaginary axis? Give one example to show why.

3. What is the geometric effect of $T(z) = z + (4 - 2i)$?

Exit Ticket Sample Solutions

1. Given $z = 3 + 2i$ and $w = -2 - i$, plot the following in the complex plane:
   a. z
   b. w
   c. $z - 2$
   d. $w + 3i$
   e. $w + z$

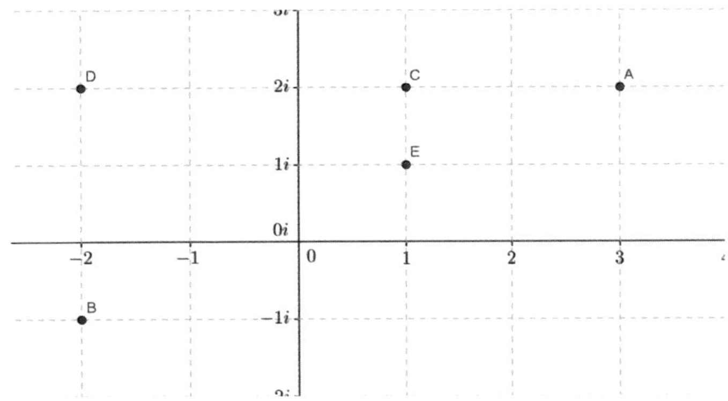

2. Given $z = a + bi$, what complex number represents the reflection of z about the imaginary axis? Give one example to show why.

   $-\bar{z}$, the negative conjugate of z. For example, $z = 2 + 3i$, $-\bar{z} = -(2 - 3i) = -2 + 3i$, which is reflected about the imaginary axis.

3. What is the geometric effect of $T(z) = z + (4 - 2i)$?

   $T(z)$ shifts 4 units to the right and the 2 units downward.

Problem Set Sample Solutions

1. Given the complex numbers $w = 2 - 3i$ and $z = -3 + 2i$, graph each of the following:
   a. $w - 2$

      $w - 2 = 2 - 3i - 2 = -3i$

   b. $z + 2$

      $z + 2 = -3 + 2i + 2 = -1 + 2i$

   c. $w + 2i$

      $w + 2i = 2 - 3i + 2i = 2 - i$

   d. $z - 3i$

      $z - 3i = -3 + 2i - 3i = -3 - i$

   e. $w + z$

      $w + z = 2 - 3i + (-3 + 2i) = -1 - i$

   f. $z - w$

      $z - w = -3 + 2i - (2 - 3i) = -5 + 5i$

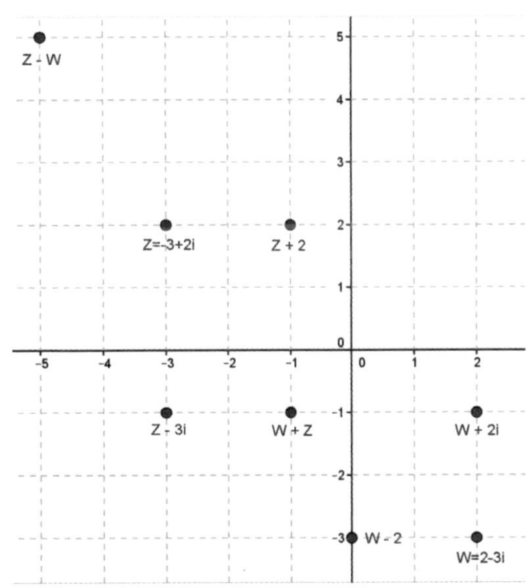

2. Let $z = 5 - 2i$. Find w for each case.

   a. z is a $90°$ counterclockwise rotation about the origin of w.

      $w \cdot i = z$; therefore, $w = \frac{z}{i} = \frac{5-2i}{i} = \frac{2+5i}{-1} = -2 - 5i$.

   b. z is reflected about the imaginary axis from w.

      $w = -\bar{z}$; therefore, $w = -(5+2i) = -5 - 2i$.

   c. z is reflected about the real axis from w.

      $w = \bar{z}$; therefore, $w = 5 + 2i$.

3. Let $z = -1 + 2i$, $w = 4 - i$. Simplify the following expressions.

   a. $z + \bar{w}$

      $z + \bar{w} = -1 + 2i + 4 + i = 3 + 3i$

   b. $|w - \bar{z}|$

      $|w - \bar{z}| = |4 - i - (-1 - 2i)| = |4 - i + 1 + 2i| = |5 + i| = \sqrt{(5)^2 + (1)^2} = \sqrt{26}$

   c. $2z - 3w$

      $2z - 3w = -2 + 4i - (12 - 3i) = -2 + 4i - 12 + 3i = -14 + 7i$

   d. $\frac{z}{w}$

      $\frac{z}{w} = \frac{-1+2i}{4-i} = \frac{(-1+2i)(4+i)}{(4-i)(4+i)} = \frac{-6+7i}{16+1} = \frac{-6}{17} + \frac{7i}{17}$

4. Given the complex number z, find a complex number w where $z + w$ is shifted:

   a. $2\sqrt{2}$ units in a northeast direction.

      $x^2 + x^2 = (2\sqrt{2})^2$, $2x^2 = 8$, $x = \pm 2$. Therefore, $w = 2 + 2i$.

   b. $5\sqrt{2}$ units in a southeast direction.

      $x^2 + x^2 = (5\sqrt{2})^2$, $2x^2 = 50$, $x = \pm 5$. Therefore, $w = 5 - 5i$.

Lesson 10: The Geometric Effect of Some Complex Arithmetic

Student Outcomes

- Students represent multiplication of complex numbers geometrically on the complex plane.

Lesson Notes

This is the second of a two-day lesson in which students continue to explore the geometric interpretations of complex arithmetic. In Lesson 9, students studied the geometric effects of adding and subtracting complex numbers. Lesson 10 focuses on multiplication of complex numbers and the geometric effect. Students revisit the concept of linearity in this lesson. Students need graph paper for each exercise and example.

Classwork

Opening Exercises (8 minutes)

Lesson 8 introduced the geometric effect of adding complex numbers. This opening exercise helps students solidify that concept as well as the geometric effect of multiplying by i and taking the complex conjugate. Students need a firm grasp of all of these topics to understand the geometric effect of multiplying by a complex number. Students should work in small groups or pairs. Each student needs a piece of graph paper.

> *Scaffolding:*
>
> Provide graphs with grid lines or with original points or points and images plotted for students with eye-hand or spatial difficulties.

Opening Exercises

1. **Given $z = 3 - 2i$, plot and label the following, and describe the geometric effect of the operation.**

   a. z

   3 units on the real axis to the right of the origin and 2 units on the imaginary axis below the origin

   b. $z - 2$

   2 units on the real axis to the left of z

   c. $z + 4i$

   4 units on the imaginary axis above z

   d. $z + (-2 + 4i)$

   2 units on the real axis to the left of z and 4 units on the imaginary axis above z

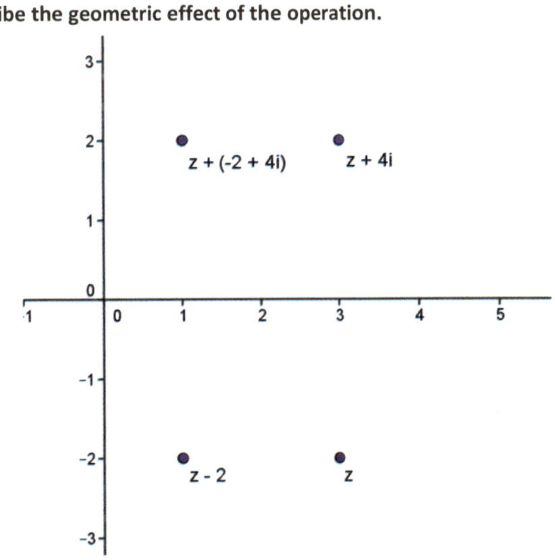

2. Describe the geometric effect of the following:

   a. Multiplying by i

      A $90°$ rotation about the origin

   b. Taking the complex conjugate

      Reflects the imaginary number across the real axis

   c. What operation reflects a complex number across the imaginary axis?

      The opposite of the conjugate ($-\bar{z}$) reflects a complex number across the imaginary axis.

Example 1 (8 minutes)

In Example 1, students multiply complex numbers by a constant to discover the geometric effect. Allow students to work in pairs or in small groups. We confirm students' conjectures that multiplying a complex number by a constant creates a dilation with the constant as the scale factor by calculating the length of the segment from the point to the origin before and after transformation. Students need graph paper.

Scaffolding:

- When confirming dilations in Example 1, assign struggling students with points on the axes or with coefficients of 1 and a constant that is a small whole number.
- Assign advanced groups points with larger and/or negative coefficients and fractional or decimal constants.

Example 1

Plot the given points, and then plot the image $L(z) = 2z$.

a. $z_1 = 3$

   $\boldsymbol{2z_1 = 6}$

b. $z_2 = 2i$

   $\boldsymbol{2z_2 = 4i}$

c. $z_3 = 1 + i$

   $\boldsymbol{2z_3 = 2 + 2i}$

d. $z_4 = -4 + 3i$

   $\boldsymbol{2z_4 = -8 + 6i}$

e. $z_5 = 2 - 5i$

   $\boldsymbol{2z_5 = 4 - 10i}$

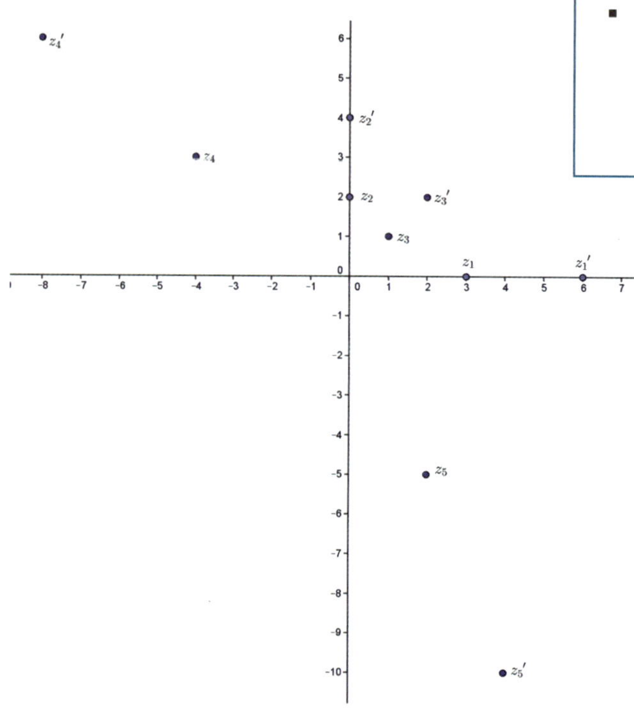

- Algebraically, how did $L(z) = 2z$ affect z?
  - *The coefficients doubled.*
- Geometrically, how did $L(z) = 2z$ affect z?
  - *There was a dilation with scale factor* 2.
- Draw a segment from the origin to one point and its image. Confirm your conjecture by finding the length of the segments.

Assign different groups different pairs of points so as a class, all pairs are checked. Students confirm that every segment from the origin to the image is double the length of the segment from the origin to the original point.

- Do you think the same would be true if we multiplied by a constant other than 2? Confirm your answer.
  - *This is true for any constant.*

Assign different groups different points by which to multiply different constants, and then have groups report their findings. This would be a good activity to use for differentiating instruction.

Exercises 1–7 (6 minutes)

Students already know that multiplying by i produces a $90°$ counterclockwise rotation about the origin. This lesson solidifies that concept in preparation for Example 2 and future lessons. This exercise can be completed in groups or pairs. Students need graph paper.

Exercises

Plot the given points, and then plot the image $L(z) = iz$.

1. $z_1 = 3$

   $iz_1 = 3i$

2. $z_2 = 2i$

   $iz_2 = -2$

3. $z_3 = 1 + i$

   $iz_3 = i(1 + i) = i - 1 = -1 + i$

4. $z_4 = -4 + 3i$

   $iz_4 = i(-4 + 3i) = -4i - 3 = -3 - 4i$

5. $z_5 = 2 - 5i$

   $iz_5 = i(2 - 5i) = 2i + 5 = 5 + 2i$

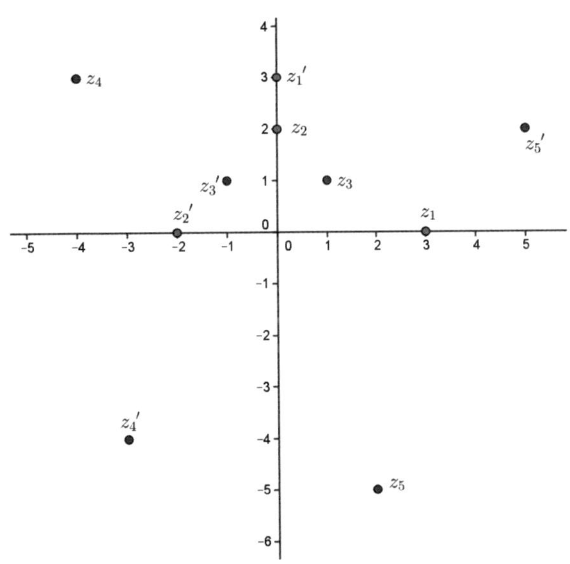

MP.7

6. **What is the geometric effect of the transformation? Confirm your conjecture using the slope of the segment joining the origin to the point and then to its image.**

Multiplying by i rotates the point $90°$ counterclockwise about the origin. This is confirmed because the slopes of the segments joining the origin and the original points and the slopes of the segments joining the origin and the image of those points are opposite reciprocals, which means the segments are perpendicular.

7. **Is $L(z)$ a linear transformation? Explain how you know.**

Yes, $L(z+w) = L(z) + L(w)$ and $kL(z) = L(kz)$.

Example 2 (15 minutes)

Students have discovered the geometric effect of multiplying a complex number by a constant and by i. In Example 2, we multiply by a complex number with a real and an imaginary part. Students may struggle with this, but in this lesson, we are just looking at the transformations graphically and letting students think about the geometric effect. We want students to understand that there is a rotation and a dilation, but the exact effect is studied in Lessons 10–12. Students should work in small groups or pairs and need graph paper.

Scaffolding:

Allow advanced learners to do Example 2 on their own and then change $L(z) = (3+4i)z$, or give them this problem instead of the one listed in the example.

Example 2

Describe the geometric effect of $L(z) = (1+i)z$ given the following. Plot the images on graph paper, and describe the geometric effect in words.

a. $z_1 = 1$

$L(z_1) = (1+i)(1) = 1+i$, ***no change***

b. $z_2 = i$

$L(z_2) = (1+i)i = i - 1 = -1 + i$, ***a*** $90°$ ***counterclockwise rotation about the origin***

MP.3

c. $z_3 = 1 + i$

$L(z_3) = (1+i)(1+i) = 1 + 2i - 1 = 2i$, ***a*** $45°$ ***counterclockwise rotation about the origin and a dilation with a scale factor of*** $\sqrt{2}$

d. $z_4 = 4 + 6i$

$L(z_4) = (1+i)(4+6i) = 4 + 10i - 6 = -2 + 10i$, ***a clockwise rotation about the origin of some angle measure θ and a dilation with a scale factor greater than*** 1

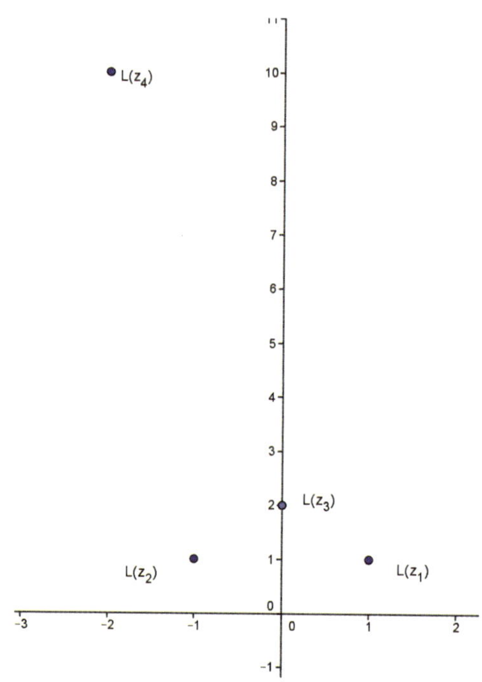

- What was the geometric result of multiplying the complex number by 1?
  - *There was no change.*
- Are you surprised by this result?
  - *No, multiplying by 1 always results in the number you started with. It is the multiplicative identity.*
- What was the geometric result of multiplying by i?
  - *The result was a $90°$ counterclockwise rotation about the origin. The point was reflected across the imaginary axis.*
- Did this result surprise you?
  - *No, multiplying by i always results in a $90°$ counterclockwise rotation about the origin, but it is not always a reflection across the imaginary axis.*
- What would always create a reflection across the real axis?
  - *Taking the complex conjugate, $1 - i$*
- What would create a reflection across the imaginary axis?
  - *The opposite of the complex conjugate, $-1 + i$*
- What was the geometric result of multiplying by $1 + i$?
  - *There was a clockwise rotation and a dilation.*
- Can you guess how many degrees the point rotated? Explain.

It is ok if students do not yet fully grasp the degree of rotation; they study this in great detail in future lessons.

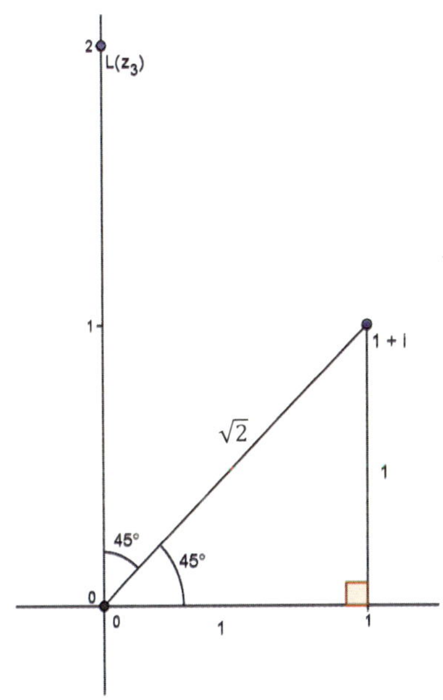

  - *$45°$. The angle of the segment joining the origin to the original point is $45°$ because if you draw a right triangle, each leg is 1 unit, so the triangle is a special $45°$–$45°$–$90°$ triangle. The image is on the y-axis, so that is a $90°$. If the original point was at a $45°$ angle and the reflection is at an angle of $90°$, the angle that the point rotated was $45°$.*
- What was the scale factor of the dilation? Explain.
  - *The segment joining the origin and the original point was the hypotenuse of a $45°$–$45°$–$90°$ with legs of length 1 unit, so it was $\sqrt{2}$ units in length by the Pythagorean theorem. The segment joining the origin and the image is 2 units in length. This means that the original length was multiplied by a scale factor of $\sqrt{2}$ to get the length of the segment containing the image.*
- What was the result of multiplying by $4 + 6i$?
  - *A clockwise rotation and a dilation greater than 1*
- Can you determine the angle of rotation and the dilation scale factor?

Students are probably not able to determine the answers for this result, but this is studied in detail later. It can be left for them to think about.

  - *The scale factor is $\sqrt{26}$, and $\theta = \pi - \tan^{-1}(5) = 1.768 \text{ radians} = 101.31°$.*

Closing (3 minutes)

Have students explain the following questions to a neighbor, and then bring the class back together for a debrief. Encourage students to draw diagrams to support their responses.

- Explain the geometric effect of multiplying a complex number by the following:
  1. 1
     - *There is no change. The number stays where it was.*
  2. i
     - *This produces a counterclockwise rotation of* $90°$ *about the origin.*
  3. $a + bi$
     - *The result is a rotation about the origin and a dilation.*

Exit Ticket (5 minutes)

Name ______________________________ Date ______________

Lesson 10: The Geometric Effect of Some Complex Arithmetic

Exit Ticket

1. Given $T(z) = z$, describe the geometric effect of the following:
   a. $T(z) = 5z$

   b. $T(z) = \frac{z}{2}$

   c. $T(z) = i \cdot z$

2. If $z = -2 + 3i$ is the result of a $90°$ counterclockwise rotation about the origin from w, find w. Plot z and w in the complex plane.

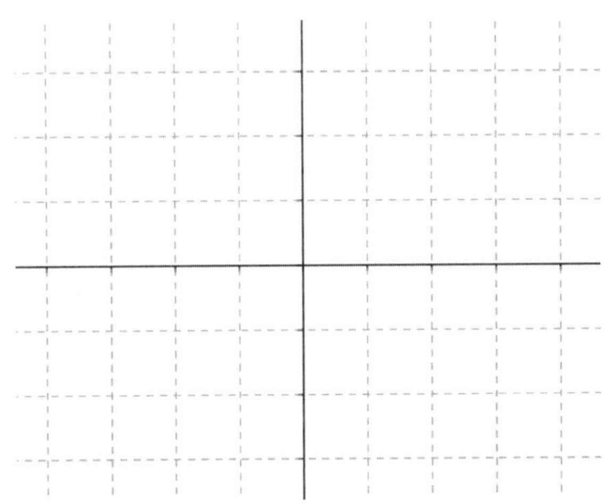

3. Explain the geometric effect of z if you multiply z by w, where $w = 1 + i$.

Exit Ticket Sample Solutions

1. Given $T(z) = z$, describe the geometric effect of the following:

   a. $T(z) = 5z$

   It has a dilation with a scale factor of 5.

   b. $T(z) = \frac{z}{2}$

   It has a dilation with a scale factor of $\frac{1}{2}$.

   c. $T(z) = iz$

   It has a $90°$ *counterclockwise rotation about the origin.*

2. If $z = -2 + 3i$ is the result of a $90°$ counterclockwise rotation about the origin from w, find w. Plot z and w in the complex plane.

$$z = i \cdot w$$
$$-2 + 3i = i \cdot w$$
$$w = \frac{-2 + 3i}{i}$$
$$w = \frac{(-2 + 3i) \cdot i}{i \cdot i}$$
$$w = \frac{-3 - 2i}{-1}$$
$$w = 3 + 2i$$

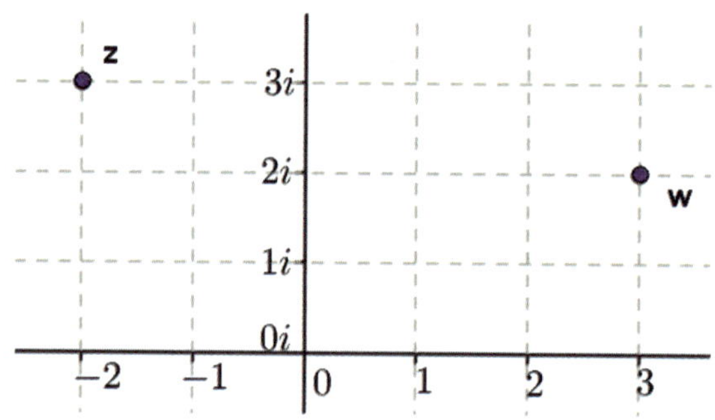

3. Explain the geometric effect of z if you multiply z by w, where $w = 1 + i$.

   It has a $45°$ *counterclockwise rotation about the origin and a dilation with a scale factor of* $\sqrt{2}$.

Problem Set Sample Solutions

1. Let $z = -4 + 2i$. Simplify the following, and describe the geometric effect of the operation. Plot the result in the complex plane.

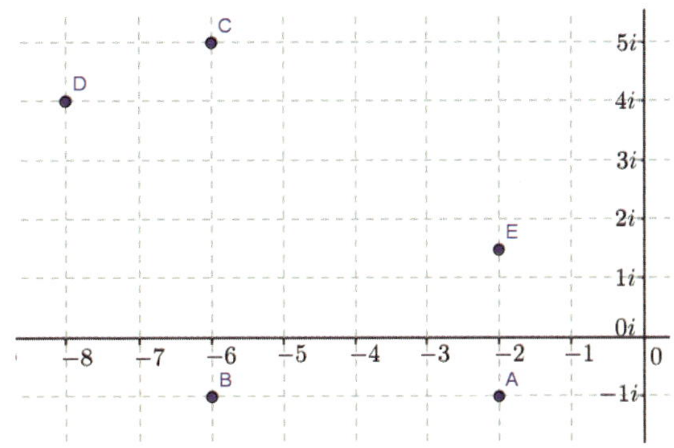

a. $z + 2 - 3i$

$-4 + 2i + 2 - 3i = -2 - i$

z is shifted 2 units to the right and 3 units downward.

b. $z - 2 - 3i$

$-4 + 2i - 2 - 3i = -6 - i$

z is shifted 2 units to the left and 3 units downward.

c. $z - (2 - 3i)$

$-4 + 2i - (2 - 3i) = -4 + 2i - 2 + 3i = -6 + 5i$

z is shifted 2 units to the left and 3 units upward.

d. $2z$

$2(-4 + 2i) = -8 + 4i$

z has a dilation with a scale factor of 2.

e. $\frac{z}{2}$

$\frac{-4 + 2i}{2} = -2 + \frac{3}{2}i$

z has a dilation with a scale factor of $\frac{1}{2}$.

2. Let $z = 1 + 2i$. Simplify the following, and describe the geometric effect of the operation.

a. iz

$iz = -2 + i$

z is rotated $90°$ counterclockwise.

b. i^2z

$i^2z = (-1)(1 + 2i) = -1 - 2i$

OR

$i^2z = i \cdot i \cdot (1 + 2i) = i(i - 2) = -1 - 2i$

z is rotated $180°$ counterclockwise.

c. $\bar{z}$

$\bar{z} = 1 - 2i$

z is reflected about the real axis.

d. $-\bar{z}$

$-\bar{z} = -(1 - 2i) = -1 + 2i$

z is reflected about the imaginary axis.

e. $i\bar{z}$

$i\bar{z} = i(1 - 2i) = 2 + i$

z is reflected about the real axis first and then is rotated $90°$ counterclockwise.

f. $2iz$

$2iz = 2i(1 + 2i) = -4 + 2i$

z is rotated $90°$ counterclockwise and then has a dilation with a scale factor of 2.

g. $iz + 5 - 3i$

$iz + 5 - 3i = i(1 + 2i) + 5 - 3i = -2 + i + 5 - 3i = 3 - 2i$

z is rotated $90°$ counterclockwise first and then shifted 5 units to the right and 3 units downward.

3. Simplify the following expressions.

a. $(4 - 2i)(5 - 3i)$

$14 - 22i$

b. $(-2 + 3i)(-2 - 3i)$

13

c. $(1 + i)^2$

$2i$

d. $(1 + i)^{10}$ (Hint: $b^{nm} = (b^n)^m$)

$(1 + i)^{10} = ((1 + i)^2)^5 = (2i)^5 = 32i$

e. $\frac{-1+2i}{1-2i}$

$\frac{(-1 + 2i)(1 + 2i)}{(1 - 2i)(1 + 2i)} = \frac{-1 - 2i + 2i - 4}{1 + 4} = \frac{-5}{5} = -1$ *or* $\frac{-1 + 2i}{1 - 2i} = \frac{-(1 - 2i)}{1 - 2i} = -1$

f. $\frac{x^2+4}{x-2i}$, provided $x \neq 2i$

$$\frac{x^2+4}{x-2i} = \frac{(x+2i)(x-2i)}{x-2i} = x+2i$$

4. Given $z = 2 + i$, describe the geometric effect of the following. Plot the result.

a. $z(1+i)$

z is rotated $45°$ counterclockwise and has a scale factor $\sqrt{2}$ multiplying to $|z|$.

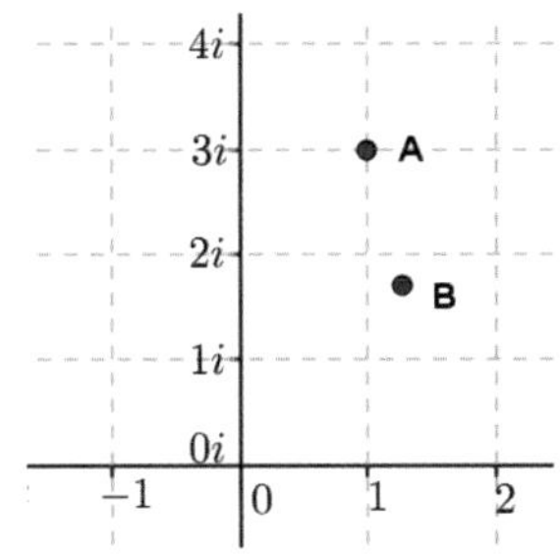

b. $z\left(\frac{\sqrt{3}}{2} + \frac{1}{2}i\right)$

z is rotated $30°$ counterclockwise and has a scale factor 1 multiplying to $|z|$.

5. We learned that multiplying by i produces a $90°$ counterclockwise rotation about the origin. What do we need to multiply by to produce a $90°$ clockwise rotation about the origin?

We need to multiply by $\frac{1}{i}$. If $w = iz$ (w is a $90°$ counterclockwise rotation about the origin from z), $z = \frac{w}{i}$, which means that if we divide w by i, we will get to z, which will create a $90°$ clockwise rotation about the origin from w. Another response: $90°$ clockwise is $270°$ counterclockwise, so you could multiply by i^3 to map (a, b) to $(b, -a)$.

6. Given z is a complex number $a + bi$, determine if $L(z)$ is a linear transformation. Explain why or why not.

a. $L(z) = i^3 z$

Yes. $L(z) = -iz$, so $L(z+w) = -iz + -iw$ and $L(z) + L(w) = -iz + -iw$.

$kL(z) = k(-iz)$ and $L(kz) = -i(kz)$.

Since $L(z+w) = L(z) + L(w)$ and $kL(z) = L(kz)$, the function is a linear transformation.

b. $L(z) = z + 4i$

No. $L(z+w) = z + w + 4i$ and $L(z) + L(w) = z + 4i + w + 4i$. $kL(z) = k(z+4i)$ and $L(kz) = kz + 4i$. Since $L(z+w) = L(z) + L(w)$ and $kL(z) \neq L(kz)$, the function is not a linear transformation.

Lesson 11: Distance and Complex Numbers

Student Outcomes

- Students calculate distances between complex numbers as the modulus of the difference.
- Students calculate the midpoint of a segment as the average of the numbers at its endpoints.

Lesson Notes

In Topic A, students saw that complex numbers have geometric interpretations associated with them since points in the complex plane seem analogous to points in the coordinate plane. In Lesson 6, students considered complex numbers as vectors and learned to add them by the tip-to-tail method. In Lessons 8 and 9, students explored the idea that every complex operation must have some geometric interpretation, eventually coming to the realization that complex addition and subtraction have the geometric effect of performing a translation to points in the complex plane. The geometric interpretation of complex multiplication was left unresolved as students realized it was not readily obvious. Later in the module, students continue to explore the question, "What is the geometric action of multiplication by a complex number w on all the points in the complex plane?" To understand this, students first explore the connection between geometry and complex numbers. The coordinate geometry studied in Geometry was about points in the coordinate plane, whereas now the thinking is about complex numbers in the complex plane.

Classwork

Opening Exercise (5 minutes)

Give students time to work independently on the Opening Exercise before discussing as a class.

Opening Exercise

a. **Plot the complex number $z = 2 + 3i$ on the complex plane. Plot the ordered pair $(2, 3)$ on the coordinate plane.**

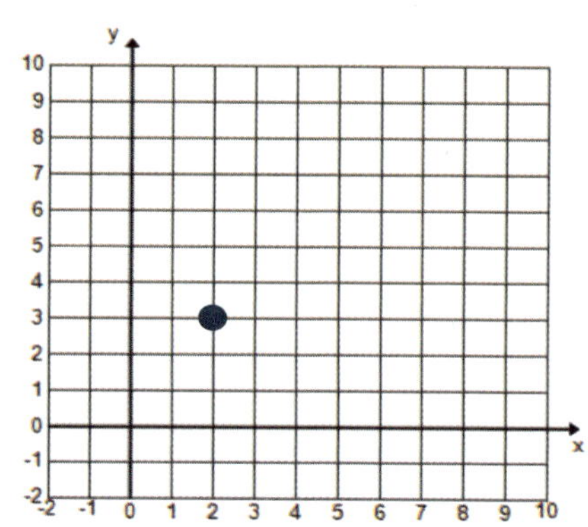

b. In what way are complex numbers *points?*

When a complex number is plotted on a complex plane, it looks just like the corresponding ordered pair plotted on a coordinate plane. For example, when $2+3i$ is plotted on the complex plane, it looks exactly the same as when the ordered pair $(2,3)$ is plotted on a coordinate plane. We can interchangeably think of a complex number $x+yi$ in the complex plane as a point (x,y) in the coordinate plane, and vice versa.

c. What point on the coordinate plane corresponds to the complex number $-1+8i$?

$(-1, 8)$

d. What complex number corresponds to the point located at coordinate $(0,-9)$?

$0-9i$ *or* $-9i$

Discussion (7 minutes)

Draw the following on the board.

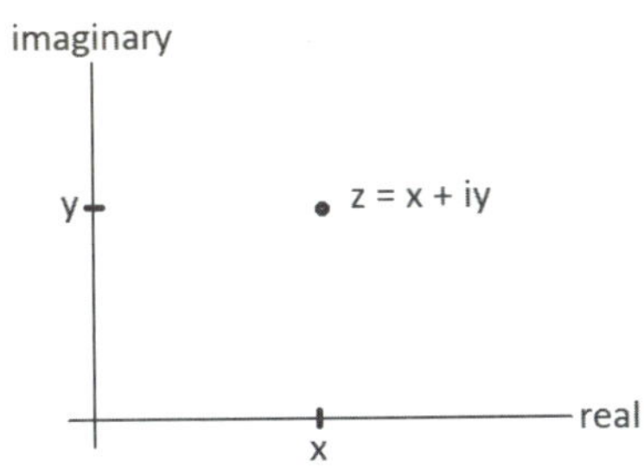

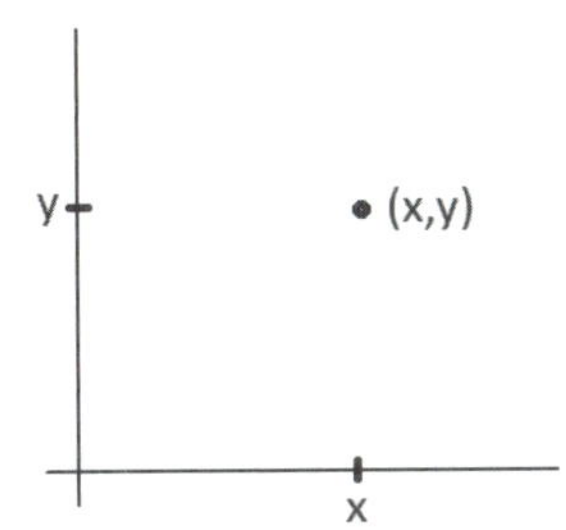

- When we say that complex numbers are points in the complex plane, what do we really mean?
  - *When a complex number $x+yi$ is plotted on the complex plane, it looks exactly the same as when the ordered pair (x,y) is plotted on a coordinate plane.*
- What does this mean in terms of connecting the ideas we learned in Geometry to complex numbers?
  - *Since we can interchangeably think of a complex number $x+yi$ in the complex plane as a point (x,y) in the coordinate plane, and vice versa, all the work we did in Geometry can be translated into the language of complex numbers, and vice versa. Therefore, any work we do with complex numbers should translate back to results from Geometry.*
- In Geometry, it did not make sense to add two points together. If $A(x_1,y_1)$ and $B(x_2,y_2)$ are points, what would the geometric meaning of $A+B$ be?
  - *It does not seem to have any meaning.*
- It does make sense to add two complex numbers together. If $A=x_1+y_1i$ and $B=x_2+y_2i$, then what is $A+B$?
  - $(x_1+x_2)+(y_1+y_2)i$

MP.7

- What is the geometric effect of transforming A with the function $f(z)=z+B$ for constant complex number B?
  - *Applying the transformation to A has the geometric effect of performing a translation. So, adding B to A will shift point A right x_2 units and up y_2.*
- If we view points in the plane as complex numbers, then we can add points in geometry.

> *Scaffolding:*
>
> If needed, provide students with an example using specific numbers rather than general parameters.
>
> - If $A=3+4i$ and $B=-1+6i$, then what is $A+B$?
>   - $(3+(-1))+(4+6)i$ or $2+10i$

Exercise 1 (3 minutes)

Have students work on Exercise 1 independently and then share results with a partner. If students do not recall how to find the midpoint, have them draw the line segment and locate the midpoint from the graph rather than providing them with the midpoint formula.

> **Exercise 1**
>
> **The endpoints of $\overline{AB}$ are $A(1,8)$ and $B(-5,3)$. What is the midpoint of $\overline{AB}$?**
>
> *The midpoint of $\overline{AB}$ is $\left(-2, \frac{11}{2}\right)$.*

- How do you find the midpoint of $\overline{AB}$?
  - *You find the average of the x-coordinates and the y-coordinates to find the halfway point.*
- In Geometry, we learned that for two points $A(x_1, y_1)$ and $B(x_2, y_2)$, the midpoint of $\overline{AB}$ is $\left(\frac{x_1 + x_2}{2}, \frac{y_1 + y_2}{2}\right)$.
- Now, view these points as complex numbers: $A = x_1 + y_1 i$ and $B = x_2 + y_2 i$.

Exercise 2 (7 minutes)

Allow students time to work on part (a) independently and then share results with a partner. Have them work in partners on part (b) before discussing as a class.

MP.2

> **Exercise 2**
>
> a. **What is the midpoint of $A = 1 + 8i$ and $B = -5 + 3i$?**
>
> $$M = \frac{1 + -5}{2} + \frac{(8+3)}{2}i = -2 + \frac{11}{2}i$$
>
> b. **Using $A = x_1 + y_1 i$ and $B = x_2 + y_2 i$, show that, in general, the midpoint of points A and B is $\frac{A+B}{2}$, the arithmetic average of the two numbers.**
>
> $$M = \frac{x_1 + x_2}{2} + \frac{y_1 + y_2}{2}i = \frac{x_1 + y_1 i + x_2 + y_2 i}{2} = \frac{A+B}{2}$$

Scaffolding:

- Provide visual learners with a graph on the complex plane.

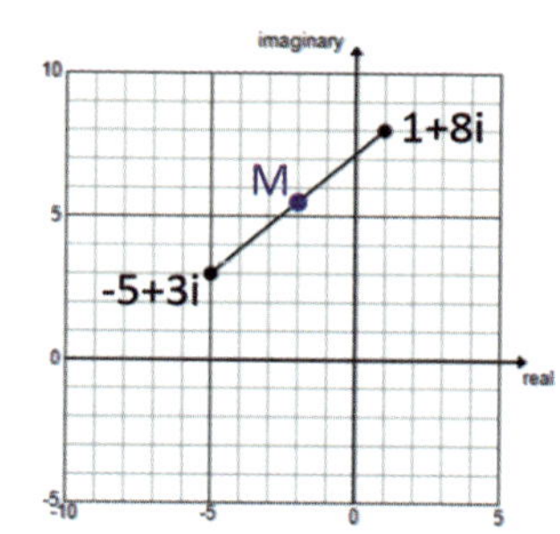

- If students need additional practice, use this example before completing part (b) of Exercise 2:

  What is the midpoint of $A = -3 + 2i$ and $B = 12 - 10i$?
  - $M = \frac{9}{2} - 4i$

Exercise 3 (5 minutes)

As with Exercise 1, have students work on Exercise 3 independently and then share results with a partner. If students do not recall how to find the length of a line segment, have them draw the line segment, and instruct them to think of it as the hypotenuse of a right triangle rather than providing them with the distance formula.

> **Exercise 3**
>
> **The endpoints of $\overline{AB}$ are $A(1,8)$ and $B(-5,3)$. What is the length of $\overline{AB}$?**
>
> ***The length of $\overline{AB}$ is $\sqrt{61}$.***

- How do you find the length of $\overline{AB}$?
  - *You use the Pythagorean theorem, which can be written as $AB = \sqrt{(x_2 - x_1)^2 + (y_2 - y_1)^2}$ for two points $A(x_1, y_1)$ and $B(x_2, y_2)$.*
- As we did previously, view these points as complex numbers: $A = x_1 + y_1 i$ and $B = x_2 + y_2 i$.

Exercise 4 (7 minutes)

As with Exercise 2, allow students time to work on part (a) independently and then share results with a partner. Have them work in partners on part (b) before discussing as a class.

> *Scaffolding:*
> - Provide visual learners with a graph on the complex plane.
>
> 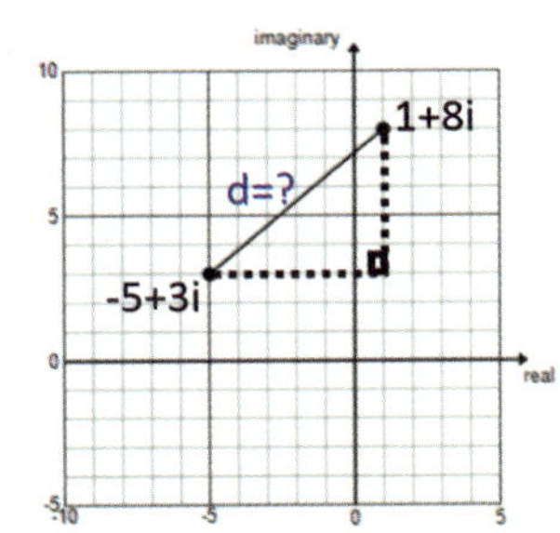
>
>
> - If students need additional practice, use this example before completing part (b) of Exercise 4:
>
>   What is the distance between $A = -3 + 2i$ and $B = 12 - 10i$?
>   - $d = \sqrt{369} = 3\sqrt{41}$

MP.2

> **Exercise 4**
>
> **a. What is the distance between $A = 1 + 8i$ and $B = -5 + 3i$?**
>
> $$d = \sqrt{(1-(-5))^2 + (8-3)^2} = \sqrt{61}$$
>
> **b. Show that, in general, the distance between $A = x_1 + y_1 i$ and $B = x_2 + y_2 i$ is the modulus of $A - B$.**
>
> $A - B = (x_1 - x_2) + (y_1 - y_2)i$
>
> $|A - B| = \sqrt{(x_1 - x_2)^2 + (y_1 - y_2)^2}$ = *distance between A and B*

Exercise 5 (3 minutes)

Allow students time to work either independently or with a partner. Circulate the room to ensure that students understand the concepts.

> *Scaffolding:*
>
> Have struggling students create a graphic organizer comparing a coordinate plane and a complex plane.
>
> | | Coordinate plane | Complex plane |
> |---|---|---|
> | | $(-3,2)$ and $(1,3)$ | $-3+2i$ and $1+3i$ |
> | Graph | | |
> | Midpoint | | |
> | Distance | | |

Exercise 5

Suppose $z = 2 + 7i$ and $w = -3 + i$.

a. **Find the midpoint m of z and w.**

$$m = -\frac{1}{2} + 4i$$

b. **Verify that $|z - m| = |w - m|$.**

$$|z - m| = \sqrt{\left(2 - \left(-\frac{1}{2}\right)\right)^2 + (7-4)^2}$$
$$= \sqrt{\frac{25}{4} + 9}$$
$$= \sqrt{\frac{61}{4}}$$

$$|w - m| = \sqrt{\left(-3 - \left(-\frac{1}{2}\right)\right)^2 + (1-4)^2}$$
$$= \sqrt{\frac{25}{4} + 9}$$
$$= \sqrt{\frac{61}{4}}$$

Closing (3 minutes)

Have students discuss each question with a partner. Then, elicit class responses.

- In what way can complex numbers be thought of as points?
  - *When a complex number is plotted on a complex plane, it looks just like the corresponding ordered pair plotted on a coordinate plane.*
- Why is it helpful to interchange between complex numbers and points on a plane?
  - *Unlike points on a plane, we can add and subtract complex numbers. Thus, we can use operations on complex numbers to find geometric measurements such as midpoint and distance.*

Lesson Summary

- Complex numbers can be thought of as points in a plane, and points in a plane can be thought of as complex numbers.
- For two complex numbers $A = x_1 + y_1 i$ and $B = x_2 + y_2 i$, the midpoint of points A and B is $\frac{A+B}{2}$.
- The distance between two complex numbers $A = x_1 + y_1 i$ and $B = x_2 + y_2 i$ is equal to $|A - B|$.

Exit Ticket (5 minutes)

Name ______________________________ Date ______________

Lesson 11: Distance and Complex Numbers

Exit Ticket

1. Kishore said that he can add two points in the coordinate plane like adding complex numbers in the complex plane. For example, for point $A(2, 3)$ and point $B(5, 1)$, he will get $A + B = (7, 4)$. Is he correct? Explain your reasoning.

2. Consider two complex numbers $A = -4 + 5i$ and $B = 4 - 10i$.
   a. Find the midpoint of A and B.

   b. Find the distance between A and B.

Exit Ticket Sample Solutions

1. Kishore said that he can add two points in the coordinate plane like adding complex numbers in the complex plane. For example, for point $A(2,3)$ and point $B(5,1)$, he will get $A+B=(7,4)$. Is he correct? Explain your reasoning.

   No. Kishore is not correct because we cannot add two points in the rectangular plane. However, we can add two complex numbers in the complex plane, which has the geometric effect of performing a translation to points in complex numbers.

2. Consider two complex numbers $A=-4+5i$ and $B=4-10i$.

   a. Find the midpoint of A and B.

   $$\begin{aligned} M &= \frac{A+B}{2} \\ &= \frac{-4+5i+4-10i}{2} \\ &= \frac{-5i}{2} \\ &= -\frac{5}{2}i \text{ or } 0-\frac{5}{2}i \end{aligned}$$

   b. Find the distance between A and B.

   $$d=\sqrt{\left(4-(-4)\right)^2+(-10-5)^2}=17$$

Problem Set Sample Solutions

1. Find the midpoint between the two given points in the rectangular coordinate plane.

   a. $2+4i$ and $4+8i$

   $$M=\frac{2+4}{2}+\frac{4+8}{2}i=3+6i$$

   b. $-3+7i$ and $5-i$

   $$M=\frac{-3+5}{2}+\frac{7-1}{2}i=1+3i$$

   c. $-4+3i$ and $9-4i$

   $$M=\frac{-4+9}{2}+\frac{3-4}{2}i=\frac{5}{2}-\frac{1}{2}i$$

   d. $4+i$ and $-12-7i$

   $$M=\frac{4-12}{2}+\frac{1-7}{2}i=-4-3i$$

e. $-8 - 3i$ and $3 - 4i$

$$M = \frac{-8+3}{2} + \frac{-3-4}{2}i = -\frac{5}{2} - \frac{7}{2}i$$

f. $\frac{2}{3} - \frac{5}{2}i$ and $-0.2 + 0.4i$

$$M = \frac{\frac{2}{3} - \frac{5}{2}i - \frac{1}{5} + \frac{2}{5}i}{2}$$
$$= \frac{\frac{2}{3} - \frac{1}{5}}{2} + \frac{\frac{-5}{2} + \frac{2}{5}}{2}i$$
$$= \frac{10-3}{30} + \frac{-25+4}{20}i$$
$$= \frac{7}{30} - \frac{21}{20}i$$

2. Let $A = 2 + 4i$, $B = 14 + 8i$, and suppose that C is the midpoint of A and B and that D is the midpoint of A and C.

a. Find points C and D.

$$C = \frac{A+B}{2} = \frac{2+4i+14+8i}{2} = \frac{16+12i}{2} = 8 + 6i$$
$$D = \frac{A+C}{2} = \frac{2+4i+8+6i}{2} = \frac{10+10i}{2} = 5 + 5i$$

b. Find the distance between A and B.

$$|A - B| = |2 + 4i - 14 - 8i|$$
$$= |-12 - 4i|$$
$$= \sqrt{(-12)^2 + (-4)^2}$$
$$= \sqrt{144 + 16}$$
$$= \sqrt{160}$$
$$= 4\sqrt{10}$$

c. Find the distance between A and C.

$$|A - C| = |2 + 4i - 8 - 6i|$$
$$= |-6 - 2i|$$
$$= \sqrt{(-6)^2 + (-2)^2}$$
$$= \sqrt{40}$$
$$= 2\sqrt{10}$$

d. Find the distance between C and D.

$$|C - D| = |8 + 6i - 5 - 5i|$$
$$= |3 + i|$$
$$= \sqrt{(3)^2 + (1)^2}$$
$$= \sqrt{10}$$

e. **Find the distance between D and B.**

$$|D - B| = |5 + 5i - 14 - 8i|$$
$$= |-9 - 3i|$$
$$= \sqrt{(-9)^2 + (-3)^2}$$
$$= \sqrt{90}$$
$$= 3\sqrt{10}$$

f. **Find a point one-quarter of the way along the line segment connecting segment A and B, closer to A than to B.**

The point is $D = 5 + 5i$.

g. **Terrence thinks the distance from B to C is the same as the distance from A to B. Is he correct? Explain why or why not.**

The distance from B to C is $2\sqrt{10}$, and the distance from A to B is $4\sqrt{10}$. The distances are not the same.

h. **Using your answer from part (g), if E is the midpoint of C and B, can you find the distance from E to C? Explain.**

The distance from B to C is $2\sqrt{10}$, and the distance from E to C should be half of this value, $\sqrt{10}$.

i. **Without doing any more work, can you find point E? Explain.**

B is $5 + 5i$, which is 3 units to the right of A in the real direction and 1 unit up in the imaginary direction. From C, you should move the same amount to get to E, so E would be $11 + 7i$.

Lesson 12: Distance and Complex Numbers

Student Outcomes

- Students apply distances between complex numbers and the midpoint of a segment.
- Students derive and apply a formula for finding the endpoint of a segment when given one endpoint and the midpoint.

Lesson Notes

In Lesson 10, students learned that it is possible to interchange between points on a coordinate plane and complex numbers. Therefore, all the work that was done in Geometry could be translated into the language of complex numbers, and vice versa. This lesson continues exploring the midpoint between complex numbers through an Exploratory Challenge in the form of a leapfrog game. In the Opening Exercise, students develop a formula for finding an endpoint of a segment when given one endpoint and the midpoint. Students then use this formula in the Exploratory Challenge that follows.

Classwork

Opening Exercise (5 minutes)

Allow students time to work on the Opening Exercise independently before discussing results as a class. The formula derived in part (b) is used in the Exploratory Challenge.

Opening Exercise

a. **Let $A = 2 + 3i$ and $B = -4 - 8i$. Find a complex number C so that B is the midpoint of A and C.**

$C = -10 - 19i$

b. **Given two complex numbers A and B, find a formula for a complex number C in terms of A and B so that B is the midpoint of A and C.**

$C = 2B - A$

c. **Verify that your formula is correct by using the result of part (a).**

$$\begin{aligned} C &= 2B - A \\ -10 - 19i &= 2(-4 - 8i) - (2 + 3i) \\ &= -8 - 16i - 2 - 3i \\ &= -10 - 19i \end{aligned}$$

Exercise 1 (7 minutes)

Give students time to work on the exercise in groups. Circulate the room to ensure students understand the problem. Encourage struggling students to try a graphical approach to the problem.

EUREKA MATH™

Exercise 1

Let $z = -100 + 100i$ and $w = 1000 - 1000i$.

a. Find a point one-quarter of the way along the line segment connecting z and w closer to z than to w.

Let M be the midpoint between z and $w = 450 - 450i$.

Let M' be the midpoint between z and $M = 175 - 175i$.

The complex number $175 - 175i$ represents a point on the complex plane that is one-quarter of the way from z on the segment connecting z and w.

b. Write this point in the form $\alpha z + \beta w$ for some real numbers α and β. Verify that this does in fact represent the point found in part (a).

$$\begin{aligned}\frac{3}{4}z + \frac{1}{4}w &= \frac{3}{4}(-100 + 100i) + \frac{1}{4}(1000 - 1000i)\\ &= (-75 + 250) + i(75 - 250)\\ &= 175 - 175i\end{aligned}$$

c. Describe the location of the point $\frac{2}{5}z + \frac{3}{5}w$ on this line segment.

This point is located $\frac{3}{5}$ of the way from z and $\frac{2}{5}$ of the way from w on $\overline{zw}$.

Scaffolding:

- Provide visual learners with a graph on the complex plane.

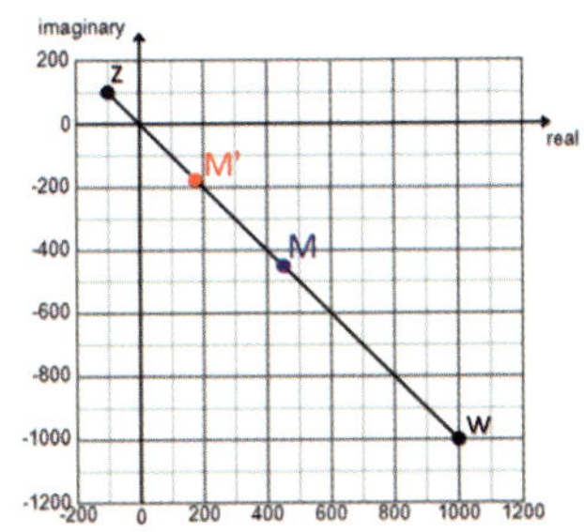

- If students need additional practice, use this example before moving on. Have some students find the answer using the midpoint and some using the result from part (b).
- Find a point one-quarter of the way along the line segment connecting segment connecting $z = 8 - 6i$ and $w = 12 + 16i$ closer to z than to w.
  - $9 - \frac{1}{2}i$

When debriefing, use the graph provided in the scaffold as needed.

- For part (b), did anyone have an answer that did not work when you tried to verify?
  - $\frac{1}{4}z + \frac{3}{4}w$ *(Note: This could be a very common incorrect answer. If nobody offers it as an answer, perhaps suggest it.)*
- Why isn't $\frac{1}{4}z + \frac{3}{4}w$ the correct answer? After all, the point is $\frac{1}{4}$ of the way from z and $\frac{3}{4}$ of the way from w.
  - *It did not work when we tried to verify it.*

$$\frac{1}{4}(-100 + 100i) + \frac{3}{4}(1000 - 1000i) = (-25 + 750) + i(25 - 750) = 725 - 725i$$

- What point would this be on the segment?
  - *It would be the point that is $\frac{3}{4}$ of the way from z and $\frac{1}{4}$ of the way from w.*
- You would think about this like a weighted average. To move the point closer to z, it must be weighted more in the calculation than w.

Exploratory Challenge 1 (15 minutes)

Have students work in groups on this challenge. Each group needs a full page of graph paper. Warn students that they need to put the three points A, B, and C fairly close together in order to stay on the page. Allow students to struggle a little with part (g), but then provide them with help getting started if needed.

Exploratory Challenge 1

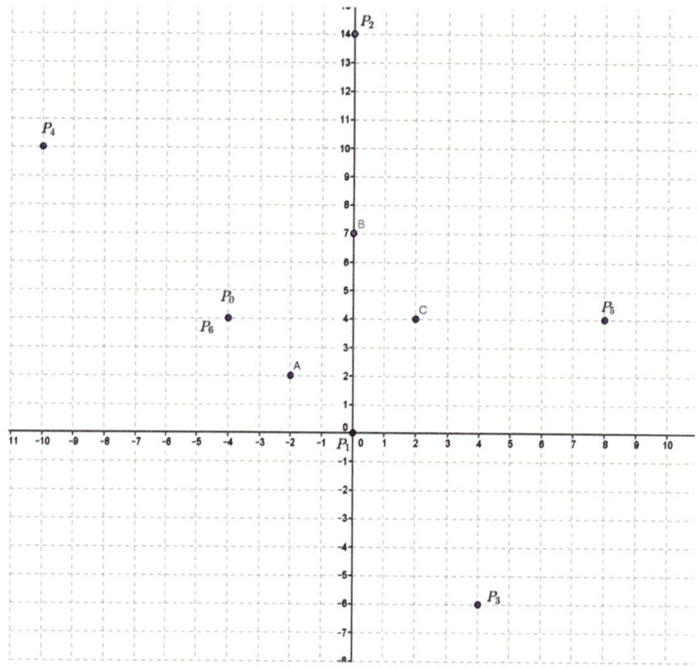

a. Draw three points A, B, and C in the plane.

b. Start at any position P_0, and leapfrog over A to a new position P_1 so that A is the midpoint of $\overline{P_0P_1}$.

c. From P_1, leapfrog over B to a new position P_2 so that B is the midpoint $\overline{P_1P_2}$.

d. From P_2, leapfrog over C to a new position P_3 so that C is the midpoint $\overline{P_2P_3}$.

e. Continue alternately leapfrogging over A, then B, and then C.

f. What eventually happens?

At the sixth jump, you end up at the initial point $P_6 = P_0$.

g. Using the formula from the Opening Exercise, part (b), show why this happens.

$P_1 = 2A - P_0$
$P_2 = 2B - P_1 = 2B - 2A + P_0$
$P_3 = 2C - P_2 = 2C - 2B + 2A - P_0$
$P_4 = 2A - P_3 = -2C + 2B + P_0$
$P_5 = 2B - P_4 = 2C - P_0$
$P_6 = 2C - P_5 = P_0$

- What happened on the sixth jump?
  - *We landed back at the starting point. (Note: Allow a couple of groups to share their graph. A sample is provided.)*

Exploratory Challenge 2 (10 minutes)

Have students continue working in groups on this challenge. Students will need a full page of graph paper for this activity.

Scaffolding:

Provide early finishers with this challenge.

- Repeat the activity with two points, A and B. But this time, leapfrog at a $90°$ angle to the left over each point. Will you return to the starting point? After how many leaps?
  - *You will return to the starting point after 4 leaps.*

Exploratory Challenge 2

a. **Plot a single point A in the plane.**

b. **What happens when you repeatedly jump over A?**

You keep alternating between landing on P_0 and landing on P_1.

c. **Using the formula from the Opening Exercise, part (b), show why this happens.**

$P_1 = 2A - P_0$

$P_2 = 2A - P_1 = 2A - 2A + P_0 = P_0$

MP.3 & MP.4

d. **Make a conjecture about what will happen if you leapfrog over two points, A and B, in the coordinate plane.**

Answers will vary.

e. **Test your conjecture by using the formula from the Opening Exercise, part (b).**

Answers will vary.

f. **Was your conjecture correct? If not, what is your new conjecture about what happens when you leapfrog over two points, A and B, in the coordinate plane?**

Answers will vary, but for the most part, students should have found that their conjecture was incorrect. In this game, you never return to the starting position. Instead, the points continue to get farther away from points A and B. This can be seen by using the formula from the Opening Exercise, part (b).

g. **Test your conjecture by actually conducting the experiment.**

Closing (3 minutes)

Discuss the results of Exploratory Challenge 2.

- What was your initial conjecture about two points?
  - *Answers will vary, but most students would have predicted that at some point you would return to the initial point.*
- How did the formula prove that it was incorrect?
  - *The formula never returned to P_0 because terms did not cancel out.*
- What happened when you leapfrogged over two points?
  - *We kept getting farther from the initial point.*

Exit Ticket (5 minutes)

Name ______________________________ Date ______________

Lesson 12: Distance and Complex Numbers

Exit Ticket

1. Find the distance between the following points.

   a. $(4, -9)$ and $(1, -5)$

   b. $4 - 9i$ and $1 - 5i$

   c. Explain why they have the same answer numerically in parts (a) and (b) but a different perspective in geometric effect.

2. Given point $A = 3 - 2i$ and point $M = -2 + i$, if M is the midpoint of A and another point B, find the coordinates of point B.

Exit Ticket Sample Solutions

1. **Find the distance between the following points.**
   a. $(4,-9)$ **and** $(1,-5)$

$$\begin{aligned} AB &= \sqrt{(4-1)^2+(-9+5)^2} \\ &= \sqrt{9+16} \\ &= 5 \end{aligned}$$

   b. $A = 4-9i$ **and** $B = 1-5i$

$$\begin{aligned} |A-B| &= |4-9i-1+5i| \\ &= |3-4i| \\ &= \sqrt{(3)^2+(-4)^2} \\ &= \sqrt{25} \\ &= 5 \end{aligned}$$

   c. **Explain why they have the same answer numerically in parts (a) and (b) but a different perspective in geometric effect**

   The length of the line segment connecting points A and B is 5.

   To find the distance between two complex numbers A and B, we need to calculate $A-B = 4-9i-1+5i$, which has the geometric effect of performing a translation—shifting one unit to the left and 5 units upward from point $A = 4-9i$. The result is $3-4i$. And $|3-4i|$ is the distance from the origin to the complex number $-4i$, which is not exactly the same as $\overline{AB}$ in terms of their position. However, they all have the same numerical value in terms of distance, which is 5.

2. **Given point $A = 3-2i$ and point $M = -2+i$, if M is the midpoint of A and another point B, find the coordinates of point B.**

$$\begin{aligned} -2+i &= \frac{(3-2i)+M}{2} \\ -4+2i &= 3-2i+M \\ -7+4i &= M \end{aligned}$$

Problem Set Sample Solutions

1. **Find the distance between the following points.**
   a. **Point $A(2,3)$ and point $B(6,6)$**

$$\begin{aligned} AB &= \sqrt{(2-6)^2+(3-6)^2} \\ &= \sqrt{16+9} \\ &= 5 \end{aligned}$$

b. $A = 2 + 3i$ and $B = 6 + 6i$

$$\begin{aligned}|A - B| &= |2 + 3i - 6 - 6i| \\ &= |-4 - 3i| \\ &= \sqrt{(-4)^2 + (-3)^2} \\ &= \sqrt{16 + 9} \\ &= 5\end{aligned}$$

c. $A = -1 + 5i$ and $B = 5 + 11i$

$$\begin{aligned}|A - B| &= |-1 + 5i - 5 - 11i| \\ &= |-6 - 6i| \\ &= \sqrt{(-6)^2 + (-6)^2} \\ &= \sqrt{2(6)^2} \\ &= 6\sqrt{2}\end{aligned}$$

d. $A = 1 - 2i$ and $B = -2 + 3i$

$$\begin{aligned}|A - B| &= |1 - 2i + 2 - 3i| \\ &= |3 - 5i| \\ &= \sqrt{(3)^2 + (-5)^2} \\ &= \sqrt{9 + 25} \\ &= \sqrt{34}\end{aligned}$$

e. $A = \frac{1}{2} - \frac{1}{2}i$ and $B = -\frac{2}{3} + \frac{1}{3}i$

$$\begin{aligned}|A - B| &= \left|\frac{1}{2} - \frac{1}{2}i + \frac{2}{3} - \frac{1}{3}i\right| \\ &= \left|\frac{7}{6} - \frac{5}{6}i\right| \\ &= \sqrt{\left(\frac{7}{6}\right)^2 + \left(\frac{-5}{6}\right)^2} \\ &= \sqrt{\frac{49 + 25}{(6)^2}} \\ &= \frac{\sqrt{74}}{6}\end{aligned}$$

2. Given three points A, B, C, where C is the midpoint of A and B.

a. If $A = -5 + 2i$ and $C = 3 + 4i$, find B.

$$C = \frac{A + B}{2}$$

$$\begin{aligned}B &= 2C - A \\ &= 2(3 + 4i) - (-5 + 2i) \\ &= 6 + 8i + 5 - 2i \\ &= 11 + 6i\end{aligned}$$

b. If $B = 1 + 11i$ and $C = -5 + 3i$, find A.

$$C = \frac{A + B}{2}$$
$$A = 2C - B$$
$$= 2(-5 + 3i) - (1 + 11i)$$
$$= -10 + 6i - 1 - 11i$$
$$= -11 - 5i$$

3. Point C is the midpoint between $A = 4 + 3i$ and $B = -6 - 5i$. Find the distance between points C and D for each point D provided below.

a. $2D = -6 + 8i$

$$C = \frac{4 - 6}{2} + \frac{3 - 5}{2}i = -1 - i$$
$$D = -3 + 4i$$
$$|C - D| = |-1 - i + 3 - 4i|$$
$$= |-2 \quad 5i|$$
$$= \sqrt{(-2)^2 + (5)^2}$$
$$= \sqrt{29}$$

b. $D = -\overline{B}$

$$D = 6 - 5i$$
$$|C - D| = |-1 - i - 6 + 5i|$$
$$= |-7 + 4i|$$
$$= \sqrt{(-7)^2 + (4)^2}$$
$$= \sqrt{65}$$

4. The distance between points $A = 1 + 1i$ and $B = a + bi$ is 5. Find the point B for each value provided below.

a. $a = 4$

$$5 = \sqrt{(1 - 4)^2 + (1 - b)^2}$$
$$25 = 9 + (1 - b)^2$$
$$1 - b = \pm 4$$
$$b = -3, 5$$

$B = 4 - 3i$ *or* $B = 4 + 5i$

b. $b = 6$

$$5 = \sqrt{(1 - a)^2 + (1 - 6)^2}$$
$$25 = (1 - a)^2 + 25$$
$$(1 - a)^2 = 0$$
$$1 - a = 0$$
$$a = 1$$

$B = 1 + 6i$

5. Draw five points in the plane A, B, C, D, E. Start at any position, P_0, and leapfrog over A to a new position, P_1 (so, A is the midpoint of $\overline{P_0P_1}$). Then, leapfrog over B, then C, then D, then E, then A, then B, then C, then D, then E, then A again, and so on. How many jumps will it take to get back to the starting position, P_0?

It takes 10 jumps to return to the starting position.

$P_1 = 2A - P_0$

$P_2 = 2B - P_1 = 2B - 2A + P_0$

$P_3 = 2C - P_2 = 2C - 2B + 2A - P_0$

$P_4 = 2D - P_3 = 2D - 2C + 2B - 2A + P_0$

$P_5 = 2E - P_4 = 2E - 2D + 2C - 2B + 2A - P_0$

$P_6 = 2A - P_5 = 2A - 2E + 2D - 2C + 2B - 2A + P_0 = -2E + 2D - 2C + 2B + P_0$

$P_7 = 2B - P_6 = 2B + 2E - 2D + 2C - 2B - P_0 = 2E - 2D + 2C - P_0$

$P_8 = 2C - P_7 = 2C - 2E + 2D - 2C + P_0 = -2E + 2D + P_0$

$P_9 = 2D - P_8 = 2D + 2E - 2D - P_0 = 2E - P_0$

$P_{10} = 2E - P_9 = 2E - 2E + P_0 = P_0$

MP.3 & MP.4

6. For the leapfrog puzzle problems in both Exploratory Challenge 1 and Problem 5, we are given an odd number of points to leapfrog over. What if we leapfrog over an even number of points? Let $A = 2$, $B = 2 + i$, and $P_0 = i$. Will P_n ever return to the starting position, P_0? Explain how you know.

No, we cannot get back to the starting position. For example, if we leapfrog over two given even points, A and B.

$P_1 = 2A - P_0$

$P_2 = 2B - P_1 = 2B - 2A + P_0$

$P_3 = 2A - P_2 = 2A - 2B + 2A - P_0 = 4A - 2B - P_0$

$P_4 = 2B - P_3 = 2B - 4A + 2B + P_0 = 4B - 4A + P_0$

$P_5 = 2A - P_4 = 2A - 4B + 4A - P_0 = 6A - 4B - P_0$

If n is even, $P_n = nB - nA + P_0 = n(B - A) + P_0$. Then, if $P_0 = P_n$, we have $0 = n(B - A)$, which would mean that $B = A$, which we know to be false. Thus, for even values of n, P_n will never return to P_0.

If n is odd, $P_n = (n + 1)A - (n - 1)B - P_0$. Then, if $P_0 = P_n$ we have

$$\begin{aligned} 2P_0 &= (n+1)A - (n-1)B \\ &= (n+1)2 - (n-1)(2+i) \\ &= -4 - (n-1)i \\ 2i &= -4 - (n-1)i \\ (n+1)i &= -4. \end{aligned}$$

Since $(n + 1)i$ is an imaginary number and -4 is a real number, it is impossible for $(n + 1)i$ to equal -4. Thus, for odd values of n, P_n will never return to P_0.

Therefore, it is not possible for P_n to ever coincide with P_0 for these values of A, B, and P_0.

Lesson 13: Trigonometry and Complex Numbers

Student Outcomes

- Students represent complex numbers in polar form and convert between rectangular and polar representations.
- Students explain why the rectangular and polar forms of a given complex number represent the same number.

Lesson Notes

This lesson introduces the polar form of a complex number and defines the argument of a complex number in terms of a rotation. This definition aligns with the definitions of the sine and cosine functions introduced in Algebra II, Module 2 and ties into work with right-triangle trigonometry from Geometry. This lesson continues to emphasize the usefulness of representing complex numbers as transformations. Analysis of the angle of rotation and the scale of the dilation brings a return to topics in trigonometry first introduced in Geometry (**G-SRT.C.6**, **G-SRT.C.7**, **G-SRT.C.8**) and expanded on in Algebra II (**FTF.A.1**, **F-TF.A.2**, **F-TF.C.8**). This lesson reinforces the geometric interpretation of the modulus of a complex number and introduces the notion of the argument of a complex number. When representing a complex number in polar form, it is apparent that every complex number can be thought of simply as a rotation and dilation of the real number 1. In addition to representing numbers in polar form and converting between them, be sure to provide opportunities for students to explain why polar and rectangular forms of a given complex number represent the same number (**N-CN.B.4**).

It may be necessary to spend some time reviewing with students the work they did in previous courses, particularly as it relates to right-triangle trigonometry, special right triangles, and the sine and cosine functions. Sample problems have been provided after the Problem Set for this lesson. Specific areas for additional review and practice include the following:

- Using proportional reasoning to determine the other two sides in a special right triangle when given one side (Geometry, Module 2),
- Finding the acute angle in a right triangle using the arctangent function (Geometry, Module 2),
- Describing rotations in both degrees and radians (Algebra II, Module 2),
- Evaluating the sine and cosine functions at special angles in both degrees and radians (Algebra II, Module 2), and
- Evaluating the sine and cosine functions at any angle using a calculator (Geometry, Module 2 and Algebra II, Module 2).

Classwork

Opening (3 minutes)

Ask students to recall the special right triangles they studied in Geometry and revisited in Algebra II by showing them a diagram with the angles labeled but the side measurements missing. Have them fill in the missing side lengths and explain their reasoning to a partner. Check for understanding by adding the side lengths to the diagrams on the board, and then direct students to record these diagrams in their notes. Display these diagrams prominently in the classroom for student reference. Announce that these relationships are very helpful as they work through today's lesson.

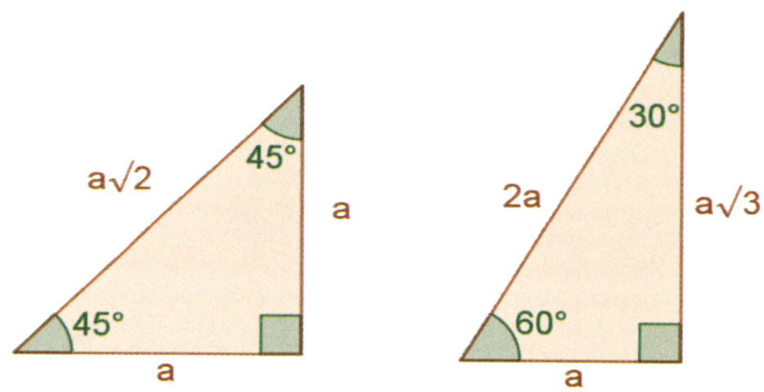

Opening Exercise (5 minutes)

The Opening Exercise reviews two key concepts from the previous lesson: (1) Each complex number $a + bi$ corresponds to a point (a, b), and (2) the modulus of a complex number is given by $\sqrt{a^2 + b^2}$. The last part of the Opening Exercise asks students to think about a rotation that takes a ray from the origin initially containing the real number 1 to its image from the origin passing through the point (a, b). The measurements of the rotation for the different points representing the different numbers should be fairly obvious to students. However, as students work, it may be necessary to remind them of the special right triangles they just discussed. These exercises should be done with a partner or with a small group. Use the discussion questions that follow to guide students as they work.

- Describe the location of the ray from the origin containing the real number 1.
    - *It lies along the positive x-axis with endpoint at the origin.*
- How can you determine the amount of rotation for z_1 and z_2?
    - *The points along the axes were one-fourth and one-half a complete rotation, which is $360°$.*
- How can you determine the amount of rotation for z_3 and z_4?
    - *The values of a and b formed the legs of a special right triangle. From there, since z_3 was located in the first quadrant, the rotation was just $45°$. For z_4, you would need to subtract $60°$ from $360°$ to give a positive counterclockwise rotation of $300°$, or use a clockwise rotation of $-60°$.*
- How did you determine the modulus?
    - *The modulus is given by the expression $\sqrt{a^2 + b^2}$.*

Scaffolding:

- Provide additional practice problems working with special right triangles.
- Project the diagram on the board, and draw in the rays and sides of a right triangle to help students see the geometric relationships for z_3 and z_4.
- Encourage students to label vertical and horizontal distances using the values of a and b.

When students are finished, have one or two of them present their solutions for each complex number. Emphasize the use of special triangles to determine the degrees of rotation for complex numbers not located along an axis. For additional scaffolding, it may be necessary to draw in the ray from the origin containing the real number 1 and the rotated ray from the origin that contains the point (a, b) for each complex number.

Opening Exercise

For each complex number shown below, answer the following questions. Record your answers in the table.

a. **What are the coordinates (a, b) that correspond to this complex number?**

b. **What is the modulus of the complex number?**

c. **Suppose a ray from the origin that contains the real number 1 is rotated $\theta°$ so it passes through the point (a, b). What is a value of θ?**

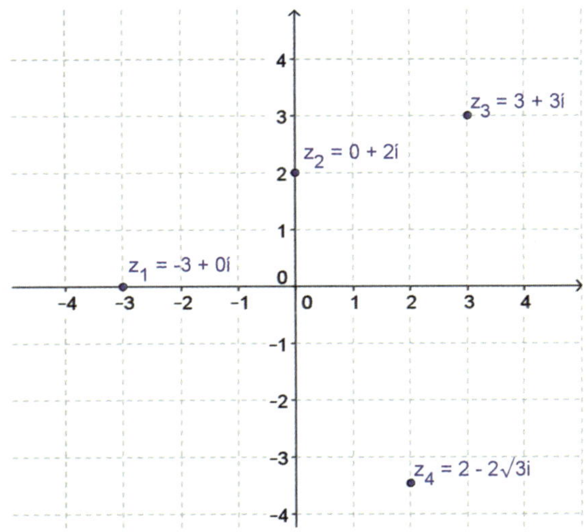

Complex Number	(a, b)	Modulus	Degrees of Rotation $\theta°$
$z_1 = -3 + 0i$	$(-3, 0)$	3	$180°$
$z_2 = 0 + 2i$	$(0, 2)$	2	$90°$
$z_3 = 3 + 3i$	$(3, 3)$	$3\sqrt{2}$	$45°$
$z_4 = 2 - 2\sqrt{3}i$	$(2, -2\sqrt{3})$	4	$300°$

MP.3

As students present their solutions, ask if anyone has a different answer for the number of degrees of the rotation, and lead a discussion so students understand that the degrees of rotation have more than one possible answer and, in fact, there are infinitely many possible answers.

- Another student said that a clockwise rotation of $270°$ would work for z_2. Do you agree or disagree? Explain.
  - *I agree. This rotation also takes the initial ray from the origin to a ray containing the point $0 + 2i$. If a complete rotation is $360°$, then $270°$ clockwise would be the same as $90°$ counterclockwise.*

At this point, remind students that positive rotations are counterclockwise and that rotation in the opposite direction is denoted using negative numbers.

Exercises 1–2 (5 minutes)

These exercises help students understand why the values of the argument of a complex number are limited to real numbers on the interval $0° \le \theta < 360°$. College-level mathematics courses make a distinction between the argument of a complex number between 0 and 360 and the set of all possible arguments of a given complex number.

Scaffolding:

Help students to generalize the expression by organizing the angles into a table (MP.8).

n	Degrees of Rotation
0	45
1	$45 + 360$
2	$45 + 360(2)$
3	$45 + 360(3)$

Exercises 1–2

1. **Can you find at least two additional rotations that would map a ray from the origin through the real number 1 to a ray from the origin passing through the point $(3, 3)$?**

   This is the number $z_3 = 3 + 3i$ from the Opening Exercise. Additional rotations could be $45° + 360° = 405°$ or $45° - 360° = -315°$.

2. **How are the rotations you found in Exercise 1 related?**

   All rotations that take the initial ray to the ray described above must be of the form $45° + 360°n$ for integer values of n.

After reviewing possible solutions to the questions above, pose this next question. It may be written on the board. Give students a few minutes to think about their responses individually, and then have them discuss them with their partners or group members before sharing responses as a whole class.

- Do you think it is possible to describe a complex number in terms of its modulus and the degrees of rotation of a ray from the origin containing the real number 1? Justify your reasoning.
  - *Student responses will vary. In general, the response should be yes, but careful students should note the difficulty of uniquely defining degrees of rotation. The modulus will be a distance from the origin, and if we want to be precise, we may need to limit the possible degrees of rotation to a subset of the real numbers such as $0° \le \theta < 360°$.*

Discussion (3 minutes)

Exercises 2 and 3 show that the rotation that maps a ray from the origin containing the real number 1 to a ray containing a given complex number is not unique. If you know one rotation, you can write an expression that represents all the rotations of a given complex number z. However, if we limit the rotations to an interval that comprises one full rotation of the initial ray, then we can *still* describe every complex number in terms of its modulus and a rotation.

Introduce the modulus and argument of a complex number.

Every complex number $z = x + yi$ appears as a point on the complex plane with coordinates (x, y) as a point in the coordinate plane.

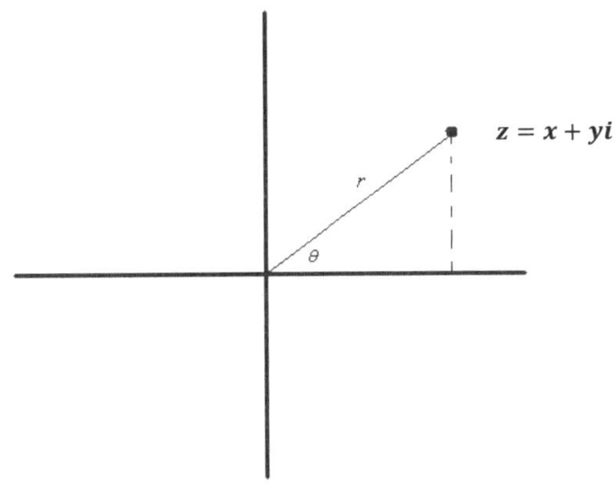

In the diagram above, notice that each complex number z has a distance r from the origin to the point (x, y) and a rotation of $\theta°$ that maps the ray from the origin along the positive real axis to the ray passing through the point (x, y).

ARGUMENT OF THE COMPLEX NUMBER z: The *argument of the complex number z* is the radian (or degree) measure of the counterclockwise rotation of the complex plane about the origin that maps the initial ray (i.e., the ray corresponding to the positive real axis) to the ray from the origin through the complex number z in the complex plane. The argument of z is denoted $\arg(z)$.

MODULUS OF A COMPLEX NUMBER z: The *modulus of a complex number z*, denoted $|z|$, is the distance from the origin to the point corresponding to z in the complex plane. If $z = a + bi$, then $|z| = \sqrt{a^2 + b^2}$.

- Is "modulus" indeed the right word? Does $r = |z|$ as we defined it in previous lessons?
  - *Yes, since r is the distance from the origin to the point (x, y), which is $\sqrt{x^2 + y^2}$, which is also how we define the modulus of a complex number.*
- Why are we limiting the argument to a subset of the real numbers?
  - *We only need these angles to sweep through all possible points in the coordinate plane. If we allowed the argument to be any real number, there would be many possible arguments for any given complex number.*

Example 1 (4 minutes): The Polar Form of a Complex Number

This example models how the polar form of a complex number is derived using the sine and cosine functions that students studied in Algebra II. Use the questions on the student materials to guide the discussion. The definitions from Algebra II are provided below for teacher reference.

SINE FUNCTION: The *sine function*, $\sin : \mathbb{R} \to \mathbb{R}$, can be defined as follows:

Let θ be any real number. In the Cartesian plane, rotate the initial ray by θ radians about the origin. Intersect the resulting terminal ray with the unit circle to get a point (x_θ, y_θ). The value of $\sin(\theta)$ is y_θ.

COSINE FUNCTION: The *cosine function*, $\cos : \mathbb{R} \to \mathbb{R}$, can be defined as follows:

Let θ be any real number. In the Cartesian plane, rotate the initial ray by θ radians about the origin. Intersect the resulting terminal ray with the unit circle to get a point (x_θ, y_θ). The value of $\cos(\theta)$ is x_θ.

- What do you recall about the definitions of the sine function and the cosine function from Algebra II?
  - *The sine function is the y-coordinate of a point, and the cosine function is the x-coordinate of the intersection point of a ray rotated θ radians about the origin and the unit circle.*
- How can the sine and cosine functions help us to relate the point (x, y) to modulus r and the argument θ?
  - *The coordinates (x, y) can be expressed in terms of the cosine and sine using the definition of the sine and cosine functions and dilating them along the terminal ray by a factor of r.*
- Why would it make sense to use these functions to relate a complex number in $a + bi$ form to one described by its modulus and argument?
  - *The modulus is a distance from the origin to the point (a, b), and the argument is basically the same type of rotation described in the definitions of the sine and cosine functions.*

MP.2 & MP.7

Example 1: The Polar Form of a Complex Number

Derive a formula for a complex number in terms of its modulus r and argument θ.

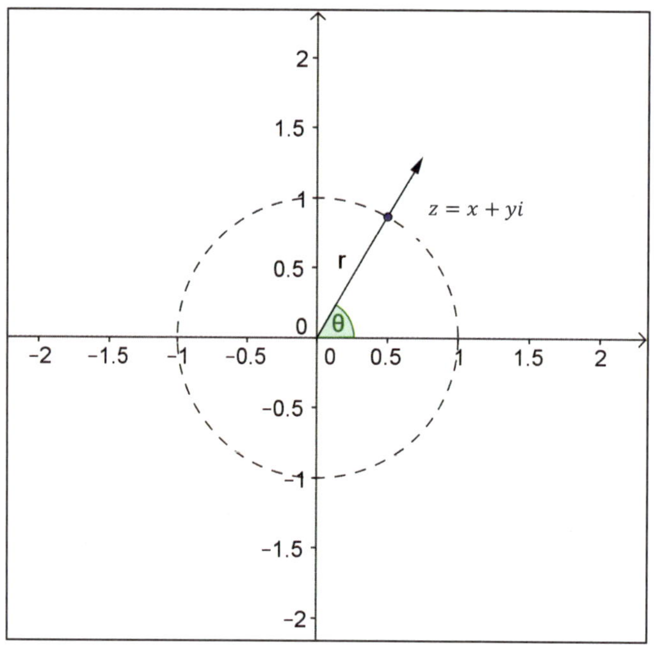

Suppose that z has coordinates (x, y) that lie on the unit circle as shown.

a. **What is the value of r, and what are the coordinates of the point (x, y) in terms of θ? Explain how you know.**

   The value of r is 1. The coordinates of the point are $(\cos(\theta), \sin(\theta))$. The definition of the sine and cosine function says that a point on the unit circle where a rotated ray intersects the unit circle has these coordinates.

b. **If $r = 2$, what would be the coordinates of the point (x, y)? Explain how you know.**

   The coordinates would be $(2\cos(\theta), 2\sin(\theta))$ because the point lies along the same ray but are just dilated by a scale factor of two along the ray from the origin compared to when $r = 1$.

c. **If $r = 20$, what would be the coordinates of the point (x, y)? Explain how you know.**

The coordinates would be $(20\cos(\theta), 20\sin(\theta))$ because a circle of radius 20 units would be similar to a circle with radius 1 but dilated by a factor of 20.

d. **Use the definitions of sine and cosine to write coordinates of the point (x, y) in terms of cosine and sine for any $r \geq 0$ and real number θ.**

$x = r\cos(\theta)$ *and* $y = r\sin(\theta)$

e. **Use your answer to part (d) to express $z = x + yi$ in terms of r and θ.**

$z = x + yi = r\cos(\theta) + ir\sin(\theta) = r(\cos(\theta) + i\sin(\theta))$

MP.2 & MP.7

Monitor students as they work in small groups to derive the polar form of a complex number from the rectangular form. After a few minutes, ask for a few volunteers to share their ideas, and then make sure to have students record the derivation shown below in their notes and revise their work to be accurate and precise.

Annotate the diagram above showing that the x- and y-values correspond to the points on a circle of radius r that is a dilation of the unit circle. Thus, the point (x, y) can be represented as $(r\cos(\theta), r\sin(\theta))$.

- The diagram shown above makes us recall the definitions of sine and cosine. We see from the following diagram:

$$x = r\cos(\theta) \text{ and } y = r\sin(\theta).$$

- Which means that every complex number can be written in the form:

$$z = x + iy = r\cos(\theta) + ir\sin(\theta) = r(\cos(\theta) + i\sin(\theta)).$$

Review the definition shown below, and then have students work in small groups to answer Exercises 3–6.

POLAR FORM OF A COMPLEX NUMBER: The *polar form of a complex number* z is $r(\cos(\theta) + i\sin(\theta))$, where $r = |z|$ and $\theta = \arg(z)$.

RECTANGULAR FORM OF A COMPLEX NUMBER: The *rectangular form of a complex number* z is $a + bi$, where z corresponds to the point (a, b) in the complex plane, and i is the imaginary unit. The number a is called the *real part* of $a + bi$, and the number b is called the *imaginary part* of $a + bi$.

Use the graphic organizer below to help students make sense of this definition. A blank version is included in the student materials. The graphic organizer has space for up to three examples of complex numbers that can either be completed as a class or assigned to students. Have students work with a partner to provide the polar and rectangular forms of both numbers. Have partners take turns explaining why the polar and rectangular forms of the examples represent the same number.

General Form	Polar Form $z = r(\cos(\theta) + i\sin(\theta))$	Rectangular Form $z = a + bi$
Examples	$3(\cos(60°) + i\sin(60°))$ $2\left(\cos\left(\frac{\pi}{2}\right) + i\sin\left(\frac{\pi}{2}\right)\right)$	$\frac{3}{2} + \frac{3\sqrt{3}}{2}i$ $0 + 2i$
Key Features	Modulus r Argument θ Coordinate $(r\cos(\theta), r\sin(\theta))$	Modulus $\sqrt{a^2 + b^2}$ Coordinate (a, b) $a = r\cos(\theta)$ $b = r\sin(\theta)$

Explain to students that this form of a complex number is particularly useful when considering geometric representations of complex numbers. This form clearly shows that every complex number z can be described as a rotation of $\theta°$ and a dilation by a factor of r of the real number 1.

Exercises 3–6 (8 minutes)

Students should complete these exercises with a partner or in small groups. Monitor progress as students work, and offer suggestions if they are struggling to work with the new representation of a complex number.

Exercises 3–6

3. **Write each complex number from the Opening Exercise in polar form.**

Rectangular	Polar Form
$z_1 = -3 + 0i$	$3(\cos(180°) + i\sin(180°))$
$z_2 = 0 + 2i$	$2(\cos(90°) + i\sin(90°))$
$z_3 = 3 + 3i$	$3\sqrt{2}(\cos(45°) + i\sin(45°))$
$z_4 = 2 - 2\sqrt{3}i$	$4(\cos(300°) + i\sin(300°))$

4. **Use a graph to help you answer these questions.**

   a. **What is the modulus of the complex number $2 - 2i$?**

   If you graph the point $(2, -2)$, then the distance between the origin and the point is given by the distance formula, so the modulus would be $\sqrt{(2)^2 + (-2)^2} = 2\sqrt{2}$.

   b. **What is the argument of the number $2 - 2i$?**

   If you graph the point $(2, -2)$, then the rotation that will take the ray from the origin through the real number 1 to a ray containing that point will be $315°$ because the point lies on a line from the origin in Quadrant IV that is exactly in between the two axes. The argument would be $315°$. We choose that rotation because we defined the argument to be a number between 0 and 360.

c. **Write the complex number in polar form.**

$$2\sqrt{2}(\cos(315^\circ) + i\sin(315^\circ))$$

d. **Arguments can be measured in radians. Express your answer to part (c) using radians.**

In radians, 315° is $\frac{7\pi}{4}$; the number would be

$$2\sqrt{2}\left(\cos\left(\frac{7\pi}{4}\right) + i\sin\left(\frac{7\pi}{4}\right)\right).$$

e. **Explain why the polar and rectangular forms of a complex number represent the same number.**

$2 - 2i$ is thought of as a point with coordinates $(2, -2)$ in the complex plane. The point can also be located by thinking of the ray extending from the origin rotated 315°. The distance from the origin to the point along that ray is the modulus, which is $2\sqrt{2}$ units.

Debrief Exercises 3 and 4 by having one or two students volunteer their solutions. On Exercise 4, some students may use right-triangle trigonometry while others take a more geometric approach and reason out the value of the argument from the graph and their knowledge of special right triangles. Pause and review radian measure if students are struggling to answer Exercise 4, part (d). When reviewing these first two exercises, be sure to emphasize why the work from Example 1 validates that the polar and rectangular forms of a complex number represent the same number.

Next, give students a few minutes to work individually on using this new form of a complex number. They need to approximate the location of a few of these rotations unless provided with a protractor. If the class is struggling to evaluate trigonometric functions of special angles, students may use a calculator, a copy of the unit circle, or their knowledge of special triangles to determine the values of a and b. Students need a calculator to answer Exercise 6, part (c).

5. **State the modulus and argument of each complex number, and then graph it using the modulus and argument.**

 a. $4(\cos(120^\circ) + i\sin(120^\circ))$

 $r = 4, \theta = 120^\circ$

 b. $5\left(\cos\left(\frac{\pi}{4}\right) + i\sin\left(\frac{\pi}{4}\right)\right)$

 $r = 5, \theta = \frac{\pi}{4}$

 c. $3(\cos(190^\circ) + i\sin(190^\circ))$

 $r = 3, \theta = 190^\circ$

6. **Evaluate the sine and cosine functions for the given values of θ, and then express each complex number in rectangular form, $z = a + bi$. Explain why the polar and rectangular forms represent the same number.**

 a. $4(\cos(120^\circ) + i\sin(120^\circ))$

$$4\left(-\frac{1}{2} + \frac{\sqrt{3}i}{2}\right) = -2 + 2\sqrt{3}i$$

 The polar form of a complex number and the rectangular form represent the same number because they both give you the same coordinates of a point that represents the complex number. In this example, 4 units along a ray from the origin rotated 120° corresponds to the coordinate $(-2, 2\sqrt{3})$.

b. $5\left(\cos\left(\frac{\pi}{4}\right) + i\sin\left(\frac{\pi}{4}\right)\right)$

$$5\left(\frac{\sqrt{2}}{2} + \frac{\sqrt{2}i}{2}\right) = \frac{5\sqrt{2}}{2} + \frac{5\sqrt{2}i}{2}$$

The polar form and rectangular form represent the same number because the values of $5\cos\left(\frac{\pi}{4}\right)$ *and* $5\sin\left(\frac{\pi}{4}\right)$ *are exactly* $\frac{5\sqrt{2}}{2}$.

c. $3(\cos(190°) + i\sin(190°))$

Rounded to two decimal places, the rectangular form is $-2.95 - 0.52i$***. This form of the number is close to, but not exactly, the same as the number expressed in polar form because the values of the trigonometric functions are rounded to the nearest hundredth.***

Review the solutions to these exercises with the entire class to check for understanding before moving on to Example 2. Make sure students understand that in Exercise 6, they rewrote each complex number given in polar form as an equivalent complex number written in rectangular form. Emphasize that in part (c), the rectangular form is an approximation of the polar form.

Example 2 (8 minutes): Writing a Complex Number in Polar Form

This example gives students a way to convert any complex number in rectangular form to its polar form using the inverse tangent function. To be consistent with work from previous grades, discussions of inverse tangent must be limited to the work students did in Geometry where they solved problems involving right triangles only. Students develop the inverse trigonometric functions in Module 3.

Ask students to recall what they did in the Opening Exercise and Exercise 5 to determine the argument.

- How were you able to determine the argument in the Opening Exercises and in Exercise 5?
  - *The complex numbers were on an axis or had coordinates that corresponded to lengths of sides in special right triangles, so we could recognize the proper degrees of rotation.*

The problems in the Opening Exercise and Exercise 5 were fairly easy because of special right-triangle relationships or the fact that the rotations coincided with an axis.

- How can you express any complex number given in rectangular form in polar form?
  - *The modulus, r, is given by $\sqrt{a^2 + b^2}$. To determine the angle, we would need a way to figure out the rotation based on the location of the point (a, b).*

Model how to construct a right triangle and use right-triangle trigonometry relationships to determine a value of an acute angle, which can then be used to determine the argument of the complex number.

- What did you learn in Geometry about finding an angle in a right triangle if you know two of the side measures?
  - *We applied the arctan, arcsin, or arccos to the ratios of the known side lengths.*

Example 2: Writing a Complex Number in Polar Form

a. **Convert $3 + 4i$ to polar form.**

$5(\cos(53.1°) + i \sin(53.1°))$

b. **Convert $3 - 4i$ to polar form.**

$5(\cos(306.9°) + i \sin(306.9°))$

- What is the modulus of $3 + 4i$?
  - *The modulus is* 5.

Draw a diagram like the one shown below, and use trigonometry ratios to help you determine the argument. Plot the point $(3, 4)$, and draw a line segment perpendicular to the x-axis from the point to the x-axis. Draw the ray from the origin through the point $(3, 4)$ and the ray from the origin through the real number 1. Label the acute angle between these rays θ.

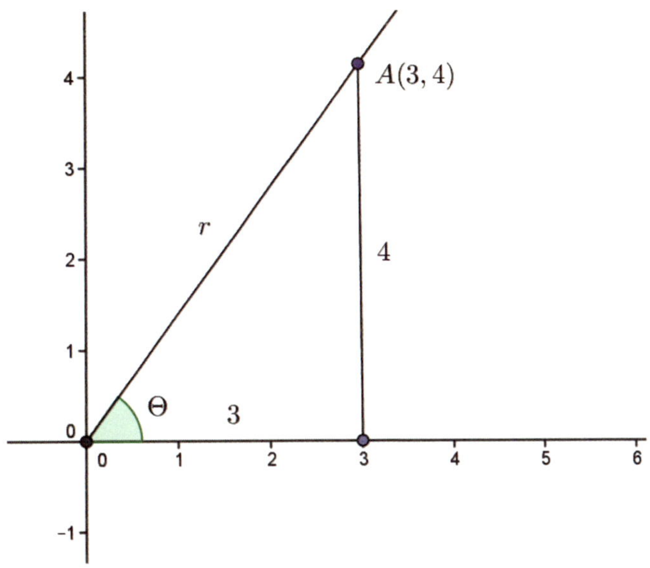

- The line segment from $(0, 0)$ to $(3, 0)$, the line segment we just drew, and the segment from the origin to the point form a right triangle. What is the tangent ratio of the acute angle whose vertex is at the origin?
  - *The tangent is* $\tan(\theta) = \frac{4}{3}$.
- Use a calculator to estimate the measure of this angle. What is the argument of $3 + 4i$?
  - *We can use* $\theta = \arctan\left(\frac{4}{3}\right)$. *Rounded to the nearest hundredth,* $\theta = 53.1°$.
- Write $3 + 4i$ in polar form.
  - *The polar form is* $5(\cos(53.1°) + i \sin(53.1°))$ *with the angle rounded to the nearest tenth.*

Part (b) of this example shows how the process above needs to be tweaked when the complex number is not located in the first quadrant.

The modulus is 5. When we plot the point $(3, -4)$ and draw a line segment perpendicular to the x-axis, we can see that the acute angle at the origin in this triangle will still have a measure equal to $\arctan\left(\frac{4}{3}\right) = 53.1°$.

Model how to draw this diagram so students see how to use the arctangent function to find the measure of the acute angle at the origin in the triangle they constructed.

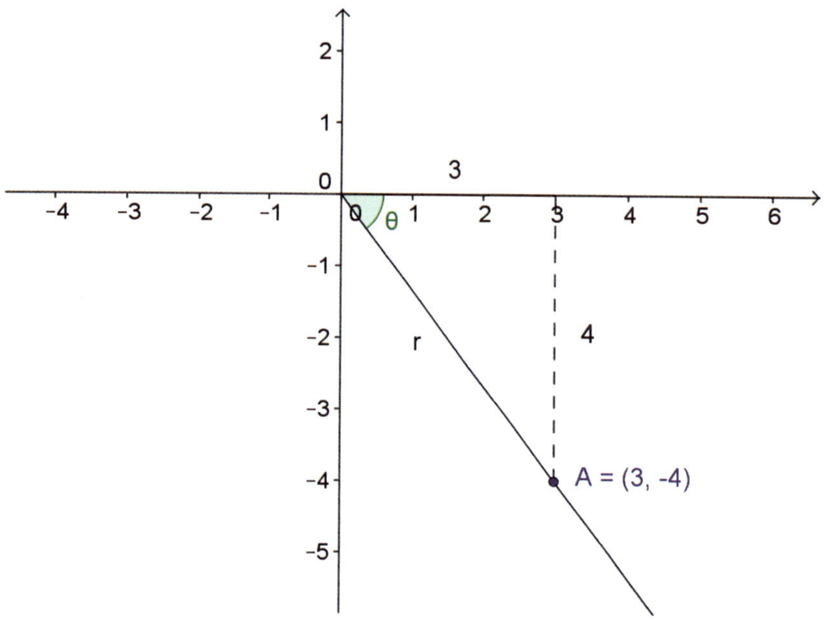

- Use your knowledge of angles to determine the argument of $3-4i$. Explain your reasoning.
  - *An argument of $3-4i$ would be $360° - 53.1° = 306.9°$. The positive rotation of ray from the origin containing the real number 1 that maps to a ray passing through this point would be $53°$ less than a full rotation of $360°$.*
- What is the polar form of $3-4i$?
  - *The polar form is $5(\cos(306.9°) + i\sin(306.9°))$.*
- Why do the polar and rectangular forms of a complex number represent the same number?
  - *$3-4i$ can be thought of as the point $(3,-4)$ in the complex plane. The point can be located by extending the ray from the origin rotated $306.9°$. The point is a distance of 5 units (the modulus) from the origin along that ray.*

Exercise 7 (4 minutes)

Have students practice the methods just demonstrated in Example 2. They can work individually or with a partner. Review the solutions to these problems with the whole class before moving on to the lesson closing.

Exercise 7

7. **Express each complex number in polar form. State the arguments in radians rounded to the nearest thousandth.**
   a. $2+5i$

   $\mathbf{arg(2+5i) = \tan^{-1}\left(\frac{5}{2}\right) \approx 1.190}$

   $\mathbf{|2+5i| = \sqrt{4+25} = \sqrt{29}}$

   $\mathbf{2+5i \approx \sqrt{29}(\cos(1.190) + i\sin(1.190))}$

b. $-6+i$

$\arg(z) = \pi - \tan^{-1}\left(\frac{1}{6}\right) \approx 2.976$

$|-6+i| = \sqrt{37}$

$-6+i = \sqrt{37}(\cos(2.976) + i\sin(2.976))$

Closing (3 minutes)

Review the Lesson Summary, and then ask students to describe to a partner the geometric meaning of the modulus and argument of a complex number. Then, have the partner describe the steps required to convert a complex number in rectangular form to polar form. Encourage students to refer back to their work in this lesson as they discuss what they learned with their partners.

Lesson Summary

The polar form of a complex number $z = r(\cos(\theta) + i\sin(\theta))$ where θ is the argument of z and r is the modulus of z. The rectangular form of a complex number is $z = a + bi$.

The polar and rectangular forms of a complex number are related by the formulas $a = r\cos(\theta)$, $b = r\sin(\theta)$, and $r = \sqrt{a^2 + b^2}$.

The notation for modulus is $|z|$, and the notation for argument is $\arg(z)$.

Exit Ticket (5 minutes)

Name ________________________________ Date ____________________

Lesson 13: Trigonometry and Complex Numbers

Exit Ticket

1. State the modulus and argument of each complex number. Explain how you know.
   a. $4 + 0i$

   b. $-2 + 2i$

2. Write each number from Problem 1 in polar form.

3. Explain why $5\left(\cos\left(\frac{\pi}{6}\right) + i\sin\left(\frac{\pi}{6}\right)\right)$ and $\frac{5\sqrt{3}}{2} + \frac{5}{2}i$ represent the same complex number.

Exit Ticket Sample Solutions

1. State the modulus and argument of each complex number. Explain how you know.

 a. $4 + 0i$

 The modulus is 4, and the argument is $0°$. The real number 4 is 4 units from the origin and lies in the same position as a ray from the origin containing the real number 1, so the rotation is $0°$.

 b. $-2 + 2i$

 The modulus is $2\sqrt{2}$, and the argument is $135°$. The values of a and b correspond to sides of a $45°$–$45°$–$90°$ right triangle, so the modulus would be $2\sqrt{2}$, and the rotation is $45°$ less than $180°$.

2. Write each number from Problem 1 in polar form.

 a. $4(\cos(0°) + i\sin(0°))$

 b. $2\sqrt{2}(\cos(135°) + i\sin(135°))$

3. Explain why $5\left(\cos\left(\frac{\pi}{6}\right) + i\sin\left(\frac{\pi}{6}\right)\right)$ and $\frac{5\sqrt{3}}{2} + \frac{5}{2}i$ represent the same complex number.

$$5\left(\cos\left(\frac{\pi}{6}\right) + i\sin\left(\frac{\pi}{6}\right)\right)$$

If you evaluate $5\cos\left(\frac{\pi}{6}\right)$ and $5\sin\left(\frac{\pi}{6}\right)$, you get $\frac{5\sqrt{3}}{2}$ and $\frac{5}{2}$, respectively.

$$5\left(\cos\left(\frac{\pi}{6}\right) + i\sin\left(\frac{\pi}{6}\right)\right) = \frac{5\sqrt{3}}{2} + \frac{5}{2}i$$

$\frac{5\sqrt{3}}{2} + \frac{5}{2}i$ is thought of as a point with coordinates $\left(\frac{5\sqrt{3}}{2}, \frac{5}{2}\right)$ in the complex plane. The point can also be located by thinking of the ray extending from the origin rotated $\frac{\pi}{6}$ radians. The distance from the origin to the point along that ray is the modulus, which is 5 units.

Problem Set Sample Solutions

1. Explain why the complex numbers $z_1 = 1 - \sqrt{3}i$, $z_2 = 2 - 2\sqrt{3}i$, and $z_3 = 5 - 5\sqrt{3}i$ can all have the same argument. Draw a diagram to support your answer.

 They all lie on the same ray from the origin that represents a $300°$ rotation.

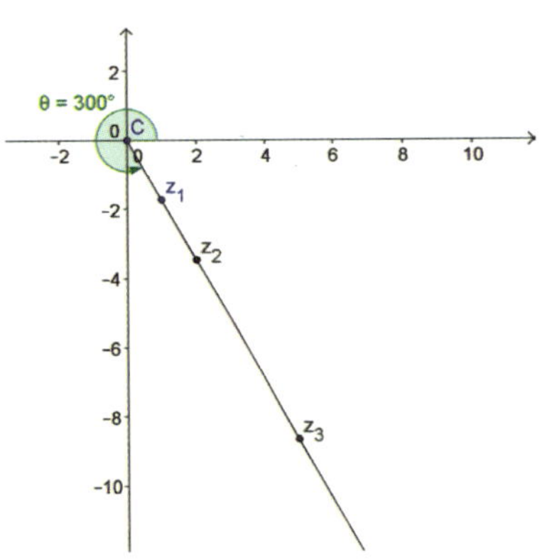

2. What is the modulus of each of the complex numbers z_1, z_2, and z_3 given in Problem 1 above?

The moduli are 2, 4, and 10.

3. Write the complex numbers from Exercise 1 in polar form.

$$z_1 = 2(\cos(300°) + i\sin(300°))$$
$$z_2 = 4(\cos(300°) + i\sin(300°))$$
$$z_3 = 10(\cos(300°) + i\sin(300°))$$

4. Explain why $1 - \sqrt{3}i$ and $2(\cos(300°) + i\sin(300°))$ represent the same number.

The point $(1, -\sqrt{3})$ lies on a ray from the origin that has been rotated $300°$ rotation from the initial ray. The distance of this point from the origin along this ray is 2 units (the modulus). Using the definitions of sine and cosine, any point along that ray will have coordinates $(2\cos(300°), 2\sin(300°))$.

5. Julien stated that a given modulus and a given argument uniquely determine a complex number. Confirm or refute Julien's reasoning.

Julien's reasoning is correct. If you rotate a ray from the origin containing the real number 1 and then locate a point a fixed number of units along that ray from the origin, it will give you a unique point in the plane.

6. Identify the modulus and argument of the complex number in polar form, convert it to rectangular form, and sketch the complex number in the complex plane. $0° \le \arg(z) \le 360°$ or $0 \le \arg(z) \le 2\pi$ (radians)

a. $z = \cos(30°) + i\sin(30°)$

$r = 1,\ \arg(z) = 30°$

$z = \dfrac{\sqrt{3}}{2} + \dfrac{1}{2}i$

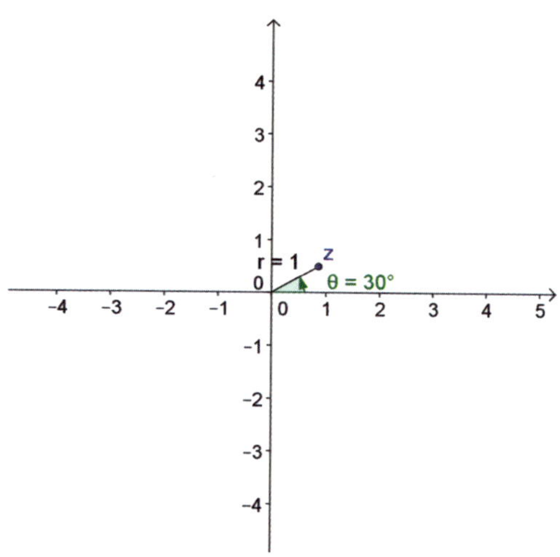

b. $z = 2\left(\cos\left(\frac{\pi}{4}\right) + i\sin\left(\frac{\pi}{4}\right)\right)$

$r = 2, \ \arg(z) = \frac{\pi}{4}$ *radians*

$z = \sqrt{2} + \sqrt{2}i$

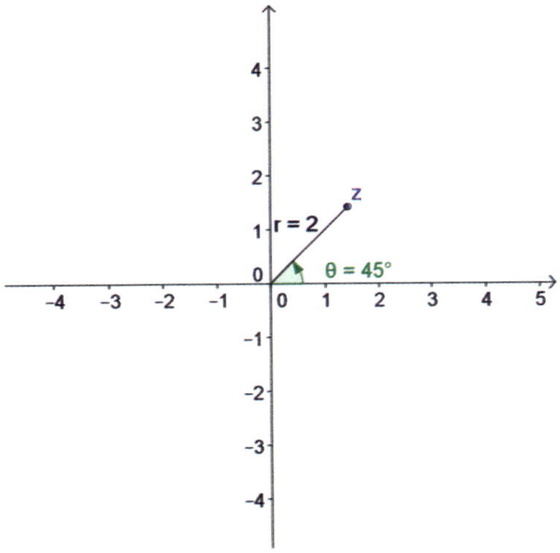

c. $z = 4\left(\cos\left(\frac{\pi}{3}\right) + i\sin\left(\frac{\pi}{3}\right)\right)$

$r = 4, \ \arg(z) = \frac{\pi}{3}$ *radians*

$z = 2 + 2\sqrt{3}i$

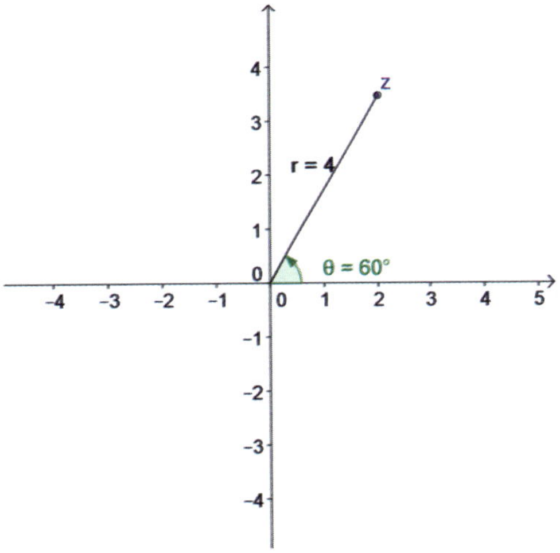

d. $z = 2\sqrt{3}\left(\cos\left(\frac{5\pi}{6}\right) + i\sin\left(\frac{5\pi}{6}\right)\right)$

$r = 2\sqrt{3}, \ \arg(z) = \frac{5\pi}{6}$ *radians*

$z = -3 + \sqrt{3}i$

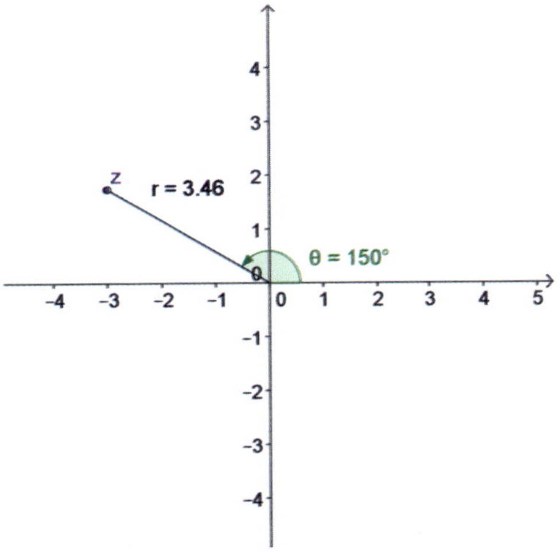

e. $z = 5(\cos(5.637) + i\sin(5.637))$

$r = 5,\ \arg(z) = 5.637$ *radians*

$z = 4 - 3i$

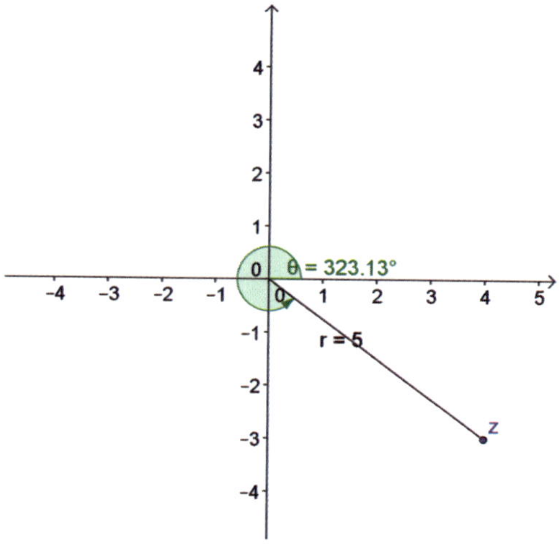

f. $z = 5(\cos(2.498) + i\sin(2.498))$

$r = 5,\ \arg(z) = 2.498$ *radians*

$z = -4 + 3i$

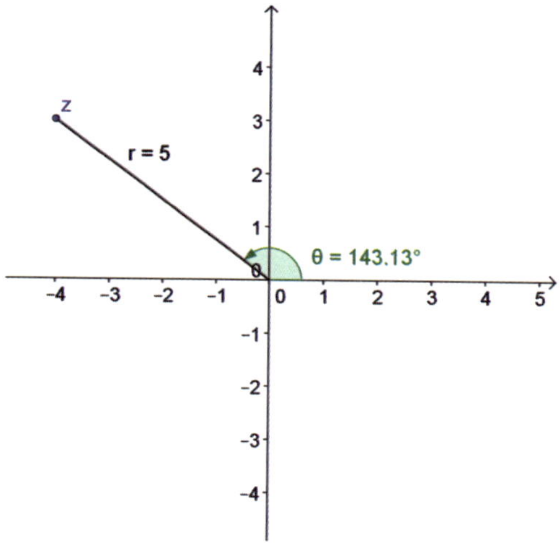

g. $z = \sqrt{34}(\cos(3.682) + i\sin(3.682))$

$r = \sqrt{34},\ \arg(z) = 3.682$

$z = -5 - 3i$

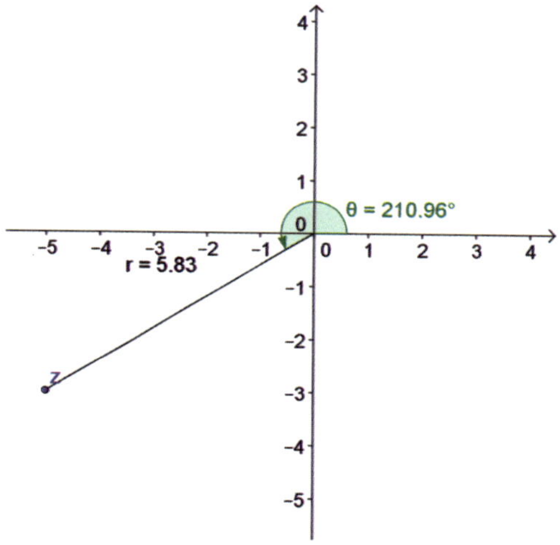

h. $z = 4\sqrt{3}\left(\cos\left(\frac{5\pi}{3}\right) + i\sin\left(\frac{5\pi}{3}\right)\right)$

$r = 4\sqrt{3},\ \arg(z) = \frac{5\pi}{3}$

$z = 2\sqrt{3} - 6i$

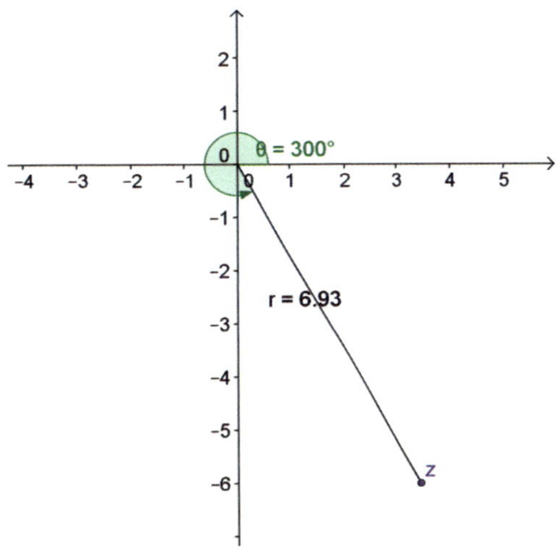

7. Convert the complex numbers in rectangular form to polar form. If the argument is a multiple of $\frac{\pi}{6}, \frac{\pi}{4}, \frac{\pi}{3}$, or $\frac{\pi}{2}$, express your answer exactly. If not, use $\arctan\left(\frac{b}{a}\right)$ to find $\arg(z)$ rounded to the nearest thousandth, $0 \le \arg(z) \le 2\pi$ (radians).

a. $z = \sqrt{3} + i$

$\arg(z)$ *is in Quadrant I.*

$$\arg(z) = \arctan\left(\frac{b}{a}\right) = \arctan\left(\frac{1}{\sqrt{3}}\right) = \frac{\pi}{6}$$

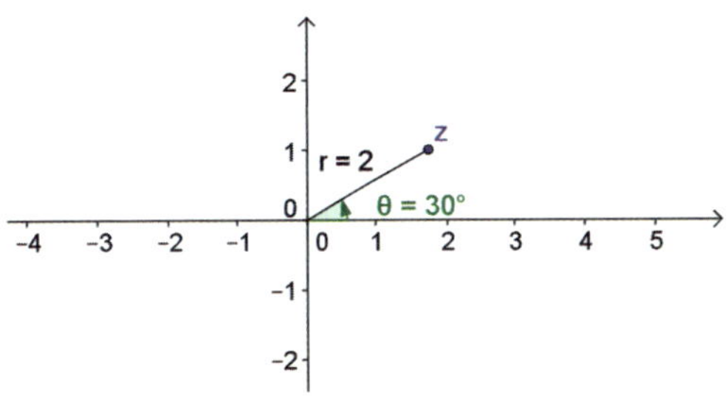

$$r = |z| = \sqrt{\left(\sqrt{3}\right)^2 + (1)^2} = 2$$

$z = 2\left(\cos\left(\frac{\pi}{6}\right) + i\sin\left(\frac{\pi}{6}\right)\right)$

b. $z = -3 + 3i$

$\arg(z)$ *is in Quadrant II.*

$$\arg(z) = \pi - \arctan\left(\frac{b}{a}\right)$$
$$= \pi - \arctan\left(\frac{3}{3}\right)$$
$$= \pi - \frac{\pi}{4}$$
$$= \frac{3\pi}{4}$$

$$r = |z|$$
$$= \sqrt{(-3)^2 + (3)^2}$$
$$= 3\sqrt{2}$$

$$z = 3\sqrt{2}\left(\cos\left(\frac{3\pi}{4}\right) + i\sin\left(\frac{3\pi}{4}\right)\right)$$

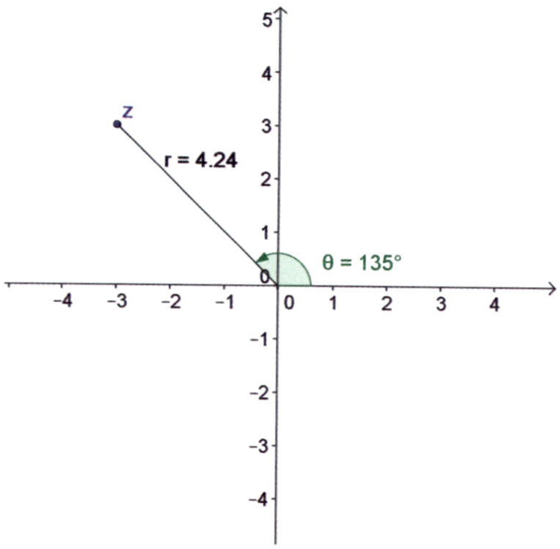

c. $z = 2 - 2\sqrt{3}i$

$\arg(z)$ *is in Quadrant IV.*

$$\arg(z) = 2\pi - \arctan\left(\frac{b}{a}\right)$$
$$= 2\pi - \arctan\left(\frac{2\sqrt{3}}{2}\right)$$
$$= 2\pi - \frac{\pi}{3}$$
$$= \frac{5\pi}{3} \text{ radians}$$

$$r = |z|$$
$$= \sqrt{(2)^2 + \left(-2\sqrt{3}\right)^2}$$
$$= 4$$

$$z = 4\left(\cos\left(\frac{5\pi}{3}\right) + i\sin\left(\frac{5\pi}{3}\right)\right)$$

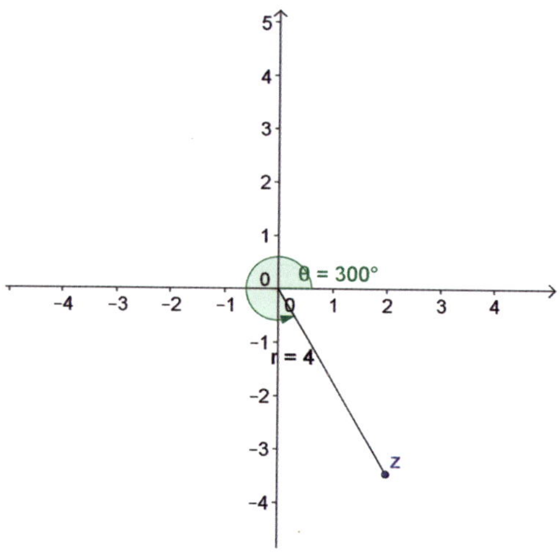

d. $z = -12 - 5i$

$\arg(z)$ *is in Quadrant III.*

$$\arg(z) = \pi + \arctan\left(\frac{5}{12}\right)$$
$$= \pi + \arctan\left(\frac{1}{\sqrt{3}}\right)$$
$$\approx 3.536 \text{ radians}$$

$$r = |z|$$
$$= \sqrt{(-12)^2 + (-5)^2}$$
$$= 13$$

$$z = 13(\cos(3.536) + i\sin(3.536))$$

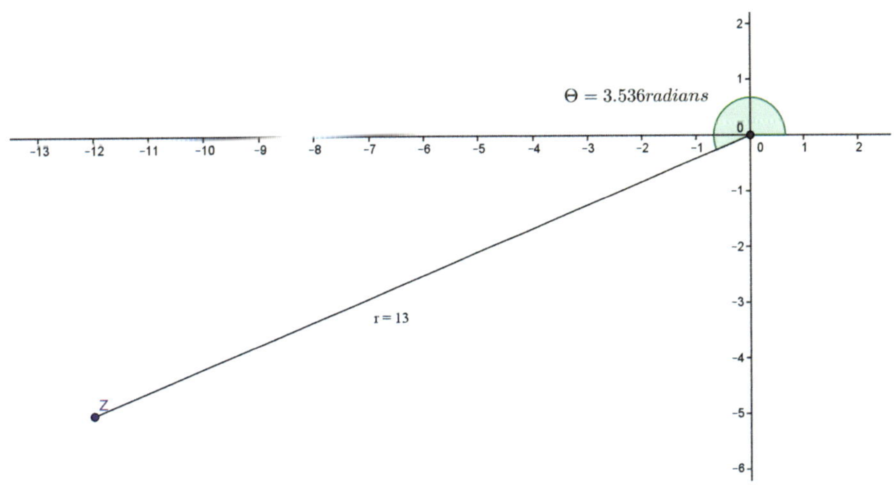

e. $z = 7 - 24i$

$\arg(z)$ *is in Quadrant IV.*

$$\arg(z) = 2\pi - \arctan\left(\frac{b}{a}\right)$$
$$= 2\pi - \arctan\left(\frac{24}{7}\right)$$
$$\approx 4.996 \text{ radians}$$

$$r = |z|$$
$$= \sqrt{(7)^2 + (-24)^2}$$
$$= 25$$

$$z = 25(\cos(4.996) + i\sin(4.996))$$

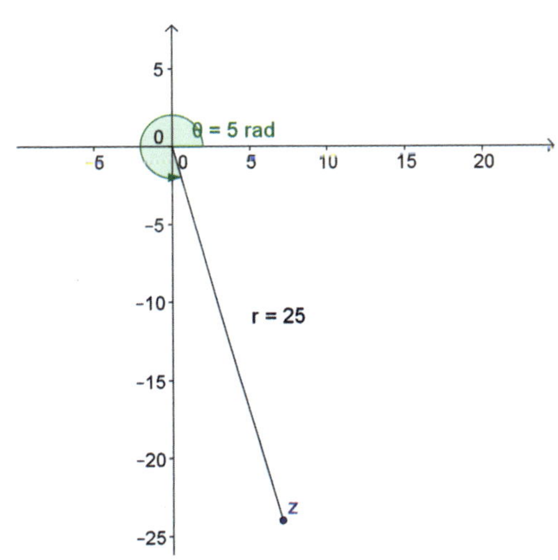

8. Show that the following complex numbers have the same argument.

   a. $z_1 = 3 + 3\sqrt{3}i$ and $z_2 = 1 + \sqrt{3}i$

   $\arg(z_1) = \arctan\left(\frac{3\sqrt{3}}{3}\right) = \frac{\pi}{3}$ *and* $\arg(z_2) = (\sqrt{3}) = \frac{\pi}{3}$

   b. $z_1 = 1 + i$ and $z_2 = 4 + 4i$

   $\arg(z_1) = \arctan\left(\frac{1}{1}\right) = \frac{\pi}{4}$ *and* $\arg(z_2) = \arctan\left(\frac{4}{4}\right) = \frac{\pi}{4}$

9. A square with side length of one unit is shown below. Identify a complex number in polar form that corresponds to each point on the square.

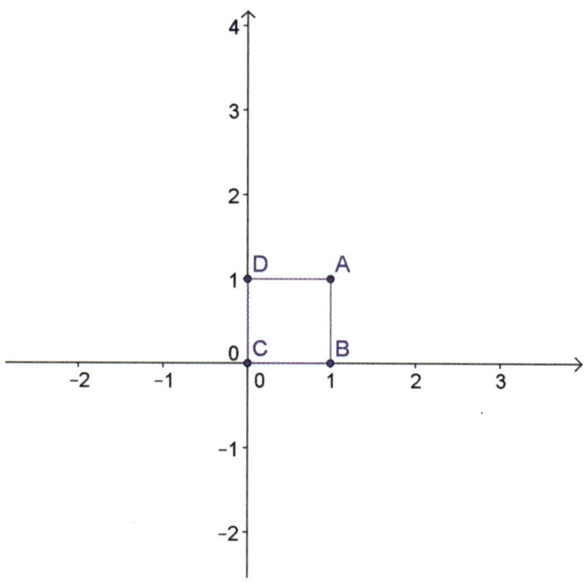

$A = \sqrt{2}(\cos(45°) + i\sin(45°))$

$B = 1(\cos(0°) + i\sin(0°))$

$C = 0(\cos(0°) + i\sin(0°))$

$D = 1(\cos(90°) + i\sin(90°))$

10. Determine complex numbers in polar form whose coordinates are the vertices of the square shown below.

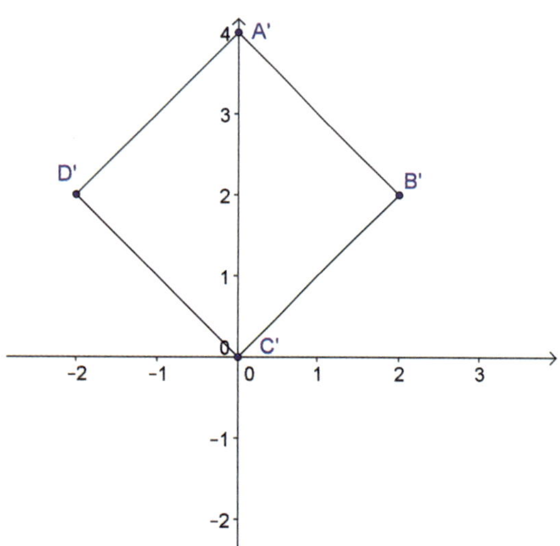

$A' = 4(\cos(90°) + i\sin(90°))$

$B' = 2\sqrt{2}(\cos(45°) + i\sin(45°))$

$C' = 0(\cos(0°) + i\sin(0°))$

$D' = 2\sqrt{2}(\cos(135°) + i\sin(135°))$

11. How do the modulus and argument of coordinate A in Problem 9 correspond to the modulus and argument of point A' in Problem 10? Does a similar relationship exist when you compare B to B', C to C', and D to D'? Explain why you think this relationship exists.

The modulus multiplied by a factor of $2\sqrt{2}$ and the argument is $45°$ more. The same is true when you compare B to B' and D to D'. The relationship could also be true for C and C', although the argument of C and C' can really be any number since the modulus is 0.

12. Describe the transformations that map $ABCD$ to $A'B'C'D'$.

Rotate by $45°$ counterclockwise, and then dilate by a factor of $2\sqrt{2}$.

Trigonometry Review: Additional Resources

1. Evaluate the following.
   a. $\sin(30°)$
   b. $\cos\left(\frac{\pi}{3}\right)$
   c. $\sin(225°)$
   d. $\cos\left(\frac{5\pi}{6}\right)$
   e. $\sin\left(\frac{5\pi}{3}\right)$
   f. $\cos(330°)$

2. Solve for the acute angle θ, both in radians and degrees, in a right triangle if you are given the opposite side, O, and adjacent side, A. Round to the nearest thousandth.
   a. $O = 3$ and $A = 4$
   b. $O = 6$ and $A = 1$
   c. $O = 3\sqrt{3}$ and $A = 2$

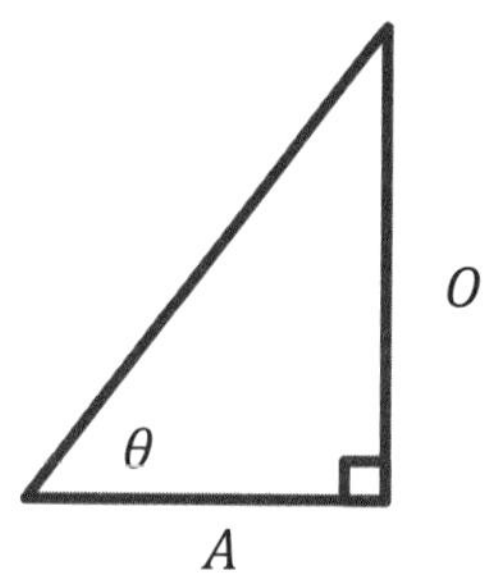

3. Convert angles in degrees to radians, and convert angles in radians to degrees.
   a. $150°$
   b. $\frac{4\pi}{3}$
   c. $\frac{3\pi}{4}$

Trigonometry Review: Additional Resources

1. **Evaluate the following.**

a. $\sin(30°)$

$\frac{1}{2}$

b. $\cos\left(\frac{\pi}{3}\right)$

$\frac{1}{2}$

c. $\sin(225°)$

$-\frac{\sqrt{2}}{2}$

d. $\cos\left(\frac{5\pi}{6}\right)$

$-\frac{\sqrt{3}}{2}$

e. $\sin\left(\frac{5\pi}{3}\right)$

$-\frac{\sqrt{3}}{2}$

f. $\cos(330°)$

$\frac{\sqrt{3}}{2}$

2. **Solve for the acute angle θ, both in radians and degrees, in a right triangle if you are given the opposite side, O, and adjacent side, A. Round to the nearest thousandth.**

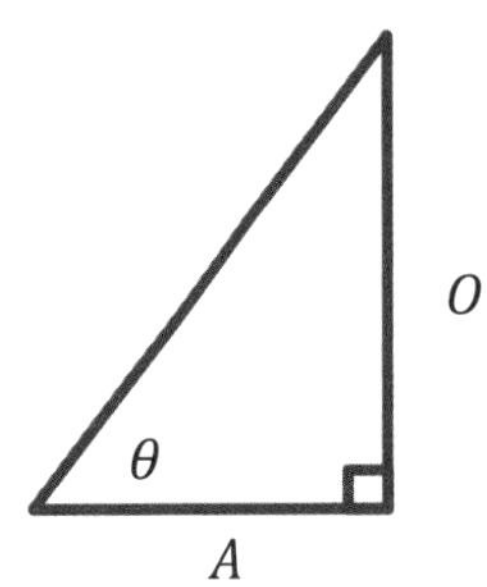

a. $O = 3$ and $A = 4$

$\arctan\left(\frac{3}{4}\right) \approx 0.644$ *radians* $= 36.898°$

b. $O = 6$ and $A = 1$

$\arctan\left(\frac{6}{1}\right) \approx 1.406$ *radians* $= 80.558°$

c. $O = 3\sqrt{3}$ and $A = 2$

$\arctan\left(\frac{3\sqrt{3}}{2}\right) \approx 1.203$ *radians* $= 68.948°$

3. **Convert angles in degrees to radians, and convert in radians to degrees.**

a. $150°$

$\frac{5\pi}{6}$

b. $\frac{4\pi}{3}$

$240°$

c. $\frac{3\pi}{4}$

$135°$

Lesson 14: Discovering the Geometric Effect of Complex Multiplication

Student Outcomes

- Students determine the geometric effects of transformations of the form $L(z) = az$, $L(z) = (bi)z$, and $L(z) = (a + bi)z$ for real numbers a and b.

Lesson Notes

In this lesson, students observe the geometric effect of transformations of the form $L(z) = (a + bi)z$ on a unit square and formulate conjectures (**G-CO.A.2**). Today's observations are mathematically established in the following lesson. As in the previous lessons, this lesson continues to associate points (a, b) in the coordinate plane with complex numbers $a + bi$, where a and b are real numbers (**N-CN.B.4**). The Problem Set includes another chance to revisit the definition and the idea of a linear transformation. Showing that these transformations are linear also provides algebraic fluency practice with complex numbers.

Classwork

Exercises 1–5 (10 minutes)

Exercises

The vertices $A(0,0)$, $B(1,0)$, $C(1,1)$, and $D(0,1)$ of a unit square can be represented by the complex numbers $A = 0$, $B = 1$, $C = 1 + i$, and $D = i$.

1. **Let $L_1(z) = -z$.**
   a. **Calculate $A' = L_1(A)$, $B' = L_1(B)$, $C' = L_1(C)$, and $D' = L_1(D)$. Plot these four points on the axes.**

   b. **Describe the geometric effect of the linear transformation $L_1(z) = -z$ on the square $ABCD$.**

      Transformation L_1 rotates the square $ABCD$ by $180°$ about the origin.

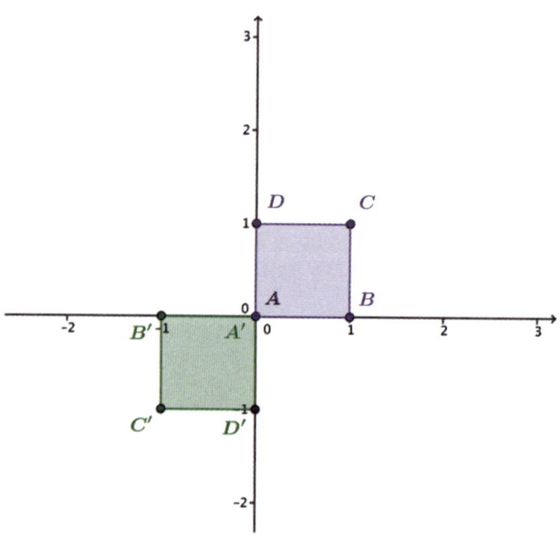

EUREKA MATH

2. Let $L_2(z) = 2z$.

   a. Calculate $A' = L_2(A)$, $B' = L_2(B)$, $C' = L_2(C)$, and $D = L_2(D)$. Plot these four points on the axes.

   b. Describe the geometric effect of the linear transformation $L_2(z) = 2z$ on the square $ABCD$.

   Transformation L_2 dilates the square $ABCD$ by a factor of 2.

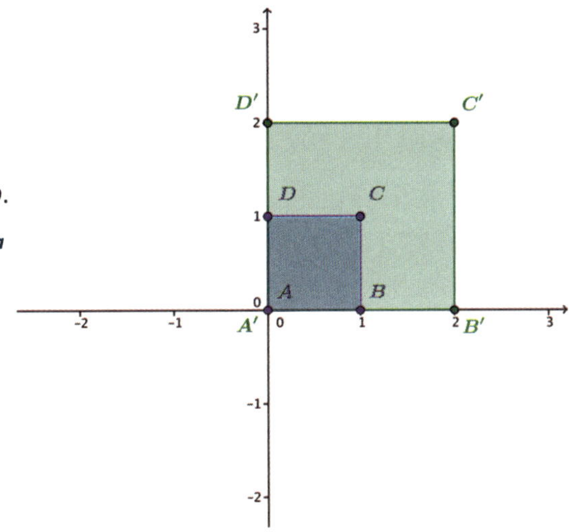

3. Let $L_3(z) = iz$.

   a. Calculate $A' = L_3(A)$, $B' = L_3(B)$, $C' = L_3(C)$, and $D' = L_3(D)$. Plot these four points on the axes.

   b. Describe the geometric effect of the linear transformation $L_3(z) = iz$ on the square $ABCD$.

   Transformation L_3 rotates the square $ABCD$ by $90°$ counterclockwise about the origin.

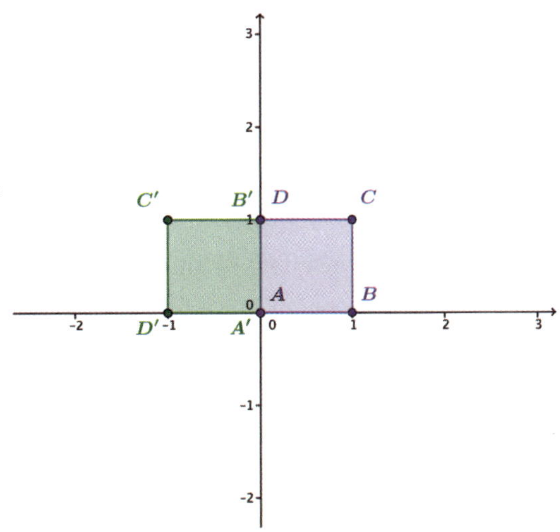

4. Let $L_4(z) = (2i)z$.

   a. Calculate $A' = L_4(A)$, $B' = L_4(B)$, $C' = L_4(C)$, and $D' = L_4(D)$. Plot these four points on the axes.

   b. Describe the geometric effect of the linear transformation $L_4(z) = (2i)z$ on the square $ABCD$.

   Transformation L_4 rotates the square $ABCD$ by $90°$ counterclockwise about the origin and dilates by a factor of 2.

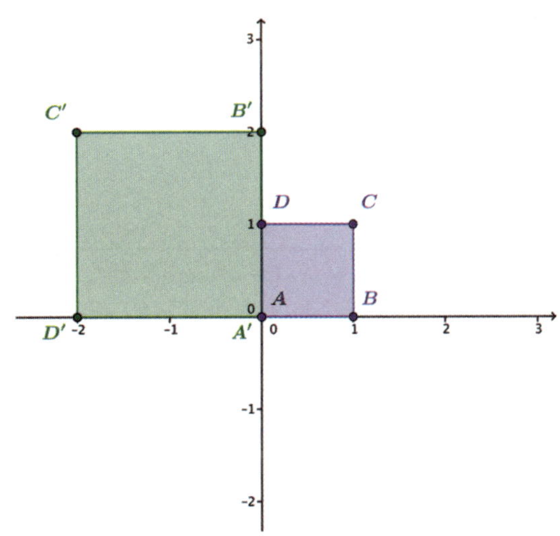

MP.7

5. **Explain how transformations L_2, L_3, and L_4 are related.**

Transformation L_4 is the result of doing transformations L_2 and L_3 (in either order).

Discussion (8 minutes)

- What is the geometric effect of the transformation $L(z) = az$ for a real number $a > 0$?
  - *The effect of L is dilation by the factor a.*
- What happens to a unit square in this case?
  - *The orientation of the square does not change; it is not reflected or rotated, but the sides of the square are dilated by a.*
- What is the effect on the square if $a > 1$?
  - *The sides of the square will get larger.*
- What is the effect on the square if $0 < a < 1$?
  - *The sides of the square will get smaller.*
- What is the geometric effect of the transformation $L(z) = az$ if $a = 0$?
  - *If $a = 0$, then $L(z) = 0$ for every complex number z. This transformation essentially shrinks the square down to the point at the origin.*
- What is the geometric effect of the transformation $L(z) = az$ for a real number $a < 0$?
  - *If $a < 0$, then $L(z) = az = -|a|z$, so L is a dilation by $|a|$ and a rotation by $180°$. This transformation will dilate the original unit square and then rotate it about point A into the third quadrant.*
- What is the geometric effect of the transformation $L(z) = (bi)z$ for a real number $b > 0$?
  - *The transformation L dilates by b and rotates by $90°$ counterclockwise.*
- What is the effect on the unit square if $b > 1$?
  - *The sides of the square will get larger.*
- What is the effect on the unit square if $0 < b < 1$?
  - *The sides of the square will get smaller.*
- What is the effect on the unit square if $b < 0$?
  - *If $b < 0$, then $L(z) = (bi)z = i(bz)$, so L is a dilation by $|b|$ and a rotation by $180°$, followed by a rotation by $90°$. This transformation will rotate and dilate the original unit square and then rotate it about point A to the fourth quadrant.*

Exercise 6 (6 minutes)

6. **We will continue to use the unit square $ABCD$ with $A = 0, B = 1, C = 1 + i, D = i$ for this exercise.**

   a. **What is the geometric effect of the transformation $L(z) = 5z$ on the unit square?**

   By our work in the first five exercises and the previous discussion, we know that this transformation dilates the unit square by a factor of 5.

EUREKA MATH

b. **What is the geometric effect of the transformation $L(z) = (5i)z$ on the unit square?**

By our work in the first five exercises, this transformation will dilate the unit square by a factor of 5 and rotate it $90°$ counterclockwise about the origin.

c. **What is the geometric effect of the transformation $L(z) = (5i^2)z$ on the unit square?**

Since $i^2 = -1$, this transformation is $L(z) = -5z$, which will dilate the unit square by 5 and rotate it $180°$ about the origin.

d. **What is the geometric effect of the transformation $L(z) = (5i^3)z$ on the unit square?**

Since $i^3 = -i$, this transformation is $L(z) = (-5i)z$, which will dilate the unit square by a factor of 5 and rotate it $270°$ counterclockwise about the origin.

e. **What is the geometric effect of the transformation $L(z) = (5i^4)z$ on the unit square?**

Since $i^4 = (i^2)^2 = (-1)^2 = 1$, this transformation is $L(z) = 5z$, which is the same transformation as in part (a). Thus, this transformation dilates the unit square by a factor of 5.

f. **What is the geometric effect of the transformation $L(z) = (5i^5)z$ on the unit square?**

Since $i^5 = i^4 \cdot i = i$, this is the same transformation as in part (b). This transformation will dilate the unit square by a factor of 5 and rotate it $90°$ counterclockwise about the origin.

g. **What is the geometric effect of the transformation $L(z) = (5i^n)z$ on the unit square, for some integer $n \geq 0$?**

If n is a multiple of 4, then $L(z) = (5i^n)z = 5z$ will dilate the unit square by a factor of 5.

If n is one more than a multiple of 4, then $L(z) = (5i^n)z = (5i)z$ will dilate the unit square by a factor of 5 and rotate it $90°$ counterclockwise about the origin.

If n is two more than a multiple of 4, then $L(z) = (5i^n)z = -5z$ will dilate the unit square by 5 and rotate it $180°$ about the origin.

If n is three more than a multiple of 4, then $L(z) = (5i^n)z = (-5i)z$ will dilate the unit square by a factor of 5 and rotate it $270°$ counterclockwise about the origin.

Exploratory Challenge (12 minutes)

Divide students into at least eight groups of two or three students each. Assign each group to the 1-team, 2-team, 3-team, or 4-team. There should be at least two groups on each team so that students can check their answers against another group when the results are shared at the end of the exercises. Each team explores a different transformation of the form $L(z) = (a + bi)z$.

Before students begin working on the Exploratory Challenge, ask the following:

- What is the geometric effect of the transformation $L(z) = 3z$?
  - *This transformation will dilate by a factor of three.*
- What is the geometric effect of the transformation $L(z) = -3z$?
  - *This transformation will dilate by a factor of three and rotate by $180°$ about the origin.*

Scaffolding:

- For struggling students, accompany this discussion with a visual representation of each transformation on the unit square $ABCD$.
- Omit this discussion for advanced students.

- What is the geometric effect of the transformation $L(z) = 4iz$?
  - *This transformation will dilate by a factor of four and rotate by $90°$ about the origin.*
- What is the geometric effect of the transformation $L(z) = -4iz$?
  - *This transformation will dilate by a factor of four and rotate by $270°$ about the origin.*

> **Exploratory Challenge**
>
> **Your group has been assigned either to the 1-team, 2-team, 3-team, or 4-team. Each team will answer the questions below for the transformation that corresponds to their team number:**
>
> $$L_1(z) = (3 + 4i)z$$
> $$L_2(z) = (-3 + 4i)z$$
> $$L_3(z) = (-3 - 4i)z$$
> $$L_4(z) = (3 - 4i)z.$$
>
> **The unit square $ABCD$ with $A = 0$, $B = 1$, $C = 1 + i$, $D = i$ is shown below. Apply your transformation to the vertices of the square $ABCD$, and plot the transformed points A', B', C', and D' on the same coordinate axes.**

The solution shown below is for transformation L_1. The transformed square for L_2, L_3, and L_4 will be rotated $90°$, $180°$, and $270°$ counterclockwise about the origin from the one shown, respectively.

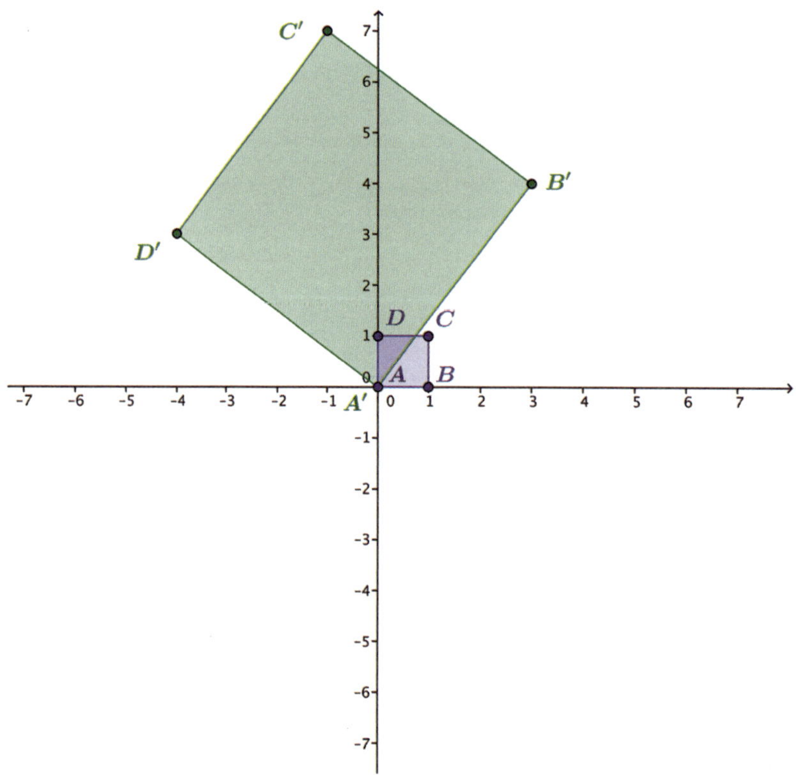

> **For the 1-team:**
>
> a. **Why is $B' = 3 + 4i$?**
>
> ***Because $B = 1$, we have $B' = L_1(B) = (3 + 4i)(1) = 3 + 4i$.***

b. What is the argument of $3 + 4i$?

The argument of $3 + 4i$ is the amount of counterclockwise rotation between the positive x-axis and the ray connecting the origin and the point $(3, 4)$.

c. What is the modulus of $3 + 4i$?

The modulus of $3 + 4i$ is $|3 + 4i| = \sqrt{3^2 + 4^2} = \sqrt{25} = 5$.

For the 2-team:

a. Why is $B' = -3 + 4i$?

Because $B = 1$, we have $B' = L_2(B) = (-3 + 4i)(1) = -3 + 4i$.

b. What is the argument of $-3 + 4i$?

The argument of $-3 + 4i$ is the amount of counterclockwise rotation between the positive x-axis and the ray connecting the origin and the point $(-3, 4)$.

c. What is the modulus of $-3 + 4i$?

The modulus of $-3 + 4i$ is $|-3 + 4i| = \sqrt{(-3)^2 + 4^2} = \sqrt{25} = 5$.

For the 3-team:

a. Why is $B' = -3 - 4i$?

Because $B = 1$, we have $B' = L_3(B) = (-3 - 4i)(1) = -3 - 4i$.

b. What is the argument of $-3 - 4i$?

The argument of $-3 - 4i$ is the amount of counterclockwise rotation between the positive x-axis and the ray connecting the origin and the point $(-3, -4)$.

c. What is the modulus of $-3 - 4i$?

The modulus of $-3 - 4i$ is $|-3 - 4i| = \sqrt{(-3)^2 + (-4)^2} = \sqrt{25} = 5$.

For the 4-team:

a. Why is $B' = 3 - 4i$?

Because $B = 1$, we have $B' = L_4(B) = (3 - 4i)(1) = 3 - 4i$.

b. What is the argument of $3 - 4i$?

The argument of $3 - 4i$ is the amount of counterclockwise rotation between the positive x-axis and the ray connecting the origin and the point $(3, -4)$.

c. What is the modulus of $3 - 4i$?

The modulus of $3 - 4i$ is $|3 - 4i| = \sqrt{3^2 + (-4)^2} = \sqrt{25} = 5$.

All groups should also answer the following:

a. Describe the amount the square has been rotated counterclockwise.

The square has been rotated the amount of counterclockwise rotation between the positive x-axis and ray $\overrightarrow{AB'}$.

b. What is the dilation factor of the square? Explain how you know.

First, we need to calculate the length of one side of the square. The length AB' is given by $AB' = \sqrt{(4-0)^2 + (3-0)^2} = 5$. Then, the dilation factor of the square is 5 because the final square has sides that are five times longer than the sides of the original square.

MP.7

c. What is the geometric effect of your transformation L_1, L_2, L_3, or L_4 on the unit square $ABCD$?

(Answered for transformation L_1.) The transformation rotates the square counterclockwise by the argument of $(3+4i)$ and dilates it by a factor of the modulus of $3+4i$.

d. Make a conjecture: What do you expect to be the geometric effect of the transformation $L(z) = (2+i)z$ on the unit square $ABCD$?

This transformation should rotate the square counterclockwise by the argument of $2+i$ and dilate it by a factor of $|2+i| = \sqrt{2^2+1^2} = \sqrt{5}$.

e. Test your conjecture with the unit square on the axes below.

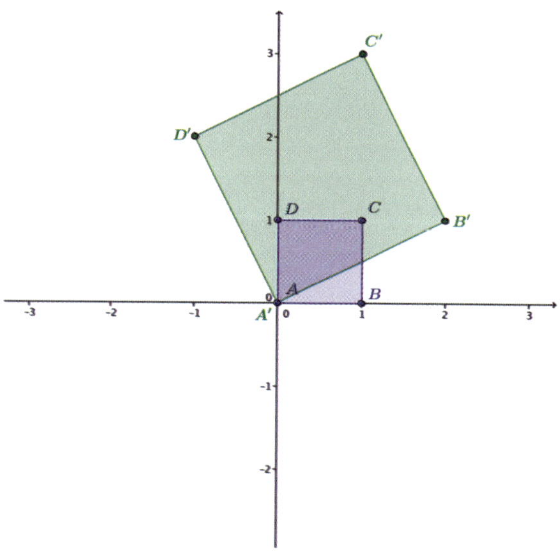

Closing (5 minutes)

Ask one group from each team to share their results from the Exploratory Challenge at the front of the class. Be sure that each group has made the connection that if the transformation is given by $L(z) = (a+bi)z$, then the geometric effect of the transformation is to dilate by $|a+bi|$ and to rotate by $\arg(a+bi)$.

Exit Ticket (4 minutes)

Name ____________________ Date ____________

Lesson 14: Discovering the Geometric Effect of Complex Multiplication

Exit Ticket

1. Identify the linear transformation L that takes square $ABCD$ to square $A'B'C'D'$ as shown in the figure on the right.

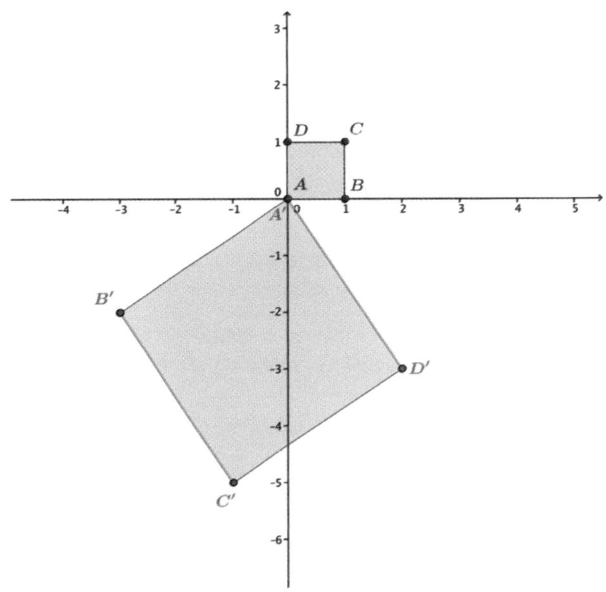

2. Describe the geometric effect of the transformation $L(z) = (1 - 3i)z$ on the unit square $ABCD$, where $A = 0, B = 1, C = 1 + i$, and $D = i$. Sketch the unit square transformed by L on the axes on the right.

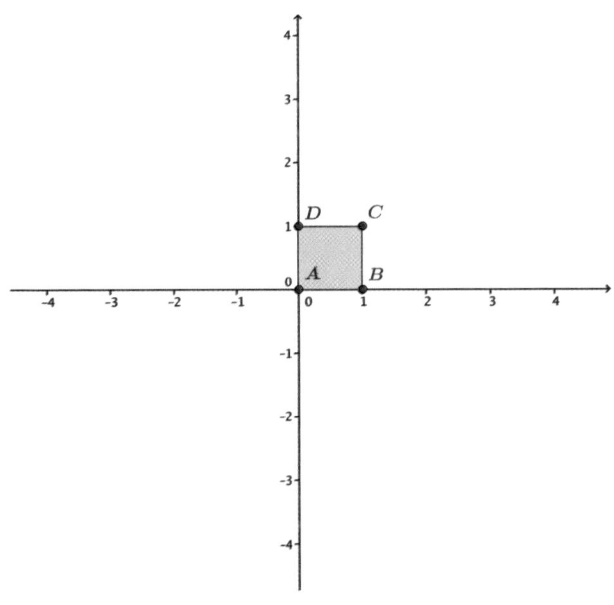

Exit Ticket Sample Solutions

1. **Identify the linear transformation L that takes square $ABCD$ to square $A'B'C'D'$ as shown in the figure on the right.**

   The transformation L takes the point $B = 1$ to the point $B' = -3 - 2i$, so this transformation is given by $L(z) = (-3 - 2i)z$.

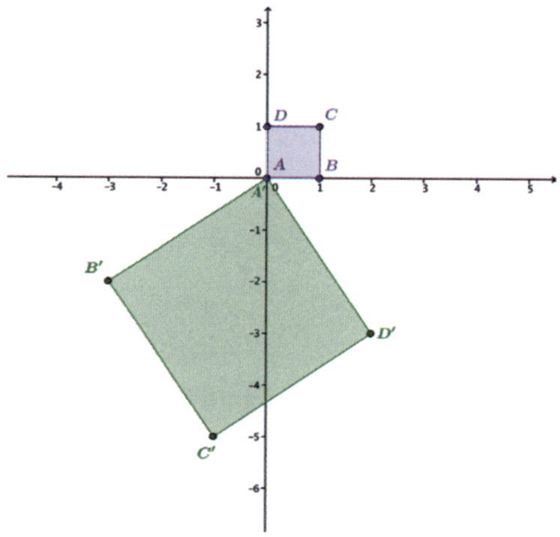

2. **Describe the geometric effect of the transformation $L(z) = (1 - 3i)z$ on the unit square $ABCD$, where $A = 0$, $B = 1$, $C = 1 + i$, and $D = i$. Sketch the unit square transformed by L on the axes on the right.**

   This transformation dilates by $|1 - 3i| = \sqrt{1^2 + 3^2} = \sqrt{10}$ and rotates counterclockwise by $\arg(1 - 3i)$.

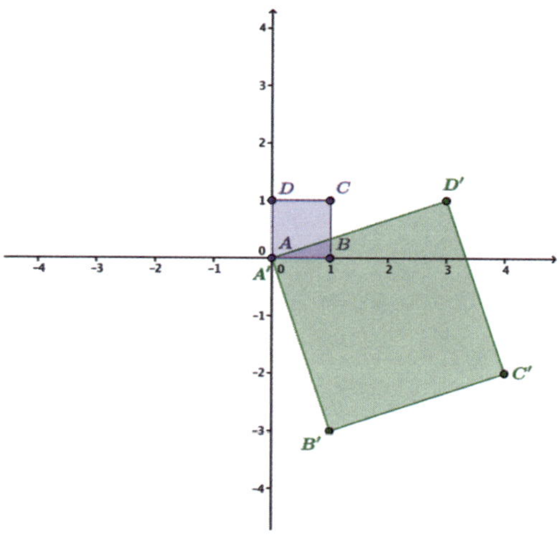

Problem Set Sample Solutions

3. **Find the modulus and argument for each of the following complex numbers.**

   a. $z_1 = \frac{\sqrt{3}}{2} + \frac{1}{2}i$

   $\left|\frac{\sqrt{3}}{2} + \frac{1}{2}i\right| = 1$, ***$z_1$ is in Quadrant I; thus,*** $\arg(z_1) = \arctan\left(\frac{\frac{1}{2}}{\frac{\sqrt{3}}{2}}\right) = 30° = \frac{\pi}{6}\text{rad}.$

   b. $z_2 = 2 + 2\sqrt{3}i$

   $|2 + 2\sqrt{3}i| = 4$, ***z_2 is in Quadrant I; thus,*** $\arg(z_2) = \arctan\left(\frac{2\sqrt{3}}{2}\right) = 60° = \frac{\pi}{3}$ rad.

c. $z_3 = -3 + 5i$

$|3 + 5i| = \sqrt{34}$, z_3 is in Quadrant II; thus, $\arg(z) = \pi - \arctan\left(\frac{5}{3}\right) \approx \pi - 1.030 \approx 2.112$ rad.

d. $z_4 = -2 - 2i$

$|-2 - 2i| = 2\sqrt{2}$, z_4 is in Quadrant III; thus, $\arg(z_4) = \pi + \arctan\left(\frac{2}{2}\right) = \pi + \frac{\pi}{4} = \frac{5\pi}{4}$ rad.

e. $z_5 = 4 - 4i$

$|4 + 4i| = 4\sqrt{2}$, z_5 is in Quadrant IV; thus, $\arg(z_5) = 2\pi - \arctan\left(\frac{4}{4}\right) = 2\pi - \frac{\pi}{4} = \frac{7\pi}{4}$ rad.

f. $z_6 = 3 - 6i$

$|3 - 6i| = 3\sqrt{5}$, z_6 is in Quadrant IV; thus, $\arg(z_6) = 2\pi - \arctan\left(\frac{6}{3}\right) = 2\pi - 1.107 = 5.176$ rad.

4. For parts (a)–(c), determine the geometric effect of the specified transformation.

a. $L(z) = -3z$

The transformation L dilates by 3 and rotates by $180°$ about the origin.

b. $L(z) = -100z$

The transformation L dilates by 100 and rotates by $180°$ about the origin.

c. $L(z) = -\frac{1}{3}z$

The transformation L dilates by $\frac{1}{3}$ and rotates by $180°$ about the origin.

d. Describe the geometric effect of the transformation $L(z) = az$ for any negative real number a.

The transformation L dilates by $|a|$ and rotates by $180°$ about the origin.

5. For parts (a)–(c), determine the geometric effect of the specified transformation.

a. $L(z) = (-3i)z$

The transformation L dilates by 3 and rotates counterclockwise by $270°$ about the origin.

b. $L(z) = (-100i)z$

The transformation L dilates by 100 and rotates by $270°$ about the origin.

c. $L(z) = \left(-\frac{1}{3}i\right)z$

The transformation L dilates by $\frac{1}{3}$ and rotates counterclockwise by $270°$ about the origin.

d. Describe the geometric effect of the transformation $L(z) = (bi)z$ for any negative real number b.

The transformation L dilates by $|b|$ and rotates by $270°$ counterclockwise about the origin.

6. Suppose that we have two linear transformations, $L_1(z) = 3z$ and $L_2(z) = (5i)z$.

 a. What is the geometric effect of first performing transformation L_1 and then performing transformation L_2?

 The transformation L_1 dilates by 3, dilates by 5, and rotates by $90°$ counterclockwise about the origin.

 b. What is the geometric effect of first performing transformation L_2 and then performing transformation L_1?

 The transformation L_1 dilates by 5, rotates by $90°$ counterclockwise about the origin, and then dilates by 3.

 c. Are your answers to parts (a) and (b) the same or different? Explain how you know.

 The answers are the same.

 $L_2(L_1(z)) = (5i)L_1(z) = (5i)(3z) = (15i)z$ $\qquad$ $L_1(L_2(z)) = 3L_2(z) = 3((5i)z) = (15i)z$

 For example, let $z = 2 - 3i$.

 $L_1 = 3(2 - 3i) = 6 - 9i$

 $L_2 = (5i)(2 - 3i) = 15 + 10i$

 $L_2(L_1) = (5i)(6 - 9i) = 45 + 30i$

 $L_1(L_2) = 3(15 + 10i) = 45 + 30i$

7. Suppose that we have two linear transformations, $L_1(z) = (4 + 3i)z$ and $L_2(z) = -z$. What is the geometric effect of first performing transformation L_1 and then performing transformation L_2?

 We have $|4 + 3i| = 5$, and the argument of $4 + 3i$ is $\arctan\left(\frac{3}{4}\right) \approx 0.644$ radians, which is about $36.87°$. Therefore, the transformation L_1 followed by L_2 dilates with scale factor 5, rotates by approximately $36.87°$ counterclockwise, and then rotates by $180°$.

8. Suppose that we have two linear transformations, $L_1(z) = (3 - 4i)z$ and $L_2(z) = -z$. What is the geometric effect of first performing transformation L_1 and then performing transformation L_2?

 We see that $|3 - 4i| = 5$, and the argument of $3 - 4i$ is $\arctan\left(\frac{4}{3}\right) \approx 2\pi - 5.356$ radians, which is about $306.87°$. Therefore, the transformation L_1 followed by L_2 dilates with scale factor 5, rotates by approximately $306.87°$ counterclockwise, and then rotates by $180°$.

9. Explain the geometric effect of the linear transformation $L(z) = (a - bi)z$, where a and b are positive real numbers.

 Note that the complex number $a - bi$ is represented by a point in the fourth quadrant. The transformation L dilates with scale factor $|a - bi|$ and rotates counterclockwise by $2\pi - \arctan\left(\frac{b}{a}\right)$.

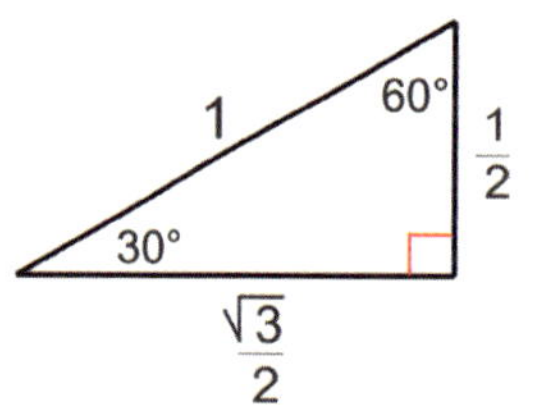

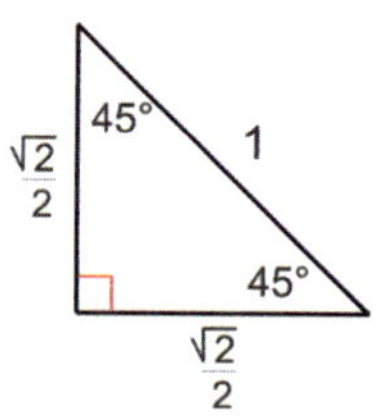

10. In Geometry, we learned the special angles of a right triangle whose hypotenuse is 1 unit. The figures are shown above. Describe the geometric effect of the following transformations.

a. $L_1(z) = \left(\frac{\sqrt{3}}{2} + \frac{1}{2}i\right)z$

$\left|\frac{\sqrt{3}}{2} + \frac{1}{2}i\right| = 1, \ \arg(z) = 30° = \frac{\pi}{6}\text{rad}$

The transformation L_1 rotates counterclockwise by $30°$.

b. $L_2(z) = (2 + 2\sqrt{3}i)z$

$|2 + 2\sqrt{3}i| = 4, \ \arg(z) = 60° = \frac{\pi}{3}\text{ rad}$

The transformation L_2 dilates with scale factor 4 and rotates counterclockwise by $60°$.

c. $L_3(z) = \left(\frac{\sqrt{2}}{2} + \frac{\sqrt{2}}{2}i\right)z$

$\left|\frac{\sqrt{2}}{2} + \frac{\sqrt{2}}{2}i\right| = 1, \ \arg(z) = 45° = \frac{\pi}{4}\text{ rad}$

The transformation L_3 dilates by 1 and rotates counterclockwise by $45°$.

d. $L_4(z) = (4 + 4i)z$

$|4 + 4i| = 4\sqrt{2}, \ \arg(z) = 45° = \frac{\pi}{4}\text{ rad}$

The transformation L_4 dilates with scale factor $4\sqrt{2}$ and rotates counterclockwise by $45°$.

11. Recall that a function L is a linear transformation if all z and w in the domain of L and all constants a meet the following two conditions:

i. $L(z + w) = L(z) + L(w)$

ii. $L(az) = aL(z)$

Show that the following functions meet the definition of a linear transformation.

a. $L_1(z) = 4z$

$L_1(z + w) = 4(z + w) = 4z + 4w = L_1(z) + L_1(w)$

$L_1(az) = 4(az) = 4az = a(4z) = aL_1(z)$

b. $L_2(z) = iz$

$L_2(z + w) = i(z + w) = iz + iw = L_2(z) + L_1(w)$

$L_2(az) = i(az) = iaz = a(iz) = aL_2(z)$

c. $L_3(z) = (4 + i)z$

$L_3(z + w) = (4 + i)(z + w) = (4 + i)z + (4 + i)w = L_3(z) + L_3(w)$

$L_3(az) = (4 + i)(az) = (4 + i)az = a((4 + i)z) = aL_3(z)$

12. The vertices $A(0,0)$, $B(1,0)$, $C(1,1)$, $D(0,1)$ of a unit square can be represented by the complex numbers $A = 0$, $B = 1$, $C = 1 + i$, $D = i$. We learned that multiplication of those complex numbers by i rotates the unit square by $90°$ counterclockwise. What do you need to multiply by so that the unit square will be rotated by $90°$ clockwise?

We need to multiply by $i^3 = -i$.

Lesson 15: Justifying the Geometric Effect of Complex Multiplication

Student Outcomes

- Students understand why the geometric transformation effect of the linear transformation $L(z) = wz$ is dilation by $|w|$ and rotation by the argument of w.

Lesson Notes (optional)

In Lesson 13, students observed that the transformation $L(z) = (3 + 4i)z$ has the geometric effect of a rotation by the argument of $3 + 4i$ and a dilation by the modulus $|3 + 4i| = 5$. In this lesson, this result is generalized to a linear transformation $L(z) = wz$ for a complex number w, using the geometric representation of a complex number as a point in the complex plane. However, before they begin thinking about the transformation L, students first need to represent multiplication of complex numbers geometrically on the complex plane (**N-CN.B.5**), so that is where this lesson begins.

This lesson covers one of nine cases for the geometric position of the complex scalar w in the coordinate plane, and the remaining cases are carefully scaffolded in the Problem Set. Consider extending this to a two-day lesson and having students work in groups on these remaining cases during the second day of class. Consider having groups present the remaining eight cases to the rest of the class.

Classwork

Opening Exercise (8 minutes)

In the Opening Exercise, students review complex multiplication and consider it geometrically to justify the geometric effect of a linear transformation $L(z) = (a + bi)z$ discovered in Lesson 13.

Opening Exercise

For each exercise below, compute the product wz. Then, plot the complex numbers z, w, and wz on the axes provided.

a. $z = 3 + i, w = 1 + 2i$

$$\begin{aligned} wz &= (3 + i)(1 + 2i) \\ &= 3 + 6i + i + 2i^2 \\ &= 3 - 2 + 7i \\ &= 1 + 7i \end{aligned}$$

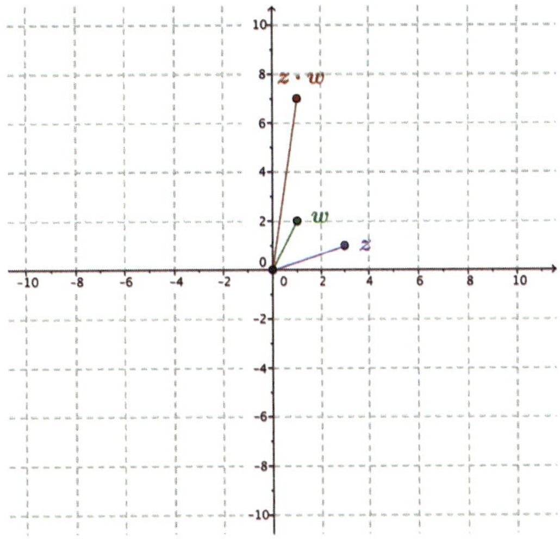

b. $z = 1 + 2i, w = -1 + 4i$

$$\begin{aligned} wz &= (1 + 2i)(-1 + 4i) \\ &= -1 + 4i - 2i + 8i^2 \\ &= -1 - 8 + 2i \\ &= -9 + 2i \end{aligned}$$

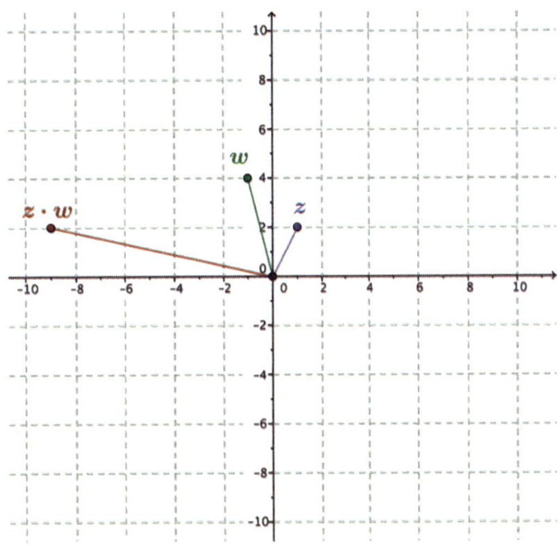

c. $z = -1 + i, w = -2 - i$

$$\begin{aligned} wz &= (-1 + i)(-2 - i) \\ &= 2 - 2i + i - i^2 \\ &= 2 + 1 - i \\ &= 3 - i \end{aligned}$$

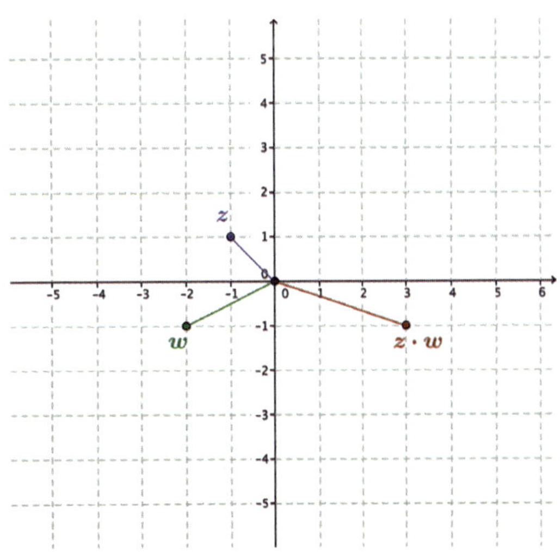

MP.8

d. **For each part (a), (b), and (c), draw line segments connecting each point z, w, and wz to the origin. Determine a relationship between the arguments of the complex numbers z, w, and wz.**

It appears that the argument of wz is the sum of the arguments of z and w.

Discussion (5 minutes)

This Discussion outlines the point of the lesson. It is claimed that the geometric effect of the linear transformation $L(z) = wz$ for complex numbers w is twofold: a dilation by $|w|$ and a rotation by the argument of w. The teacher then leads students through the justification for why these observations hold in every case. The observation was made in Lesson 13 using the particular examples $L_1(z) = (3+4i)z$, $L_2(z) = (-3+4i)z$, $L_3(z) = (-3-4i)z$, and $L_4(z) = (3-4i)z$. In the lesson itself, only the case of $L(z) = (a+bi)z$ where $a > 0$ and $b > 0$ is addressed. The remaining cases are included in the Problem Set.

- At the end of Lesson 13, what did you discover about the geometric effects of the transformations $L_1(z) = (3+4i)z$, $L_2(z) = (-3+4i)z$, $L_3(z) = (-3-4i)z$, and $L_4(z) = (3-4i)z$?
  - *These transformations had the geometric effect of dilation by $|3+4i| = 5$ and rotation by the argument of $3+4i$ (or $3-4i$, $-3-4i$, $-3+4i$, as appropriate).*
- Can we generalize this result to any linear transformation $L(z) = wz$, for a complex number w? Why or why not?
  - *Yes, it seems that we can generalize this. We tried it for $L(z) = (2+i)z$, and it worked.*
- For a general linear transformation $L(z) = wz$, what do we need to establish in order to generalize what we discovered in Lesson 13?

MP.6

Students may struggle with stating these ideas using proper mathematical terminology. Allow them time to grapple with the phrasing before providing the correct terminology.

  - *We need to show that the modulus of $L(z)$ is equal to the product of the modulus of w and the modulus of z. That is, we need to show that $|L(z)| = |w| \cdot |z|$.*
  - *We need to show that the angle made by the ray through the origin and z is a rotation of the ray through the origin and $L(z)$ by $\arg(w)$. That is, we need to show that $\arg(L(z)) = \arg(w) + \arg(z)$.*

Exercises 1–2 (5 minutes)

Exercises

1. **Let $w = a + bi$ and $z = c + di$.**

   a. **Calculate the product wz.**

   $$\begin{aligned} wz &= (a+bi)(c+di) \\ &= (ac-bd) + (ad+bc)i \end{aligned}$$

Scaffolding:

- Allow struggling students to complete these exercises for concrete values of z and w, such as $z = 4 - 3i$ and $w = 5 + 12i$.
- Ask advanced students to think about the relationship between the arguments of w and z.

b. **Calculate the moduli $|w|$, $|z|$, and $|wz|$.**

$$|w| = \sqrt{a^2 + b^2}$$
$$|z| = \sqrt{c^2 + d^2}$$
$$\begin{aligned}|wz| &= \sqrt{(ac - bd)^2 + (ad + bc)^2} \\ &= \sqrt{a^2c^2 - 2abcd + b^2d^2 + a^2d^2 + 2abcd + b^2c^2} \\ &= \sqrt{a^2(c^2 + d^2) + b^2(c^2 + d^2)} \\ &= \sqrt{(a^2 + b^2)(c^2 + d^2)}\end{aligned}$$

c. **What can you conclude about the quantities $|w|$, $|z|$, and $|wz|$?**

From part (b), we can see that $|wz| = |w| \cdot |z|$.

2. **What does the result of Exercise 1 tell us about the geometric effect of the transformation $L(z) = wz$?**

We see that $|L(z)| = |wz| = |w| \cdot |z|$, so the transformation L dilates by a factor of $|w|$.

Discussion (15 minutes)

In this Discussion, lead students through the geometric argument that $\arg(wz)$ is the sum of $\arg(w)$ and $\arg(z)$. The images presented here show one of many cases, but the mathematics is not dependent on the case. The remaining cases are addressed in the Problem Set.

- We have established half of what we need to show today, that is, that one geometric effect of the transformation $L(z) = wz$ is a dilation by the modulus of w, $|w|$. Now, we will demonstrate that another geometric effect of this transformation is a rotation by the argument of w.
- Let $w = a + bi$, where a and b are real numbers. Representations of the complex numbers z and w as points in the coordinate plane are shown below.

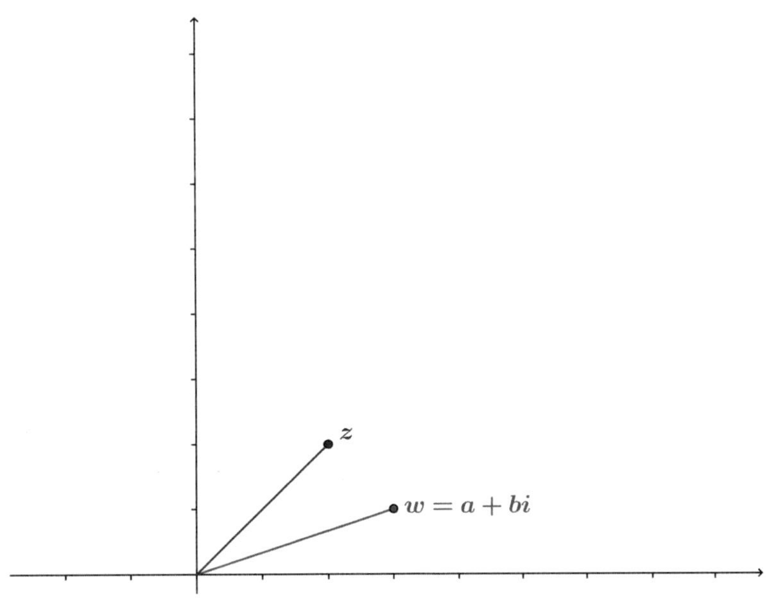

- Then, $wz = (a + bi)z = az + (bi)z$. Recall from Lesson 13 that az is a dilation of z by a, and $(bi)z$ is a dilation of z by b and a rotation by $90°$. Let's add the points az and $(bi)z$ to the figure.

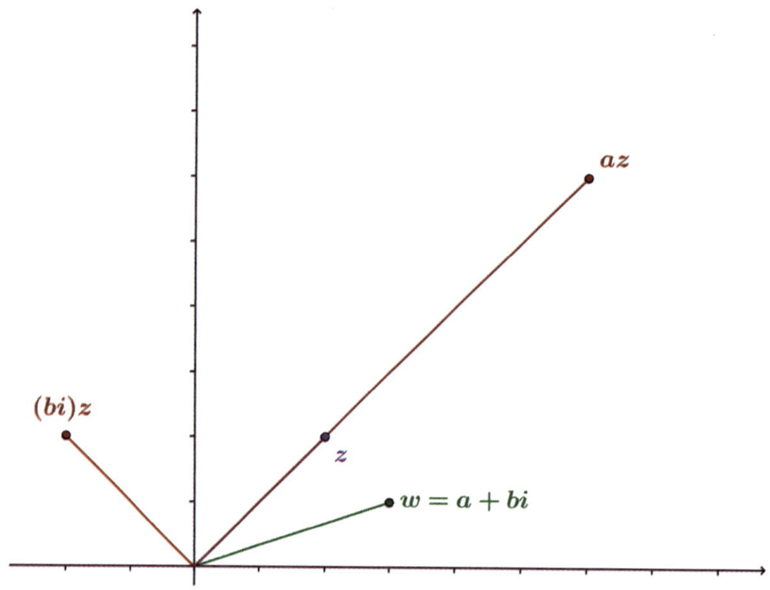

- We know that $wz = az + (bi)z$, so we can find the location of wz in the plane by adding $az + (bi)z$ geometrically. (We do not need to find a formula for the coordinates of wz.)

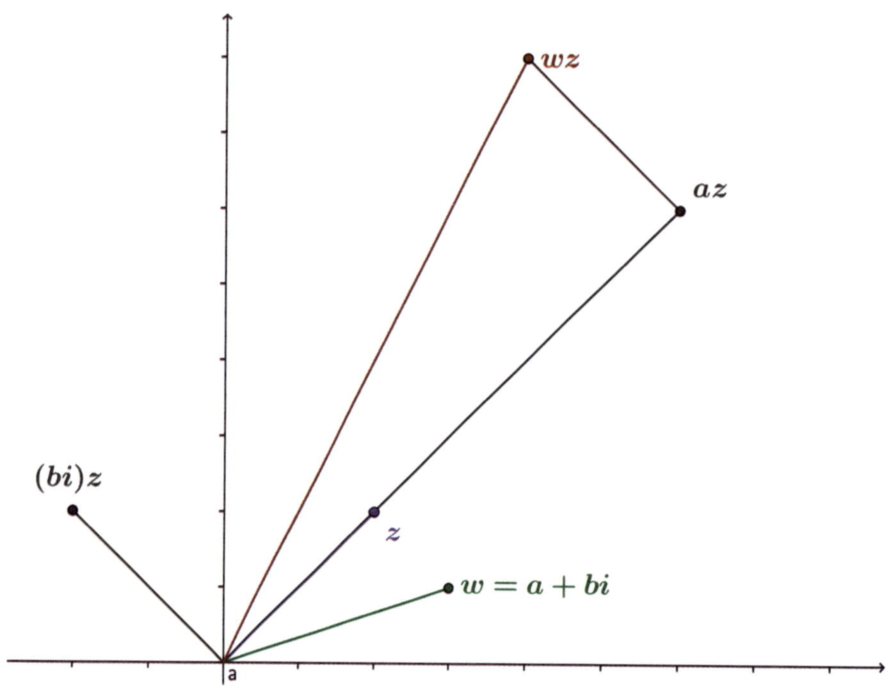

- Now, we can build a triangle with vertices at the origin, az and wz. And we can build another triangle with vertices at the origin, w and a.

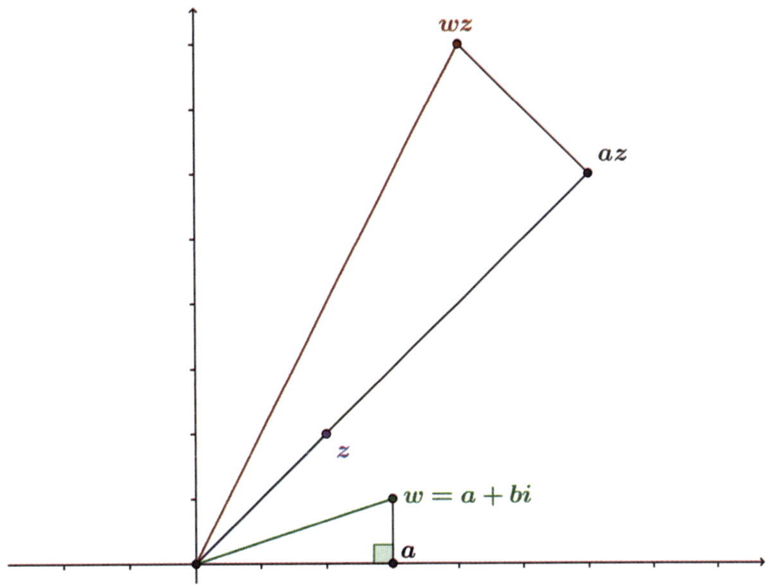

- What do we notice about these two triangles?
  - *They appear to both be right triangles. They appear to be similar.*
- For simplicity's sake, let's label the vertices of these triangles. Denote the origin by O, and let $P = w$, $Q = a$, $R = wz$, and $S = az$.

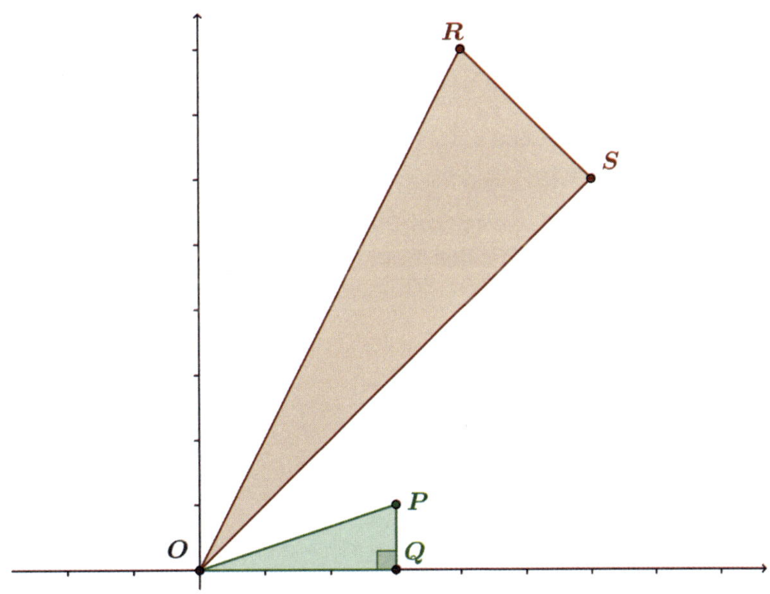

- What are the lengths of the sides of the small triangle, $\triangle OPQ$?
  - *We have*

$$OP = |w|$$
$$OQ = |a|$$
$$PQ = |b|.$$

- What are the lengths of the sides of the large triangle, $\triangle ORS$?
  - *We have*

$$\begin{aligned} OR &= |wz| = |w| \cdot |z| \\ OS &= |az| = |a| \cdot |z| \\ RS &= |wz - az| \\ &= |az + (bi)z - az| \\ &= |(bi)z| \\ &= |bi| \cdot |z| \\ &= |b| \cdot |i| \cdot |z| \\ &= |b| \cdot |z|. \end{aligned}$$

- How do the side lengths of $\triangle ORS$ and $\triangle OPQ$ relate?
  - *We see that*

$$\frac{OR}{OP} = \frac{|w| \cdot |z|}{|w|} = |z|,$$
$$\frac{OS}{OQ} = \frac{|a| \cdot |z|}{|a|} = |z|,$$
$$\frac{RS}{PQ} = \frac{|b| \cdot |z|}{|b|} = |z|.$$

- What can we conclude about $\triangle ORS$ and $\triangle OPQ$?
  - *We can conclude that $\triangle ORS \sim \triangle OPQ$ by SSS similarity.*
- Now that we know $\triangle ORS \sim \triangle OPQ$, we can conclude that $\angle ROS \cong \angle POQ$. So, how can we use this angle congruence to help us answer the original question?

EUREKA MATH™

- Where are $\arg(z)$, $\arg(w)$, and $\arg(wz)$ in our diagrams? How do they relate to the angles in the triangles?

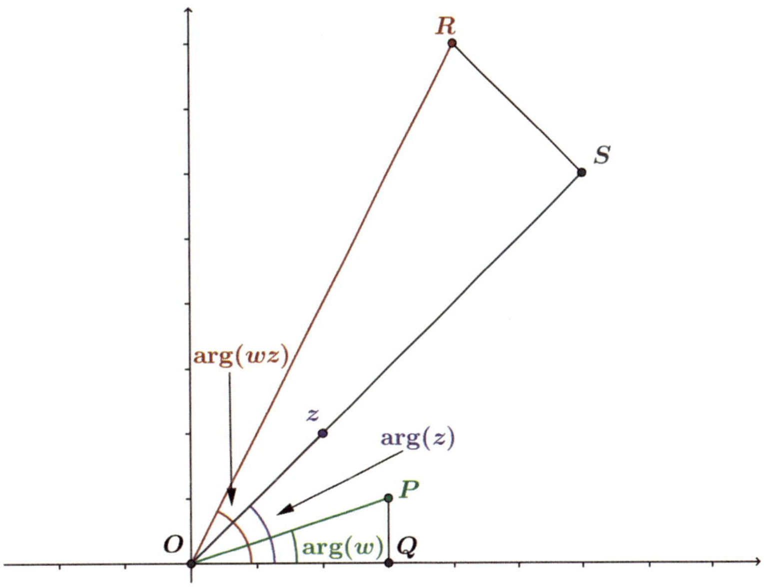

  - *From the diagram,* $\arg(w) = m\angle POQ$, $\arg(z) = m\angle SOQ$, *and* $\arg(wz) = m\angle ROQ$.

- However, we have shown that $m\angle POQ = m\angle ROS$.

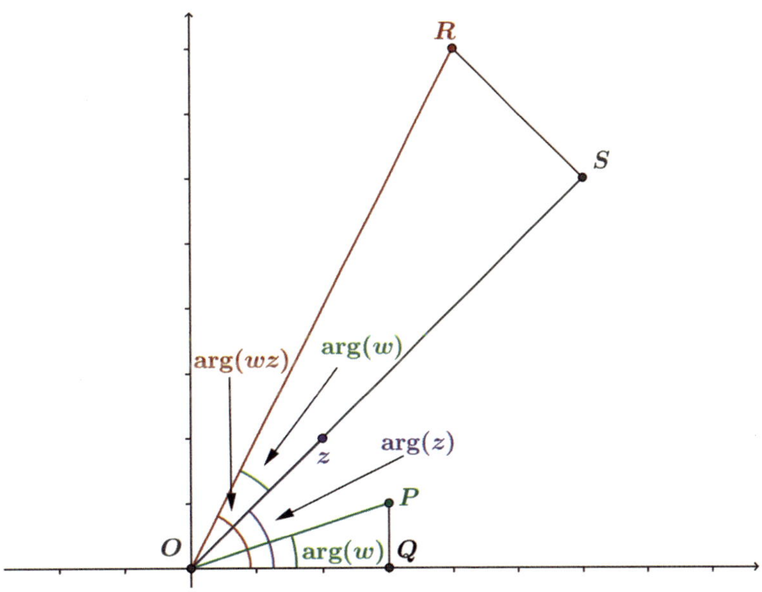

- We see that

$$\begin{aligned}\arg(wz) &= m\angle ROQ\\ &= m\angle ROS + m\angle SOQ\\ &= \arg(z) + \arg(w).\end{aligned}$$

- Then, since $\arg(L(z)) = \arg(wz) = \arg(z) + \arg(w)$, the point $L(z) = w(z)$ is the image of z under rotation by $\arg(w)$ about the origin. Thus, the transformation $L(z) = wz$ also has the geometric effect of rotation by $\arg(w)$.
- While our Discussion only addressed the case where w is represented by a point in the first quadrant, the result holds for any complex number w. You will consider the other cases for w in the Problem Set.

Exercise 3 (4 minutes)

3. If z and w are the complex numbers with the specified arguments and moduli, locate the point that represents the product wz on the provided coordinate axes.

a. $|w| = 3,\ \arg(w) = \frac{\pi}{4}$

$|z| = \frac{2}{3},\ \arg(z) = -\frac{\pi}{2}$

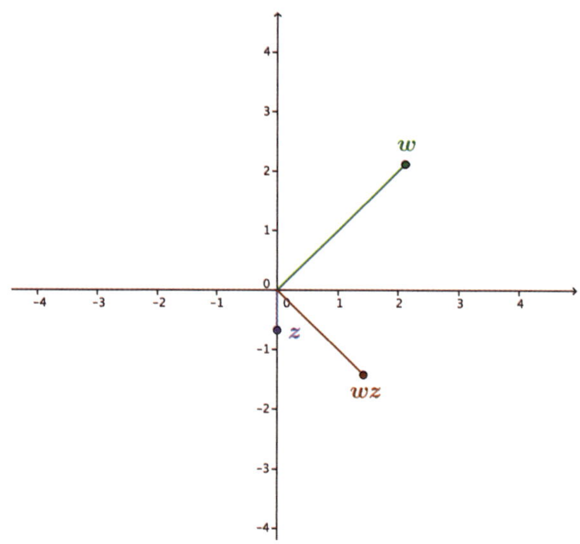

b. $|w| = 2,\ \arg(w) = \pi$

$|z| = 1,\ \arg(z) = \frac{\pi}{4}$

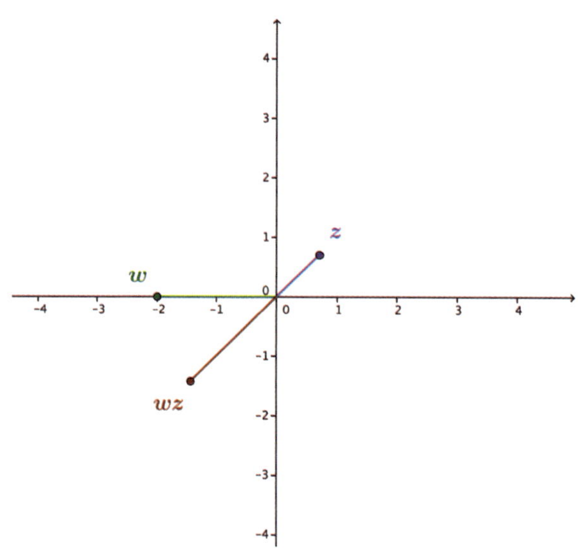

c. $|w| = \frac{1}{2}$, $\arg(w) = \frac{4\pi}{3}$
$|z| = 4$, $\arg(z) = -\frac{\pi}{6}$

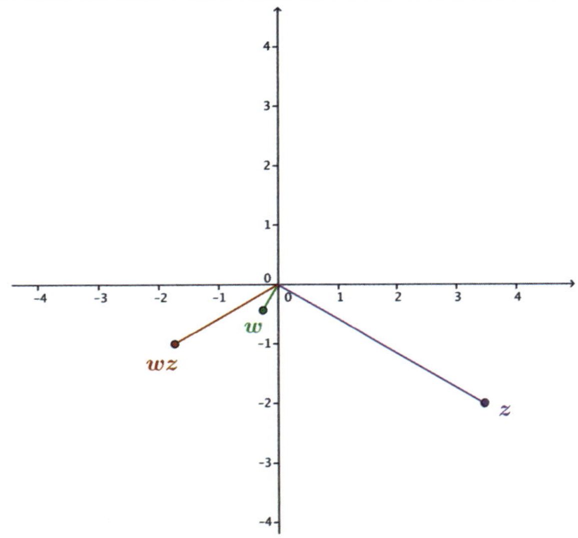

Closing (4 minutes)

Ask students to write in their journals or notebooks to explain the process for geometrically describing the product of two complex numbers. Students should mention the following key points.

- For complex numbers z and w, the modulus of the product is the product of the moduli:

$$|wz| = |w| \cdot |z|.$$

- For complex numbers z and w, the argument of the product is the sum of the arguments:

$$\arg(wz) = \arg(w) + \arg(z).$$

Lesson Summary

For complex numbers z and w,

- **The modulus of the product is the product of the moduli:**

$$|wz| = |w| \cdot |z|.$$

- **The argument of the product is the sum of the arguments:**

$$\arg(wz) = \arg(w) + \arg(z).$$

Exit Ticket (4 minutes)

Name ______________________________ Date ______________

Lesson 15: Justifying the Geometric Effect of Complex Multiplication

Exit Ticket

1. What is the geometric effect of the transformation $L(z) = (-6 + 8i)z$?

2. Suppose that w is a complex number with $|w| = \frac{3}{2}$ and $\arg(w) = \frac{5\pi}{6}$, and z is a complex number with $|z| = 2$ and $\arg(z) = \frac{\pi}{3}$.

   a. Explain how you can geometrically locate the point that represents the product wz in the coordinate plane.

   b. Plot w, z, and wz on the coordinate grid.

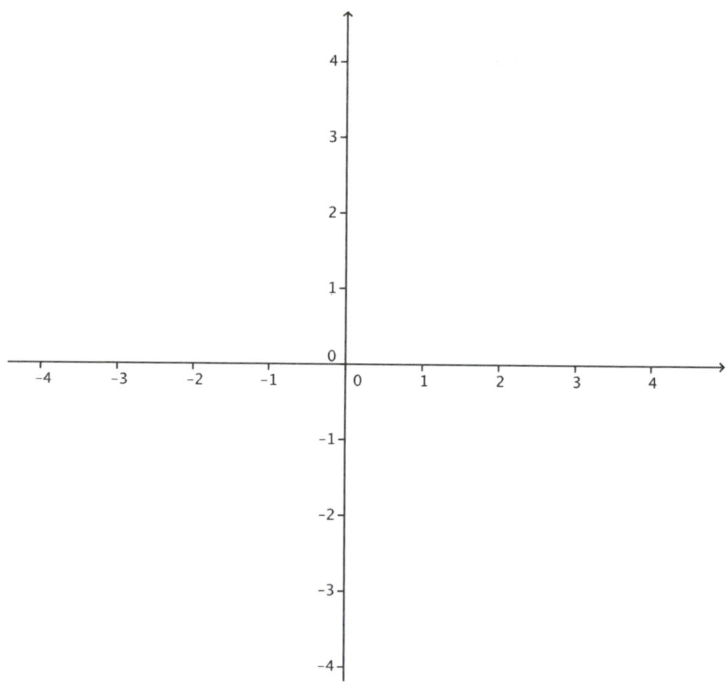

Exit Ticket Sample Solutions

1. What is the geometric effect of the transformation $L(z) = (-6 + 8i)z$?

   For this transformation, $w = -6 + 8i$, so $|w| = \sqrt{(-6)^2 + 8^2} = \sqrt{100} = 10$. The transformation L dilates by a factor of 10 and rotates counterclockwise by $\arg(-6 + 8i)$.

2. Suppose that w is a complex number with $|w| = \frac{3}{2}$ and $\arg(w) = \frac{5\pi}{6}$, and z is a complex number with $|z| = 2$ and $\arg(z) = \frac{\pi}{3}$.

   a. Explain how you can geometrically locate the point that represents the product wz in the coordinate plane.

   The product wz has argument $\frac{5\pi}{6} + \frac{\pi}{3} = \frac{7\pi}{6}$ and modulus $\frac{3}{2} \cdot 2 = 3$. So, we find the point that is distance 3 units from the origin on the ray that has been rotated $\frac{7\pi}{6}$ radians from the positive x-axis.

   b. Plot w, z, and wz on the coordinate grid.

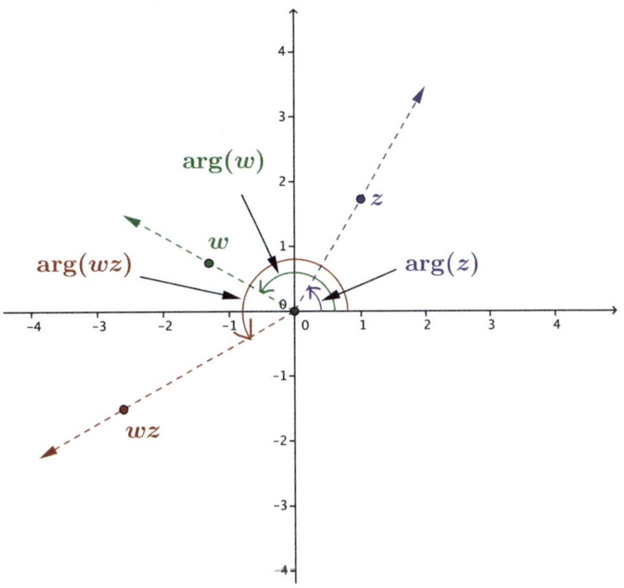

Problem Set Sample Solutions

Problems 1 and 2 establish that any linear transformation of the form $L(z) = wz$ has the geometric effect of a rotation by $\arg(w)$ and dilation by $|w|$. Problems 3 and 4 lead to the development in the next lesson in which students build new transformations from ones they already know.

1. **In the lesson, we justified our observation that the geometric effect of a transformation $L(z) = wz$ is a rotation by $\arg(w)$ and a dilation by $|w|$ for a complex number w that is represented by a point in the first quadrant of the coordinate plane. In this exercise, we will verify that this observation is valid for any complex number w. For a complex number $w = a + bi$, we only considered the case where $a > 0$ and $b > 0$. There are eight additional possibilities we need to consider.**

    a. **Case 1: The point representing w is the origin. That is, $a = 0$ and $b = 0$.**

    In this case, explain why $L(z) = (a + bi)z$ has the geometric effect of rotation by $\arg(a + bi)$ and dilation by $|a + bi|$.

    If $a + bi = 0 + 0i = 0$, then $\arg(a + bi) = 0$, and $|a + bi| = 0$. Rotating a point z by $0°$ does not change the location of z, and dilation by 0 sends each point to the origin. Since $L(z) = 0z = 0$ for every complex number z, we can say that L dilates by 0 and rotates by 0, so L rotates counterclockwise by $\arg(a + bi)$ and dilates by $|a + bi|$.

    b. **Case 2: The point representing w lies on the positive real axis. That is, $a > 0$ and $b = 0$.**

    In this case, explain why $L(z) = (a + bi)z$ has the geometric effect of rotation by $\arg(a + bi)$ and dilation by $|a + bi|$.

    If $b = 0$, then $L(z) = az$, which dilates z by a factor of a and does not rotate z. Since $a + bi$ lies on the positive real axis, $\arg(a + bi) = 0$. Also, $|a + bi| = |a| = a$, since $a > 0$. Thus, L dilates by $|a + bi|$ and rotates counterclockwise by $\arg(a + bi)$.

    c. **Case 3: The point representing w lies on the negative real axis. That is, $a < 0$ and $b = 0$.**

    In this case, explain why $L(z) = (a + bi)z$ has the geometric effect of rotation by $\arg(a + bi)$ and dilation by $|a + bi|$.

    If $b = 0$, then $L(z) = az$, which dilates z by a factor of $|a|$ and rotates z by $180°$. Since $a + bi$ lies on the negative real axis, $\arg(a + bi) = 180°$. Also, $|a + bi| = |a|$. Thus, L dilates by $|a + bi|$ and rotates counterclockwise by $\arg(a + bi)$.

    d. **Case 4: The point representing w lies on the positive imaginary axis. That is, $a = 0$ and $b > 0$.**

    In this case, explain why $L(z) = (a + bi)z$ has the geometric effect of rotation by $\arg(a + bi)$ and dilation by $|a + bi|$.

    If $a = 0$, then $L(z) = (bi)z$, which dilates z by a factor of b and rotates z by $90°$ counterclockwise. Since $a + bi$ lies on the positive imaginary axis, $\arg(a + bi) = 90°$. Also, $|a + bi| = b$. Thus, L dilates by $|a + bi|$ and rotates counterclockwise by $\arg(a + bi)$.

    e. **Case 5: The point representing w lies on the negative imaginary axis. That is, $a = 0$ and $b < 0$.**

    In this case, explain why $L(z) = (a + bi)z$ has the geometric effect of rotation by $\arg(a + bi)$ and dilation by $|a + bi|$.

    If $a = 0$, then $L(z) = (bi)z$, which dilates z by a factor of $|b|$ and rotates z by $270°$ counterclockwise. Since $a + bi$ lies on the negative imaginary axis, $\arg(a + bi) = 270°$. Also, $|a + bi| = |b|$. Thus, L dilates by $|a + bi|$ and rotates counterclockwise by $\arg(a + bi)$.

f. Case 6: The point representing $w = a + bi$ lies in the second quadrant. That is, $a < 0$ and $b > 0$. Points representing w, z, az, $(bi)z$, and $wz = az + (bi)z$ are shown in the figure below.

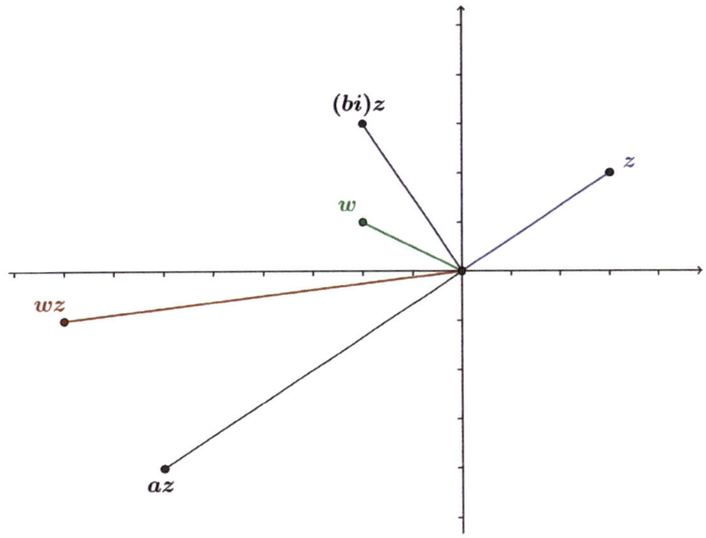

For convenience, rename the origin O, and let $P = w$, $Q = a$, $R = wz$, $S = az$, and $T = z$, as shown below. Let $m\angle POQ = \theta$.

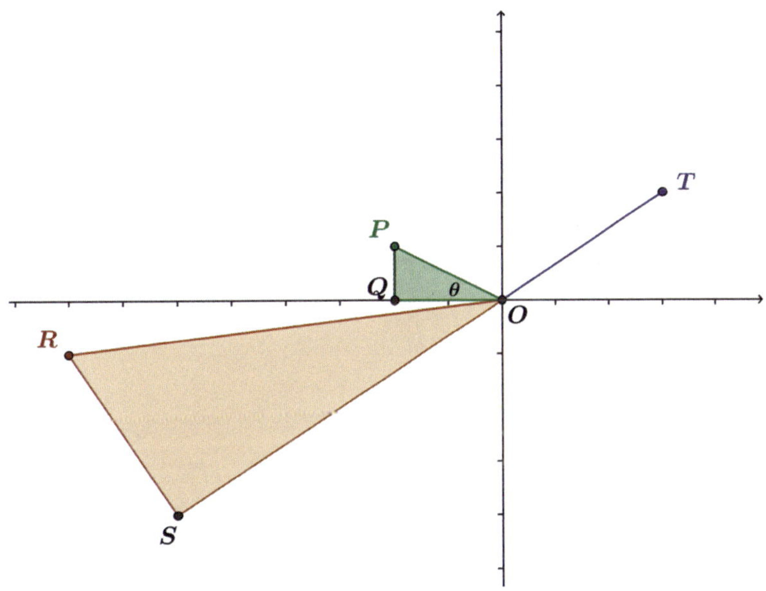

i. Argue that $\triangle OPQ \sim \triangle ORS$.

The lengths of the sides of the triangles are the following:

$OP = |w|$ $\quad$ $OR = |w| \cdot |z|$

$OQ = |a|$ $\quad$ $OS = |a| \cdot |z|$

$PQ = b$ $\quad$ $RS = b \cdot |z|$

Thus, $\frac{OR}{OP} = \frac{OS}{OQ} = \frac{RS}{PQ} = |z|$, *so* $\triangle OPQ \sim \triangle ORS$.

ii. Express the argument of az in terms of $\arg(z)$.

$\arg(az) = 180° + \arg(z)$

iii. Express $\arg(w)$ in terms of θ, where $\theta = m\angle POQ$.

$\arg(w) = 180° - \theta$

iv. Explain why $\arg(wz) = \arg(az) - \theta$.

Because $\triangle OPQ \sim \triangle ORS$, $m\angle ROS = m\angle POQ = \theta$.

$$\begin{aligned}\arg(wz) &= \arg(az) - m\angle ROW\\ &= \arg(az) - m\angle POQ\\ &= \arg(az) - \theta\end{aligned}$$

v. Combine your responses from parts (ii), (iii), and (iv) to express $\arg(wz)$ in terms of $\arg(z)$ and $\arg(w)$.

$$\begin{aligned}\arg(wz) &= \arg(az) - \theta\\ &= \left(180° + \arg(z)\right) - \left(180° - \arg(w)\right)\\ &= \arg(z) + \arg(w)\end{aligned}$$

g. Case 7: The point representing $w = a + bi$ lies in the third quadrant. That is, $a < 0$ and $b < 0$. Points representing w, z, az, $(bi)z$, and $wz = az + (bi)z$ are shown in the figure below.

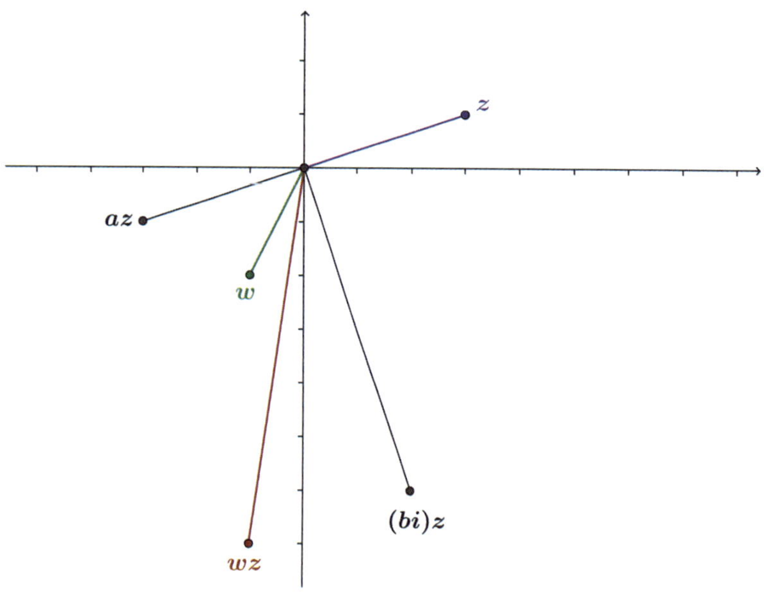

For convenience, rename the origin O, and let $P = w$, $Q = a$, $R = wz$, $S = az$, and $T = z$, as shown below. Let $m\angle POQ = \theta$.

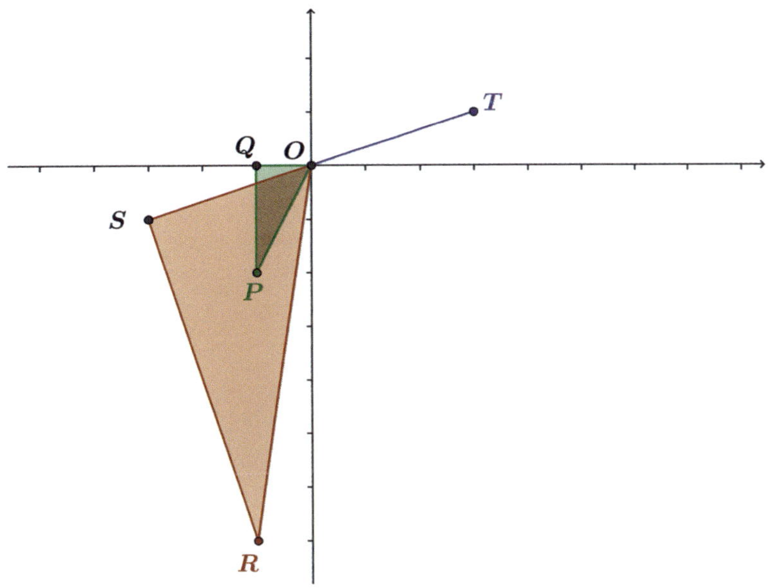

i. Argue that $\triangle OPQ \sim \triangle ORS$.

The lengths of the sides of the triangles are as follows:

$OP = |w|$ $\quad$ $OR = |w| \cdot |z|$

$OQ = |a|$ $\quad$ $OS = |a| \cdot |z|$

$PQ = |b|$ $\quad$ $RS = |b| \cdot |z|$

Thus, $\frac{OR}{OP} = \frac{OS}{OQ} = \frac{RS}{PQ} = |z|$, *so* $\triangle OPQ \sim \triangle ORS$.

ii. Express the argument of az in terms of $\arg(z)$.

$\arg(az) = 180° + \arg(z)$

iii. Express $\arg(w)$ in terms of θ, where $\theta = m\angle POQ$.

$\arg(w) = 180° + \theta$

iv. Explain why $\arg(wz) = \arg(az) + \theta$.

Because $\triangle OPQ \sim \triangle ORS$, $m\angle ROS = m\angle POQ = \theta$.

$$\begin{aligned}\arg(wz) &= \arg(az) + m\angle ROS \\ &= \arg(az) + m\angle POQ \\ &= \arg(az) + \theta\end{aligned}$$

v. **Combine your responses from parts (ii), (iii), and (iv) to express $\arg(wz)$ in terms of $\arg(z)$ and $\arg(w)$.**

$$\begin{aligned}\arg(wz) &= \arg(az) + \theta \\ &= \left(180° + \arg(z)\right) + (\arg(w) - 180°) \\ &= \arg(z) + \arg(w)\end{aligned}$$

h. **Case 8: The point representing $w = a + bi$ lies in the fourth quadrant. That is, $a > 0$ and $b < 0$. Points representing w, z, az, $(bi)z$, and $wz = az + (bi)z$ are shown in the figure below.**

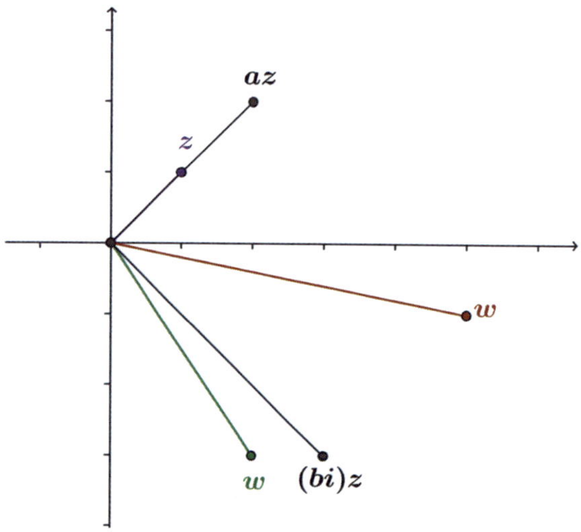

For convenience, rename the origin O, and let $P = w$, $Q = a$, $R = wz$, $S = az$, and $T = z$, as shown below. Let $m\angle POQ = \theta$.

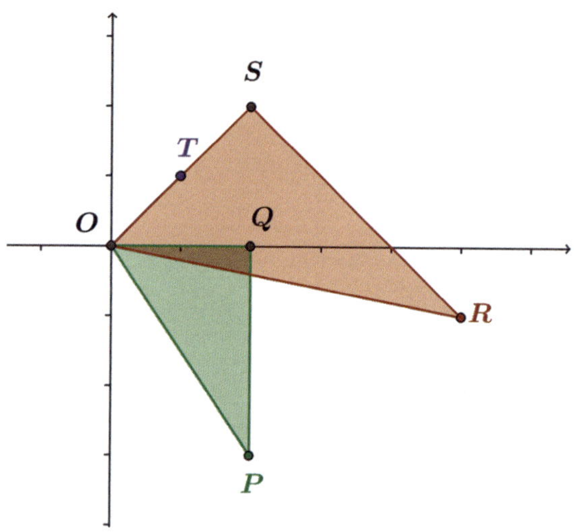

i. Argue that $\triangle OPQ \sim \triangle ORS$.

The lengths of the sides of the triangles are the following:

$$OP = |w| \qquad OR = |w| \cdot |z|$$
$$OQ = a \qquad OS = a \cdot |z|$$
$$PQ = |b| \qquad RS = |b| \cdot |z|$$

Thus, $\frac{OR}{OP} = \frac{OS}{OQ} = \frac{RS}{PQ} = |z|$, *so* $\triangle OPQ \sim \triangle ORS$.

ii. Express $\arg(w)$ in terms of θ, where $\theta = m\angle POQ$.

$\arg(w) = 360° - \theta$

iii. Explain why $m\angle QOR = \theta - \arg(z)$.

Because $\triangle OPQ \sim \triangle ORS$, $m\angle SOR = m\angle POQ = \theta$.

$$\begin{aligned} m\angle QOR &= m\angle SOR - m\angle SOQ \\ &= \theta - m\angle SOQ \\ &= \theta - \arg(z) \end{aligned}$$

iv. Express $\arg(wz)$ in terms of $m\angle QOR$.

Because $\triangle OPQ \sim \triangle ORS$, $m\angle ROS = m\angle POQ = \theta$.

$$\begin{aligned} \arg(wz) &= \arg(az) + m\angle ROS \\ &= \arg(az) + m\angle POQ \\ &= \arg(az) + \theta \end{aligned}$$

v. Combine your responses from parts (ii), (iii), and (iv) to express $\arg(wz)$ in terms of $\arg(z)$ and $\arg(w)$.

$$\begin{aligned} \arg(wz) &= 360° - m\angle QOR \\ &= 360° - (\theta - \arg(z)) \\ &= (360° - \theta) + \arg(z) \\ &= \arg(w) + \arg(z) \end{aligned}$$

2. Summarize the results of Problem 1, parts (a)–(h), and the lesson.

For any complex number w, the transformation $L(z) = wz$ has the geometric effect of rotation by $\arg(w)$ and dilation by $|w|$.

3. Find a linear transformation L that will have the geometric effect of rotation by the specified amount without dilating.

a. $45°$ counterclockwise

We need to find a complex number w so that $|w| = 1$ and $\arg(w) = 45°$. Then, w can be represented by a point on the unit circle such that the ray through the origin and w is the terminal ray of the positive x-axis rotated by $45°$. Then, the x-coordinate of w is $\cos(45°)$, and the y-coordinate of w is $\sin(45°)$, so we have $w = \cos(45°) + i\sin(45°) = \frac{\sqrt{2}}{2} + \frac{i\sqrt{2}}{2}$. Then, $L(z) = \left(\frac{\sqrt{2}}{2} + \frac{i\sqrt{2}}{2}\right)z = \frac{\sqrt{2}}{2}(1+i)z$.

b. 60° counterclockwise

$$L(z) = (\cos(60°) + i\sin(60°))z$$
$$= \left(\frac{1}{2} + \frac{i\sqrt{3}}{2}\right)z$$

c. 180° counterclockwise

$$L(z) = (\cos(180°) + i\sin(180°))z$$
$$= -z$$

d. 120° counterclockwise

$$L(z) = (\cos(120°) + i\sin(120°))z$$
$$= \left(-\frac{1}{2} + \frac{i\sqrt{3}}{2}\right)z$$

e. 30° clockwise

$$L(z) = (\cos(-30°) + i\sin(-30°))z$$
$$= \left(\frac{\sqrt{3}}{2} - \frac{1}{2}i\right)z$$

f. 90° clockwise

$$L(z) = (\cos(-90°) + i\sin(-90°))z$$
$$= -iz$$

g. 180° clockwise

$$L(z) = (\cos(-180°) + i\sin(-180°))z$$
$$= -z$$

h. 135° clockwise

$$L(z) = (\cos(-135°) + i\sin(-135°))z$$
$$= \left(-\frac{\sqrt{2}}{2} - \frac{i\sqrt{2}}{2}\right)z$$
$$= -\frac{\sqrt{2}}{2}(1+i)z$$

4. Suppose that we have linear transformations L_1 and L_2 as specified below. Find a formula for $L_2(L_1(z))$ for complex numbers z.

a. $L_1(z) = (1+i)z$ and $L_2(z) = (1-i)z$

$$L_2(L_1(z)) = L_2((1+i)z)$$
$$= (1-i)((1+i)z)$$
$$= (1-i)(1+i)z$$
$$= 2z$$

b. $L_1(z) = (3 - 2i)z$ and $L_2(z) = (2 + 3i)z$

$$\begin{aligned} L_2(L_1(z)) &= L_2((3-2i)z) \\ &= (2+3i)((3-2i)z) \\ &= (2+3i)(3-2i)z \\ &= (12+5i)z \end{aligned}$$

c. $L_1(z) = (-4 + 3i)z$ and $L_2(z) = (-3 - i)z$

$$\begin{aligned} L_2(L_1(z)) &= L_2((-4+3i)z) \\ &= (-3-i)((-4+3i)z) \\ &= (-3-i)(-4+3i)z \\ &= (15-5i)z \end{aligned}$$

d. $L_1(z) = (a + bi)z$ and $L_2(z) = (c + di)z$ for real numbers a, b, c, and d.

$$\begin{aligned} L_2(L_1(z)) &= L_2((a+bi)z) \\ &= (c+di)((a+bi)z) \\ &= (a+bi)(c+di)z \end{aligned}$$

Lesson 16: Representing Reflections with Transformations

Student Outcomes

- Students create a sequence of transformations that produce the geometric effect of reflection across a given line through the origin.

Lesson Notes

In this lesson, students apply complex multiplication from Lesson 14 to construct a transformation of the plane that reflects across a given line. So far, students have only looked at linear transformations of the form $L: \mathbb{C} \rightarrow \mathbb{C}$ by $L(z) = wz$ for a complex number w, and all such linear transformations have the geometric effect of rotation by $\arg(w)$ and dilation by $|w|$. In later lessons, when matrices are used to define transformations, students see that reflection can be represented by a transformation of the form $L\left(\begin{bmatrix} x \\ y \end{bmatrix}\right) = A\begin{bmatrix} x \\ y \end{bmatrix}$ for a matrix A, which better fits the form that they are used to for linear transformations. This lesson relies upon the foundational standards **G-CO.A.2**, **G-CO.A.4**, and **G.CO.A.5** and strengthens student understanding of **N-CN.A.3** and **N-CN.B.4**. Students may need to be reminded of the following notations for transformations of the plane from Geometry:

- A rotation by θ degrees about the origin is denoted by $R_{(0,\theta°)}$.
- A reflection across line ℓ is denoted by r_ℓ.

Classwork

Opening Exercise (6 minutes)

Students should work in pairs or small groups for these exercises. Students did problems identical or nearly identical to parts (a) and (b) in the Problem Set for Lesson 14, and they learned in Lesson 6 that taking the conjugate of z produces the reflection of z across the real axis.

Opening Exercise

a. **Find a transformation $R_{(0,45°)}: \mathbb{C} \rightarrow \mathbb{C}$ that rotates a point represented by the complex number z by $45°$ counterclockwise in the coordinate plane but does not produce a dilation.**

$$L(z) = \left(\frac{\sqrt{2}}{2} + i\frac{\sqrt{2}}{2}\right)z$$

b. **Find a transformation $R_{(0,-45°)}: \mathbb{C} \rightarrow \mathbb{C}$ that rotates a point represented by the complex number z by $45°$ clockwise in the coordinate plane but does not produce a dilation.**

$$L(z) = \left(\frac{\sqrt{2}}{2} - i\frac{\sqrt{2}}{2}\right)z$$

c. **Find a transformation $r_{x\text{-axis}}: \mathbb{C} \rightarrow \mathbb{C}$ that reflects a point represented by the complex number z across the x-axis.**

$L(z) = \bar{z}$, the conjugate of z

Discussion (15 minutes)

This Discussion sets up the problem for the day, which is finding a linear transformation that reflects across a line through the origin. For familiarity and ease of calculation, students begin with a reflection across the line with equation $y = x$. Students need to know the results of the Opening Exercise, so be sure to verify that all groups got the correct answers before proceeding with the Discussion. The circle with radius z is shown lightly in the figure to help with performing transformations accurately.

Discussion

We want to find a transformation $r_\ell: \mathbb{C} \rightarrow \mathbb{C}$ that reflects a point representing a complex number z across the diagonal line ℓ with equation $y = x$.

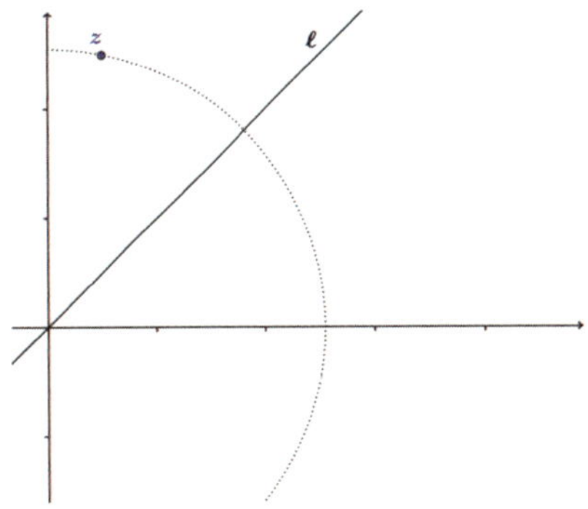

Recall from Algebra II, Module 3 that the transformation $(x, y) \rightarrow (y, x)$ accomplishes this reflection across the diagonal line in the coordinate plane, but students are now looking for a formula that produces this result for the complex number $x + yi$. If students mention this transformation, praise them for making the connection to past work, and ask them to keep this response in mind for verifying the answer they get with the new approach. The steps outlined below demonstrate that the reflection across a line other than the x-axis or y-axis can be accomplished by a sequence of rotations so that the line of reflection aligns with the x-axis, reflects across the x-axis, and rotates so that the line is back in its original position.

Display or reproduce the image above to guide students through this Discussion as they take notes. Ask students to draw a point $r_\ell(z)$ where they think the reflection of z across line ℓ is. Draw it on the teacher version also. Walk through the sequence of transformations geometrically before introducing the analytical formulas.

Scaffolding:

For struggling students, use transparency sheets to model the sequence of rotating by $-45°$, reflecting across the real axis, and then rotating by $45°$.

We know how to find transformations that produce the effect of rotating by a certain amount around the origin, dilating by a certain scale factor, and reflecting across the x-axis or the y-axis. Which of these transformations help us to reflect across the diagonal line? Allow students to make suggestions or conjectures.

- How much do we have to rotate around the origin to have line ℓ align with the positive x-axis?
  - *We need to rotate $-45°$.*

Draw the image of z after rotation by $-45°$ about the origin. Label the new point z_1. Give students a quick minute to draw z_1 on their version before displaying the teacher version.

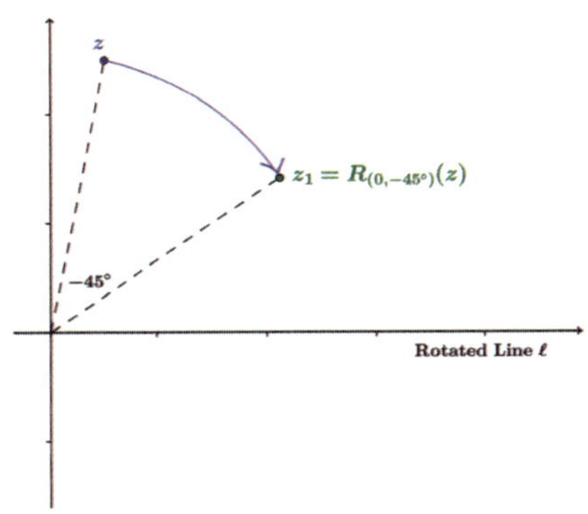

- Where is the original line ℓ now?
  - *It coincides with the positive x-axis.*

- We know how to reflect across the x-axis. Draw the reflection of point z_1 across the x-axis, and label it z_2.

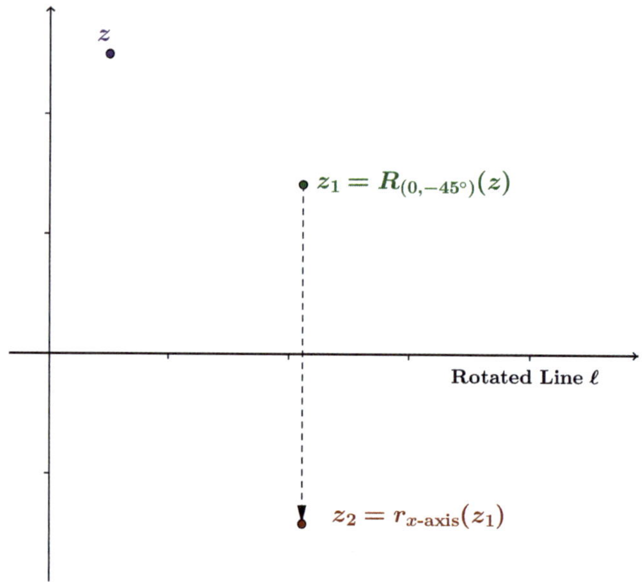

- Now that we have done a reflection, we need to rotate back to where we started. How much do we have to rotate around the origin to put line ℓ back where it originally was?
  - *We need to rotate* $45°$*.*
- Draw the image of z_2 under rotation by $45°$ about the origin. The image should coincide with the original estimate of $r_\ell(z)$.

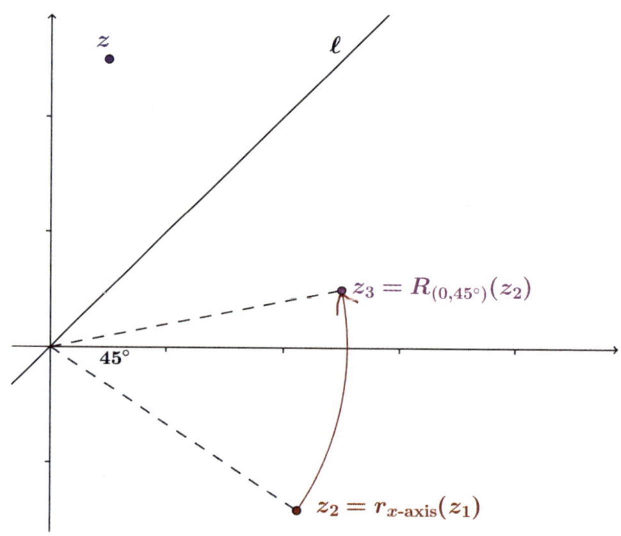

Now that students have had a chance to think about the geometric steps involved in reflecting z across diagonal line ℓ, repeat the process using the formulas for the three transformations.

- What is the transformation that accomplishes rotation by $-45°$? Students answered this question in the Opening Exercise.
  - *The transformation is* $L(z) = \left(\frac{\sqrt{2}}{2} - i\frac{\sqrt{2}}{2}\right)z$*.*
- We will refer to this transformation using the notation we used in Geometry. We can also factor out the constant $\frac{\sqrt{2}}{2}$: $R_{(0,-45°)}(z) = \frac{\sqrt{2}}{2}(1-i)z$.
- After we rotate the plane so that line ℓ lies along the x-axis, we reflect the new point z across the x-axis. What is the formula for the transformation $r_{x\text{-axis}}$ we use to accomplish the reflection?
  - *We use the conjugate of* z*, so we have* $r_{x\text{-axis}}(z) = \bar{z}$*.*

- Now, we can rotate the plane back to its original position by rotating by $45°$ counterclockwise around the origin. What is the formula for this rotation?
  - *From the Opening Exercise, using the notation from Geometry, we have* $R_{(0,45°)}(z) = \frac{\sqrt{2}}{2}(1+i)z$.
- We then have

$$\begin{aligned} z_1 &= R_{(0,-45°)}(z) = \frac{\sqrt{2}}{2}(1-i)z \\ z_2 &= r_{x\text{-axis}}(z_1) = \overline{z_1} \\ z_3 &= R_{(0,45°)}(z_2) = \frac{\sqrt{2}}{2}(1+i)z_2. \end{aligned}$$

- Putting the formulas together, we have

$$\begin{aligned} z_3 &= R_{(0,45°)}(z_2) \\ &= R_{(0,45°)}\big(r_{x\text{-axis}}(z_1)\big) \\ &= R_{(0,45°)}\Big(r_{x\text{-axis}}\Big(R_{(0,-45°)}(z)\Big)\Big). \end{aligned}$$

Stop here before going forward with the analytic equations, and ensure that all students understand that this formula means that they are first rotating point z by $-45°$ about the origin, then reflecting across the x-axis, and then rotating by $45°$ about the origin. Remind students that the innermost transformations happen first.

- Applying the formulas, we have

$$\begin{aligned} z_3 &= R_{(0,45°)}\Big(r_{x\text{-axis}}\Big(R_{(0,-45°)}(z)\Big)\Big) \\ &= R_{(0,45°)}\left(r_{x\text{-axis}}\left(\frac{\sqrt{2}}{2}(1-i)z\right)\right) \\ &= R_{(0,45°)}\left(\overline{\frac{\sqrt{2}}{2}(1-i)z}\right) \\ &= R_{(0,15°)}\left(\overline{\frac{\sqrt{2}}{2}(1-i)} \cdot \bar{z}\right) \\ &= R_{(0,45°)}\left(\frac{\sqrt{2}}{2}(1+i)\bar{z}\right) \\ &= \frac{\sqrt{2}}{2}(1+i)\left(\frac{\sqrt{2}}{2}(1+i)\bar{z}\right) \\ &= \frac{1}{2}(1+i)^2\bar{z} \\ &= i\bar{z}. \end{aligned}$$

Then, the transformation $r_{\ell}(z) = i\bar{z}$ has the geometric effect of reflection across the diagonal line ℓ with equation $y = x$.

Exercises 1–2 (5 minutes)

Exercises

1. **The number z in the figure used in the Discussion above is the complex number $1 + 5i$. Compute $r_\ell(1 + 5i)$, and plot it below.**

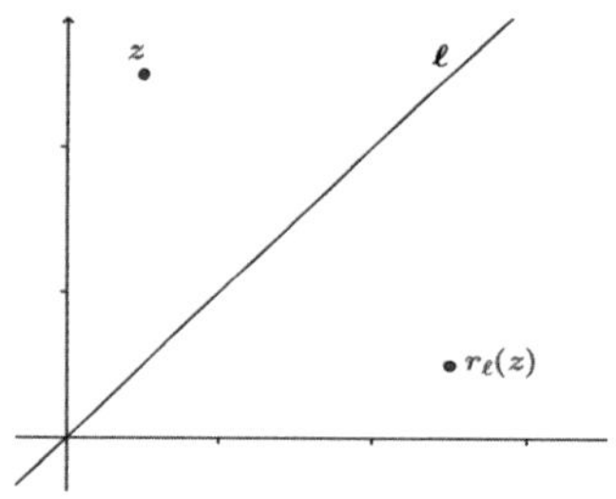

2. **We know from previous courses that the reflection of a point (x, y) across the line with equation $y = x$ is the point (y, x). Does this agree with our result from the previous discussion?**

Yes, we can represent the point (x, y) by $z = x + iy$. Then,

$$\begin{aligned} r_\ell(z) &= r_\ell(x + iy) \\ &= i\,\overline{(x + iy)} \\ &= i\,(x - iy) \\ &= y + ix, \end{aligned}$$

which corresponds to the point (y, x).

Exercise 3 (10 minutes)

In this exercise, students repeat the previous calculations to find an analytic formula for reflection across the line ℓ that makes a $60°$ angle with the positive x-axis.

3. **We now want to find a formula for the transformation of reflection across the line ℓ that makes a $60°$ angle with the positive x-axis. Find formulas to represent each component of the transformation, and use them to find one formula that represents the overall transformation.**

The transformation consists of rotating line ℓ so that it coincides with the x-axis, reflecting across the x-axis, and rotating the x-axis back to the original position of line ℓ. The components of the transformation can be represented by these formulas:

$$\begin{aligned} z_1 &= R_{(0,-60°)}(z) = \left(\frac{1}{2} - \frac{\sqrt{3}}{2}i\right)z \\ z_2 &= r_{x\text{-axis}}(z_1) = \overline{z_1} \\ z_3 &= R_{(0,60°)}(z_2) = \left(\frac{1}{2} + \frac{\sqrt{3}}{2}i\right)z_2. \end{aligned}$$

Putting the formulas together, we have

$$\begin{aligned} z_3 &= R_{(0,60°)}(z_2) \\ &= R_{(0,60°)}\left(r_{x\text{-axis}}(z_1)\right) \\ &= R_{(0,60°)}\left(r_{x\text{-axis}}\left(R_{(0,-60°)}(z)\right)\right). \end{aligned}$$

Scaffolding:

Ask struggling students the following questions to guide their work in Exercise 3.

- What transformation will rotate line ℓ so that it coincides with the x-axis?
  - $-60°$
- What transformation will reflect across the x-axis?
  - The conjugate
- What transformation will rotate the x-axis back to the original position of line ℓ?
  - $60°$

Stop here before going forward with the analytic equations, and ensure that all students understand that this formula means that we are first rotating point z by $-60°$ about the origin, then reflecting across the x-axis, and then rotating by $60°$ about the origin. Remind students that the innermost transformations happen first.

Applying the formulas, we have

$$\begin{aligned} r_{\ell}(z) &= R_{(0,60°)}\left(r_{x\text{-axis}}\left(R_{(0,-60°)}(z)\right)\right) \\ &= R_{(0,60°)}\left(r_{x\text{-axis}}\left(\left(\frac{1}{2}-\frac{\sqrt{3}}{2}i\right)z\right)\right) \\ &= R_{(0,60°)}\left(\overline{\left(\frac{1}{2}-\frac{\sqrt{3}}{2}i\right)z}\right) \\ &= R_{(0,60°)}\left(\overline{\left(\frac{1}{2}-\frac{\sqrt{3}}{2}i\right)}\cdot\overline{z}\right) \\ &= R_{(0,60°)}\left(\left(\frac{1}{2}+\frac{\sqrt{3}}{2}i\right)\overline{z}\right) \\ &= \left(\frac{1}{2}+\frac{\sqrt{3}}{2}i\right)\left(\left(\frac{1}{2}+\frac{\sqrt{3}}{2}i\right)\overline{z}\right) \\ &= \left(-\frac{1}{2}+\frac{\sqrt{3}}{2}i\right)\overline{z}. \end{aligned}$$

Then, the transformation $r_{\ell}(z) = \left(-\frac{1}{2}+\frac{\sqrt{3}}{2}i\right)\overline{z}$ has the geometric effect of reflection across the line ℓ that makes a $60°$ angle with the positive x-axis.

> *Scaffolding:*
>
> Have early finishers repeat Exercise 3 for the line ℓ that makes a $-30°$ angle with the positive x-axis.

Closing (4 minutes)

Ask students to write in their journals or notebooks to explain the sequence of transformation that produces reflection across a line ℓ through the origin that contains the terminal ray of a rotation of the x-axis by θ. Key points are summarized in the box below.

> **Lesson Summary**
>
> **Let ℓ be a line through the origin that contains the terminal ray of a rotation of the x-axis by θ. Then, reflection across line ℓ can be done by the following sequence of transformations:**
>
> - **Rotation by $-\theta$ about the origin.**
> - **Reflection across the x-axis.**
> - **Rotation by θ about the origin.**

Exit Ticket (5 minutes)

Name ______________________________ Date ______________

Lesson 16: Representing Reflections with Transformations

Exit Ticket

Explain the process used in the lesson to locate the reflection of a point z across the diagonal line with equation $y = x$. Include figures in your explanation.

Exit Ticket Sample Solutions

Explain the process used in the lesson to locate the reflection of a point z across the diagonal line with equation $y = x$. Include figures in your explanation.

First, we rotated the point z by $-45°$ to align the diagonal line with equation $y = x$ with the x-axis to get a new point z_1.

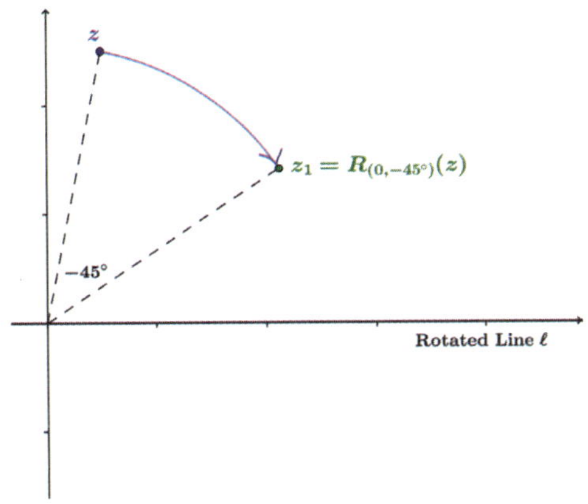

Then, we reflected the point z_1 across the real axis to find point z_2.

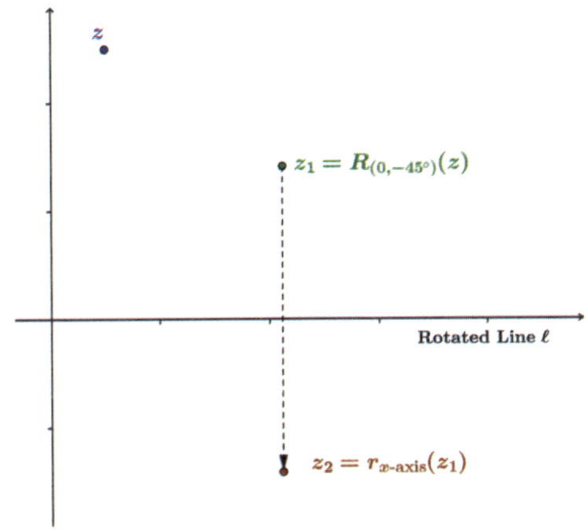

Finally, we rotated everything back by $45°$ to find the final point $z_3 = r_\ell(z)$.

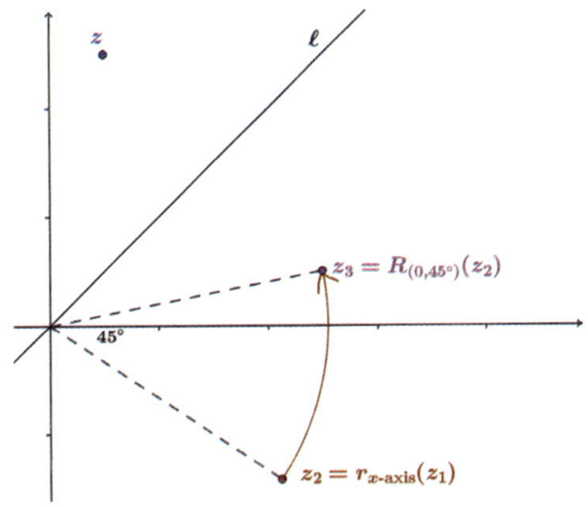

Problem Set Sample Solutions

1. Find a formula for the transformation of reflection across the line ℓ with equation $y = -x$.

   $z_1 = R_{(0,-135^\circ)}(z) = \left(-\frac{\sqrt{2}}{2} - \frac{\sqrt{2}}{2}i\right)z$; *if students cannot see it, you can say that:*

   $$R_{(0,-135^\circ)}(z) = R_{(0,-45^\circ)}\left(R_{(0,-45^\circ)}\left(R_{(0,-45^\circ)}z\right)\right) = \left(\frac{\sqrt{2}}{2} - \frac{\sqrt{2}}{2}i\right)^3 z = \left(-\frac{\sqrt{2}}{2} - \frac{\sqrt{2}}{2}i\right)z$$

   $$z_2 = r_{x\text{-axis}}(z_1) = \overline{\left(-\frac{\sqrt{2}}{2} - \frac{\sqrt{2}}{2}i\right)z} = \left(\overline{-\frac{\sqrt{2}}{2} - \frac{\sqrt{2}}{2}i}\right)\bar{z} = \left(-\frac{\sqrt{2}}{2} + \frac{\sqrt{2}}{2}i\right)\bar{z}$$

   $$z_3 = R_{(0,135^\circ)}(z_2) = \left(-\frac{\sqrt{2}}{2} + \frac{\sqrt{2}}{2}i\right)(z_2)$$

   $$z_3 = R_{(0,135^\circ)}(r_{x\text{-axis}}(z_1)) = R_{(0,135^\circ)}\left(r_{x\text{-axis}}\left(R_{(0,-145^\circ)}(z)\right)\right) = R_{(0,135^\circ)}\left(r_{x\text{-axis}}\left(\left(-\frac{\sqrt{2}}{2} - \frac{\sqrt{2}}{2}i\right)z\right)\right)$$

   $$= R_{(0,135^\circ)}\left(\overline{\left(-\frac{\sqrt{2}}{2} - \frac{\sqrt{2}}{2}i\right)z}\right) = R_{(0,135^\circ)}\left(\left(-\frac{\sqrt{2}}{2} + \frac{\sqrt{2}}{2}i\right)\bar{z}\right) = \left(-\frac{\sqrt{2}}{2} + \frac{\sqrt{2}}{2}i\right)\left(-\frac{\sqrt{2}}{2} + \frac{\sqrt{2}}{2}i\right)\bar{z}$$

   $$= -i\bar{z}$$

2. Find the formula for the sequence of transformations comprising reflection across the line with equation $y = x$ and then rotation by 180° about the origin.

   $$z_1 = R_{(0,-45^\circ)}(z) = \left(\frac{\sqrt{2}}{2} - \frac{\sqrt{2}}{2}i\right)z$$

   $$z_2 = r_{x\text{-axis}}(z_1) = \overline{\left(\frac{\sqrt{2}}{2} - \frac{\sqrt{2}}{2}i\right)z} = \left(\overline{\frac{\sqrt{2}}{2} - \frac{\sqrt{2}}{2}i}\right)\bar{z} = \left(\frac{\sqrt{2}}{2} + \frac{\sqrt{2}}{2}i\right)\bar{z}$$

   $$z_3 = R_{(0,45^\circ)}(z_2) = \left(\frac{\sqrt{2}}{2} + \frac{\sqrt{2}}{2}i\right)(z_2)$$

   $$z_3 = R_{(0,45^\circ)}(z_2) = \left(\frac{\sqrt{2}}{2} + \frac{\sqrt{2}}{2}i\right)(z_2) = \left(\frac{\sqrt{2}}{2} + \frac{\sqrt{2}}{2}i\right)\left(\frac{\sqrt{2}}{2} + \frac{\sqrt{2}}{2}i\right)\bar{z} = i\bar{z}$$

   $$z_4 = -z_3 = -i\bar{z}$$

3. Compare your answers to Problems 1 and 2. Explain what you find.

   They have the same answer/formula that will produce the same transformation of z.

4. Find a formula for the transformation of reflection across the line ℓ that makes a -30° angle with the positive x-axis.

   $$z_1 = R_{(0,30^\circ)}(z) = \left(\frac{\sqrt{3}}{2} + \frac{1}{2}i\right)z$$

   $$z_2 = r_{x\text{-axis}}(z_1) = \overline{\left(\frac{\sqrt{3}}{2} + \frac{1}{2}i\right)z} = \left(\overline{\frac{\sqrt{3}}{2} + \frac{1}{2}i}\right)\bar{z} = \left(\frac{\sqrt{3}}{2} - \frac{1}{2}i\right)\bar{z}$$

   $$z_3 = R_{(0,-30^\circ)}(z_2) = \left(\frac{\sqrt{3}}{2} - \frac{1}{2}i\right)(z_2)$$

   $$z_3 = R_{(0,-30^\circ)}(z_2) = \left(\frac{\sqrt{3}}{2} - \frac{1}{2}i\right)(z_2) = \left(\frac{\sqrt{3}}{2} - \frac{1}{2}i\right)\left(\frac{\sqrt{3}}{2} - \frac{1}{2}i\right)\bar{z} = \left(\frac{1}{2} - \frac{\sqrt{3}}{2}i\right)\bar{z}$$

5. Max observed that when reflecting a complex number, $z = a + bi$ about the line $y = x$, a and b are reversed, which is similar to how we learned to find an inverse function. Will Max's observation also be true when the line $y = -x$ is used, where $a = -b$ and $b = -a$? Give an example to show his assumption is either correct or incorrect.

Yes, to reflect a complex number $z = a + bi$ about the line $y = -x$, we need to do $R_{0,-135^\circ}$, $r_{x\text{-axis}}$, and then $R_{0,135^\circ}$, which will produce the answer to be $z = b + ai$.

The examples vary. This example will work: $z = 1 + i$.

$z = a + bi$

$$z_3 = R_{0,135^\circ}(\mathrm{r}_{x\text{-axis}}\left(R_{0,-135^\circ}(z)\right) = R_{0,135^\circ}(r_{x\text{-axis}}\left(\left(-\frac{\sqrt{2}}{2}-\frac{\sqrt{2}}{2}i\right)z\right) = R_{0,135^\circ}\left(\frac{\sqrt{2}}{2}\overline{(-1-i)z}\right)$$

$$= R_{0,135^\circ}\left(\frac{\sqrt{2}}{2}(-1+i)\bar{z}\right) = \frac{\sqrt{2}}{2}(-1+i)\frac{\sqrt{2}}{2}(-1+i)\bar{z} = \frac{2}{4}(-2i)\bar{z} = -i\bar{z} = -i(a+bi) = b - a$$

$z = 1 + i, z_3 = -i(1 - i) = -1 - i$

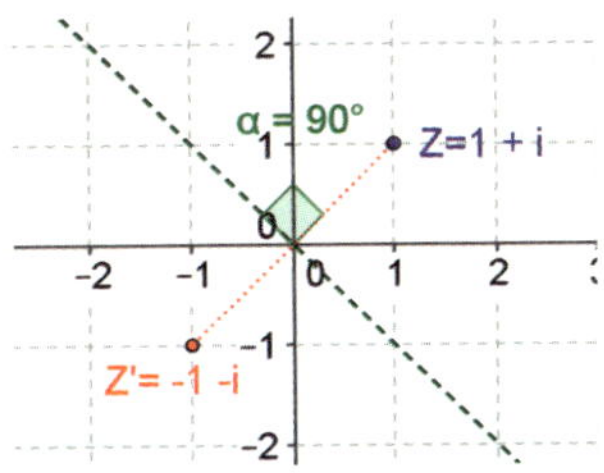

6. For reflecting a complex number, $z = a + bi$ about the line $y = 2x$, will Max's idea work if he makes $b = 2a$ and $a = \frac{b}{2}$? Use $z = 1 + 4i$ as an example to show whether or not it works.

No, it will not work based on the example shown. $z_1 = \frac{b}{2} + 2ai = \frac{4}{2} + 2 \times 1i = 2 + 2i$ *Since the angle $\alpha \neq 90°$, this is not a reflection.*

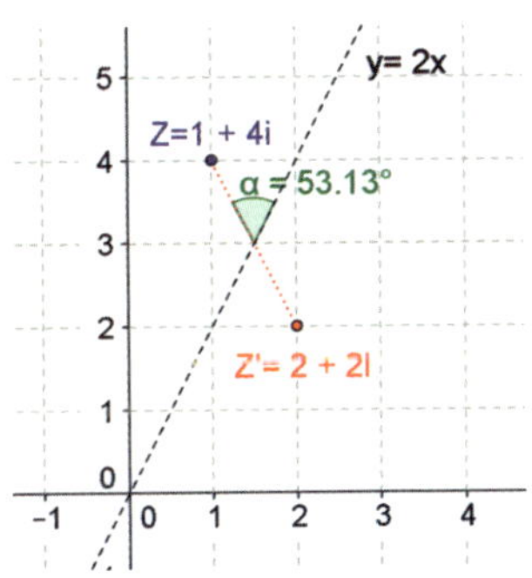

7. What would the formula look like if you want to reflect a complex number about the line $y = mx$, where $m > 0$?

For reflecting a complex number or a point about the line going through the origin, we need to know the angle of the line with respect to the positive x-axis to do rotations. So, we can use the slope of the line to find the angle that we need to rotate, which is $\arctan(m)$. Now we can come up with a general formula that can be applied to reflecting about the line $y = mx$, where $m > 0$.

$z_3 = R_{0,\arctan(m)}(r_{x\text{-axis}}\left(R_{0,\arctan(-m)}(z)\right)$,

where $R_{0,\arctan(-m)} = \cos(\arctan(-m)) + i \cdot \sin(\arctan(-m))$

$R_{0,\arctan(m)} = \cos(\arctan(m)) + i \cdot \sin(\arctan(m))$

Lesson 17: The Geometric Effect of Multiplying by a Reciprocal

Student Outcomes

- Students apply their knowledge to understand that multiplication by the reciprocal provides the inverse geometric operation to a rotation and dilation.
- Students understand the geometric effects of multiple operations with complex numbers.

Lesson Notes

This lesson explores the geometric effect of multiplication by the reciprocal to construct a transformation that undoes multiplication. In this lesson, students verify that the transformation of multiplication by the reciprocal produces the same result geometrically as it does algebraically. This lesson ties together the work of Lessons 13–15 on linear transformations of the form $L: \mathbb{C} \rightarrow \mathbb{C}$ by $L(z) = wz$ for a complex number w and all such linear transformations having the geometric effect of rotation by $\arg(w)$ and dilation by $|w|$ to the work done in Lessons 6 and 7 on complex number division. In later lessons, when matrices are used to define transformations, these topics are revisited and extended. This lesson relies upon the foundational standards **G-CO.A.2**, **G-CO.A.4**, and **G.CO.A.5** and strengthens student understanding of **N-CN.A.3**, **N-CN.B.4**, and **N-CN.B.5.**

This lesson is structured as a series of exploratory challenges that are scaffolded to allow students to make sense of the connections between algebraic operations with complex numbers and the corresponding transformations. The lesson concludes with students considering all the operations (and their related transformations) together and working with combinations of operations and describing them as a series of transformations of a complex number. In the Problem Set, students connect the work of this module back to linear transformations that they studied in Lessons 1 and 2.

Scaffolding:

Use these concrete examples to scaffold the opening as needed for students:

- In $x + 2$, how do you undo adding 2?
  - *You would subtract* 2.
    $x + 2 - 2 = x$
- In $5x$, how do you undo multiplication by 5?
  - *You would divide by* 5.
    $\frac{5x}{5} = x$
- In x^3, how do you undo the operation of cubing?
  - *You would take the cube root.*
    $\sqrt[3]{x^3} = x$

Classwork

Opening (2 minutes)

Ask students to brainstorm real-world operations that *undo* each other, for example, putting their shoes on and taking them off. Have students briefly share ideas with their group mates. Next, have them think of mathematical operations that undo each other. For example, division by 3 undoes multiplication by 3. Remind students that the word *inverse* is often used when talking about operations that undo each other. During this lesson, be sure to correct students who confuse the words *opposite, reciprocal,* and *inverse*.

EUREKA MATH

Opening Exercise (3 minutes)

Students work with this complex number in the subsequent Exploratory Challenge. If students struggle to find the argument and modulus of $1+i$ in this exercise, take time to review notation and methods for determining the argument of a complex number written in rectangular form. All the actual complex numbers in this lesson correspond to *friendly* rotations such as $\frac{\pi}{4}, \frac{\pi}{3}, \frac{\pi}{2}$, etc.

> *Scaffolding:*
> Throughout this lesson, students work with friendly rotations and use their knowledge of special right triangles and proportional reasoning. In the classroom, display prominent visual reminders such as drawings of special triangles (see Lesson 12 of this module), a unit circle with benchmark rotations labeled in degrees and radians (see Algebra II, Module 2), etc.

Opening Exercise

Given $w = 1 + i$, what is $\arg(w)$ and $|w|$? Explain how you got your answer.

$\arg(w) = \frac{\pi}{4}$ and $|w| = \sqrt{2}$. I used the formula $|w| = \sqrt{a^2+b^2}$ to determine the modulus, and since the point $(1, 1)$ lies along a ray from the origin that has been rotated $45°$ from the ray through the origin that contains the real number 1, the argument must be $\frac{\pi}{4}$.

Exploratory Challenge 1/Exercises 1–9 (10 minutes)

Students should complete the next nine exercises in small groups of three to four students. As teams work on these problems, circulate around the room to monitor progress. Some groups may get stuck on Exercise 3. Since the argument of a complex number on an interval $0 \le \arg(z) < 2\pi$ has been defined, students need to figure a positive rotation on this interval that is equivalent to $-\arg(z)$. Lead a whole-class discussion at this point if needed before moving the groups on to complete the rest of the exercises in this Exploratory Challenge. After Exercise 6, have one or two students report their group's response to this item.

Exploratory Challenge 1/Exercises 1–9

1. **Describe the geometric effect of the transformation $L(z) = (1+i)z$.**

   The transformation rotates the complex number about the origin through $45°$ and dilates the number by a scale factor of $\sqrt{2}$.

2. **Describe a way to undo the effect of the transformation $L(z) = (1+i)z$.**

   Geometrically, we need to rotate in the opposite direction, $-45°$, and dilate by a factor of $\frac{1}{\sqrt{2}}$.

3. **Given that $0 \le \arg(z) < 2\pi$ for any complex number, how could you describe any clockwise rotation of θ as an argument of a complex number?**

   $\arg(z) = 2\pi - \theta$ would result in the same rotation as a clockwise rotation of θ.

4. **Write a complex number in polar form that describes a rotation and dilation that will undo multiplication by $(1+i)$, and then convert it to rectangular form.**

   $$\frac{1}{\sqrt{2}}(\cos(315°) + i\sin(315°)) = \frac{1}{\sqrt{2}}\cdot\frac{1}{\sqrt{2}} + i\cdot\frac{1}{\sqrt{2}}\cdot -\frac{1}{\sqrt{2}} = \frac{1}{2} - \frac{1}{2}i$$

> *Scaffolding:*
> Work with specific angle measures to help struggling students understand the answer to Exercise 3.
> - Name a positive and negative rotation that take a ray from the origin containing the real number 1 through each point.
>
> $(1,1)$
> $(0,2)$
> $(-1,\sqrt{3})$
> $(-3,0)$
> $(-2,-2)$

5. In a previous lesson, you learned that to undo multiplication by $1+i$, you would multiply by the reciprocal $\frac{1}{1+i}$. Write the complex number $\frac{1}{1+i}$ in rectangular form $z = a + bi$ where a and b are real numbers.

$$\frac{1}{1+i} = \frac{1-i}{(1+i)(1-i)} = \frac{1-i}{2} = \frac{1}{2} - \frac{1}{2}i$$

6. How do your answers to Exercises 4 and 5 compare? What does that tell you about the geometric effect of multiplication by the reciprocal of a complex number?

The geometric effect of rotation by $2\pi - \arg(z)$ and dilation by $\frac{1}{|z|}$ appears to be the same as multiplication by the reciprocal when the problem is solved algebraically.

MP.3

7. Jimmy states the following:

Multiplication by $\frac{1}{a+bi}$ has the reverse geometric effect of multiplication by $+bi$.

Do you agree or disagree? Use your work on the previous exercises to support your reasoning.

Geometrically undoing the effect of multiplication by $a + bi$ by rotating in the opposite direction by the argument and dilating by the reciprocal of the modulus gave us the same results as when we rewrote $\frac{1}{a+bi}$ in rectangular form. This statement appears to be true. In each case, we got the same complex number.

8. Show that the following statement is true when $z = 2 - 2\sqrt{3}i$:

The reciprocal of a complex number z with modulus r and argument θ is $\frac{1}{z}$ with modulus $\frac{1}{r}$ and argument $2\pi - \theta$.

Since $2 - 2\sqrt{3}i$ has modulus 4 and argument $\frac{5\pi}{3}$, we must have

$$\frac{1}{2-2\sqrt{3}i} = \frac{1}{4}\left(\cos\left(\frac{\pi}{3}\right) + i\sin\left(\frac{\pi}{3}\right)\right) = \frac{1}{4}\left(\frac{1}{2} + \frac{\sqrt{3}}{2}i\right) = \frac{1}{8} + \frac{\sqrt{3}}{8}i.$$

Multiplying by 1 using the conjugate, we have

$$\frac{1}{2-2\sqrt{3}i} = \frac{(2+2\sqrt{3}i)}{(2-2\sqrt{3}i)(2+2\sqrt{3}i)} = \frac{2+2\sqrt{3}i}{16} = \frac{1}{8} + \frac{\sqrt{3}}{8}i.$$

Since both methods produce equivalent complex numbers, this statement is true when $z = 2 - 2\sqrt{3}i$.

9. Explain using transformations why $z \cdot \frac{1}{z} = 1$.

The complex number z can be thought of as a rotation of the real number 1 by $\arg(z)$ and a dilation by $|z|$. If we multiply this number by its reciprocal, then we rotate $\arg(z)$ in the opposite direction and dilate by a factor of $\frac{1}{|z|}$. This will return the rotation to 0 and the modulus to 1, which describes the real number 1.

Debrief this section by making sure that students are clear on the geometric effect of multiplication by the reciprocal of a complex number. Explain that this allows us to understand division of complex numbers as transformations as well. A proof that the geometric effect of multiplication by the reciprocal is the same as $\frac{1}{a+bi}$ is provided below.

Let $a + bi = r(\cos(\theta) + i\sin(\theta))$. Then, a complex number whose modulus is $\frac{1}{r}$ and whose argument is $2\pi - \theta$ would be $\frac{1}{r}(\cos(2\pi - \theta) + i\sin(2\pi - \theta))$. We need to show that

$$\frac{1}{a+bi} = \frac{1}{r(\cos(\theta) + i\sin(\theta))}$$

is equivalent to $\frac{1}{r}(\cos(2\pi - \theta) + i\sin(2\pi - \theta))$.

$$\begin{aligned}\frac{1}{a+bi} &= \frac{1}{r(\cos(\theta) + i\sin(\theta))} \\ &= \frac{1}{r(\cos(\theta) + i\sin(\theta))} \cdot \frac{r(\cos(\theta) - i\sin(\theta))}{r(\cos(\theta) - i\sin(\theta))} \\ &= \frac{r(\cos(\theta) - i\sin(\theta))}{r^2(\cos^2(\theta) - i^2\sin^2(\theta))} \\ &= \frac{1}{r} \cdot \frac{(\cos(\theta) - i\sin(\theta))}{\cos^2(\theta) + \sin^2(\theta)}\end{aligned}$$

By the Pythagorean identity,

$$\frac{1}{a+bi} = \frac{1}{r}(\cos(\theta) - i\sin(\theta)).$$

By using identities $\cos(-\theta) = \cos(\theta)$ and $\sin(-\theta) = -\sin(\theta)$ and $\cos(2\pi - \theta) = \cos(-\theta)$ and $\sin(2\pi - \theta) = \sin(-\theta)$, we substitute to get

$$\frac{1}{a+bi} = \frac{1}{r}(\cos(2\pi - \theta) + i\sin(2\pi - \theta)).$$

Scaffolding:

As an alternative to presenting this proof, have students verify the geometric effect of multiplying by the reciprocal of a complex number with specific examples.

- Let $z = 2 + 2i$. For each number below, find $\frac{z}{w}$.

  $w = 1 - i$

  $w = 3i$

  $w = -4 - 4i$

  $w = -5$

Exploratory Challenge 2/Exercise 10 (15 minutes)

This second challenge is a culminating activity that gives students the opportunity to review all the transformations and operations on complex numbers studied in this module. Students should continue working in groups on this activity. After groups have completed the graphic organizer, invite representatives from the different groups to summarize their findings one row per group. If time is an issue, each group may work on only one row. Depending on the size of the class, more than one group may be assigned the same row. Have students prepare and present a poster summarizing their work on their assigned operation. While each group is presenting, students can take notes.

MP.2 & MP.7

In this section, students are making connections between the algebraic structure of complex numbers and the related geometric representations and transformations, which is MP.2 and MP.7.

MP.2 & MP.7

Exploratory Challenge 2/Exercise 10

10. Complete the graphic organizer below to summarize your work with complex numbers so far.

Operation	Geometric Transformation	Example. Illustrate algebraically and geometrically. Let $z = 3 - 3i$ and $w = -2i$.
Addition $z + w$	*Translation of z by w*	$3 - 3i + (-2i) = 3 - 5i$ *You can see that the point $(3, -3)$ has been translated down 2 units.* 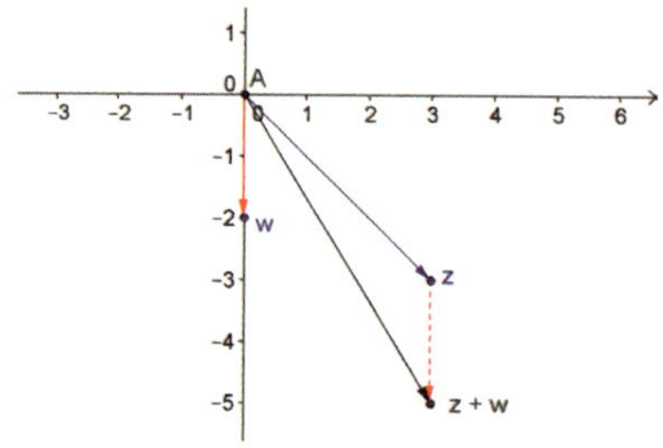
Subtraction $z - w$	*Translation of z by w*	$3 - 3i - (-2i) = 3 - i$ *You can see that the point $(3, -3)$ has been translated up 2 units.* 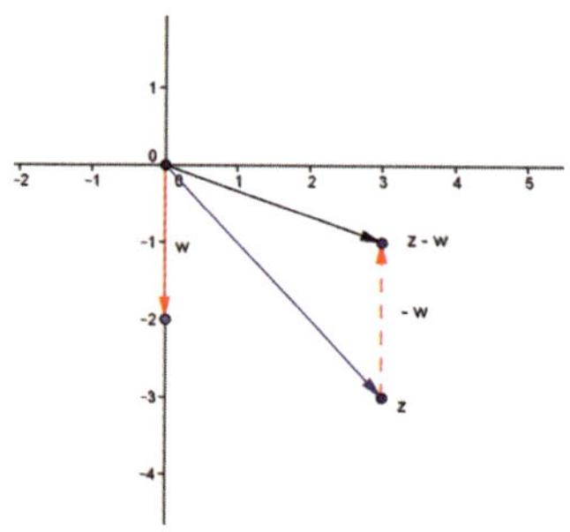
Conjugate of z	*Reflection of z across the real axis*	$\bar{z} = 3 + 3i$. *The point $(3, -3)$ is reflected across the real axis to the point $(3, 3)$.* 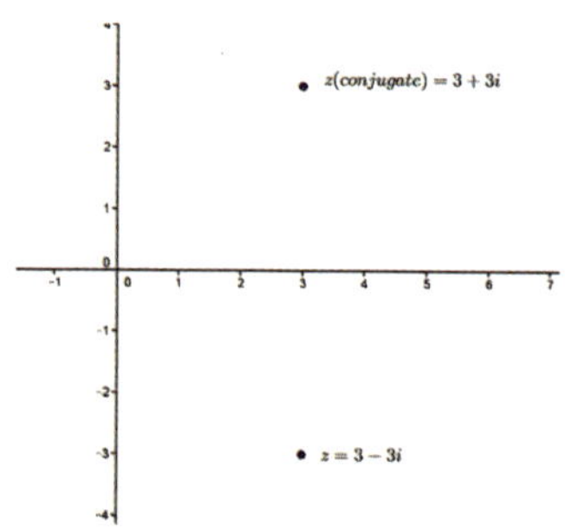

EUREKA MATH

MP.2 & MP.7

Multiplication $w \cdot z$	*Rotation of z by $\arg(w)$ followed by dilation by a factor of $\lvert w \rvert$*	*$\arg(w) = \frac{3\pi}{2}$ and $\lvert w \rvert = 2$. Thus, wz is $3 - 3i$ rotated $\frac{3\pi}{2}$ and dilated by a factor of 2.* *$3 - 3i = 3\sqrt{2}\left(\cos\left(\frac{7\pi}{4}\right) + i\sin\left(\frac{7\pi}{4}\right)\right)$, so the new modulus will be $6\sqrt{2}$, and the new argument will be a number between 0 and 2π that corresponds to a rotation of $\frac{7\pi}{4} + \frac{3\pi}{2} = \frac{13\pi}{4}$. The argument would be $\frac{5\pi}{4}$.* *$-2i(3 - 3i) = -6i - 6 = -6 - 6i$; that does indeed have modulus $6\sqrt{2}$ and argument $\frac{5\pi}{4}$.* 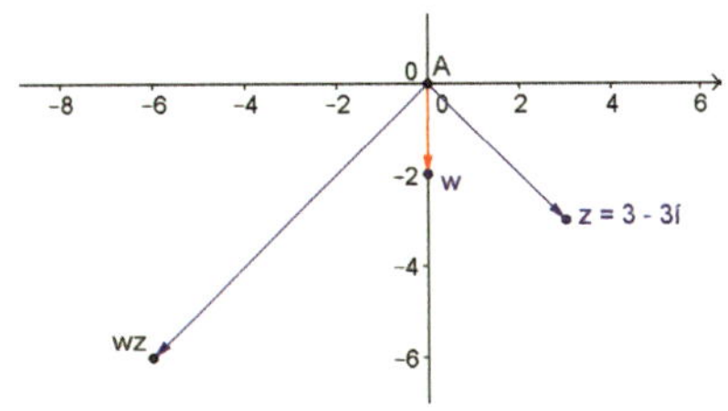
Reciprocal of z	*Rotates the real number 1 the opposite rotation of z and a dilation by the reciprocal of the modulus of z*	*$3 - 3i$ is a rotation of $\frac{7\pi}{4}$ and a dilation by $3\sqrt{2}$ of the real number 1. The reciprocal would be a rotation of $-\frac{7\pi}{4}$ and a dilation by $\frac{1}{3\sqrt{2}}$. The argument of the reciprocal would be $\frac{\pi}{4}$, and the modulus would be $\frac{\sqrt{2}}{6}$.* $\frac{1}{3-3i} = \frac{3+3i}{(3-3i)(3+3i)} = \frac{3+3i}{18} = \frac{1}{6} + \frac{1}{6}i$ *with* $\arg(z) = \frac{\pi}{4}$ *and* $\lvert z \rvert = \frac{1}{6}\sqrt{2}$ 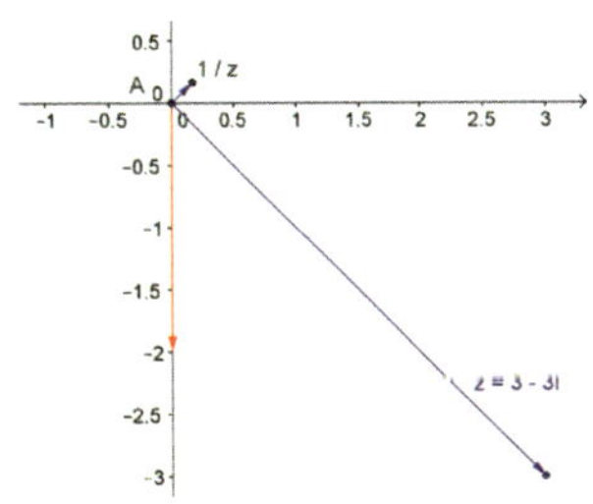

Division $\frac{w}{z}$	*Rotates w by $2\pi - arg(z)$ and followed by a dilation of $\frac{1}{\lvert z \rvert}$*	*Rotation of w by $2\pi - \arg(z) = \frac{\pi}{4}$ and dilation by $\frac{\sqrt{2}}{6}$ would result in a complex number whose argument was $\frac{3\pi}{2} + \frac{\pi}{4} = \frac{7\pi}{4}$ and a modulus of $\frac{\sqrt{2}}{6} \cdot 2 = \frac{\sqrt{2}}{3}$.* $\frac{w}{z} = \frac{-2i}{3-3i} = \frac{-2i(3+3i)}{(3-3i)(3+3i)} = \frac{-6i+6}{18} = \frac{1}{3} - \frac{1}{3}i$ $\left\lvert \frac{w}{z} \right\rvert = \frac{1}{3}\sqrt{2}$ *and* $\arg\left(\frac{w}{z}\right) = \frac{7\pi}{4}$

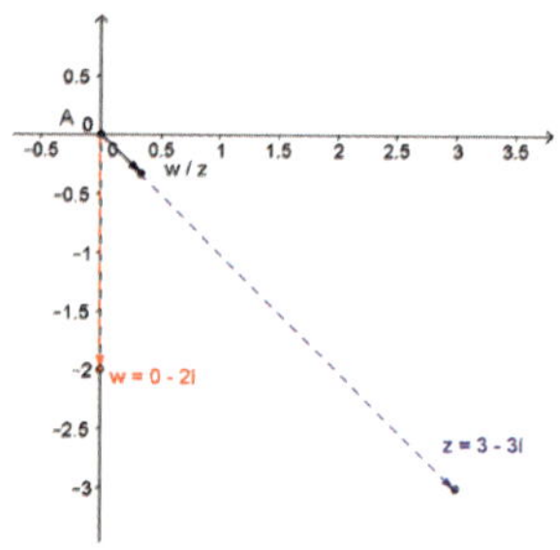

Exercises 11–13 (7 minutes)

Students should work on these problems with a partner. Ask each student to explain one problem to their partners to check for understanding. Then, invite one or two students to share their results on the board.

Exercises 11–13

Let $z = -1 + i$, and let $w = 2i$. Describe each complex number as a transformation of z, and then write the number in rectangular form.

11. $w\bar{z}$

z is reflected across the real axis; then, that number is rotated by $\arg(w)$ and dilated by $\lvert w \rvert$.

$\arg(z) = \frac{3\pi}{4}$ and $\lvert z \rvert = \sqrt{2}$. $\arg(\bar{z}) = \frac{5\pi}{4}$ with the same modulus as z. Rotation by $\arg(w) = \frac{\pi}{2}$ and dilation by 2 would give a complex number with argument of $\frac{5\pi}{4} + \frac{\pi}{2} = \frac{7\pi}{4}$ and modulus of $2\sqrt{2}$, which is the modulus and argument of the number shown below.

$$w\bar{z} = 2i(-1 - i) = 2 - 2i$$

12. $\frac{1}{\bar{z}}$

z is reflected across the real axis and then rotated $2\pi - \arg(\bar{z})$ and dilated by $\frac{1}{\lvert \bar{z} \rvert}$. The result is a dilation of z.

Reflection of z across the real axis results in a complex number with argument $\frac{5\pi}{4}$ and modulus of $\sqrt{2}$. The reciprocal has argument $2\pi - \frac{5\pi}{4} = \frac{3\pi}{4}$ and modulus $\frac{1}{\sqrt{2}} = \frac{\sqrt{2}}{2}$. This number has the same argument as z and is a dilation by a factor of $\frac{1}{2}$.

$$\frac{1}{\bar{z}} = \frac{1}{-1-i} = \frac{-1+i}{(-1-i)(-1+i)} = \frac{-1+i}{2} = -\frac{1}{2} + \frac{i}{2}$$

> 13. $\overline{w+z}$
>
> ***z is translated by w vertically 2 units up since the real part of w is 0 and the imaginary part is 2. This new number is reflected across the real axis.***
>
> $$w+z=2i-1+i=-1+3i$$
>
> $$\overline{w+z}=-1-3i$$

Closing (3 minutes)

The graphic organizer students made in Exploratory Challenge 2 function as a summary for this lesson. Invite students to answer the following questions in writing or to discuss them with a partner.

> *Scaffolding:*
>
> If needed, make the closing questions more concrete by specifying specific complex numbers for z and w.

- What is the geometric effect of complex number division (multiplication of z by $\frac{1}{w}$)?
  - *The number z is rotated $2\pi - \arg(w)$ and dilated by $\frac{1}{|w|}$.*
- How are the modulus and argument of the complex number $\frac{1}{z}$ related to the complex number z?
  - *The modulus of $\frac{1}{z}$ is $\frac{1}{r}$, and the argument of $\frac{1}{z}$ is $2\pi - \arg(z)$, which is the same as rotation of $-\arg(z)$.*

Exit Ticket (5 minutes)

Name ______________________________ Date ______________

Lesson 17: The Geometric Effect of Multiplying by a Reciprocal

Exit Ticket

Let $z = 1 + \sqrt{3}i$ and $w = \sqrt{3} - i$. Describe each complex number as a transformation of z, and then write the number in rectangular form, and identify its modulus and argument.

1. $\dfrac{z}{w}$

2. $\dfrac{1}{wz}$

Exit Ticket Sample Solutions

Let $z = 1 + \sqrt{3}i$ and $w = \sqrt{3} - i$. Describe each complex number as a transformation of z, and then write the number in rectangular form, and identify its modulus and argument.

1. $\dfrac{z}{w}$

z is rotated by $-\arg(w)$ and dilated by $\dfrac{1}{|w|}$.

$\arg(z) = \dfrac{\pi}{3}$ and $2\pi - \arg(w) = 2\pi - \dfrac{11\pi}{6} = \dfrac{\pi}{6}$. So, division by w should rotate z to $\dfrac{\pi}{2}$. $|z| = 2$, and $\dfrac{1}{|w|} = \dfrac{1}{2}$, so the modulus of $\dfrac{z}{w}$ should be $2 \cdot \dfrac{1}{2} = 1$. This rotation and dilation describe the complex number i. Algebraically, we get the same number.

$$\frac{z}{w} = \frac{1+\sqrt{3}i}{\sqrt{3}-i} = \frac{1+\sqrt{3}i}{\sqrt{3}-i} \cdot \frac{\sqrt{3}+i}{\sqrt{3}+i} = \frac{4i}{4} = i$$

2. $\dfrac{1}{wz}$

z is rotated $\arg(w)$ and dilated by $|w|$ and then rotated $-\arg(wz)$ and dilated by $\dfrac{1}{|wz|}$. For the given values of z and w, this transformation results in a dilation of w by a factor of $\dfrac{1}{4}$.

$\arg(z) = \dfrac{\pi}{3}$ and $\arg(w) = \dfrac{11\pi}{6}$. Adding these arguments and finding an equivalent rotation between 0 and 2π gives a rotation of $\dfrac{\pi}{6}$ and $|wz| = 2 \cdot 2 = 4$. This describes the complex number $2\sqrt{3} + 2i$. The reciprocal of this number has argument $\dfrac{11\pi}{6}$ and modulus $\dfrac{1}{4}$, which describes the complex number $\dfrac{\sqrt{3}}{8} - \dfrac{1}{8}i$.

$$\frac{1}{wz} = \frac{1}{(\sqrt{3}-i)(1+\sqrt{3}i)} = \frac{1}{2\sqrt{3}+2i} = \frac{1}{2\sqrt{3}+2i} \cdot \frac{2\sqrt{3}-2i}{2\sqrt{3}-2i} = \frac{2\sqrt{3}-2i}{16} = \frac{\sqrt{3}}{8} - \frac{1}{8}i$$

Problem Set Sample Solutions

1. **Describe the geometric effect of multiplying z by the reciprocal of each complex number listed below.**

   a. $w_1 = 3i$

   $\arg(w_1) = \dfrac{\pi}{2}$ and $|w_1| = 3$

   z is rotated by $2\pi - \arg(w_1)$, which is $2\pi - \dfrac{\pi}{2} = \dfrac{3\pi}{2}$, and dilated by $\dfrac{1}{3}$.

   b. $w_2 = -2$

   $\arg(w_2) = \pi$ and $|w_2| = 2$

   z is rotated by $2\pi - \arg(w_2)$, which is $2\pi - \pi = \pi$, and dilated by $\dfrac{1}{2}$.

c. $w_3 = \sqrt{3} + i$

$\arg(w_3) = \frac{\pi}{6}$ *and* $|w_3| = 2$

z is rotated by $2\pi - \arg(w_3)$, *which is* $2\pi - \frac{\pi}{6} = \frac{11\pi}{6}$, *and dilated by* $\frac{1}{2}$.

d. $w_4 = 1 - \sqrt{3}i$

$\arg(w_4) = \frac{5\pi}{3}$ *and* $|w_4| = 2$

z is rotated by $2\pi - \arg(w_4)$, *which is* $2\pi - \left(\frac{5\pi}{3}\right) = \frac{\pi}{3}$, *and dilated by* $\frac{1}{2}$.

2. **Let $z = -2 - 2\sqrt{3}i$. Show that the geometric transformations you described in Problem 1 really produce the correct complex number by performing the indicated operation and determining the argument and modulus of each number.**

a. $\frac{-2-2\sqrt{3}i}{w_1}$

$$\frac{z}{w_1} = \frac{-2-2\sqrt{3}i}{3i} = \frac{-2-2\sqrt{3}i}{3i} \cdot \frac{-3i}{-3i} = \frac{-6\sqrt{3}+6i}{9} = -\frac{2\sqrt{3}}{3} + \frac{2}{3}i, \left|\frac{z}{w_1}\right| = \frac{4}{3}, \arg\left(\frac{z}{w_1}\right) = \frac{5\pi}{6}$$

$|z| = 4$, *so the result of division is a complex number whose modulus is* $\frac{1}{3}$ *of* 4.

$\arg(z) = \frac{4\pi}{3}$, *so the result of division by a complex number is whose argument represents the same rotation as* $\frac{4\pi}{3} + \frac{3\pi}{2} = \frac{17\pi}{6}$, *which would be* $\frac{5\pi}{6}$.

b. $\frac{-2-2\sqrt{3}i}{w_2}$

$$\frac{z}{w_2} = \frac{-2-2\sqrt{3}i}{-2} = 1 + \sqrt{3}i, \left|\frac{z}{w_2}\right| = 2, \arg\left(\frac{z}{w_2}\right) = \frac{\pi}{3}$$

$|z| = 4$ *and* $\frac{1}{2}$ *of* 4 *is* 2.

$\arg(z) = \frac{4\pi}{3}$, *so the result of division will rotate z by* $-\frac{\pi}{2}$ *and* $\frac{4\pi}{3} - \frac{\pi}{2} = \frac{5\pi}{6}$.

c. $\frac{-2-2\sqrt{3}i}{w_3}$

$$\frac{z}{w_3} = \frac{-2-2\sqrt{3}i}{\sqrt{3}+i} = \frac{-2\sqrt{3}-6i+2i-2\sqrt{3}}{4} = -\sqrt{3} - i, \left|\frac{z}{w_3}\right| = 2, \arg\left(\frac{z}{w_3}\right) = \frac{7\pi}{6}$$

$|z| = 4$ *and* $\frac{1}{2}$ *of* 4 *is* 2.

$\arg(z) = \frac{4\pi}{3}$, *so the result of division will rotate z by* $-\frac{\pi}{6}$ *and* $\frac{4\pi}{3} - \frac{\pi}{6} = \frac{7\pi}{6}$.

d. $\frac{-2-2\sqrt{3}i}{w_4}$

$\frac{z}{w_4} = \frac{-2-2\sqrt{3}i}{1-\sqrt{3}i} = \frac{-2-2\sqrt{3}i-2\sqrt{3}i+6}{4} = 1-\sqrt{3}i, \left|\frac{z}{w_4}\right| = 2, \arg\left(\frac{z}{w_4}\right) = \frac{5\pi}{3}$

$|z| = 4$, and $\frac{1}{2}$ of 4 is 2.

$\arg(z) = \frac{4\pi}{3}$, so the result of division will rotate z by $-\frac{5\pi}{3}$, and $\frac{4\pi}{3} - \frac{5\pi}{3} = -\frac{\pi}{3}$. A rotation of $-\frac{\pi}{3}$ will be the same as a rotation of $\frac{5\pi}{3}$, which is the argument of the quotient.

3. In Exercise 12 of this lesson, you described the complex number $\frac{1}{z}$ as a transformation of z for a specific complex number z. Show that this transformation always produces a dilation of $z = a + bi$.

$\bar{z} = a - bi$, and $\frac{1}{a-bi} = \frac{1}{a-bi} \cdot \frac{a+bi}{a+bi} = \frac{a+bi}{a^2+b^2} = \frac{1}{a^2+b^2}(a+bi)$. This complex number is the product of a real number and the original complex number z, so it will have the same argument as $a + bi$, but the modulus will be a different number.

4. Does $L(z) = \frac{1}{z}$ satisfy the conditions that $L(z+w) = L(z) + L(w)$ and $L(mz) = mL(z)$, which makes it a linear transformation? Justify your answer.

$\frac{1}{z} = \frac{1}{a+bi} = \frac{a-bi}{a^2+b^2}$, which is a complex number whose real part is $\frac{a}{a^2+b^2}$ and whose imaginary part is $-\frac{b}{a^2+b^2}$. Since all complex numbers satisfy the conditions that make them a linear transformation, and $\frac{1}{z}$ is a complex number, it will also be a linear transformation.

5. Show that $L(z) = w\left(\frac{1}{w}z\right)$ describes a reflection of z about the line containing the origin and w for $z = 3i$ and $w = 1 + i$.

$L(z) = (1+i)\left(\frac{1}{1+i}(3i)\right) = (1+i)\left(\frac{-3i}{1-i}\right) = \frac{-3i(1+i)(1+i)}{(1-i)(1+i)} = \frac{6}{2} = 3$, which is the image of the transformation z that is reflected about the line containing the origin and w.

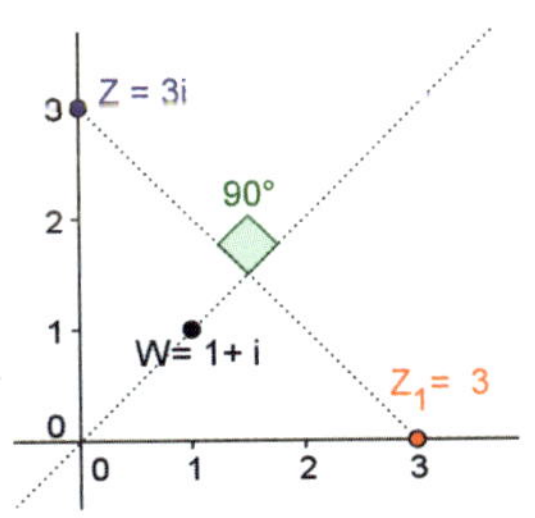

6. Describe the geometric effect of each transformation function on z where z, w, and a are complex numbers.

a. $L_1(z) = \frac{z-w}{a}$

z is translated by w, rotated by $2\pi - \arg(a)$, and dilated by $\frac{1}{|a|}$.

b. $L_2(z) = \overline{\left(\frac{z-w}{a}\right)}$

z is translated by w, reflected about the real axis, rotated by $2\pi - \arg(a)$, and dilated by $\frac{1}{|a|}$, and reflected about the real axis.

c. $L_3(z) = a\overline{\left(\frac{z-w}{a}\right)}$

z is translated by w, reflected about the real axis, rotated by $2\pi - \arg(a)$, and dilated by $\frac{1}{|a|}$, and reflected about the real axis, rotated by $\arg(a)$, and dilated by $|a|$.

d. $L_3(z) = a\overline{\left(\frac{z-w}{a}\right)} + w$

z is translated by w, reflected about the real axis, rotated by $2\pi - \arg(a)$, and dilated by $\frac{1}{|a|}$, and reflected about the real axis, rotated by $\arg(a)$, dilated by $|a|$, and translated by w.

7. Verify your answers to Problem 6 if $z = 1 - i$, $w = 2i$, and $a = -1 - i$.

a. $L_1(z) = \frac{z-w}{a}$

$|z| = \sqrt{2}, \arg(z) = \frac{7\pi}{4}, |w| = 2, \arg(w) = \pi, |a| = \sqrt{2}, \arg(a) = \frac{5\pi}{4} = 3.927$

$\frac{z-w}{a} = \frac{1-i-2i}{-1-i} = \frac{1-3i}{-1-i} = 1+2i, \left|\frac{z-w}{a}\right| = \sqrt{5}, \arg\left(\frac{z-w}{a}\right) = 1.107$

$z - w = 1 - 3i, |z-w| = \sqrt{10}, \arg(z-w) = 5.034$

$\frac{1}{|a|} \times |z-w| = \frac{1}{\sqrt{2}} \times \sqrt{10} = \sqrt{5} = \left|\frac{z-w}{a}\right|,$

$\arg\left(\frac{z-w}{a}\right) = \arg(z-w) + (2\pi - \arg(a)) = 5.034 + 2\pi - 3.927 = 1.107 + 2\pi = 1.107$

b. $L_2(z) = \overline{\left(\frac{z-w}{a}\right)}$

$\overline{\left(\frac{z-w}{a}\right)} = \overline{\left(\frac{1-3\imath}{-1-\imath}\right)} = \overline{(1+2\imath)} = 1-2i, \left|\overline{\left(\frac{z-w}{a}\right)}\right| = \sqrt{5}, \arg\left(\overline{\left(\frac{z-w}{a}\right)}\right) = -1.107$

$\overline{\left(\frac{1}{|a|}\right)} \times \overline{|z-w|} = \frac{1}{\sqrt{2}} \times \sqrt{10} = \sqrt{5} = \left|\overline{\left(\frac{z-w}{a}\right)}\right|$

$\arg\left(\overline{\left(\frac{z-w}{a}\right)}\right) = 2\pi - \left(\arg(z-w) + (2\pi - \arg(a))\right) = 2\pi - 5.034 - 2\pi + 3.927 = -1.107$

c. $L_3(z) = a\overline{\left(\frac{z-w}{a}\right)}$

$a \times \overline{\left(\frac{z-w}{a}\right)} = (-1-i)\overline{\left(\frac{1-3i}{-1-i}\right)} = (-1-i)\overline{(1+2i)} = (-1-i)(1-2i) = -3+i,$

$\left|a \times \overline{\left(\frac{z-w}{a}\right)}\right| = \sqrt{10}, \arg\left((-1-i)\overline{\left(\frac{z-w}{a}\right)}\right) = \pi - 0.322 = 2.820$

$|a| \times \overline{\left(\frac{1}{|a|}\right)} \times \overline{|z-w|} = \sqrt{2} \times \frac{1}{\sqrt{2}} \times \sqrt{10} = \sqrt{10} = \left|(-1-i)\overline{\left(\frac{z-w}{a}\right)}\right|$

$\arg\left(a \times \overline{\left(\frac{z-w}{a}\right)}\right) = \arg(a) + \left(\arg(z-w) + \left(2\pi - \arg(a)\right)\right)$

$= 3.927 + 2\pi - 5.034 - 2\pi + 3.927 = 2.280$

d. $L_3(z) = a\overline{\left(\frac{z-w}{a}\right)} + w$

$a \times \overline{\left(\frac{z-w}{a}\right)} + w = (-1-i)\overline{\left(\frac{1-3i}{-1-i}\right)} + 2i = -3+i+2i = -3+3i,$

$\left|a \times \overline{\left(\frac{z-w}{a}\right)} + w\right| = 3\sqrt{2}, \arg\left((-1-i)\overline{\left(\frac{z-w}{a}\right)} + w\right) = \frac{3\pi}{4} = 2.356$

$\left|a \times \overline{\left(\frac{z-w}{a}\right)} + w\right| = |-3+i+2i| = 3\sqrt{2} =,$

$\arg\left(a \times \overline{\left(\frac{z-w}{a}\right)} + w\right) = \arg(-3+i+2i) = \arg(-3+3i) = 2.356$

Name ______________________________ Date ______________

1. Given $z = 3 - 4i$ and $w = -1 + 5i$:

   a. Find the distance between z and w.

   b. Find the midpoint of the segment joining z and w.

2. Let $z_1 = 2 - 2i$ and $z_2 = (1 - i) + \sqrt{3}(1 + i)$.

   a. What is the modulus and argument of z_1?

   b. Write z_1 in polar form. Explain why the polar and rectangular forms of a given complex number represent the same number.

c. Find a complex number w, written in the form $w = a + ib$, such that $wz_1 = z_2$.

d. What is the modulus and argument of w?

e. Write w in polar form.

f. When the points z_1 and z_2 are plotted in the complex plane, explain why the angle between z_1 and z_2 measures $\arg(w)$.

g. What type of triangle is formed by the origin and the two points represented by the complex numbers z_1 and z_2? Explain how you know.

h. Find the complex number, v, closest to the origin that lies on the line segment connecting z_1 and z_2. Write v in rectangular form.

3. Let z be the complex number $2 + 3i$ lying in the complex plane.

   a. What is the conjugate of z? Explain how it is related geometrically to z.

   b. Write down the complex number that is the reflection of z across the vertical axis. Explain how you determined your answer.

Let m be the line through the origin of slope $\frac{1}{2}$ in the complex plane.

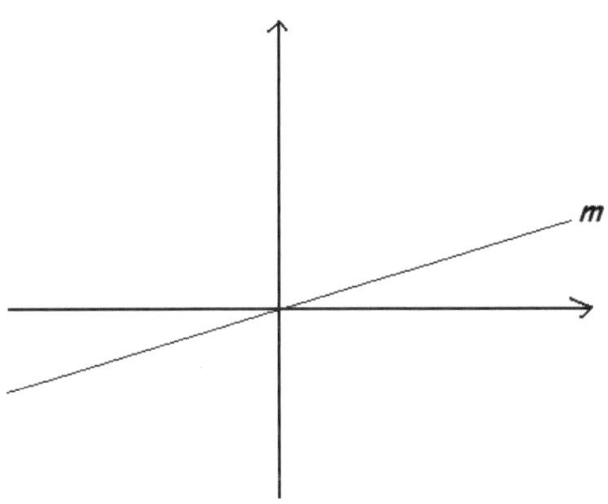

   c. Write down a complex number, w, of modulus 1 that lies on m in the first quadrant in rectangular form.

d. What is the modulus of wz?

e. Explain the relationship between wz and z. First, use the properties of modulus to answer this question, and then give an explanation involving transformations.

f. When asked,
"What is the argument of $\frac{1}{w}z$?"

Paul gave the answer $\arctan\left(\frac{3}{2}\right) - \arctan\left(\frac{1}{2}\right)$, which he then computed to two decimal places.

Provide a geometric explanation that yields Paul's answer.

g. When asked,

"What is the argument of $\frac{1}{w}z$?"

Mable did the complex number arithmetic and computed $z \div w$.

She then gave an answer in the form $\arctan\left(\frac{a}{b}\right)$ for some fraction $\frac{a}{b}$. What fraction did Mable find? Up to two decimal places, is Mable's final answer the same as Paul's?

A Progression Toward Mastery					
Assessment Task Item		**STEP 1** Missing or incorrect answer and little evidence of reasoning or application of mathematics to solve the problem.	**STEP 2** Missing or incorrect answer but evidence of some reasoning or application of mathematics to solve the problem.	**STEP 3** A correct answer with some evidence of reasoning or application of mathematics to solve the problem, OR an incorrect answer with substantial evidence of solid reasoning or application of mathematics to solve the problem.	**STEP 4** A correct answer supported by substantial evidence of solid reasoning or application of mathematics to solve the problem.
1	**a** N-CN.B.6	Student provides an incorrect response, and there is no evidence to support that student understands how to compute the distance.	Student shows some knowledge of the distance formula but does not use the formula correctly.	Student uses the formula correctly but makes minor mathematical mistakes. OR Student computes distance correctly with no supporting work.	Student computes the distance correctly with supporting work shown.
	b N-CN.B.6	Student provides an incorrect response, and there is no evidence to support that student understands how to compute the midpoint.	Student shows some knowledge of the midpoint formula but does not use the formula correctly.	Student uses the formula correctly but makes minor mathematical mistakes. OR Student computes the midpoint correctly with no supporting work.	Student computes the midpoint correctly with supporting work shown.
2	**a** N-CN.B.4	Student provides an incorrect response, and there is no evidence to support that student understands how to compute the modulus and argument.	Student uses the correct method and answer for either the modulus or the argument. OR Student uses the correct method for both but makes minor errors.	Student uses the correct methods for both, but either the modulus or argument is incorrect due to a minor error. OR Student gives correct answers for both with no supporting work shown.	Student computes modulus and argument correctly. Work is shown to support the answer.
	b N-CN.B.4	Student shows no knowledge of polar form.	Student attempts to put the complex number into polar form but makes errors.	Student writes the correct polar form but does not explain why polar and rectangular forms represent the same number.	Student writes the correct polar form and correctly explains why polar and rectangular forms represent the same number.

	c **N-CN.A.3**	Student does not attempt to divide z_2 by z_1 to find w.	Student attempts to divide z_2 by z_1 without applying the correct algorithm involving multiplication by 1 in the form of the conjugate z_1 divided by the conjugate of z_1.	Student applies the division algorithm correctly and shows work but has minor mathematical errors leading to an incorrect final answer.	Student applies the division algorithm correctly and gives a correct answer with sufficient work shown to demonstrate understanding of the process.
	d **N-CN.B.5** **N-CN.B.6**	Student does not compute either answer correctly, nor is there evidence to support that student understands how to compute the modulus and argument.	Student computes the modulus or argument correctly. OR Student uses correct methods for both but arrives at incorrect answers due to minor errors.	Student uses correct methods for both, but either the modulus or the argument is incorrect due to a minor error. OR Student gives correct answers for both with no supporting work shown.	Student computes modulus and argument correctly for the answer to part (b), and work is shown to support answer. Note: Student can earn full points for this even if the answer to part (b) is incorrect.
	e **N-CN.B.4**	Student shows no knowledge of polar form.	Student shows some knowledge of polar form but does make major mathematical errors.	Student shows knowledge of polar form but makes minor mathematical errors.	Student writes the correct polar form of the number.
	f **N-CN.B.4** **N-CN.B.5**	Student makes little or no attempt to identify the angle measure.	Student cannot determine the angle between the two complex numbers whose vertex is at the origin. Student provides an explanation that fails to connect multiplication with transformations. The explanation may include a comparison of the moduli of z_1 and z_2. The answer may include a sketch to support the answer.	Student may not indicate the requested angle measurement or gives an incorrect angle measurement. Student provides an explanation that does not fully address the connection between multiplication and transformations but may include a comparison of the moduli of z_1 and z_2. The answer may be supported with a sketch.	Student explanation clearly connects multiplication with the correct transformations by explaining that z_2 is the image of z_1 achieved by rotating z_1 by the $\arg(w)$ and no dilation since $\|w\| = 1$. The answer may be supported with a sketch.
	g **N-CN.B.4** **N-CN.B.5**	Student makes little or no attempt to identify the triangle.	Student may identify the triangle as isosceles but with little or no explanation.	Student identifies the triangle as isosceles or equilateral but gives an incomplete or incorrect explanation.	Student identifies the triangle as equilateral and gives a complete explanation.

	h **N-CN.B.4** **N-CN.B.6**	Student makes little or no attempt to find v.	Student may attempt to sketch the situation, but more than one misconception or mathematical error leads to an incorrect or incomplete solution.	Student finds v correctly, but there is little or no explanation or work explaining why v is the midpoint of the line segment. OR Student identifies that v would be at the midpoint but fails to compute it correctly.	Student averages z_1 and z_2 to find v and clearly explains why using a geometric argument regarding v's location on the perpendicular bisector of the triangle.
3	**a** **N-CN.A.3** **N-CN.B.5**	Student incorrectly answers both the real and the imaginary part.	Student incorrectly answers either the real or the imaginary part.	Student gives the correct answer but does not give an explanation.	Student gives the correct answer and includes the correct explanation of how it is geometrically related to z.
	b **N-CN.A.3** **N-CN.B.5**	Student incorrectly answers both the real and the imaginary part.	Student incorrectly answers either the real or the imaginary part.	Student gives the correct answer but does not give an explanation.	Student gives the correct answer and explains that the real part is the opposite but that the imaginary part stays the same.
	c **N-CN.B.4**	Student provides an answer that is not a complex number. OR Student provides an answer that is a complex number with both an incorrect modulus and argument. There is little or no supporting work shown.	Student gives an incorrect answer with little evidence of correct reasoning (solution may fail to address modulus of 1 but be a complex number on the line m or may have a modulus of 1 but not be a complex number on the line m).	Student gives the correct answer with limited reasoning or work to support the answer. OR Student uses correct reasoning, but minor mathematical errors lead to an incorrect solution.	Student gives a correct answer with work shown to support approach. Student reasoning could use the polar form of a complex number or apply proportional reasoning to find w with the correct argument and modulus.
	d **N-CN.B.4**	Student shows no knowledge of wz or the modulus.	Student shows some knowledge of calculating the modulus, but the answer is incorrect with little supporting work shown.	Student makes minor mathematical errors in calculating the modulus.	Student gives a correct answer with supporting work shown clearly.
	e **N-CN.B.4**	Student does not explain the relationship between z and wz.	Student gives minimal explanation with some mistakes.	Student explains connection but does not include transformations that occur.	Student gives clear and correct explanation that includes transformations that occur.

	f **N-CN.B.5** **N-CN.B.6**	Student shows little or no work. The explanation fails to address transformations in a meaningful way.	Student provides an explanation that may include references to rotations and dilations but does not clearly address the fact that multiplication by $\frac{1}{w}$ represents a clockwise rotation of $\arg(w)$.	Student identifies $\arg(z)$ and $\arg(w)$ and explains that multiplication by $\frac{1}{w}$ would create a clockwise rotation, but the explanation lacks a clear reason why the two should be subtracted or contains other minor errors.	Student explains both the rotation and lack of dilation correctly since $\lvert w \rvert = 1$. In the explanation, student identifies $\arg(z)$ and $\arg(w)$ and supports the reason for the difference. A sketch may be included.
	g **N-CN.A.3** **N-CN.B.5**	Student answers both $\frac{z}{w}$ and $\arg\left(\frac{z}{w}\right)$ incorrectly due to major misconceptions or calculation errors. Student may argue incorrectly that the arguments in parts (e) and (f) should be different.	Student answers either $\frac{z}{w}$ or $\arg\left(\frac{z}{w}\right)$ incorrectly due to mathematical errors. Student indicates that both arguments should be the same.	Student correctly computes $\frac{z}{w}$ and $\arg\left(\frac{z}{w}\right)$ but fails to compare the arguments in parts (d) and (e) or explain why they should be the same.	Student correctly computes both arguments to two decimal places. Work shown uses appropriate notation and sufficient steps to follow the solution.

Name ______________________________ Date ______________

1. Given $z = 3 - 4i$ and $w = -1 + 5i$:

   a. Find the distance between z and w.

$$\begin{aligned} \text{Distance} &= \sqrt{(3-(-1))^2 - (-1) + ((-4) - 5)^2} \\ &= \sqrt{4^2 + (-9)^2} \\ &= \sqrt{16+81} \\ &= \sqrt{97} \end{aligned}$$

   b. Find the midpoint of the segment joining z and w.

$$\begin{aligned} \text{Midpoint} &= \frac{3+(-1)}{2} + \frac{(-4)+5}{2}i \\ &= \frac{2}{2} + \frac{1}{2}i \\ &= 1 + \frac{1}{2}i \end{aligned}$$

2. Let $z_1 = 2 - 2i$ and $z_2 = (1 - i) + \sqrt{3}(1 + i)$.

   a. What is the modulus and argument of z_1?

$$|z_1| = \sqrt{(2)^2 + (-2)^2} = \sqrt{8} = 2\sqrt{2}$$

$$\arg(z_1) = \tan^{-1}\left(\frac{-2}{2}\right) = \frac{-\pi}{4}$$

   b. Write z_1 in polar form. Explain why the polar and rectangular forms of a given complex number represent the same number.

$$z_1 = 2\sqrt{2}\left[\cos\left(\frac{-\pi}{4}\right) + i\sin\left(\frac{-\pi}{4}\right)\right]$$

The modulus represents the distance from the origin to the point. The degree of rotation is the angle from the x-axis. When the polar form is expanded, the result is the rectangular form of a complex number.

c. Find a complex number w, written in the form $w = a + ib$, such that $wz_1 = z_2$.

$$z_2 = 1 - i + \sqrt{3} + i\sqrt{3} = (\sqrt{3} + 1) + (\sqrt{3} - 1)i$$

$wz_1 = w_2$ *implies that*

$$w = \frac{z_2}{z_1} = \frac{[(\sqrt{3}+1) + (\sqrt{3}-1)i]}{(2-2i)} \times \frac{(2+2i)}{(2+2i)}$$
$$= \frac{2\sqrt{3} + 2 + 2i\sqrt{3} + 2i + 2i\sqrt{3} - 2i - 2\sqrt{3} + 2}{4+4}$$
$$= \frac{4 + 4i\sqrt{3}}{8}$$
$$= \frac{1}{2} + \frac{\sqrt{3}}{2}i.$$

d. What is the modulus and argument of w?

$$|w| = \sqrt{\left(\frac{1}{2}\right)^2 + \left(\frac{\sqrt{3}}{2}\right)^2} = \sqrt{\frac{1}{4} + \frac{3}{4}} = 1$$

$$arg(w) = tan^{-1}\left(\frac{\frac{\sqrt{3}}{2}}{\frac{1}{2}}\right) = \frac{\pi}{3}$$

e. Write w in polar form.

$$w = 1\left[cos\left(\frac{\pi}{3}\right) + i\,sin\left(\frac{\pi}{3}\right)\right]$$

f. When the points z_1 and z_2 are plotted in the complex plane, explain why the angle between z_1 and z_2 measures $\arg(w)$.

Since $z_2 = wz_1$, *then* z_2 *is the transformation of* z_1 *rotated counterclockwise by* $arg(w)$, *which is* $\frac{\pi}{3}$.

g. What type of triangle is formed by the origin and the two points represented by the complex numbers z_1 and z_2? Explain how you know.

Since $|w| = 1$ and $\arg(w) = \frac{\pi}{3}$, the triangle formed by the origin and the points representing z_1 and z_2 will be equilateral. All of the angles are $60°$ in this triangle.

h. Find the complex number, v, closest to the origin that lies on the line segment connecting z_1 and z_2. Write v in rectangular form.

The point that represents v is the midpoint of the segment connecting z_1 and z_2 since it must be on the perpendicular bisector of the triangle with vertex at the origin.

To find the midpoint, average z_1 and z_2.

$$v = \frac{\sqrt{3}+1+2}{2} + \frac{\sqrt{3}-1-2}{2}i = \frac{\sqrt{3}+3}{2} + \frac{\sqrt{3}-3}{2}i$$

3. Let z be the complex number $2 + 3i$ lying in the complex plane.

   a. What is the conjugate of z? Explain how it is related geometrically to z.

   $$\bar{z} = 2 - 3i$$

   This number is the conjugate of z and the reflection of z across the horizontal axis.

   b. Write down the complex number that is the reflection of z across the vertical axis. Explain how you determined your answer.

   This number is $-2+3i$. The real coordinate has the opposite sign, but the imaginary part keeps the same sign. This means a reflection across the imaginary (vertical) axis.

Let m be the line through the origin of slope $\frac{1}{2}$ in the complex plane.

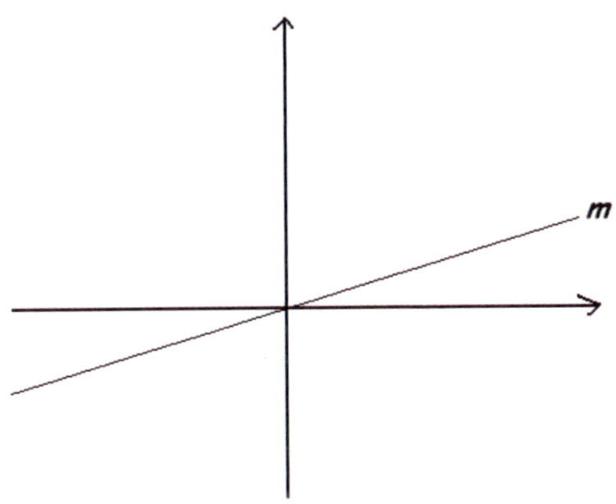

   c. Write down a complex number, w, of modulus 1 that lies on m in the first quadrant in rectangular form.

   Because the slope of m is $\frac{1}{2}$, the argument of w is $\tan^{-1}\left(\frac{1}{2}\right)$.

   Using the polar form of w,

   $w = 1\left[\cos\left(\tan^{-1}\left(\frac{1}{2}\right)\right) + i\sin\left(\tan^{-1}\left(\frac{1}{2}\right)\right)\right]$. From the triangle shown below, $w = \frac{2}{\sqrt{5}} + \frac{1}{\sqrt{5}}i$.

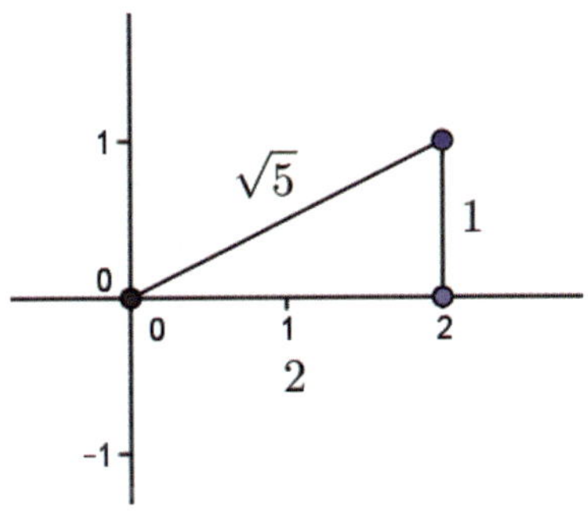

d. What is the modulus of wz?

The modulus of wz is $\sqrt{13}$.

e. Explain the relationship between wz and z. First, use the properties of modulus to answer this question, and then give an explanation involving transformations.

The modulus of wz is $\sqrt{13}$ and is the same as $|z|$.

Using the properties of modulus,

$$|wz| = |w| \times |z| = 1 \times |z| = 1 \times \sqrt{2^2 + 3^2} = \sqrt{13}.$$

Geometrically, multiplying by w will rotate z by the $\arg(w)$ and dilate z by $|w|$. Since $|w| = 1$, the transformation is a rotation only, so both w and z are the same distance from the origin.

f. When asked,

"What is the argument of $\frac{1}{w}z$?"

Paul gave the answer $\arctan\left(\frac{3}{2}\right) - \arctan\left(\frac{1}{2}\right)$, which he then computed to two decimal places. Provide a geometric explanation that yields Paul's answer.

The product $\frac{1}{w}z$ would result in a clockwise rotation of z by the $\arg(w)$. There would be no dilation since $|w| = 1$.

$$\arg(z) = \tan^{-1}\left(\frac{3}{2}\right)$$

$$\arg(w) = \tan^{-1}\left(\frac{1}{2}\right)$$

$$\arg\left(\frac{1}{w}z\right) = \tan^{-1}\left(\frac{3}{2}\right) - \tan^{-1}\left(\frac{1}{2}\right)$$

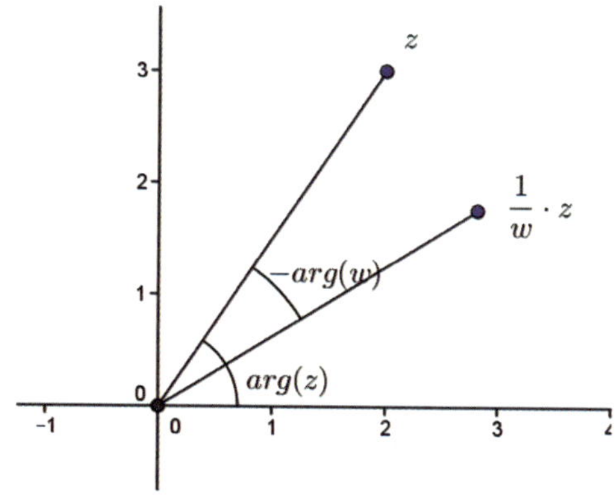

g. When asked,

"What is the argument of $\frac{1}{w}z$?"

Mable did the complex number arithmetic and computed $z \div w$. She then gave an answer in the form $\arctan\left(\frac{a}{b}\right)$ for some fraction $\frac{a}{b}$. What fraction did Mable find? Up to two decimal places, is Mable's final answer the same as Paul's?

$$\frac{z}{w} = \frac{z\overline{w}}{w\overline{w}} = \frac{z\overline{w}}{|w|} = \frac{(2+3i)\times\left(\frac{2}{\sqrt{5}} - \frac{1}{\sqrt{5}}i\right)}{1}$$

$$= \frac{4}{\sqrt{5}} + \frac{6i}{\sqrt{5}} - \frac{2i}{\sqrt{5}} + \frac{3}{\sqrt{5}}$$

$$= \frac{7}{\sqrt{5}} + \frac{4}{\sqrt{5}}i$$

Comparing these angles shows they are the same.

$$\tan^{-1}\left(\frac{4}{7}\right) \approx 0.52$$

$$\tan^{-1}\left(\frac{3}{2}\right) - \tan^{-1}\left(\frac{1}{2}\right) \approx 0.52$$

Topic C

The Power of the Right Notation

N-CN.B.4, N-CN.B.5, N-VM.C.8, N-VM.C.10, N-VM.C.11, N-VM.C.12

Focus Standards:	N-CN.B.4	(+) Represent complex numbers on the complex plane in rectangular and polar form (including real and imaginary numbers), and explain why the rectangular and polar forms of a given complex number represent the same number.
	N-CN.B.5	(+) Represent addition, subtraction, multiplication, and conjugation of complex numbers geometrically on the complex plane; use properties of this representation for computation. *For example,* $(-1+\sqrt{3}i)^3 = 8$ *because* $(-1+\sqrt{3}i)$ *has modulus* 2 *and argument* $120°$.
	N-VM.C.8	(+) Add, subtract, and multiply matrices of appropriate dimensions.
	N-VM.C.10	(+) Understand that the zero and identity matrices play a role in matrix addition and multiplication similar to the role of 0 and 1 in the real numbers. The determinant of a square matrix is nonzero if and only if the matrix has a multiplicative inverse.
	N-VM.C.11	(+) Multiply a vector (regarded as a matrix with one column) by a matrix of suitable dimensions to produce another vector. Work with matrices as transformations of vectors.
	N-VM.C.12	(+) Work with 2×2 matrices as transformations of the plane, and interpret the absolute value of the determinant in terms of area.
Instructional Days:	13	
Lessons 18–19:	Exploiting the Connection to Trigonometry (E, P)[1]	
Lesson 20:	Exploiting the Connection to Cartesian Coordinates (S)	
Lesson 21:	The Hunt for Better Notation (P)	
Lessons 22–23:	Modeling Video Game Motion with Matrices (P, P)	
Lesson 24:	Matrix Notation Encompasses New Transformations! (P)	
Lesson 25:	Matrix Multiplication and Addition (P)	
Lessons 26–27:	Getting a Handle on New Transformations (E, P)	
Lessons 28–30:	When Can We Reverse a Transformation? (P, E, P)	

[1]Lesson Structure Key: **P**-Problem Set Lesson, **M**-Modeling Cycle Lesson, **E**-Exploration Lesson, **S**-Socratic Lesson

The theme of Topic C is to highlight the effectiveness of changing notations and the power provided by certain notations such as matrices. Lessons 18 and 19 exploit the connection to trigonometry, as students see how much of complex arithmetic is simplified (**N-CN.B.4**, **N-CN.B.5**). Students use the connection to trigonometry to solve problems such as find the three cube roots of -1. In Lesson 20, complex numbers are regarded as points in the Cartesian plane. If $w = a + ib$, then the modulus is $r = \sqrt{a^2 + b^2}$, and the argument is $\alpha = \arctan\left(\frac{b}{a}\right)$. Students begin to write analytic formulas for translations, rotations, and dilations in the plane and revisit the ideas of Geometry (**G-CO.A.2**, **G-CO.A.4**, **G-CO.A.5**) in this light. In Lesson 21, students discover a better notation (i.e., matrices) and develop the 2×2 matrix notation for planar transformations represented by complex number arithmetic. This work leads to Lessons 22 and 23 as students discover how geometry software and video games efficiently perform rigid motion calculations. Students discover the flexibility of 2×2 matrix notation in Lessons 24 and 25 as they add matrices and multiply by the identity matrix and the zero matrix (**N-VM.C.8**, **N-VM.C.11**). Students understand that multiplying matrix A by the identity matrix results in matrix A and connect the multiplicative identity matrix to the role of 1, the multiplicative identity, in the real number system. This is extended as students see that the identity matrix does not transform the unit square. Students then add matrices and conclude that the zero matrix added to matrix A results in matrix A and is similar to 0 in the real number system. They extend this concept to transformations on the unit square and see that adding the zero matrix has no effect, but multiplying by the zero matrix collapses the unit square to zero. This allows for the study of additional matrix transformations (shears, for example) in Lessons 26 and 27, multiplying matrices, and the meaning of the determinant of a 2×2 matrix (**N-VM.C.10**, **N-VM.C.12**). Lessons 28–30 conclude Topic C and Module 1 as students discover the inverse matrix (matrix A is called an *inverse matrix* to a matrix B if $AB = I$ and $BA = I$) and determine when matrices do not have inverses. Students begin to think and reason abstractly about the geometric effects of the operations of complex numbers (MP.2) as they see the connection to trigonometry and the Cartesian plane.

The study of vectors and matrices is only introduced in Module 1 through a coherent connection to transformations and complex numbers. Further and more formal study of multiplication of matrices occurs in Module 2. **N-M.C.8** is assessed secondarily, in the context of other standards, but not directly on Mid- and End-of-Module Assessments until Module 2.

Lesson 18: Exploiting the Connection to Trigonometry

Student Outcomes

- Students derive the formula for $z^n = r^n(\cos(n\theta) + i\sin(n\theta))$ and use it to calculate powers of a complex number.

Lesson Notes

This lesson builds on the concepts from Topic B by asking students to extend their thinking about the geometric effect of multiplication of two complex numbers to the geometric effect of raising a complex number to an integer exponent (**N-CN.B.5**). This lesson is part of a two-day lesson that gives students another opportunity to work with the polar form of a complex number, to see its usefulness in certain situations, and to exploit that form to quickly calculate powers of a complex number. Students compare and convert between polar and rectangular form and graph complex numbers represented both ways (**N-CN.B.4**). On the second day, students examine graphs of powers of complex numbers in a polar grid and then reverse the process from Day 1 to calculate n^{th} roots of a complex number (**N-CN.B.5**). Throughout the lesson, students construct and justify arguments (MP.3), looking for patterns in repeated reasoning (MP.8), and use the structure of expressions and visual representations to make sense of the mathematics (MP.7).

Classwork

Opening (5 minutes)

Display two complex numbers on the board: $1 + i$ and $\sqrt{2}\left(\cos\left(\frac{\pi}{4}\right) i\sin\left(\frac{\pi}{4}\right)\right)$.

MP.3

- Do these represent the same number? Explain why or why not.
  - $\sqrt{2}\left(\frac{\sqrt{2}}{2} + i\frac{\sqrt{2}}{2}\right) = 1 + i$; *yes, they are the same number. When you expand* $\sqrt{2}\left(\cos\left(\frac{\pi}{4}\right) i\sin\left(\frac{\pi}{4}\right)\right)$, *you get* $1 + i$.
- What are the advantages of writing a complex number in polar form? What are the disadvantages?
  - *In polar form, you can see the modulus and argument. It is easy to multiply the numbers because you multiply the modulus and add the arguments. It can be difficult to graph the numbers because you have to use a compass and protractor to graph them accurately. If you are unfamiliar with the rotations and evaluating sine and cosine functions, then converting to rectangular is difficult. It is not so easy to add complex numbers in polar form unless you have a calculator and convert them to rectangular form.*
- What are the advantages of writing a complex number in rectangular form? What are the disadvantages?
  - *They are easy to graph; addition and multiplication are not too difficult either. It is difficult to understand the geometric effect of multiplication when written in rectangular form. It is not so easy to calculate the argument of the number, and you have to use a formula to calculate the modulus.*

Opening Exercise (5 minutes)

Tell students that in this lesson they are going to begin to exploit the advantages of writing a number in polar form, and have them quickly do the Opening Exercises. Students should work these problems individually. These exercises also serve as a check for understanding. If students are struggling to complete these exercises quickly and accurately, consider providing some additional practice in the form of Sprints.

Scaffolding:

- For struggling students, encourage them to work from a copy of a unit circle to quickly identify the sine and cosine function values.
- On Opening Exercise part (b), help students recall how to graph complex numbers, construct a triangle, and use special triangle ratios to determine the argument.

Opening Exercise

a. **Identify the modulus and argument of each complex number, and then rewrite it in rectangular form.**

i. $2\left(\cos\left(\frac{\pi}{4}\right) + i\sin\left(\frac{\pi}{4}\right)\right)$

The modulus is 2, and the argument is $\frac{\pi}{4}$. The number is $\sqrt{2} + i\sqrt{2}$.

ii. $5\left(\cos\left(\frac{2\pi}{3}\right) + i\sin\left(\frac{2\pi}{3}\right)\right)$

The modulus is 5, and the argument is $\frac{2\pi}{3}$. The number is $-\frac{5}{2} + i\frac{5\sqrt{3}}{2}$.

iii. $3\sqrt{2}\left(\cos\left(\frac{7\pi}{4}\right) + i\sin\left(\frac{7\pi}{4}\right)\right)$

The modulus is $3\sqrt{2}$, and the argument is $\frac{7\pi}{4}$. The number is $3 - 3i$.

iv. $3\left(\cos\left(\frac{7\pi}{6}\right) + i\sin\left(\frac{7\pi}{6}\right)\right)$

The modulus is 3, and the argument is $\frac{7\pi}{6}$. The number is $-\frac{3\sqrt{3}}{2} - \frac{3}{2}i$.

v. $1(\cos(\pi) + i\sin(\pi))$

The modulus is 1, and the argument is π. The number is -1.

b. **What is the argument and modulus of each complex number? Explain how you know.**

i. $2 \quad 2i$

We have $|2 - 2i| = 2\sqrt{2}$, and $\arg(2 - 2i) = \frac{7\pi}{4}$. The point $(2, -2)$ is located in the fourth quadrant. The ray from the origin containing the point is a rotation of $\frac{7\pi}{4}$ from the ray through the origin containing the real number 1.

ii. $3\sqrt{3} + 3i$

We have $\left|3\sqrt{3} + 3i\right| = 6$, and $\arg\left(3\sqrt{3} + 3i\right) = \frac{\pi}{6}$. The point $(3\sqrt{3}, 3)$ is located in the first quadrant. The ray from the origin containing the point is a rotation of $\frac{\pi}{6}$ from the ray through the origin containing the real number 1.

iii. $-1-\sqrt{3}i$

We have $|-1-\sqrt{3}i| = 2$, *and* $\arg(-1-\sqrt{3}i) = \frac{4\pi}{3}$. *The point* $(-1, -\sqrt{3})$ *is located in the third quadrant. The ray from the origin containing the point is a rotation of* $\frac{4\pi}{3}$ *from the ray through the origin containing the real number* 1.

iv. $-5i$

We have $|-5i| = 5$, *and* $\arg(-5i) = \frac{3\pi}{2}$. *The point* $(0, -5)$ *is located on the imaginary axis. The ray from the origin containing the point is a rotation of* $\frac{3\pi}{2}$ *from the ray through the origin containing the real number* 1.

v. 1

We have $|1| = 1$, *and* $\arg(1) = 0$. *This is the real number* 1.

Exploratory Challenge/Exercises 1–12 (20 minutes)

Students investigate and ultimately generalize a formula for quickly calculating the value of z^n. The class should work on these problems in teams of three to four students each. Use the discussion questions to help move individual groups forward as they work through the exercises in this exploration. Each group should have graph paper for each group member and access to a calculator to check calculations if needed.

In Exercise 3, most groups probably expand the number and perform the calculation in rectangular form. Here polar form offers little advantage. Perhaps when the exponent is a 4, a case could be made that polar form is more efficient for calculating a power of a complex number.

Be sure to pause and debrief with the entire class after Exercise 5. All students need to have observed the patterns in the table in order to continue to make progress discovering the relationships about powers of a complex number.

Exploratory Challenge/Exercises 1–12

1. **Rewrite each expression as a complex number in rectangular form.**

   a. $(1+i)^2$

   $(1+i)(1+i) = 1 + 2i + i^2 = 1 + 2i - 1 = 2i$

   b. $(1+i)^3$

   $(1+i)^3 = (1+i)^2(1+i) = 2i(1+i) = 2i + 2i^2 = -2 + 2i$

   c. $(1+i)^4$

   $(1+i)^4 = (1+i)^2(1+i)^2 = 2i \cdot 2i = 4i^2 = -4$

2. **Complete the table below showing the rectangular form of each number and its modulus and argument.**

Power of $(1+i)$	Rectangular Form	Modulus	Argument
$(1+i)^0$	1	1	0
$(1+i)^1$	$1+i$	$\sqrt{2}$	$\frac{\pi}{4}$
$(1+i)^2$	$2i$	2	$\frac{\pi}{2}$
$(1+i)^3$	$-2+2i$	$2\sqrt{2}$	$\frac{3\pi}{4}$
$(1+i)^4$	-4	4	π

MP.8

3. **What patterns do you notice each time you multiply by another factor of $(1+i)$?**

The argument increases by $\frac{\pi}{4}$. The modulus is multiplied by $\sqrt{2}$.

Before proceeding to the rest of the exercises in this Exploratory Challenge, check to make sure each group observed the patterns in the table required for them to make the connection that repeatedly multiplying by the same complex number causes repeated rotation by the argument, dilation, and modulus of the number.

Debrief the first five exercises by having one or two groups present their findings on the board or document camera.

4. **Graph each power of $1+i$ shown in the table on the same coordinate grid. Describe the location of these numbers in relation to one another using transformations.**

Starting with $(1+i)^0$, each subsequent complex number is a $45°$ rotation and a dilation by a factor of $\sqrt{2}$ of the previous one. The graph shows the graphs of $z_n = (1+i)^n$ for $n = 0, 1, 2, 3, 4, 5$.

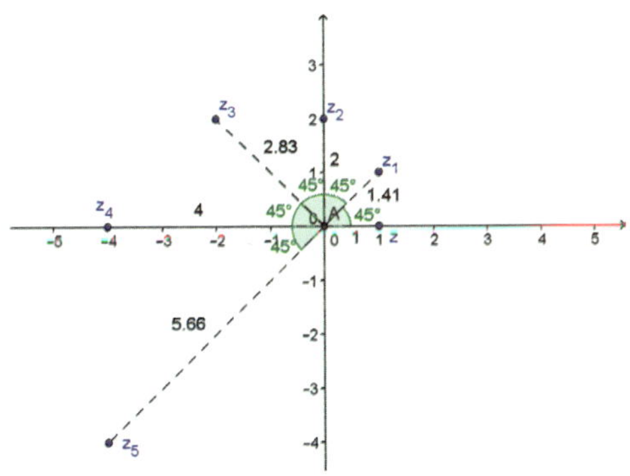

5. **Predict what the modulus and argument of $(1+i)^5$ would be without actually performing the multiplication. Explain how you made your prediction.**

The modulus would be $4\sqrt{2}$, and the argument would be $\pi + \frac{\pi}{4} = \frac{5\pi}{4}$.

6. **Graph $(1+i)^5$ in the complex plane using the transformations you described in Exercise 5.**

See the solution to Exercises 4 and 5.

7. **Write each number in polar form using the modulus and argument you calculated in Exercise 4.**

$(1+i)^0 = 1(\cos(0) + i\sin(0))$

$(1+i)^1 = \sqrt{2}\left(\cos\left(\frac{\pi}{4}\right) + i\sin\left(\frac{\pi}{4}\right)\right)$

$(1+i)^2 = 2\left(\cos\left(\frac{\pi}{2}\right) + i\sin\left(\frac{\pi}{2}\right)\right)$

$(1+i)^3 = 2\sqrt{2}\left(\cos\left(\frac{3\pi}{4}\right) + i\sin\left(\frac{3\pi}{4}\right)\right)$

$(1+i)^4 = 4(\cos(\pi) + i\sin(\pi))$

8. **Use the patterns you have observed to write $(1+i)^5$ in polar form, and then convert it to rectangular form.**

$(1+i)^5 = 4\sqrt{2}\left(\cos\left(\frac{5\pi}{4}\right) + i\sin\left(\frac{5\pi}{4}\right)\right)$

9. **What is the polar form of $(1+i)^{20}$? What is the modulus of $(1+i)^{20}$? What is its argument? Explain why $(1+i)^{20}$ is a real number.**

In polar form, the number would be $(\sqrt{2})^{20}\left(\cos\left(20\cdot\frac{\pi}{4}\right) + i\sin\left(20\cdot\frac{\pi}{4}\right)\right)$. The modulus is $(\sqrt{2})^{20} = 2^{10} = 1024$. The argument is the rotation between 0 and 2π that corresponds to a rotation of $20\cdot\frac{\pi}{4} = 5\pi$. The argument is π. This rotation takes the number 1 to the negative real-axis and dilates it by a factor of $1,024$, resulting in the number $-1,024$, which is a real number.

Pause here to discuss the advantages of considering the geometric effect of multiplication by a complex number when raising a complex number to a large integer exponent. Lead a discussion so students understand that the polar form of a complex number makes this type of multiplication very efficient.

- How do you represent multiplication by a complex number when written in polar form?
  - *The product of two complex numbers has a modulus that is the product of the two factors' moduli and an argument that is the sum of the two factors' arguments.*
- How does understanding the geometric effect of multiplication by a complex number make solving Exercises 10 and 11 easier than repeatedly multiplying by the rectangular form of the number?
  - *If you know the modulus and argument of the complex number, and you want to calculate z^n, then the argument will be n times the argument, and the modulus will be the modulus raised to the n.*
- In these exercises, you worked with powers of $1+i$. Do you think the patterns you observed can be generalized to any complex number raised to a positive integer exponent? Explain your reasoning.
  - *Since the patterns we observed are based on repeatedly multiplying by the same complex number, and since the geometric effect of multiplication always involves a rotation and dilation, this process should apply to all complex numbers.*
- How can you quickly raise any complex number of a large integer exponent?
  - *Determine the modulus and argument of the complex number. Then, multiply the argument by the exponent, and raise the modulus to the exponent. Then, you can write the number easily in polar and then rectangular form.*

This exploration largely relies on students using inductive reasoning to observe patterns in powers of complex numbers. The formula they write in Exercise 11 is known as DeMoivre's formula (or DeMoivre's theorem). More information and a proof by mathematical induction that this relationship holds can be found at http://en.wikipedia.org/wiki/De_Moivre's_formula.

If students have been struggling with this exploration, lead a whole-class discussion on the next several exercises, or groups can proceed to work through the rest of this Exploratory Challenge on their own. Be sure to monitor groups, and keep referring them back to the patterns they observed in the tables and graphs as they make their generalizations. Before students begin, announce that they are generalizing the patterns they observed in the previous exercises. Make sure they understand that the goal is a formula or process for quickly raising a complex number to an integer exponent. Observe groups, and encourage students to explain to one another how they are seeing the formula as they work through these exercises.

10. **If z has modulus r and argument θ, what are the modulus and argument of z^2? Write the number z^2 in polar form.**

    The modulus would be r^2, and the argument would be a rotation between 0 and 2π that is equivalent to 2θ.

    $$z^2 = r^2(\cos(2\theta) + i\sin(2\theta))$$

11. **If z has modulus r and argument θ, what are the modulus and argument of z^n where n is a nonnegative integer? Write the number z^n in polar form. Explain how you got your answer.**

    The modulus would be r^n, and the argument would be a rotation between 0 and 2π that is equivalent to $n\theta$.

    $$z^n = r^n(\cos(n\theta) + i\sin(n\theta))$$

12. **Recall that $\frac{1}{z} = \frac{1}{r}(\cos(-\theta) + i\sin(-\theta))$. Explain why it would make sense that the formula holds for all integer values of n.**

    Since $\frac{1}{z} = z^{-1}$, it would make sense that the formula would hold for negative integers as well. If you plot $\frac{1}{z^2}$, $\frac{1}{z^3}$, etc., you can see the pattern holds.

In Exercise 14, students must consider why this formula holds for negative integers as well. Ask them how they could verify graphically or algebraically that these formulae could be extended to include negative integer exponents. Consider demonstrating this using graphing software such as Geogebra or Desmos.

Close this section by recording the formula shown below on the board. Ask students to summarize to a partner how to use this formula with the number $(1+i)^{10}$ and to record it in their notes.

- Given a complex number z with modulus r and argument θ, the nth power of z is given by $z^n = r^n(\cos(n\theta) + i\sin(n\theta))$ where n is an integer.

Exercises 13–14 (5 minutes)

Students should work these exercises in their small groups or with a partner. After a few minutes, review the solutions, and discuss any problems students had with their calculations.

Exercises 13–14

13. Compute $\left(\frac{1-i}{\sqrt{2}}\right)^7$, and write it as a complex number in the form $a+bi$ where a and b are real numbers.

The modulus of $\frac{1-i}{\sqrt{2}}$ is 1, and the argument is $\frac{7\pi}{4}$. The polar form of the number is

$1^7\left(\cos\left(7\cdot\frac{7\pi}{4}\right)+i\sin\left(7\cdot\frac{7\pi}{4}\right)\right).$

Converting this number to rectangular form by evaluating the sine and cosine values produces $\frac{\sqrt{2}}{2}+\frac{\sqrt{2}}{2}i$.

14. Write $(1+\sqrt{3}i)^6$, and write it as a complex number in the form $a+bi$ where a and b are real numbers.

The modulus of $1+\sqrt{3}i$ is 2, and the argument is $\frac{\pi}{3}$. The polar form of the number is

$2^6\left(\cos\left(6\cdot\frac{\pi}{3}\right)+i\sin\left(6\cdot\frac{\pi}{3}\right)\right).$

Converting this number to rectangular form by evaluating the sine and cosine values produces
$64(1+0\cdot i)=64.$

Closing (5 minutes)

Revisit one of the questions from the beginning of the lesson. Students can write their responses or discuss them with a partner.

- Describe an additional advantage to polar form that we discovered during this lesson.
  - *When raising a complex number to an integer exponent, the polar form gives a quick way to express the repeated transformations of the number and quickly determine its location in the complex plane. This then leads to quick conversion to rectangular form.*

Review the relationship that students discovered in this lesson.

Lesson Summary

Given a complex number z with modulus r and argument θ, the nth power of z is given by

$z^n=r^n(\cos(n\theta)+i\sin(n\theta))$ where n is an integer.

Exit Ticket (5 minutes)

Name ______________________________ Date ______________

Lesson 18: Exploiting the Connection to Trigonometry

Exit Ticket

1. Write $(2 + 2i)^8$ as a complex number in the form $a + bi$ where a and b are real numbers.

2. Explain why a complex number of the form $(a + ai)^n$ will either be a pure imaginary or a real number when n is an even number.

Exit Ticket Sample Solutions

1. Write $(2+2i)^8$ as a complex number in the form $a+bi$ where a and b are real numbers.

   We have $|2+2i| = 2\sqrt{2}$ *and* $\arg(2+2i) = \frac{\pi}{4}$.

   Thus, $(2+2i)^8 = \left(2\sqrt{2}\right)^8\left(\cos\left(8\cdot\frac{\pi}{4}\right) + i\sin\left(8\cdot\frac{\pi}{4}\right)\right) = 2^{12}(\cos(2\pi) + i\sin(2\pi)) = 2^{12}(1+0i) = 2^{12} + 0i.$

2. Explain why a complex number of the form $(a+ai)^n$ where a is a positive real number will either be a pure imaginary or a real number when n is an even number.

   Since the argument will always be $\frac{\pi}{4}$*, any even number multiplied by this number will be a multiple of* $\frac{\pi}{2}$*. This will result in a rotation to one of the axes, which means the complex number will either be a real number or a pure imaginary number.*

Problem Set Sample Solutions

1. Write the complex number in $a+bi$ form where a and b are real numbers.

   a. $2\left(\cos\left(\frac{5\pi}{3}\right) + i\sin\left(\frac{5\pi}{3}\right)\right)$

   $$2\left(\cos\left(\frac{5\pi}{3}\right) + i\sin\left(\frac{5\pi}{3}\right)\right) = 2\left(\frac{1}{2} - \frac{\sqrt{3}}{2}i\right) = 1 - \sqrt{3}i$$

   b. $3(\cos(210^\circ) + i\sin(210^\circ))$

   $$3(\cos(210^\circ) + i\sin(210^\circ)) = 3\left(-\frac{\sqrt{3}}{2} - \frac{1}{2}i\right) = -\frac{3\sqrt{3}}{2} - \frac{3}{2}i$$

   c. $\left(\sqrt{2}\right)^{10}\left(\cos\left(\frac{15\pi}{4}\right) + i\sin\left(\frac{15\pi}{4}\right)\right)$

   $$\left(\sqrt{2}\right)^{10}\left(\cos\left(\frac{15\pi}{4}\right) + i\sin\left(\frac{15\pi}{4}\right)\right) = 32\left(\cos\left(\frac{7\pi}{4}\right) + i\sin\left(\frac{7\pi}{4}\right)\right) = 32\left(\frac{\sqrt{2}}{2} - \frac{i\sqrt{2}}{2}\right) = 16\sqrt{2} - 16\sqrt{2}i$$

   d. $\cos(9\pi) + i\sin(9\pi)$

   $$\cos(9\pi) + i\sin(9\pi) = \cos(\pi) + i\sin(\pi) = -1 + 0i = -1$$

e. $4^3\left(\cos\left(\frac{3\pi}{4}\right)+i\sin\left(\frac{3\pi}{4}\right)\right)$

$$4^3\left(\cos\left(\frac{3\pi}{4}\right)+i\sin\left(\frac{3\pi}{4}\right)\right)=64\left(-\frac{\sqrt{2}}{2}+i\frac{\sqrt{2}}{2}\right)$$
$$=-32\sqrt{2}+32\sqrt{2}i$$

f. $6(\cos(480°)+i\sin(480°))$

$$6(\cos(480°)+i\sin(480°))=6(\cos(120°)+i\sin(120°))$$
$$=6\left(-\frac{1}{2}+\frac{\sqrt{3}}{2}i\right)$$
$$=-3+3\sqrt{3}i$$

2. **Use the formula discovered in this lesson to compute each power of z. Verify that the formula works by expanding and multiplying the rectangular form and rewriting it in the form $a+bi$ where a and b are real numbers.**

a. $\left(1+\sqrt{3}i\right)^3$

Since $z=1+\sqrt{3}i$, we have $|z|=\sqrt{1+3}=2$, and $\theta=\frac{\pi}{3}$. Then,

$$\left(1+\sqrt{3}i\right)^3=2^3\left(\cos\left(3\cdot\frac{\pi}{3}\right)+i\sin\left(3\cdot\frac{\pi}{3}\right)\right)=8(\cos(\pi)+i\sin(\pi))=-8$$
$$(1+\sqrt{3}i)^3=(1+\sqrt{3}i)(1+2\sqrt{3}i-3)=(1+\sqrt{3}i)(-2+2\sqrt{3}i)=-2+2\sqrt{3}i-2\sqrt{3}i-6=-8$$

b. $(-1+i)^4$

Since $z=-1+i$, we have $|z|=\sqrt{1+1}=\sqrt{2}$, and $\theta=\frac{3\pi}{4}$. Then,

$$(-1+i)^4=\left(\sqrt{2}\right)^4\left(\cos\left(4\cdot\frac{3\pi}{4}\right)+i\sin\left(4\cdot\frac{3\pi}{4}\right)\right)=4(\cos(3\pi)+i\sin(3\pi))=-4$$
$$(-1+i)^4=(-1+i)^2(-1+i)^2=(1-2i-1)(1-2i-1)=(-2i)(-2i)=-4$$

c. $(2+2i)^5$

Since $z=2+2i$, we have $|z|=\sqrt{2^2+2^2}=2\sqrt{2}$, and $\theta=\frac{\pi}{4}$. Then,

$$(2+2i)^5=\left(2\sqrt{2}\right)^5\left(\cos\left(5\cdot\frac{\pi}{4}\right)+i\sin\left(5\cdot\frac{\pi}{4}\right)\right)=128\sqrt{2}\left(-\frac{\sqrt{2}}{2}-\frac{\sqrt{2}}{2}i\right)=-128-128i$$
$$(2+2i)^5=(2+2i)^2(2+2i)^2(2+2i)=(4+8i-4)(4+8i-4)(2+2i)=(8i)(8i)(2+2i)$$
$$=-64(2+2i)=-128-128i$$

d. $(2-2i)^{-2}$

Since $z=2-2i$, we have $|z|=\sqrt{2^2+2^2}=2\sqrt{2}$, and $\theta=\frac{7\pi}{4}$. Then,

$$(2-2i)^{-2}=\left(2\sqrt{2}\right)^{-2}\left(\cos\left(-2\cdot\frac{7\pi}{4}\right)+i\sin\left(-2\cdot\frac{7\pi}{4}\right)\right)=\frac{1}{8}(0+i)=\frac{1}{8}i$$
$$(2-2i)^{-2}=\frac{1}{(2-2i)^2}=\frac{1}{4-8i-4}=\frac{1}{-8i}\cdot\frac{i}{i}=\frac{i}{8}=\frac{1}{8}i$$

e. $\left(\sqrt{3}-i\right)^4$

Since $z=\sqrt{3}-i$, we have $|z|=\sqrt{\sqrt{3}^2+1^2}=2$, and $\theta=\frac{11\pi}{6}$. Then,

$$\left(\sqrt{3}-i\right)^4=2^4\left(\cos\left(4\cdot\frac{11\pi}{6}\right)+i\sin\left(4\cdot\frac{11\pi}{6}\right)\right)=16\left(\cos\left(\frac{22\pi}{3}\right)+i\sin\left(\frac{22\pi}{3}\right)\right)=16\left(-\frac{1}{2}-\frac{\sqrt{3}}{2}i\right)$$
$$=-8-8\sqrt{3}i$$

$$\left(\sqrt{3}-i\right)^4=\left(\sqrt{3}-i\right)^2\left(\sqrt{3}-i\right)^2=\left(3-2\sqrt{3}i-1\right)\left(3-2\sqrt{3}i-1\right)=\left(2-2\sqrt{3}i\right)\left(2-2\sqrt{3}i\right)$$
$$=4-8\sqrt{3}i-12=-8-8\sqrt{3}i$$

f. $\left(3\sqrt{3}-3i\right)^6$

Since $z=3\sqrt{3}-3i$, we have $|z|=\sqrt{(3\sqrt{3})^2+3^2}=\sqrt{36}=6$, and $\theta=\frac{11\pi}{6}$. Then,

$$\left(3\sqrt{3}-3i\right)^6=6^6\left(\cos\left(6\cdot\frac{11\pi}{6}\right)+i\sin\left(6\cdot\frac{11\pi}{6}\right)\right)=46656(\cos(11\pi)+i\sin(11\pi))=-46656$$

$$\left(3\sqrt{3}-3i\right)^6=\left(3\sqrt{3}-3i\right)^2\left(3\sqrt{3}-3i\right)^2\left(3\sqrt{3}-3i\right)^2$$
$$=\left(27-18\sqrt{3}i-9\right)\left(27-18\sqrt{3}i-9\right)\left(27-18\sqrt{3}i-9\right)$$
$$=\left(18-18\sqrt{3}i\right)\left(18-18\sqrt{3}i\right)\left(18-18\sqrt{3}i\right)$$
$$=\left(324-648\sqrt{3}i-972\right)\left(18-18\sqrt{3}i\right)=\left(-648-648\sqrt{3}i\right)\left(18-18\sqrt{3}i\right)$$
$$=-11664+11664\sqrt{3}i-11664\sqrt{3}i-34992=-46656$$

3. Given $z=-1-i$, graph the first five powers of z by applying your knowledge of the geometric effect of multiplication by a complex number. Explain how you determined the location of each in the coordinate plane.

Multiplication by $-1-i$ will dilate by $|-1-i|=\sqrt{1+1}=\sqrt{2}$ and rotate by $\arg(-1-i)=\frac{5\pi}{4}$. Then, the graph below shows $z=-1-i$, $z^2=(-1-i)^2$, $z^3=(-1-i)^3$, $z^4=(-1-i)^4$, and $z^5=(-1-i)^5$.

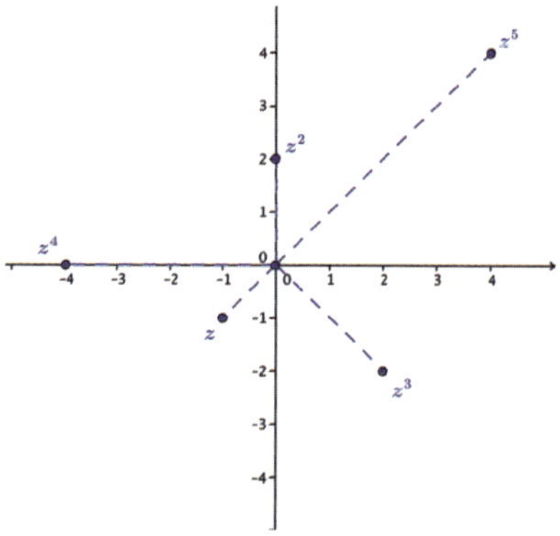

To locate each point, multiply the distance from the previous point to the origin by the modulus ($\sqrt{2}$), and rotate counterclockwise $\frac{5\pi}{4}$.

4. Use your work from Problem 3 to determine three values of n for which $(-1-i)^n$ is a multiple of $-1-i$.

Since multiplication by $-1-i$ rotates the point by $\frac{5\pi}{4}$ radians, the point $(-1-i)^n$ is a multiple of the original $z=-1-i$ every 8 iterations. Thus, $(-1-i)^9$, $(-1-i)^{17}$, $(-1-i)^{25}$ are all multiples of $(1-i)$. Answers will vary.

5. Find the indicated power of the complex number, and write your answer in form $a + bi$ where a and b are real numbers.

a. $\left[2\left(\cos\left(\frac{3\pi}{4}\right) + i\sin\left(\frac{3\pi}{4}\right)\right)\right]^3$

$$\begin{aligned}\left[2\left(\cos\left(\frac{3\pi}{4}\right) + i\sin\left(\frac{3\pi}{4}\right)\right)\right]^3 &= 2^3\left(\cos\left(3\cdot\frac{3\pi}{4}\right) + i\sin\left(3\cdot\frac{3\pi}{4}\right)\right)\\ &= 8\left(\cos\left(\frac{9\pi}{4}\right) + i\sin\left(\frac{9\pi}{4}\right)\right)\\ &= 8\left(\frac{\sqrt{2}}{2} + \frac{\sqrt{2}}{2}i\right)\\ &= 4\sqrt{2} + 4\sqrt{2}i\end{aligned}$$

b. $\left[\sqrt{2}\left(\cos\left(\frac{\pi}{4}\right) + i\sin\left(\frac{\pi}{4}\right)\right)\right]^{10}$

$$\begin{aligned}\left[\sqrt{2}\left(\cos\left(\frac{\pi}{4}\right) + i\sin\left(\frac{\pi}{4}\right)\right)\right]^{10} &= \left(\sqrt{2}\right)^{10}\left(\cos\left(\frac{10\pi}{4}\right) + i\sin\left(\frac{10\pi}{4}\right)\right)\\ &= 32(0 + 1i)\\ &= 32i\end{aligned}$$

c. $\left(\cos\left(\frac{5\pi}{6}\right) + i\sin\left(\frac{5\pi}{6}\right)\right)^6$

$$\begin{aligned}\left(\cos\left(\frac{5\pi}{6}\right) + i\sin\left(\frac{5\pi}{6}\right)\right)^6 &= \cos\left(\frac{30\pi}{6}\right) + i\sin\left(\frac{30\pi}{6}\right)\\ &= \cos(5\pi) + i\sin(5\pi)\\ &= -1\end{aligned}$$

d. $\left[\frac{1}{3}\left(\cos\left(\frac{3\pi}{2}\right) + i\sin\left(\frac{3\pi}{2}\right)\right)\right]^4$

$$\begin{aligned}\left[\frac{1}{3}\left(\cos\left(\frac{3\pi}{2}\right) + i\sin\left(\frac{3\pi}{2}\right)\right)\right]^4 &= \left(\frac{1}{3}\right)^4\left(\cos\left(4\cdot\frac{3\pi}{2}\right) + i\sin\left(4\cdot\frac{3\pi}{2}\right)\right)\\ &= \frac{1}{81}(\cos(6\pi) + i\sin(6\pi))\\ &= \frac{1}{81}\end{aligned}$$

e. $\left[4\left(\cos\left(\frac{4\pi}{3}\right) + i\sin\left(\frac{4\pi}{3}\right)\right)\right]^{-4}$

$$\begin{aligned}\left[4\left(\cos\left(\frac{4\pi}{3}\right) + i\sin\left(\frac{4\pi}{3}\right)\right)\right]^{-4} &= 4^{-4}\left(\cos\left(-4\cdot\frac{4\pi}{3}\right) + i\sin\left(4\cdot\frac{4\pi}{3}\right)\right)\\ &= \frac{1}{256}\left(\cos\left(-\frac{16\pi}{3}\right) + i\sin\left(-\frac{16\pi}{3}\right)\right)\\ &= \frac{1}{256}\left(-\frac{1}{2} + \frac{\sqrt{3}}{2}i\right)\\ &= -\frac{1}{512} + \frac{\sqrt{3}}{512}i\end{aligned}$$

Lesson 19: Exploiting the Connection to Trigonometry

Student Outcomes

- Students understand how a formula for the nth roots of a complex number is related to powers of a complex number.
- Students calculate the nth roots of a complex number.

Lesson Notes

This lesson builds on the work from Topic B by asking students to extend their thinking about the geometric effect of multiplication of two complex numbers to the geometric effect of raising a complex number to an integer exponent (**N-CN.B.5**). It is part of a two-day lesson that gives students another opportunity to work with the polar form of a complex number, to see its usefulness in certain situations, and to exploit that to quickly calculate powers of a complex number. In this lesson, students continue to work with polar and rectangular form and graph complex numbers represented both ways (**N-CN.B.4**). They examine graphs of powers of complex numbers in a polar grid and then write the nth root as a fractional exponent and reverse the process from Day 1 to calculate the nth roots of a complex number (**N-CN.B.5**). Throughout the lesson, students are constructing and justifying arguments (MP.3), using precise language (MP.6), and using the structure of expressions and visual representations to make sense of the mathematics (MP.7).

Classwork

Opening (4 minutes)

Introduce the notion of a polar grid. Representing complex numbers in polar form on a polar grid makes this lesson seem easier for students and emphasizes the geometric effect of the roots of a complex number.

Display a copy of the polar grid on the right, and model how to plot a few complex numbers in polar form to illustrate that the concentric circles make it easy to measure the modulus and the rays at equal intervals and make representing the rotation of the complex number easy as well.

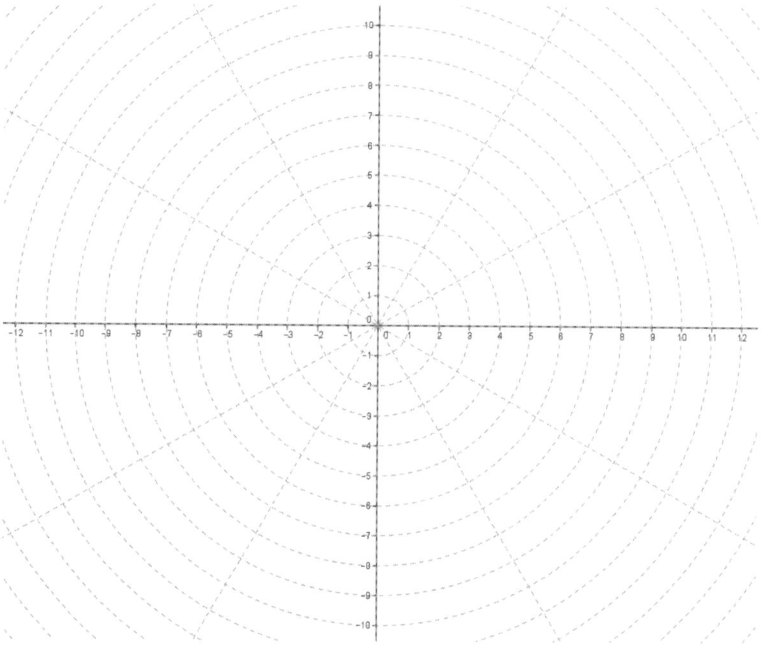

Plot a point with the given modulus and argument.

A: modulus $= 1$, argument $= 0°$

B: modulus $= 3$, argument $= 90°$

C: modulus $= 5$, argument $= 30°$

D: modulus $= 7$, argument $= 120°$

Explain to students that each circle represents a distance from the origin (the modulus). Each line represents an angle measure. To plot a point, find the angle of rotation, and then move out to the circle that represents the distance from the origin given by the modulus.

Opening Exercise (7 minutes)

These exercises give students an opportunity to practice working with a polar grid and to review their work from the previous day's lesson. Students should work individually or with a partner on these exercises. Monitor student progress to check for understanding, and provide additional support as needed.

Scaffolding:

- For struggling students, encourage them to label the rays in the polar grid with the degrees of rotation.
- Provide additional practice plotting complex numbers in polar form. Some students may find working with degrees easier than working with radians.

Opening Exercise

A polar grid is shown below. The grid is formed by rays from the origin at equal rotation intervals and concentric circles centered at the origin. The complex number $z = \sqrt{3} + i$ is graphed on this polar grid.

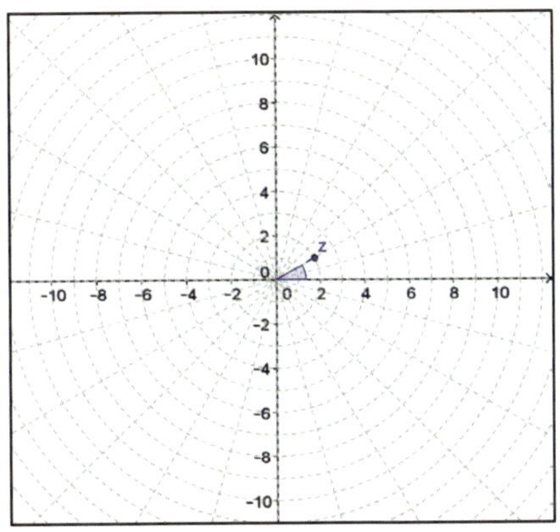

a. **Use the polar grid to identify the modulus and argument of z.**

The argument is $\frac{\pi}{6}$, and the modulus is 2.

b. **Graph the next three powers of z on the polar grid. Explain how you got your answers.**

Each power of z is another $30°$ rotation and a dilation by a factor of 2 from the previous number.

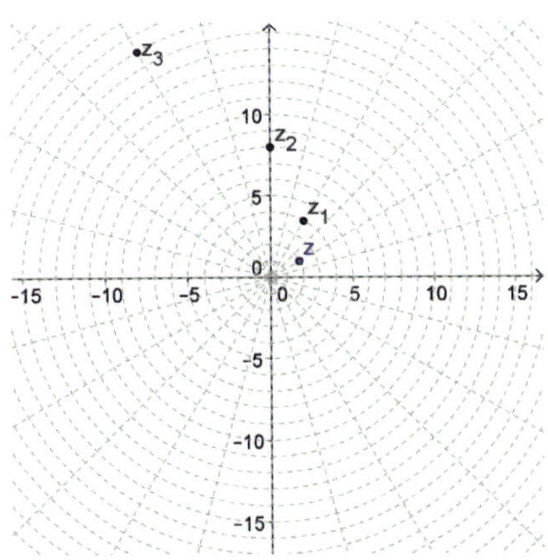

c. **Write the polar form of the number in the table below, and then rewrite it in rectangular form.**

Power of z	Polar Form	Rectangular Form
$\sqrt{3}+i$	$2\left(\cos\left(\frac{\pi}{6}\right)+i\sin\left(\frac{\pi}{6}\right)\right)$	$\sqrt{3}+i$
$\left(\sqrt{3}+i\right)^2$	$4\left(\cos\left(\frac{\pi}{3}\right)+i\sin\left(\frac{\pi}{3}\right)\right)$	$2+2\sqrt{3}i$
$\left(\sqrt{3}+i\right)^3$	$8\left(\cos\left(\frac{\pi}{2}\right)+i\sin\left(\frac{\pi}{2}\right)\right)$	$8i$
$\left(\sqrt{3}+i\right)^4$	$16\left(\cos\left(\frac{2\pi}{3}\right)+i\sin\left(\frac{2\pi}{3}\right)\right)$	$8-8\sqrt{3}i$

Have early finishers check their work by calculating one or two powers of z by expanding and then multiplying the rectangular form. Examples are shown below.

$$\left(\sqrt{3}+i\right)^2 = 3+2\sqrt{3}i+i^2 = 2+2\sqrt{3}i$$

$$\left(\sqrt{3}+i\right)^3 = \left(\sqrt{3}+i\right)\left(2+2\sqrt{3}i\right) = 2\sqrt{3}+6i+2i+2\sqrt{3}i^2 = 8i$$

Debrief by having one or two students explain their process to the class. Remind them again of the efficiency of working with complex numbers in polar form and the patterns that emerge when they graph powers of a complex number.

- Which way of expanding a power of a complex number would be quicker if you were going to expand $\left(\sqrt{3}+i\right)^{10}$?
- *Using the polar form would be far easier. It would be* $2^{10}\left(\cos\left(\frac{5\pi}{3}\right)+i\sin\left(\frac{5\pi}{3}\right)\right)$ow could you describe the pattern of the numbers if we continued graphing the powers of z?
  - *The numbers are spiraling outward as each number is on a ray rotated* $30°$ *from the previous one and farther from the origin by a factor of* 2.

Next, transition to the main focus of this lesson by giving students time to consider the next question. Have them respond in writing and discuss their answers with a partner.

- How do you think we could reverse this process, in other words, undo squaring a complex number or undo cubing a complex number?
  - *That would be like taking a square root or cube root. We would have to consider how to undo the rotation and dilation effects.*

Exercises 1–3 (7 minutes)

In these exercises, students explore one of the square roots of a complex number. Later in the lesson, show students that complex numbers have multiple roots just like a real number has two square roots (e.g., the square roots of 4 are 2 and -2). Students should work these exercises with a partner. If the class is struggling to make sense of Exercise 3, work that one as a whole class.

Exercises 1–3

The complex numbers $z_2 = \left(-1+\sqrt{3}i\right)^2$ and z_1 are graphed below.

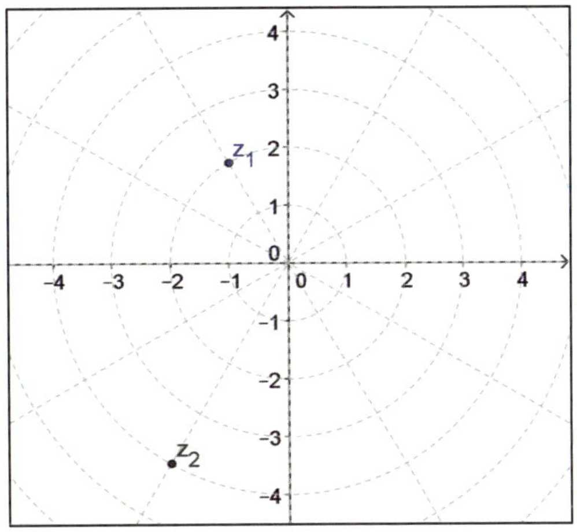

1. Use the graph to help you write the numbers in polar and rectangular form.

   $z_1 = 2\left(\cos\left(\frac{2\pi}{3}\right) + i\sin\left(\frac{2\pi}{3}\right)\right) = -1+\sqrt{3}i$

   $z_2 = 4\left(\cos\left(\frac{4\pi}{3}\right) + i\sin\left(\frac{4\pi}{3}\right)\right) = -2-2\sqrt{3}i$

2. Describe how the modulus and argument of $z_1 = -1+\sqrt{3}i$ are related to the modulus and argument of $z_2 = \left(-1+\sqrt{3}i\right)^2$.

   The modulus and argument are both cut in half.

MP.3

3. Why could we call $-1+\sqrt{3}i$ a square root of $-2-2\sqrt{3}i$?

   Clearly, $\left(-1+\sqrt{3}i\right)^2 = -2-2\sqrt{3}$. We can demonstrate this using the rectangular or polar form and verify it using transformations of the numbers when they are plotted in the complex plane. So, it would make sense then that raising both sides of this equation to the $\frac{1}{2}$ power should give us the desired result.

   Start with the equation, $\left(-1+\sqrt{3}i\right)^2 = -2-2\sqrt{3}i$.

$$\left(\left(-1+\sqrt{3}i\right)^2\right)^{\frac{1}{2}} = \left(-2-2\sqrt{3}i\right)^{\frac{1}{2}}$$

$$-1+\sqrt{3}i = \sqrt{-2-2\sqrt{3}i}$$

   Alternately, using the formula from Lesson 17, replace n with $\frac{1}{2}$.

$$z^n = r^n(\cos(n\theta) + i\sin(n\theta))$$
$$z^{\frac{1}{2}} = r^{\frac{1}{2}}\left(\cos\left(\frac{1}{2}\theta\right) + i\sin\left(\frac{1}{2}\theta\right)\right)$$

   So, a square root of $-2-2\sqrt{3}i$ would be

$$4^{\frac{1}{2}}\left(\cos\left(\frac{1}{2}\cdot\frac{4\pi}{3}\right) + i\sin\left(\frac{1}{2}\cdot\frac{4\pi}{3}\right)\right) = 2\left(\cos\left(\frac{2\pi}{3}\right) + i\sin\left(\frac{2\pi}{3}\right)\right) = -1+\sqrt{3}i.$$

After giving students a few minutes to work these exercises with a partner, make sure they understand that the modulus is cut in half because $2 \cdot 2 = 4$ shows a repeated multiplication by 2.

- How would this problem change if the modulus was 9 instead of 4? How would this problem change if the modulus was 3 instead of 4?
  - *The new modulus would have to be a number that when squared equals 9, so we would need the new modulus to be the square root of the original modulus.*

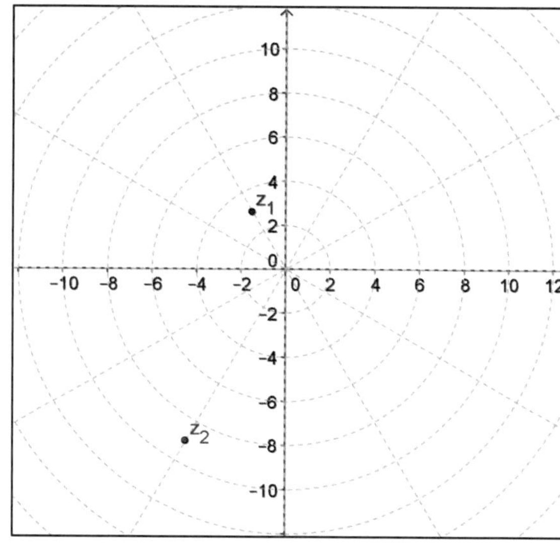

If students seem to think that the modulus would always be divided by 2 instead of it being the square root of the original modulus, model this using Geogebra. A sample screenshot is provided below showing a complex number with the same argument and a modulus of 9. Notice that the modulus of z_1 is 3 while the argument is still cut in half.

Discussion (7 minutes): The nth Roots of a Complex Number

In this Discussion, model how to derive a formula to find all the nth roots of a complex number. Begin by reminding students of the definition of square roots learned in Grade 8 and Algebra I.

- Recall that each real number has two square roots. For example, what are the two square roots of 4? The two square roots of 10? How do you know?
  - *They are 2 and -2 because $(2)^2 = 4$, and $(-2)^2 = 4$. The two square roots of 10 are $\sqrt{10}$ and $-\sqrt{10}$.*
- How many square roots do you think a complex number has? How many cube roots? Fourth roots, etc.?
  - *Since all real numbers are complex numbers, complex numbers should have two square roots as well. Since roots are solutions to an equation $x^n = r$, it would make sense that if our solution set is the complex numbers, then there would be three cube roots when $n = 3$ and four fourth roots when $n = 4$.*

Thus, complex numbers have multiple n^{th} roots when n is a positive integer. In fact, every complex number has 2 square roots, 3 cube roots, 4 fourth roots, and so on. This work relates back to Modules 1 and 3 in Algebra II where students learned that a degree n polynomial equation has n complex zeros and to previous work extending the properties of exponents to the real number exponents. Students should take notes as the work shown below is presented.

Using the formula from Lesson 17, suppose we have an nth root of z, $w = s(\cos(\alpha) + i\sin(\alpha))$. Then, for $r > 0, s > 0$, we have

$$w^n = z$$

$$s^n(\cos(n\alpha) + i\sin(n\alpha)) = r(\cos(\theta) + i\sin(\theta))$$

Equating the moduli, $s^n = r$, which implies that $s = \sqrt[n]{r}$.

Equating the arguments, $n\alpha = \theta$. However, since the sine and cosine functions are periodic functions with period 2π, this equation does not have a unique solution for α. We know that $\cos(\theta + 2\pi k) = \cos(\theta)$ and $\sin(\theta + 2\pi k) = \sin(\theta)$ for integer values of k and real numbers θ.

Therefore,

$$n\alpha = \theta + 2\pi k$$

Or, $\alpha = \frac{\theta}{n} + \frac{2\pi k}{n}$ for values of k up to $n - 1$. When $k = n$ or greater, we start repeating values for α.

Going back to your work in Example 1 and Exercise 6, we can find both roots of $-2 - 2\sqrt{3}i$ and all three cube roots of this number.

Example 1 (5 minutes): Find the Two Square Roots of a Complex Number

Example: Find the Two Square Roots of a Complex Number

Find both of the square roots of $-2 - 2\sqrt{3}i$.

The polar form of this number is $4\left(\cos\left(\frac{4\pi}{3}\right) + i\sin\left(\frac{4\pi}{3}\right)\right)$. The square roots of this number will have modulus $\sqrt{4} = 2$ and arguments given by $\alpha = \frac{\theta}{n} + \frac{2\pi k}{n}$ for $k = 0, 1$ where $\theta = \frac{4\pi}{3}$. Thus,

$$\alpha = \frac{1}{2}\left(\frac{4\pi}{3} + 2\pi \cdot 0\right) = \frac{2\pi}{3}$$

and

$$\alpha = \frac{1}{2}\left(\frac{4\pi}{3} + 2\pi \cdot 1\right) = \frac{5\pi}{3}$$

The square roots are

$$2\left(\cos\left(\frac{2\pi}{3}\right) + i\sin\left(\frac{2\pi}{3}\right)\right) = -1 + \sqrt{3}i$$

and

$$2\left(\cos\left(\frac{5\pi}{3}\right) + i\sin\left(\frac{5\pi}{3}\right)\right) = 1 - \sqrt{3}i.$$

Have students go back and add the graph of the second square root to the graph at the beginning of Exercise 1.

Exercises 4–6 (7 minutes)

Students work with the formula developed in the Discussion and presented in the Lesson Summary. Students can work individually or with a partner. If time is running short, assign these as Problem Set exercises as well.

Exercises 4–6

4. **Find the cube roots of $-2 = 2\sqrt{3}i$.**

$$-2-2\sqrt{3}i = 4\left(\cos\left(\frac{4\pi}{3}\right) + i\sin\left(\frac{4\pi}{3}\right)\right)$$

The modulus of the cube roots will be $\sqrt[3]{4}$. The arguments for $k = 0$, 1, and 2 are given by $\alpha = \frac{\theta}{n} + \frac{2\pi k}{n}$ where $\theta = \frac{4\pi}{3}$ and $n = 3$. Using this formula, the arguments are $\frac{4\pi}{9}$, $\frac{10\pi}{9}$, and $\frac{16\pi}{9}$. The three cube roots of $-2-2\sqrt{3}i$ are

$$\sqrt[3]{4}\left(\cos\left(\frac{4\pi}{9}\right) + i\sin\left(\frac{4\pi}{9}\right)\right)$$

$$\sqrt[3]{4}\left(\cos\left(\frac{10\pi}{9}\right) + i\sin\left(\frac{10\pi}{9}\right)\right)$$

$$\sqrt[3]{4}\left(\cos\left(\frac{16\pi}{9}\right) + i\sin\left(\frac{16\pi}{9}\right)\right)$$

5. **Find the square roots of $4i$.**

$$4i = 4\left(\cos\left(\frac{\pi}{2}\right) + i\sin\left(\frac{\pi}{2}\right)\right)$$

The modulus of the square roots is $\sqrt{4} = 2$. The arguments for $k = 0$ and 1 are given by $\alpha = \frac{\theta}{n} + \frac{2\pi k}{n}$ where $\theta = \frac{\pi}{2}$ and $n = 2$. Using this formula, the arguments are $\frac{\pi}{4}$ and $\frac{3\pi}{4}$. The two square roots of $4i$ are

$$2\left(\cos\left(\frac{\pi}{4}\right) + i\sin\left(\frac{\pi}{4}\right)\right) = \sqrt{2} + \sqrt{2}i$$

and

$$2\left(\cos\left(\frac{3\pi}{4}\right) + i\sin\left(\frac{3\pi}{4}\right)\right) = -\sqrt{2} + \sqrt{2}i.$$

6. **Find the cube roots of 8.**

In polar form,

$$8 = 8(\cos(0) + i\sin(0)).$$

The modulus of the cube roots is $\sqrt[3]{8} = 2$. The arguments for $k = 0$, 1, and 2 are given by $\alpha = \frac{\theta}{n} + \frac{2\pi k}{n}$ where $\theta = 0$ and $n = 3$. Using this formula, the arguments are 0, $\frac{2\pi}{3}$, and $\frac{4\pi}{3}$. The three cube roots of 8 are

$$2(\cos(0) + i\sin(0)) = 2$$

$$2\left(\cos\left(\frac{2\pi}{3}\right) + i\sin\left(\frac{2\pi}{3}\right)\right) = -1 + \sqrt{3}i$$

$$2\left(\cos\left(\frac{4\pi}{3}\right) + i\sin\left(\frac{4\pi}{3}\right)\right) = -1 - \sqrt{3}i$$

Consider pointing out to students that the answers to Exercise 9 are the solutions to the equation $x^3 - 8 = 0$.

$$x^3 - 8 = 0$$

$$(x-2)(x^2 + 2x + 4) = 0$$

One solution is -2, and the other two are solutions to $x^2 + 2x + 4 = 0$. Using the quadratic formula, the other two solutions are given by

$$x = \frac{-2 \pm \sqrt{2^2 - 4(1)(4)}}{2}.$$

This expression gives the solutions $-1 + \sqrt{3}i$ and $-1 - \sqrt{3}i$. This connection with the work in Algebra II, Module 2 is revisited in the last few exercises in the Problem Set.

Closing (3 minutes)

Ask students to respond to this question either in writing or with a partner. They can use one of the exercises above to explain the process. Then, review the formula that was derived during the Discussion portion of this lesson.

MP.6

- How do you find the nth roots of a complex number?
  - *Determine the argument and the modulus of the original number. Then, the modulus of the roots is the nth root of the original modulus. The arguments are found using the formula $\alpha = \frac{\theta}{n} + \frac{2\pi k}{n}$ for k is the integers from 0 to $n - 1$. Write the roots in polar form. If you are finding the cube roots, there will be three of them; if you are finding fourth roots, there will be four, and so on.*

Review the formula students can use to find the nth roots of a complex number.

Lesson Summary

Given a complex number z with modulus r and argument θ, the nth roots of z are given by

$$\sqrt[n]{r}\left(\cos\left(\frac{\theta}{n} + \frac{2\pi k}{n}\right) + i\sin\left(\frac{\theta}{n} + \frac{2\pi k}{n}\right)\right)$$

for integers k and n such that $n > 0$ and $0 \le k < n$.

Exit Ticket (5 minutes)

Name ______________________________ Date ______________

Lesson 19: Exploiting the Connection to Trigonometry

Exit Ticket

Find the fourth roots of $-2 - 2\sqrt{3}i$.

Exit Ticket Sample Solutions

Find the fourth roots of $-2-2\sqrt{3}i$.

The modulus is 4*, and the argument is* $\frac{4\pi}{3}$*. Using the formula, the modulus of the fourth roots will be* $\sqrt[4]{4}$*, and the arguments will be* $\frac{1}{4}\left(\frac{4\pi}{3}\right)+\frac{1}{4}(2\pi k)$ *for* $k=0,1,2,3$*. This gives the following complex numbers as the fourth roots of* $-2-2\sqrt{3}i$*.*

$$\sqrt[4]{4}\left(\cos\left(\frac{\pi}{3}\right)+i\sin\left(\frac{\pi}{3}\right)\right)=\sqrt[4]{4}\left(\frac{1}{2}+\frac{\sqrt{3}}{2}i\right)$$

$$\sqrt[4]{4}\left(\cos\left(\frac{5\pi}{6}\right)+i\sin\left(\frac{5\pi}{6}\right)\right)=\sqrt[4]{4}\left(-\frac{\sqrt{3}}{2}+\frac{1}{2}i\right)$$

$$\sqrt[4]{4}\left(\cos\left(\frac{4\pi}{3}\right)+i\sin\left(\frac{4\pi}{3}\right)\right)=\sqrt[4]{4}\left(-\frac{1}{2}-\frac{\sqrt{3}}{2}i\right)$$

$$\sqrt[4]{4}\left(\cos\left(\frac{11\pi}{6}\right)+i\sin\left(\frac{11\pi}{6}\right)\right)=\sqrt[4]{4}\left(\frac{\sqrt{3}}{2}-\frac{1}{2}i\right)$$

Problem Set Sample Solutions

1. **For each complex number, what is z^2?**
   a. $1+\sqrt{3}i$

      $-2+2\sqrt{3}i$

   b. $3-3i$

      $-18i$

   c. $4i$

      -16

   d. $-\frac{\sqrt{3}}{2}+\frac{1}{2}i$

      $\frac{1}{2}-\frac{\sqrt{3}}{2}i$

   e. $\frac{1}{9}+\frac{1}{9}i$

      $\frac{2}{81}i$

   f. -1

      1

2. For each complex number, what are the square roots of z?

a. $1 + \sqrt{3}i$

$z = 1 + \sqrt{3}i, r = 2, \arg(z) = \frac{\pi}{3},$

$\alpha = \frac{1}{2}\left(\frac{\pi}{3} + 2\pi k\right), k = 0 \text{ or } 1$ $\alpha = \frac{\pi}{6} \text{ or } \frac{7\pi}{6}$ *Let the square roots of z be w_1 and w_2.*

$w_1 = \sqrt{2}\left(\cos\frac{\pi}{6} + i \cdot \sin\frac{\pi}{6}\right) = \sqrt{2}\left(\frac{\sqrt{3}}{2} + \frac{1}{2}i\right)$

$w_2 = \sqrt{2}\left(\cos\frac{7\pi}{6} + i \cdot \sin\frac{7\pi}{6}\right) = \sqrt{2}\left(\frac{-\sqrt{3}}{2} - \frac{1}{2}i\right)$

b. $3 - 3i$

$z = 3 - 3i, r = \sqrt{18}, \arg(z) = \frac{7\pi}{4},$

$\alpha = \frac{1}{2}\left(\frac{7\pi}{4} + 2\pi k\right), k = 0 \text{ or } 1$ $\alpha = \frac{7\pi}{8} \text{ or } \frac{15\pi}{8}$ *Let the square roots of z be w_1 and w_2.*

$w_1 = \sqrt[4]{18}\left(\cos\frac{7\pi}{8} + i \cdot \sin\frac{7\pi}{8}\right)$

$w_2 = \sqrt[4]{18}\left(\cos\frac{15\pi}{8} + i \cdot \sin\frac{15\pi}{8}\right)$

c. $4i$

$z = 0 + 4i, r = 4, \arg(z) = \frac{\pi}{2},$

$\alpha = \frac{1}{2}\left(\frac{\pi}{2} + 2\pi k\right), k = 0 \text{ or } 1$ $\alpha = \frac{\pi}{4} \text{ or } \frac{5\pi}{4}$ *Let the square roots of z be w_1 and w_2.*

$w_1 = \sqrt{4}\left(\cos\frac{\pi}{4} + i \cdot \sin\frac{\pi}{4}\right) = 2\left(\frac{\sqrt{2}}{2} + \frac{\sqrt{2}}{2}i\right) = \sqrt{2} + \sqrt{2}i$

$w_2 = \sqrt{4}\left(\cos\frac{5\pi}{4} + i \cdot \sin\frac{5\pi}{4}\right) = 2\left(\frac{-\sqrt{2}}{2} + \frac{-\sqrt{2}}{2}i\right) = -\sqrt{2} - \sqrt{2}i$

d. $-\frac{\sqrt{3}}{2} + \frac{1}{2}i$

$z = \frac{-\sqrt{3}}{2} + \frac{1}{2}i, r = 1, \arg(z) = \frac{5\pi}{6},$

$\alpha = \frac{1}{2}\left(\frac{5\pi}{6} + 2\pi k\right), k = 0 \text{ or } 1$ $\alpha = \frac{5\pi}{12} \text{ or } \frac{17\pi}{12}$ *Let the square roots of z be w_1 and w_2.*

$w_1 = \sqrt{1}\left(\cos\frac{5\pi}{12} + i \cdot \sin\frac{5\pi}{12}\right) = \cos\frac{5\pi}{12} + i \cdot \sin\frac{5\pi}{12}$

$w_2 = \sqrt{1}\left(\cos\frac{5\pi}{12} + i \cdot \sin\frac{5\pi}{12}\right) = \cos\frac{17\pi}{12} + i \cdot \sin\frac{17\pi}{12}$

e. $\frac{1}{9}+\frac{1}{9}i$

$z=\frac{1}{9}+\frac{1}{9}i, r=\frac{\sqrt{2}}{9}, \arg(z)=\frac{\pi}{4},$

$\alpha=\frac{1}{2}\left(\frac{\pi}{4}+2\pi k\right), k=0$ *or* 1 $\alpha=\frac{\pi}{8}$ *or* $\frac{9\pi}{8}$ *Let the square roots of z be* w_1 *and* w_2.

$$w_1=\sqrt{\frac{\sqrt{2}}{9}}\left(\cos\frac{\pi}{8}+i\cdot\sin\frac{9\pi}{8}\right)=\frac{\sqrt[4]{2}}{3}\left(\cos\frac{\pi}{8}+i\cdot\sin\frac{9\pi}{8}\right)$$

$$w_2=\sqrt{\frac{\sqrt{2}}{9}}\left(\cos\frac{5\pi}{12}+i\cdot\sin\frac{5\pi}{12}\right)=\frac{\sqrt[4]{2}}{3}\left(\cos\frac{5\pi}{12}+i\cdot\sin\frac{5\pi}{12}\right)$$

f. -1

$z=-1+0i, r=1, \arg(z)=\pi$

$\alpha=\frac{1}{2}(\pi+2\pi k), k=0$ or 1 $\alpha=\frac{\pi}{2}$ or $\frac{3\pi}{2}$ *Let the square roots of z be* w_1 *and* w_2.

$$w_1=1\left(\cos\frac{\pi}{2}+i\cdot\sin\frac{\pi}{2}\right)=i$$

$$w_2=1\left(\cos\frac{3\pi}{2}+i\cdot\sin\frac{3\pi}{2}\right)=-i$$

3. **Graph each complex number on a polar grid.**

a. $2\left(\cos\left(\frac{\pi}{3}\right)+i\sin\left(\frac{\pi}{3}\right)\right)$

$r=2, \theta=\frac{\pi}{3}$

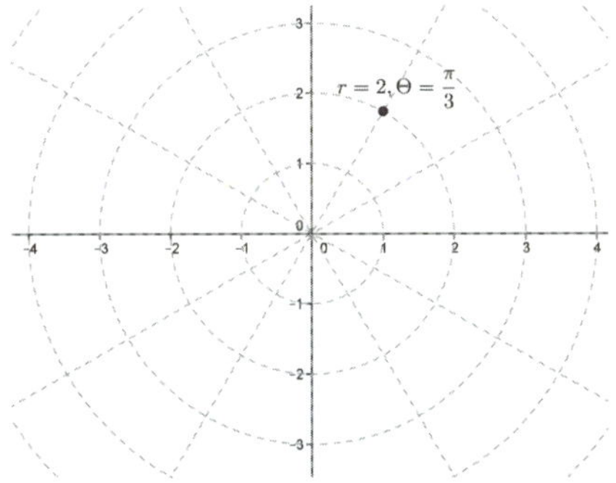

b. $3(\cos(210°) + i\sin(210°))$

$r = 3, \theta = 210°$

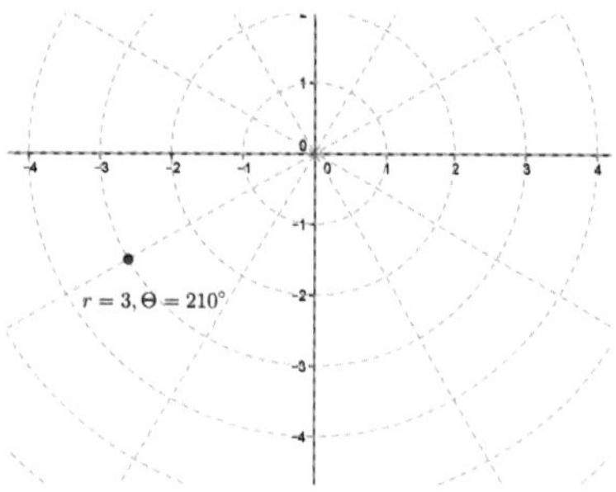

c. $2\left(\cos\left(\frac{\pi}{4}\right) + i\sin\left(\frac{\pi}{4}\right)\right)$

$r = 2, \theta = \frac{\pi}{4}$

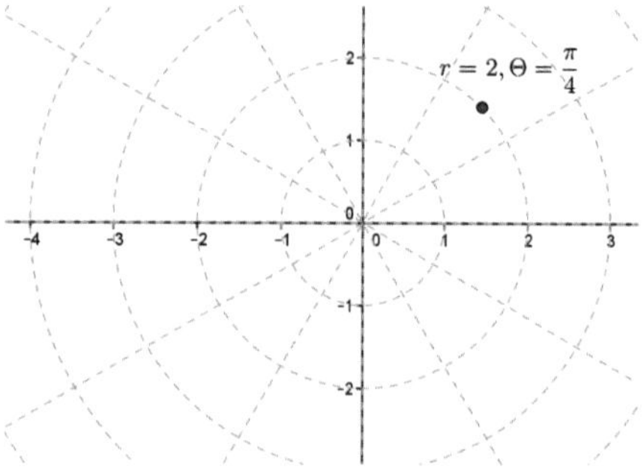

d. $\cos(\pi) + i\sin(\pi)$

$r = 1, \theta = \pi$

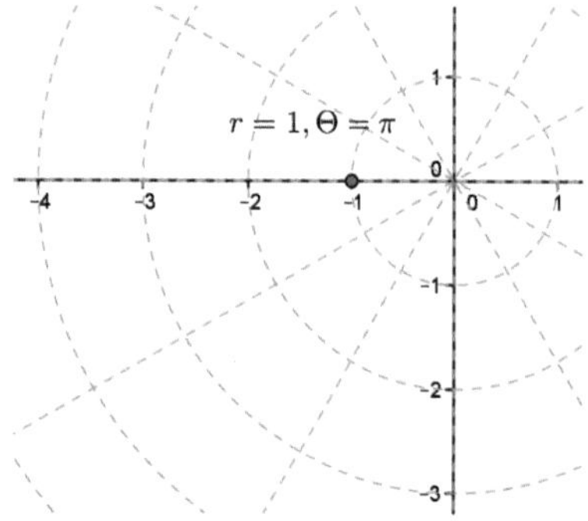

e. $4\left(\cos\left(\frac{3\pi}{4}\right) + i\sin\left(\frac{3\pi}{4}\right)\right)$

$r = 4, \theta = \frac{3\pi}{4}$

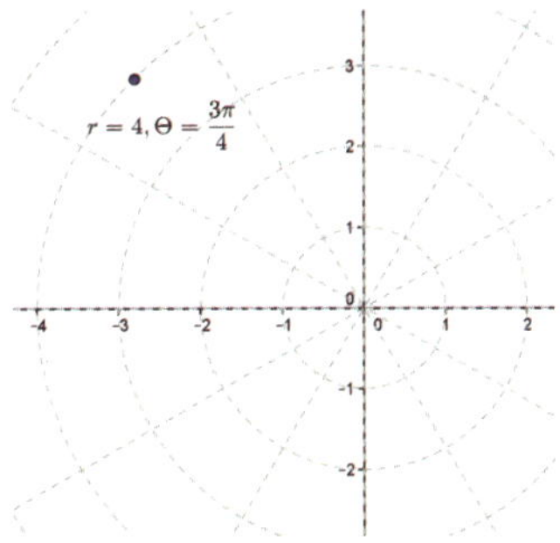

f. $\frac{1}{2}(\cos(60°) + i\sin(60°))$

$r = \frac{1}{2}, \theta = 60°$

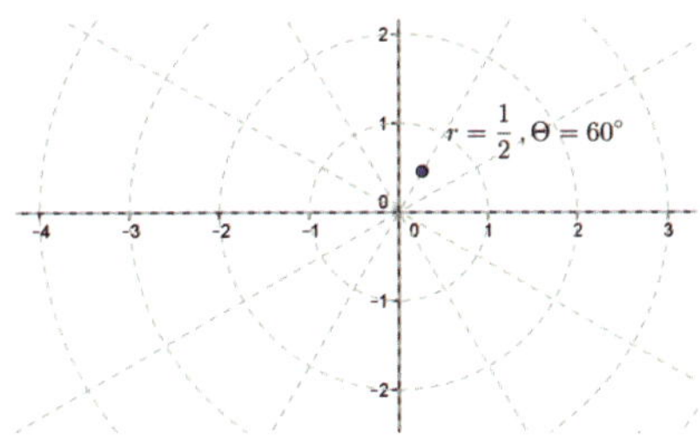

4. What are the cube roots of $-3i$?

$z = 0 - 3i, r = 3, \arg(z) = \frac{3\pi}{2}$,

$\alpha = \frac{1}{3}\left(\frac{3\pi}{2} + 2\pi k\right), k = 0, 1, or\ 2\ \alpha = \frac{\pi}{2}, \frac{7\pi}{6}\ or\ \frac{11\pi}{6}$ *Let the cube roots of z be* w_1, w_2, *and* w_3.

$w_1 = \sqrt[3]{3}\left(\cos\frac{\pi}{2} + i \cdot \sin\frac{\pi}{2}\right) = \sqrt[3]{3}(0 + i) = \sqrt[3]{3} \cdot i$

$w_2 = \sqrt[3]{3}\left(\cos\frac{7\pi}{6} + i \cdot \sin\frac{7\pi}{6}\right) = \sqrt[3]{3}\left(\frac{-\sqrt{3}}{2} - \frac{1}{2}i\right)$

$w_3 = \sqrt[3]{3}\left(\cos\frac{11\pi}{6} + i \cdot \sin\frac{11\pi}{6}\right) = \sqrt[3]{3}\left(\frac{\sqrt{3}}{2} - \frac{1}{2}i\right)$

5. What are the fourth roots of 64?

$z = 64 + 0i, r = 64, \arg(z) = 0,$

$\alpha = \frac{1}{4}(0 + 2\pi k), k = 0, 1, 2, or\ 3\ \ \alpha = 0, \frac{\pi}{2}, \pi, or\ \frac{3\pi}{2}$ *Let the fourth roots of z be* w_1, w_2, w_3, *and* w_4.

$w_1 = \sqrt[4]{64}(\cos 0 + i \cdot \sin 0) = 2\sqrt{2}(1 + 0) = 2\sqrt{2}$

$w_2 = \sqrt[4]{64}\left(\cos\frac{\pi}{2}, +i \cdot \sin\frac{\pi}{2}\right) = 2\sqrt{2}(0 + i) = 2\sqrt{2} \cdot i$

$w_3 = \sqrt[4]{64}(\cos\pi + i \cdot \sin\pi) = 2\sqrt{2}(-1 + 0) = -2\sqrt{2}$

$w_4 = \sqrt[4]{64}\left(\cos\frac{3\pi}{2} + i \cdot \sin\frac{3\pi}{2}\right) = 2\sqrt{2}(0 - i) = -2\sqrt{2} \cdot i$

6. **What are the square roots of $-4-4i$?**

$z=-4-4i, r=4\sqrt{2}, \arg(z)=\frac{5\pi}{4}$,

$\alpha=\frac{1}{2}\left(\frac{5\pi}{4}+2\pi k\right), k=0$ *or* 1 $\alpha=\frac{5\pi}{8}, \pi,$ *or* $\frac{13\pi}{8}$ *Let the square roots of z be w_1 and w_2.*

$w_1=2\sqrt[4]{2}\left(\cos\frac{5\pi}{8}+i\cdot\sin\frac{5\pi}{8}\right)$

$w_2=2\sqrt[4]{2}\left(\cos\frac{13\pi}{8}+i\cdot\sin\frac{13\pi}{8}\right)$

7. **Find the square roots of -5. Show that the square roots satisfy the equation $x^2+5=0$.**

$z=-5+0i, r=5, \arg(z)=\pi$,

$\alpha=\frac{1}{2}(\pi+2\pi k), k=0$ *or* 1 $\alpha=\frac{\pi}{2}$ *or* $\frac{3\pi}{2}$ *Let the square roots of z be w_1 and w_2.*

$w_1=\sqrt{5}\left(\cos\frac{\pi}{2}+i\cdot\sin\frac{\pi}{2}\right)=\sqrt{5}(0+i)=\sqrt{5}\cdot i$

$w_2=\sqrt{5}\left(\cos\frac{3\pi}{2}+i\cdot\sin\frac{3\pi}{2}\right)=\sqrt{5}(0-i)=-\sqrt{5}\cdot i$

$\left(\sqrt{5}\cdot i\right)^2+5=-5+5=0$

$\left(-\sqrt{5}\cdot i\right)^2+5=-5+5=0$

8. **Find the cube roots of 27. Show that the cube roots satisfy the equation $x^3-27=0$.**

$z=27+0i, r=27, \arg(z)=0$,

$\alpha=\frac{1}{3}(0+2\pi k), k=0, 1,$ *or* 2 $\alpha=0, \frac{2\pi}{3}$ *or* $\frac{4\pi}{3}$ *Let the cube roots of z be w_1, w_2, and w_3.*

$w_1=\sqrt[3]{27}(\cos 0+i\cdot\sin)=3(1+0)=3$

$w_2=\sqrt[3]{27}\left(\cos\frac{2\pi}{3}+i\cdot\sin\frac{2\pi}{3}\right)=3\left(-\frac{1}{2}+\frac{\sqrt{3}}{2}i\right)=-\frac{3}{2}+\frac{3\sqrt{3}}{2}i$

$w_3=\sqrt[3]{27}\left(\cos\frac{4\pi}{3}+i\cdot\sin\frac{4\pi}{3}\right)=3\left(-\frac{1}{2}-\frac{\sqrt{3}}{2}i\right)=-\frac{3}{2}-\frac{3\sqrt{3}}{2}i$

$(3)^3-27=0$

$$\left(3\left(-\frac{1}{2}+\frac{\sqrt{3}}{2}i\right)\right)^3-27=-27\left(\frac{-1}{8}-\frac{3\sqrt{3}}{8}i+\frac{9}{8}+\frac{3\sqrt{3}}{8}i\right)-27=-27(1)-27=0$$

$$\left(3\left(-\frac{1}{2}-\frac{\sqrt{3}}{2}i\right)\right)^3-27=-27\left(\frac{1}{8}+\frac{3\sqrt{3}}{8}i-\frac{9}{8}-\frac{3\sqrt{3}}{8}i\right)-27=-27(-1)-27=0$$

Lesson 20: Exploiting the Connection to Cartesian Coordinates

Student Outcomes

- Students interpret complex multiplication as the corresponding function of two real variables.
- Students calculate the amount of rotation and the scale factor of dilation in a transformation of the form $L(x, y) = (ax - by, bx + ay)$.

Lesson Notes

This lesson leads into the introduction of matrix notation in the next lesson. The primary purpose of this lesson is to formalize the idea that when the complex number $x + iy$ is identified with the point (x, y) in the coordinate plane, multiplication by a complex number performs a rotation and dilation in the plane. All dilations throughout this lesson and module are centered at the origin. When the formulas are written out for such rotation and dilation in terms of the real components x and y of $z = x + iy$, the formulas are rather cumbersome, leading to the need for a new notation using matrices in the next lesson. This lesson serves to solidify many of the ideas introduced in Topic B and link them to matrices. This lesson has a coherent connection to the standards within F-IF domain. For example, students connect operations with complex numbers to the language and symbols of functions.

Classwork

Opening Exercise (6 minutes)

Opening Exercise

a. **Find a complex number w so that the transformation $L_1(z) = wz$ produces a clockwise rotation by $1°$ about the origin with no dilation.**

Because there is no dilation, we need $|w| = 1$, and because there is rotation by $1°$, we need $\arg(w) = 1°$. Thus, we need to find the point where the terminal ray of a $1°$ rotation intersects the unit circle. From Algebra II, we know the coordinates of the point are

$$(x, y) = (\cos(1°), \sin(1°)),$$

so that the complex number w is

$$w = x + iy = \cos(1°) + i\sin(1°).$$

(Students may use a calculator to find the approximation $w = 0.99998 + 0.01745i$.)

Scaffolding:

- For struggling students, scaffold part (a) by first asking them to write a complex number with modulus 1 and argument $1°$, and then ask the question stated. This helps students see the connection. Do the same for part (b).
- Ask advanced students to find complex numbers w so that:

  a) the transformation $L_1(z) = wz$ produces a clockwise rotation by $\alpha°$ about the origin with no dilation, and

  b) the transformation $L_2(z) = wz$ produces a dilation with scale factor r with no rotation.

b. **Find a complex number w so that the transformation $L_2(z) = wz$ produces a dilation with scale factor 0.1 with no rotation.**

In this case, there is no rotation, so the argument of w must be 0. This means that the complex number w corresponds to a point on the positive real axis, so w has no imaginary part; this means that w is a real number, and $w = a + bi = a$. Thus, $|w| = a = 0.1$, so $w = 0.1$.

Discussion (8 minutes)

This teacher-led Discussion provides justification for why new notation needs to be developed.

- We have seen that we can use complex multiplication to perform dilation and rotation in the coordinate plane.
- By identifying the point (x, y) with the complex number $(x + yi)$, we can think of $L(z) = wz$ as a transformation in the coordinate plane. Then, complex multiplication gives us a way of finding formulas for rotation and dilation in two-dimensional geometry.
- Video game creators are very interested in the mathematics of rotation and dilation. In a first-person video game, you are centered at the origin. When you move forward in the game, the images on the screen need to undergo a translation to mimic what you see as you walk past them. As you walk closer to objects, they look larger, requiring dilation. If you turn, then the images on the screen need to rotate.
- We have established the necessary mathematics for representing rotation and dilation in two-dimensional geometry, but in video games we need to use three-dimensional geometry to mimic our three-dimensional world. Eventually, we'll need to translate our work from two dimensions into three dimensions.
- Complex numbers are inherently two-dimensional, with our association $x + iy \leftrightarrow (x, y)$. We will need some way to represent points (x, y, z) in three dimensions.
- Before we can jump to three-dimensional geometry, we need to better understand the mathematics of two-dimensional geometry.
  1. First, we will rewrite all of our work about rotation and dilation of complex numbers $x + iy$ in terms of points (x, y) in the coordinate plane and see what rotation and dilation looks like from that perspective.
  2. Then, we will see if we can generalize the mathematics of rotation and dilation of two-dimensional points (x, y) to three-dimensional points (x, y, z).
- We address point (1) in this lesson and the ones that follow and leave point (2) until the next module.
- Using the notation of complex numbers, if $w = a + bi$, then $|w| = \sqrt{a^2 + b^2}$, and $\arg(w) = \arg(a + bi)$.
- Then, how can we describe the geometric effect of multiplication by w on a complex number z?
  - *The geometric effect of multiplication wz is dilation by $|w|$ and counterclockwise rotation by $\arg(w)$ about the origin.*
- Now, let's rephrase this more explicitly as follows: Multiplying $x + yi$ by $a + bi$ rotates $x + yi$ about the origin through $\arg(a + bi)$ and dilates that point from the origin with scale factor $\sqrt{a^2 + b^2}$.
- We can further refine our statement: The transformation $(x + yi) \to (a + bi)(x + yi)$ corresponds to a rotation of the plane about the origin through $\arg(a + bi)$ and dilation with scale factor $\sqrt{a^2 + b^2}$.
- How does this transformation work on points (x, y) in the plane? Rewrite it to get a transformation in terms of coordinate points (x, y).
  - *We have $(x + iy) \to (a + bi)(x + yi)$ and $(a + bi)(x + yi) = (ax - by) + (bx + ay)i$, so we can rewrite this as the transformation $(x, y) \to (ax - by, bx + ay)$.*

- This is the formula we want for rotation and dilation of points (x, y) in the coordinate plane.
  For real numbers a and b, the transformation $L(x, y) = (ax - by, bx + ay)$ corresponds to a counterclockwise rotation by $\arg(a + bi)$ about the origin and dilation with scale factor $\sqrt{a^2 + b^2}$.
- We have just written a function in two variables. Let's practice that. If $L(x, y) = (2x, x + y)$, how can we find $L(2, 3)$? Explain this in words.
  - *We would substitute 2 in for x and 3 for y.*
- What is $L(2, 3)$?
  - $L(2, 3) = (4, 5)$
- How can $L(2, 3)$ be interpreted?
  - *When $L(2,3)$ is multiplied by $a + bi$, it is transformed to the point $(4, 5)$.*

MP.2
- Returning back to our formula, explain how the quantity $ax - by$ was derived and what it represents in the formula $L(x, y)$.
  - *When multiplying $(x + yi)$ by $(a + bi)$, the real component is $ax - by$. This represents the transformation of the x component.*

Exercises 1–4 (12 minutes)

These exercises link the geometric interpretation of rotation and dilation to the analytic formulas. Have students work on these exercises in pairs or small groups. Use these exercises to check for understanding. The exercises can be modified and/or assigned as instructionally necessary.

Exercises

1.

a. Find values of a and b so that $L_1(x, y) = (ax - by, bx + ay)$ has the effect of dilation with scale factor 2 and no rotation.

We need $\arg(a + bi) = 0$ and $\sqrt{a^2 + b^2} = 2$. Since $\arg(a + bi) = 0$, the point corresponding to $a + bi$ lies along the positive x-axis, so we know that $b = 0$ and $a > 0$. Then, we have $\sqrt{a^2 + b^2} = \sqrt{a^2} = a$, so $a = 2$. Thus, the transformation $L_1(x, y) = (2x - 0y, 0x + 2y) = (2x, 2y)$ has the geometric effect of dilation by scale factor 2.

b. Evaluate $L_1(L_1(x, y))$, and identify the resulting transformation.

$$\begin{aligned} L_1(L_1(x, y)) &= L_1(2x, 2y) \\ &= (4x, 4y) \end{aligned}$$

If we take $L_1(L_1(x, y))$, we are dilating the point (x, y) with scale factor 2 twice. This means that we are dilating with scale factor $2 \cdot 2 = 4$.

2.

a. Find values of a and b so that $L_2(x, y) = (ax - by, bx + ay)$ has the effect of rotation about the origin by $180°$ counterclockwise and no dilation.

Since there is no dilation, we have $\sqrt{a^2 + b^2} = 1$, and $\arg(a + bi) = 180°$ means that the point (a, b) lies on the negative x-axis. Then, $a < 0$ and $b = 0$, so $\sqrt{a^2 + b^2} = \sqrt{a^2} = |a| = 1$, so $a = -1$. Then, the transformation $L_2(x, y) = (-x - 0y, 0x - y) = (-x, -y)$ has the geometric effect of rotation by $180°$ without dilation.

b. **Evaluate $L_2(L_2(x,y))$, and identify the resulting transformation.**

$$\begin{aligned} L_2(L_2(x,y)) &= L_2(-x,-y) \\ &= (-(-x),-(-y)) \\ &= (x,y) \end{aligned}$$

Thus, if we take $L_2(L_2(x,y))$, we are rotating the point (x,y) by $180°$ twice, which results in a rotation of $360°$ and has the net effect of doing nothing to the point (x,y). This is the identity transformation.

3.

a. **Find values of a and b so that $L_3(x,y) = (ax - by, bx + ay)$ has the effect of rotation about the origin by $90°$ counterclockwise and no dilation.**

Since there is no dilation, we have $\sqrt{a^2+b^2} = 1$, and since the rotation is $90°$ counterclockwise, we know that $a + bi$ must lie on the positive imaginary axis. Thus, $a = 0$, and we must have $b = 1$. Then, the transformation $L_3(x,y) = (0x - y, x + 0y) = (-y, x)$ has the geometric effect of rotation by $90°$ counterclockwise with no dilation.

b. **Evaluate $L_3(L_3(x,y))$, and identify the resulting transformation.**

$$\begin{aligned} L_3(L_3(x,y)) &= L_3(-y,x) \\ &= (-x,-y) \\ &= L_2(x,y) \end{aligned}$$

Thus, if we take $L_3(L_3(x,y))$, we are rotating the point (x,y) by $90°$ twice, which results in a rotation of $180°$. This is the transformation L_2.

4.

a. **Find values of a and b so that $L_3(x,y) = (ax - by, bx + ay)$ has the effect of rotation about the origin by $45°$ counterclockwise and no dilation.**

Since there is no dilation, we have $\sqrt{a^2+b^2} = 1$, and since the rotation is $45°$ counterclockwise, we know that the point (a,b) lies on the line $y = x$, and thus, $a = b$. Then, $\sqrt{a^2+b^2} = \sqrt{a^2+a^2} = 1$, so $2a^2 = 1$, and thus $a = \frac{\sqrt{2}}{2}$, so we also have $b = \frac{\sqrt{2}}{2}$. Then, the transformation $L(x,y) = \left(\frac{\sqrt{2}}{2}x - \frac{\sqrt{2}}{2}y, \frac{\sqrt{2}}{2}x + \frac{\sqrt{2}}{2}y\right)$ has the geometric effect of rotation by $45°$ counterclockwise with no dilation.

(Students may also find the values of a and b by $+bi = \cos(45°) + i\sin(45°)$.)

b. **Evaluate $L_4(L_4(x,y))$, and identify the resulting transformation.**

We then have

$$\begin{aligned} L_4(L_4(x,y)) &= L_4\left(\frac{\sqrt{2}}{2}x - \frac{\sqrt{2}}{2}y, \frac{\sqrt{2}}{2}x + \frac{\sqrt{2}}{2}y\right) \\ &= \left(\frac{\sqrt{2}}{2}\left(\frac{\sqrt{2}}{2}x - \frac{\sqrt{2}}{2}y\right) - \frac{\sqrt{2}}{2}\left(\frac{\sqrt{2}}{2}x + \frac{\sqrt{2}}{2}y\right), \frac{\sqrt{2}}{2}\left(\frac{\sqrt{2}}{2}x - \frac{\sqrt{2}}{2}y\right) + \frac{\sqrt{2}}{2}\left(\frac{\sqrt{2}}{2}x + \frac{\sqrt{2}}{2}y\right)\right) \\ &= \left(\left(\frac{1}{2}x - \frac{1}{2}y\right) - \left(\frac{1}{2}x + \frac{1}{2}y\right), \left(\frac{1}{2}x - \frac{1}{2}y\right) + \left(\frac{1}{2}x + \frac{1}{2}y\right)\right) \\ &= (-y,x) \\ &= L_3(x,y) \end{aligned}$$

Thus, if we take $L_4(L_4(x,y))$, we are rotating the point (x,y) by $45°$ twice, which results in a rotation of $90°$. This is the transformation L_3.

Exercises 5–6 (10 minutes)

These exercises encourage students to question whether a given analytic formula represents a rotation and/or dilation. Have students work on these exercises in pairs or small groups.

5. **The figure below shows a quadrilateral with vertices $A(0,0)$, $B(1,0)$, $C(3,3)$, and $D(0,3)$.**

   a. **Transform each vertex under $L_5 = (3x + y, 3y - x)$, and plot the transformed vertices on the figure.**

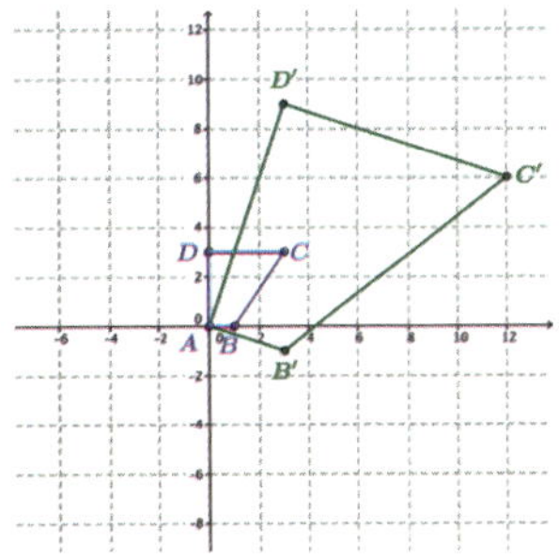

   b. **Does L_5 represent a rotation and dilation? If so, estimate the amount of rotation and the scale factor from your figure.**

   The transformed image is roughly three times larger than the original and rotated about $20°$ clockwise.

   c. **If L_5 represents a rotation and dilation, calculate the amount of rotation and the scale factor from the formula for L_5. Do your numbers agree with your estimate in part (b)? If not, explain why there are no values of a and b so that $L_5(x, y) = (ax - by, bx + ay)$.**

   From the formula, we have $a = 3$ and $b = -1$. The transformation dilates by the scale factor $|a + bi| = \sqrt{3^3 + (-1)^2} = \sqrt{10} \approx 3.16$ and rotates by $\arg(a + bi) = \arctan\left(\frac{b}{a}\right) = \arctan\left(-\frac{1}{3}\right) \approx -18.435°$.

6. **The figure below shows a figure with vertices $A(0,0)$, $B(1,0)$, $C(3,3)$, and $D(0,3)$.**

   a. **Transform each vertex under $L_6 = (2x + 2y, 2x - 2y)$, and plot the transformed vertices on the figure.**

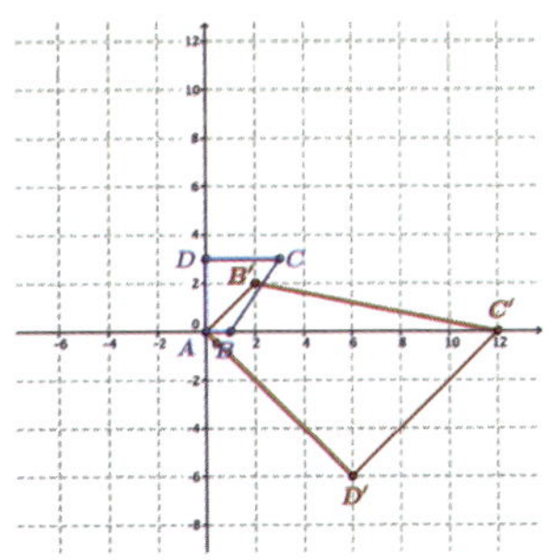

   b. **Does L_6 represent a rotation and dilation? If so, estimate the amount of rotation and the scale factor from your figure.**

   The transformed image is dilated and rotated but is also reflected, so transformation L_6 is not a rotation and dilation.

c. **If L_5 represents a rotation and dilation, calculate the amount of rotation and the scale factor from the formula for L_6. Do your numbers agree with your estimate in part (b)? If not, explain why there are no values of a and b so that $L_6(x, y) = (ax - by, bx + ay)$.**

Suppose that $(2x + 2y, 2x - 2y) = (ax - by, bx + ay)$. Then, $a = 2$ and $a = -2$, which is not possible. This transformation does not fit our formula for rotation and dilation.

Closing (4 minutes)

Ask students to summarize the lesson in writing or orally with a partner. Some key elements are summarized below.

Lesson Summary

For real numbers a and b, the transformation $L(x, y) = (ax - by, bx + ay)$ corresponds to a counterclockwise rotation by $\arg(a + bi)$ about the origin and dilation with scale factor $\sqrt{a^2 + b^2}$.

Exit Ticket (5 minutes)

Name ____________________ Date ____________

Lesson 20: Exploiting the Connection to Cartesian Coordinates

Exit Ticket

1. Find the scale factor and rotation induced by the transformation $L(x,y) = (-6x - 8y, 8x - 6y)$.

2. Explain how the transformation of complex numbers $L(x + iy) = (a + bi)(x + iy)$ leads to the transformation of points in the coordinate plane $L(x,y) = (ax - by, bx + ay)$.

Exit Ticket Sample Solutions

1. Find the scale factor and rotation induced by the transformation $L(x, y) = (-6x - 8y, 8x - 6y)$.

   This is a transformation of the form $L(x, y) = (ax - by, bx + ay)$ with $a = -6$ and $b = 8$. The scale factor is then $\sqrt{(-6)^2 + 8^2} = 10$.

   The rotation is the $\arctan\left(\frac{8}{-6}\right) \approx -51.13°$.

2. Explain how the transformation of complex numbers $L(x + iy) = (a + bi)(x + iy)$ leads to the transformation of points in the coordinate plane $L(x, y) = (ax - by, bx + ay)$.

   First, we associate the complex number $x + iy$ to the point (x, y) in the coordinate plane. Then, the point associated with the complex number $(a + bi)(x + iy) = (ax - by) + (bx + ay)i$ is $(ax - by, bx + ay)$. Thus, we can interpret the original transformation of complex numbers as the transformation of points $L(x, y) = (ax - by, bx + ay)$.

Problem Set Sample Solutions

1. Find real numbers a and b so that the transformation $L(x, y) = (ax - by, bx + ay)$ produces the specified rotation and dilation.

   a. Rotation by $270°$ counterclockwise and dilation by scale factor $\frac{1}{2}$.

   We need to find real numbers a and b so that $a + bi$ has modulus $\frac{1}{2}$ and argument $270°$. Then, (a, b) lies on the negative y-axis, so $a = 0$ and $b < 0$. We need $\frac{1}{2} = |a + bi| = |bi| = |b|$, so this means that $b = -\frac{1}{2}$. Thus, the transformation $L(x, y) = \left(\frac{1}{2}y, -\frac{1}{2}x\right)$ will rotate by $270°$ and dilate by a scale factor of $\frac{1}{2}$.

   b. Rotation by $135°$ counterclockwise and dilation by scale factor $\sqrt{2}$.

   We need to find real numbers a and b so that $a + bi$ has modulus $\sqrt{2}$ and argument $135°$. Thus, (a, b) lies in the second quadrant on the diagonal line with equation $y = -x$, so we know that $a > 0$ and $b = -a$. Since $\sqrt{2} = \sqrt{a^2 + b^2}$ and $a = -b$, we have $\sqrt{2} = \sqrt{a^2 + (-a)^2}$ so $\sqrt{2} = \sqrt{2a^2}$, and thus $a = 1$. It follows that $b = -1$. Then, the transformation $L(x, y) = (x + y, -x + y)$ rotates by $135°$ counterclockwise and dilates by a scale factor of $\sqrt{2}$.

   c. Rotation by $45°$ clockwise and dilation by scale factor 10.

   We need to find real numbers a and b so that $a + bi$ has modulus 10 and argument $45°$. Thus, (a, b) lies in the first quadrant on the line with equation $y = x$, so we know that $a = b$ and $a > 0$, $b > 0$. Since $10 = \sqrt{(a^2 + b^2)} = \sqrt{a^2 + a^2}$, we know that $2a^2 = 100$, and $a = b = 5\sqrt{2}$. Thus, the transformation $L(x, y) = (5\sqrt{2}x - 5\sqrt{2}y, 5\sqrt{2}x + 5\sqrt{2}y)$ rotates by $45°$ counterclockwise and dilates by a scale factor of 10.

d. **Rotation by 540° counterclockwise and dilation by scale factor 4.**

Rotation by 540° counterclockwise has the same effect as rotation by 180° counterclockwise. Thus, we need to find real numbers a and b so that the argument of $(a+bi)$ is 180° and $|a+bi| = \sqrt{a^2+b^2} = 4$. Since $\arg(a+bi) = 180^\circ$, we know that the point (a, b) lies on the negative x-axis, and we have $a < 0$ and $b = 0$. We then have $a = -4$ and $b = 0$, so the transformation $L(x,y) = (-4x, -4y)$ will rotate by 540° counterclockwise and dilate with scale factor 4.

2. **Determine if the following transformations represent a rotation and dilation. If so, identify the scale factor and the amount of rotation.**

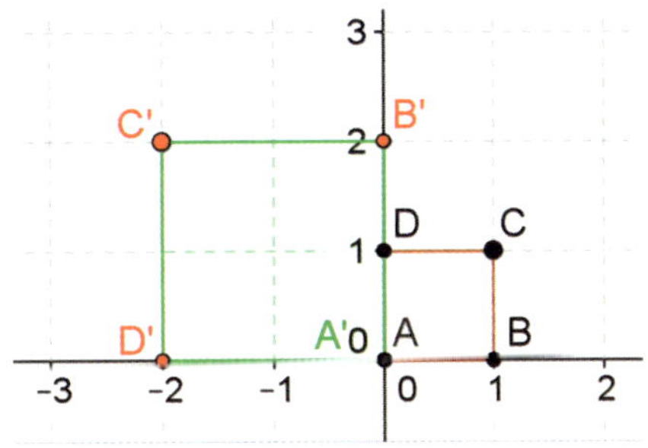

a. $L(x,y) = (3x+4y, 4x+3y)$

If $L(x,y)$ is of the form$(x,y) = (ax-by, bx+ay)$, then $a = 3$ and b must be both 3 and -3. Since this is impossible, this transformation does not consist of rotation and dilation.

b. $L(x,y) = (-5x+12y, -12x-5y)$

If we let $a = -5$ and $b = -12$, then $L(x,y)$ is of the form $(x,y) = (ax-by, bx+ay)$. Thus, this transformation does consist of rotation and dilation. The dilation has scale factor $\sqrt{(-5)^2+(-12)^2} = 13$, and the transformation rotates through $\arg(-5-12i) = \arctan\left(\frac{12}{5}\right) \approx 67.38^\circ$.

c. $L(x,y) = (3x+3y, -3y+3x)$

If we let $a = 3$ and $b = -3$, then $L(x,y)$ is of the form $(x,y) = (ax-by, bx+ay)$. Thus, the transformation does consist of rotation and dilation. The dilation has scale factor $\sqrt{(3)^2+(-3)^2} = 3\sqrt{2}$, and the transformation rotates through $\arg(3-3i) = 315^\circ$.

3. **Grace and Lily have a different point of view about the transformation on cube $ABCD$ that is shown above. Grace states that it is a reflection about the imaginary axis and a dilation of factor of 2. However, Lily argues it should be a 90° counterclockwise rotation about the origin with a dilation of a factor of 2.**

a. **Who is correct? Justify your answer.**

Lily is correct because the vertices of the cube stay the same with respect to each other.

b. **Represent the above transformation in the form $L(x,y) = (ax-by, bx+ay)$.**

Rotating 90° with a dilation of a factor of 2: $a+bi = 2(\cos(90^\circ) + i\cdot\sin(90^\circ)) = 2(0+1i) = 0+2i$. Therefore, $a = 0, b = 2, L(x,y) = (0x-2y, 2x+0y) = (-2y, 2x)$.

4. **Grace and Lily still have a different point of view on this transformation on triangle ABC shown above. Grace states that it is reflected about the real axis first, then reflected about the imaginary axis, and then dilated with a factor of 2. However, Lily asserts that it is a $180°$ counterclockwise rotation about the origin with a dilation of a factor of 2.**

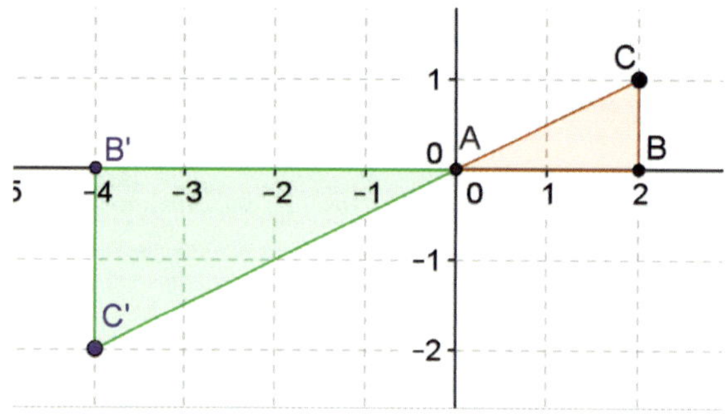

a. **Who is correct? Justify your answer.**

Both are correct. Both sequences of transformations result in the same image.

b. **Represent the above transformation in the form $L(x, y) = (ax - by, bx + ay)$.**

Rotating $180°$ with a dilation of a factor of 2:
$a + bi = 2(\cos(180°) + i \cdot \sin(180°)) = 2(-1 + 0i) = -2 + 0i.$

Therefore, $a - 2, b = 0, L(x, y) = (-2x - 0y, 0x - 2y) = (-2x, -2y).$

5. **Given $z = \sqrt{3} + i$**

a. **Find the complex number w that will cause a rotation with the same number of degrees as z without a dilation.**

$z = \sqrt{3} + i, |z| = 2, w = \frac{1}{2}(\sqrt{3} + i)$

b. **Can you come up with a general formula $L(x, y) = (ax - by, bx + ay)$ for any complex number $z = x + yi$ to represent this condition?**

$w = x + yi, |z| = \sqrt{x^2 + y^2}, a = x, b = y,$

$L(x, y) = \frac{1}{\sqrt{x^2+y^2}}(x \cdot x - y \cdot y, y \cdot x + x \cdot y) = \frac{1}{\sqrt{x^2+y^2}}(x^2 - y^2, 2xy)$

Lesson 21: The Hunt for Better Notation

Student Outcomes

- Students represent linear transformations of the form $L(x, y) = (ax + by, cx + dy)$ by matrix multiplication $L\begin{pmatrix} x \\ y \end{pmatrix} = \begin{pmatrix} a & b \\ c & d \end{pmatrix}\begin{pmatrix} x \\ y \end{pmatrix}$.
- Students recognize when a linear transformation of the form $L\begin{pmatrix} x \\ y \end{pmatrix} = \begin{pmatrix} a & b \\ c & d \end{pmatrix}\begin{pmatrix} x \\ y \end{pmatrix}$ represents rotation and dilation in the plane.
- Students multiply matrix products of the form $\begin{pmatrix} a & b \\ c & d \end{pmatrix}\begin{pmatrix} x \\ y \end{pmatrix}$.

Lesson Notes

This lesson introduces 2×2 matrices and their use for representing linear transformation through multiplication (**N-VM.C.11**, **N-VM.C.12**). Matrices provide a third method of representing rotation and dilation of the plane, as well as other linear transformations that students have not yet been exposed to in this module, such as reflection and shearing.

Classwork

Opening Exercise (5 minutes)

Have students work on this exercise in pairs or small groups. Students see how cumbersome this notation can be.

> *Scaffolding:*
> - Have struggling students evaluate $L_1(1,2)$ and $L_2(1,2)$.
> - Have advanced learners find $L_2(L_1(p,q))$ and $L_1(L_2(p,q))$ and determine values of p and q where $L_2(L_1(p,q))$ and $L_1(L_2(p,q))$ are equal.

Opening Exercise

Suppose that $L_1(x, y) = (2x - 3y, 3x + 2y)$ and $L_2(x, y) = (3x + 4y, -4y + 3x)$. Find the result of performing L_1 and then L_2 on a point (p, q). That is, find $L_2(L_1(p, q))$.

$$\begin{aligned} L_2(L_1(p,q)) &= L_2(2p - 3q, 3p + 2q) \\ &= (3(2p - 3q) + 4(3p + 2q), -4(2p - 3q) + 3(3p + 2q)) \\ &= (6p - 9q + 12p + 8q, -8p + 12q + 9p + 6q) \\ &= (18p - q, p + 18q) \end{aligned}$$

Discussion (6 minutes)

Use this Discussion to review the answer to the Opening Exercise and to motivate and introduce matrix notation.

- What answer did you get to the Opening Exercise?
  - $L_2(L_1(p, q)) = (18p - q, p + 18q)$
- How do you feel about this notation? Do you find it confusing or cumbersome?
  - *Answers will vary, but most students will find the composition confusing or cumbersome or both.*

- What if I told you there was a simpler way to find the answer? We just have to learn some new mathematics first.
- In the mid-1800s and through the early 1900s, formulas such as $L(x, y) = (ax - by, bx + ay)$ kept popping up in mathematical situations, and people were struggling to find a simpler way to work with these expressions. Mathematicians used a representation called a *matrix*. A matrix is a rectangular array of numbers that looks like $\begin{pmatrix} a \\ b \end{pmatrix}$ or $\begin{bmatrix} a \\ b \end{bmatrix}$. We can represent matrices as soft or hard brackets, but a matrix is a rectangular array of numbers. These matrices both have 1 column and 2 rows. Matrices can be any size. A square matrix has the same number of rows and columns and could look like $\begin{bmatrix} a & b \\ c & d \end{bmatrix}$. We call this a 2×2 matrix because it has 2 columns and 2 rows.
- A matrix with one column can be used to represent a point $\begin{pmatrix} x \\ y \end{pmatrix}$.
- It can also represent a vector from point A to point B. If $A(a_1, a_2)$ and $B(b_1, b_2)$, then $\overrightarrow{AB}$ can be represented as $\begin{bmatrix} b_1 - a_1 \\ b_2 - a_2 \end{bmatrix}$. This translation maps A to B.
- Explain what we have just said about a matrix and a vector to your neighbor.
- Let's think about what a transformation $L(x, y) = (ax + by, cx + dy)$ does to the components of the point (or vector) (x, y). It will be helpful to write a point (x, y) as $\begin{pmatrix} x \\ y \end{pmatrix}$. Then, the transformation becomes
$$L\begin{pmatrix} x \\ y \end{pmatrix} = \begin{pmatrix} ax + by \\ cx + dy \end{pmatrix}.$$
- The important parts of this transformation are the four coefficients a, b, c, and d. We will record them in a matrix: $\begin{pmatrix} a & b \\ c & d \end{pmatrix}$.
- A matrix is a rectangular array of numbers, symbols, or expressions arranged in rows and columns.
- We can define a new type of multiplication so that $\begin{pmatrix} a & b \\ c & d \end{pmatrix}\begin{pmatrix} x \\ y \end{pmatrix} = \begin{pmatrix} ax + by \\ cx + dy \end{pmatrix}$.

MP.7

- Based on this definition, explain how the entries in the matrix are used in the process of multiplication.
  - *When we use matrix multiplication, we think of multiplying the first row of the matrix* $(a \quad b)$ *by the column* $\begin{pmatrix} x \\ y \end{pmatrix}$ *so that* $(a \quad b) \cdot \begin{pmatrix} x \\ y \end{pmatrix} = ax + by$, *and we write that result in the first row. (This multiplication* $(a \quad b) \cdot \begin{pmatrix} x \\ y \end{pmatrix} = ax + by$ *is called a dot product. The teacher may choose whether or not to share this terminology with students.) Then, we multiply the second row of the matrix* $(c \quad d)$ *by the column* $\begin{pmatrix} x \\ y \end{pmatrix}$ *so that* $(c \quad d) \cdot \begin{pmatrix} x \\ y \end{pmatrix} = cx + dy$, *and we write that result in the second row, giving the final answer.*

Example 1 (6 minutes)

Do the following numerical examples to illustrate matrix-vector multiplication. It may be necessary to do more or fewer examples based on assessment of students' understanding.

- Evaluate the product $\begin{pmatrix} 1 & 2 \\ 3 & 4 \end{pmatrix}\begin{pmatrix} 5 \\ 6 \end{pmatrix}$.
  - $\begin{pmatrix} 1 & 2 \\ 3 & 4 \end{pmatrix}\begin{pmatrix} 5 \\ 6 \end{pmatrix} = \begin{pmatrix} 1 \cdot 5 + 2 \cdot 6 \\ 3 \cdot 5 + 4 \cdot 6 \end{pmatrix} = \begin{pmatrix} 17 \\ 39 \end{pmatrix}$

EUREKA MATH

- Evaluate the product $\begin{pmatrix}1 & 2\\3 & 4\end{pmatrix}\begin{pmatrix}2\\-5\end{pmatrix}$.
  - $\begin{pmatrix}1 & 2\\3 & 4\end{pmatrix}\begin{pmatrix}2\\-5\end{pmatrix} = \begin{pmatrix}1\cdot 2+2\cdot(-5)\\3\cdot 2+4\cdot(-5)\end{pmatrix} = \begin{pmatrix}-8\\-14\end{pmatrix}$
- Evaluate the product $\begin{pmatrix}1 & 2\\3 & 4\end{pmatrix}\begin{pmatrix}x\\y\end{pmatrix}$.
  - $\begin{pmatrix}1 & 2\\3 & 4\end{pmatrix}\begin{pmatrix}x\\y\end{pmatrix} = \begin{pmatrix}x+2y\\3x+4y\end{pmatrix}$

Exercises 1–2 (6 minutes)

Have students work these exercises in pairs or small groups.

Exercises

1. **Calculate each of the following products.**

   a. $\begin{pmatrix}3 & -2\\-1 & 4\end{pmatrix}\begin{pmatrix}1\\5\end{pmatrix}$

   $\begin{pmatrix}3-10\\-1+20\end{pmatrix} = \begin{pmatrix}-7\\19\end{pmatrix}$

   b. $\begin{pmatrix}3 & 3\\3 & 3\end{pmatrix}\begin{pmatrix}4\\-4\end{pmatrix}$

   $\begin{pmatrix}3 & 3\\3 & 3\end{pmatrix}\begin{pmatrix}4\\-4\end{pmatrix} = \begin{pmatrix}12-12\\12-12\end{pmatrix} = \begin{pmatrix}0\\0\end{pmatrix}$

   c. $\begin{pmatrix}2 & -4\\5 & -1\end{pmatrix}\begin{pmatrix}3\\-2\end{pmatrix}$

   $\begin{pmatrix}2 & -4\\5 & -1\end{pmatrix}\begin{pmatrix}3\\-2\end{pmatrix} = \begin{pmatrix}6+8\\15+2\end{pmatrix} = \begin{pmatrix}14\\17\end{pmatrix}$

2. **Find a value of k so that $\begin{pmatrix}1 & 2\\k & 1\end{pmatrix}\begin{pmatrix}3\\-1\end{pmatrix} = \begin{pmatrix}1\\11\end{pmatrix}$.**

   Multiplying this out, we have $\begin{pmatrix}1 & 2\\k & 1\end{pmatrix}\begin{pmatrix}3\\-1\end{pmatrix} = \begin{pmatrix}1\\3k-1\end{pmatrix} = \begin{pmatrix}1\\11\end{pmatrix}$, so $3k-1=11$, and thus, $k=4$.

Example 2 (6 minutes)

Use this example to connect the process of multiplying a matrix by a vector to the geometric transformations of rotation and dilation in the plane students have been doing in the past few lessons.

- We know that a linear transformation $L(x,y) = (ax-by, bx+ay)$ has the geometric effect of a counterclockwise rotation in the plane by $\arg(a+bi)$ and dilation with scale factor $|a+bi|$. How would we represent this rotation and dilation using matrix multiplication?
  - $L\begin{pmatrix}x\\y\end{pmatrix} = \begin{pmatrix}a & -b\\b & a\end{pmatrix}\begin{pmatrix}x\\y\end{pmatrix}$

MP.2

- What is the geometric effect of the transformation $L\begin{pmatrix} x \\ y \end{pmatrix} = \begin{pmatrix} 1 & -2 \\ 2 & 1 \end{pmatrix}\begin{pmatrix} x \\ y \end{pmatrix}$?
  - *This corresponds to the transformation $L(x,y) = (ax - by, bx + ay)$ with $a = 1$ and $b = 2$, so the geometric effect of this transformation is a counterclockwise rotation through $\arctan\left(\frac{2}{1}\right)$ and dilation with scale factor $|1 + 2i| = \sqrt{5}$.*
- Evaluate the product $\begin{pmatrix} 1 & -2 \\ 2 & 1 \end{pmatrix}\begin{pmatrix} 1 \\ 0 \end{pmatrix}$.
  - $\begin{pmatrix} 1 & -2 \\ 2 & 1 \end{pmatrix}\begin{pmatrix} 1 \\ 0 \end{pmatrix} = \begin{pmatrix} 1\cdot 1 + (-2)\cdot 0 \\ 2\cdot 1 + 1\cdot 0 \end{pmatrix} = \begin{pmatrix} 1 \\ 2 \end{pmatrix}$
- The points represented by $\begin{pmatrix} 1 \\ 0 \end{pmatrix}$ and $\begin{pmatrix} 1 \\ 2 \end{pmatrix}$ are shown on the axes below. We see that the point $\begin{pmatrix} 1 \\ 2 \end{pmatrix}$ is the image of the point $\begin{pmatrix} 1 \\ 0 \end{pmatrix}$ under rotation by $\arg(1 + 2i) = \arctan(2) \approx 63.435°$ and dilation by $|1 + 2i| = \sqrt{5} \approx 2.24$.

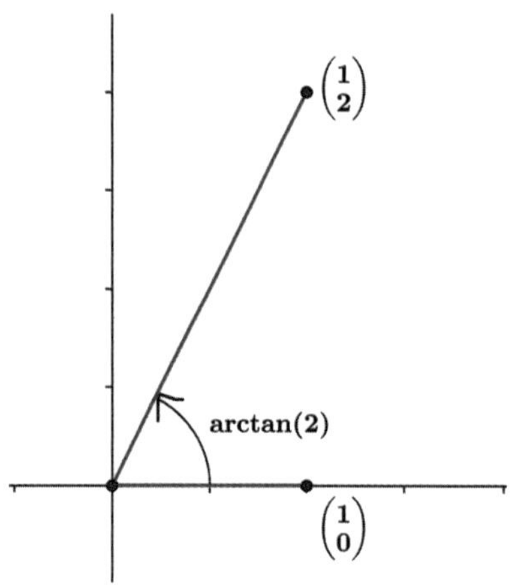

> *Scaffolding:*
> - Remember from Algebra II that $\theta = \arctan\left(\frac{b}{a}\right)$ means students are finding the angle θ such that $\tan(\theta) = \frac{b}{a}$.
> - Students know $\tan\left(\frac{\pi}{4}\right) = 1$, so $\arctan(1) = \frac{\pi}{4}$.

Exercises 3–9 (8 minutes)

Have students work in pairs or small group on these exercises.

3. Find a matrix $\begin{pmatrix} a & b \\ c & d \end{pmatrix}$ so that we can represent the transformation $L(x,y) = (2x - 3y, 3x + 2y)$ by $L\begin{pmatrix} x \\ y \end{pmatrix} = \begin{pmatrix} a & b \\ c & d \end{pmatrix}\begin{pmatrix} x \\ y \end{pmatrix}$.

The matrix is $\begin{pmatrix} 2 & -3 \\ 3 & 2 \end{pmatrix}$.

4. If a transformation $L\begin{pmatrix} x \\ y \end{pmatrix} = \begin{pmatrix} a & b \\ c & d \end{pmatrix}\begin{pmatrix} x \\ y \end{pmatrix}$ has the geometric effect of rotation and dilation, what do you know about the values $a, b, c,$ and d?

Since the transformation $L(x,y) = (ax - by, bx + ay)$ has matrix representation $L\begin{pmatrix} x \\ y \end{pmatrix} = \begin{pmatrix} a & -b \\ b & a \end{pmatrix}\begin{pmatrix} x \\ y \end{pmatrix}$, we know that $a = d$ and $c = -b$.

5. Describe the form of a matrix $\begin{pmatrix} a & b \\ c & d \end{pmatrix}$ so that the transformation $L\begin{pmatrix} x \\ y \end{pmatrix} = \begin{pmatrix} a & b \\ c & d \end{pmatrix}\begin{pmatrix} x \\ y \end{pmatrix}$ has the geometric effect of only dilation by a scale factor r.

The transformation that scales by factor r has the form $L(x, y) = r(x, y) = (rx, ry) = (rx - 0y, 0x + ry)$, so the matrix has the form $\begin{pmatrix} r & 0 \\ 0 & r \end{pmatrix}$.

6. Describe the form of a matrix $\begin{pmatrix} a & b \\ c & d \end{pmatrix}$ so that the transformation $L\begin{pmatrix} x \\ y \end{pmatrix} = \begin{pmatrix} a & b \\ c & d \end{pmatrix}\begin{pmatrix} x \\ y \end{pmatrix}$ has the geometric effect of only rotation by θ. Describe the matrix in terms of θ.

The matrix has the form $\begin{pmatrix} a & -b \\ b & a \end{pmatrix}$, where $\arg(a + bi) = \theta$. Thus, $a = \cos(\theta)$ and $b = \sin(\theta)$, so the matrix has the form $\begin{pmatrix} \cos(\theta) & -\sin(\theta) \\ \sin(\theta) & \cos(\theta) \end{pmatrix}$.

7. Describe the form of a matrix $\begin{pmatrix} a & b \\ c & d \end{pmatrix}$ so that the transformation $L\begin{pmatrix} x \\ y \end{pmatrix} = \begin{pmatrix} a & b \\ c & d \end{pmatrix}\begin{pmatrix} x \\ y \end{pmatrix}$ has the geometric effect of rotation by θ and dilation with scale factor r. Describe the matrix in terms of θ and r.

The matrix has the form $\begin{pmatrix} a & -b \\ b & a \end{pmatrix}$, where $\arg(a + bi) = \theta$ and $r = |a + bi|$. Thus, $a = r\cos(\theta)$ and $b = r\sin(\theta)$, so the matrix has the form $\begin{pmatrix} r\cos(\theta) & -r\sin(\theta) \\ r\sin(\theta) & r\cos(\theta) \end{pmatrix}$.

8. Suppose that we have a transformation $L\begin{pmatrix} x \\ y \end{pmatrix} = \begin{pmatrix} 1 & 2 \\ 0 & 1 \end{pmatrix}\begin{pmatrix} x \\ y \end{pmatrix}$.

   a. Does this transformation have the geometric effect of rotation and dilation?

   No, the matrix is not in the form $\begin{pmatrix} a & -b \\ b & a \end{pmatrix}$, so this transformation is not a rotation and dilation.

   b. Transform each of the points $A = \begin{pmatrix} 0 \\ 0 \end{pmatrix}$, $B = \begin{pmatrix} 1 \\ 0 \end{pmatrix}$, $C = \begin{pmatrix} 1 \\ 1 \end{pmatrix}$, and $D = \begin{pmatrix} 0 \\ 1 \end{pmatrix}$, and plot the images in the plane shown.

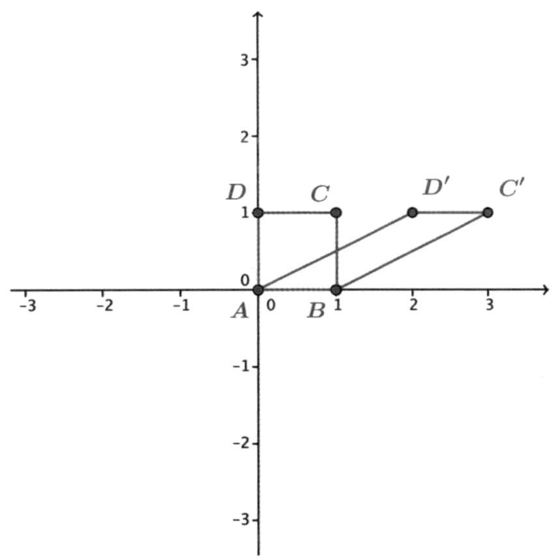

9. Describe the geometric effect of the transformation $L\begin{pmatrix}x\\y\end{pmatrix} = \begin{pmatrix}1 & 0\\0 & 1\end{pmatrix}\begin{pmatrix}x\\y\end{pmatrix}$.

This transformation does nothing to the point (x, y) in the plane; it is the identity transformation.

Closing (3 minutes)

Ask students to summarize the lesson in writing or orally with a partner. Some key elements are summarized below.

Lesson Summary

For real numbers a, b, c, and d, the transformation $L(x, y) = (ax + by, cx + dy)$ can be represented using matrix multiplication by $L\begin{pmatrix}x\\y\end{pmatrix} = \begin{pmatrix}a & b\\c & d\end{pmatrix}\begin{pmatrix}x\\y\end{pmatrix}$, where $\begin{pmatrix}a & b\\c & d\end{pmatrix}\begin{pmatrix}x\\y\end{pmatrix} = \begin{pmatrix}ax + by\\cx + dy\end{pmatrix}$ and the $\begin{pmatrix}x\\y\end{pmatrix}$ represents the point (x, y) in the plane.

- The transformation is a counterclockwise rotation by θ if and only if the matrix representation is $L\begin{pmatrix}x\\y\end{pmatrix} = \begin{pmatrix}\cos(\theta) & -\sin(\theta)\\\sin(\theta) & \cos(\theta)\end{pmatrix}\begin{pmatrix}x\\y\end{pmatrix}$.
- The transformation is a dilation with scale factor k if and only if the matrix representation is $L\begin{pmatrix}x\\y\end{pmatrix} = \begin{pmatrix}k & 0\\0 & k\end{pmatrix}\begin{pmatrix}x\\y\end{pmatrix}$.
- The transformation is a counterclockwise rotation by $\arg(a + bi)$ and dilation with scale factor $|a + bi|$ if and only if the matrix representation is $L\begin{pmatrix}x\\y\end{pmatrix} = \begin{pmatrix}a & -b\\b & a\end{pmatrix}\begin{pmatrix}x\\y\end{pmatrix}$. If we let $r = |a + bi|$ and $\theta = \arg(a + bi)$, then the matrix representation is $L\begin{pmatrix}x\\y\end{pmatrix} = \begin{pmatrix}r\cos(\theta) & -r\sin(\theta)\\r\sin(\theta) & r\cos(\theta)\end{pmatrix}\begin{pmatrix}x\\y\end{pmatrix}$.

Exit Ticket (5 minutes)

Name ______________________________ Date ______________

Lesson 21: The Hunt for Better Notation

Exit Ticket

1. Evaluate the product $\begin{pmatrix} 10 & 2 \\ -8 & -5 \end{pmatrix}\begin{pmatrix} 3 \\ -2 \end{pmatrix}$.

2. Find a matrix representation of the transformation $L(x, y) = (3x + 4y, x - 2y)$.

3. Does the transformation $L\begin{pmatrix} x \\ y \end{pmatrix} = \begin{pmatrix} 5 & 2 \\ -2 & 5 \end{pmatrix}\begin{pmatrix} x \\ y \end{pmatrix}$ represent a rotation and dilation in the plane? Explain how you know.

Exit Ticket Sample Solutions

1. Evaluate the product $\begin{pmatrix} 10 & 2 \\ -8 & -5 \end{pmatrix}\begin{pmatrix} 3 \\ -2 \end{pmatrix}$.

$$\begin{pmatrix} 10 & 2 \\ -8 & -5 \end{pmatrix}\begin{pmatrix} 3 \\ -2 \end{pmatrix} = \begin{pmatrix} 30-4 \\ -24+10 \end{pmatrix} = \begin{pmatrix} 26 \\ -14 \end{pmatrix}$$

2. Find a matrix representation of the transformation $L(x,y) = (3x+4y, x-2y)$.

$$L\begin{pmatrix} x \\ y \end{pmatrix} = \begin{pmatrix} 3 & 4 \\ 1 & -2 \end{pmatrix}\begin{pmatrix} x \\ y \end{pmatrix}$$

3. Does the transformation $L\begin{pmatrix} x \\ y \end{pmatrix} = \begin{pmatrix} 5 & 2 \\ -2 & 5 \end{pmatrix}\begin{pmatrix} x \\ y \end{pmatrix}$ represent a rotation and dilation in the plane? Explain how you know.

Yes; this transformation can also be represented as $L(x,y) = (5x-(-2)y, -2x+5y)$, which has the geometric effect of counterclockwise rotation by $\arg(5-2i)$ and dilation by $|5-2i|$.

Problem Set Sample Solutions

1. Perform the indicated multiplication.

   a. $\begin{pmatrix} 1 & 2 \\ 4 & 8 \end{pmatrix}\begin{pmatrix} 3 \\ -2 \end{pmatrix}$

   $\begin{pmatrix} -1 \\ -4 \end{pmatrix}$

   b. $\begin{pmatrix} 3 & 5 \\ -2 & -6 \end{pmatrix}\begin{pmatrix} 2 \\ 4 \end{pmatrix}$

   $\begin{pmatrix} 26 \\ -28 \end{pmatrix}$

   c. $\begin{pmatrix} 1 & 1 \\ 1 & -1 \end{pmatrix}\begin{pmatrix} 6 \\ 8 \end{pmatrix}$

   $\begin{pmatrix} 14 \\ -2 \end{pmatrix}$

   d. $\begin{pmatrix} 5 & 7 \\ 4 & 9 \end{pmatrix}\begin{pmatrix} 10 \\ 100 \end{pmatrix}$

   $\begin{pmatrix} 750 \\ 940 \end{pmatrix}$

   e. $\begin{pmatrix} 4 & 2 \\ 3 & 7 \end{pmatrix}\begin{pmatrix} -3 \\ 1 \end{pmatrix}$

   $\begin{pmatrix} -10 \\ -2 \end{pmatrix}$

EUREKA MATH™

f. $\begin{pmatrix} 6 & 4 \\ 9 & 6 \end{pmatrix}\begin{pmatrix} 2 \\ -3 \end{pmatrix}$

$\begin{pmatrix} 0 \\ 0 \end{pmatrix}$

g. $\begin{pmatrix} \cos(\theta) & -\sin(\theta) \\ \sin(\theta) & \cos(\theta) \end{pmatrix}\begin{pmatrix} x \\ y \end{pmatrix}$

$\begin{pmatrix} x\cos(\theta) - y\sin(\theta) \\ x\sin(\theta) + y\cos(\theta) \end{pmatrix}$

h. $\begin{pmatrix} \pi & 1 \\ 1 & -\pi \end{pmatrix}\begin{pmatrix} 10 \\ 7 \end{pmatrix}$

$\begin{pmatrix} 10\pi + 7 \\ 10 - 7\pi \end{pmatrix}$

2. Find a value of k so that $\begin{pmatrix} k & 3 \\ 4 & k \end{pmatrix}\begin{pmatrix} 4 \\ 5 \end{pmatrix} = \begin{pmatrix} 7 \\ 6 \end{pmatrix}$.

We have $\begin{pmatrix} k & 3 \\ 4 & k \end{pmatrix}\begin{pmatrix} 4 \\ 5 \end{pmatrix} = \begin{pmatrix} 4k + 15 \\ 16 + 5k \end{pmatrix}$, so $4k + 15 = 7$ and $16 + 5k = 6$. Thus, $4k = -8$ and $5k = -10$, so $k = -2$.

3. Find values of k and m so that $\begin{pmatrix} k & 3 \\ -2 & m \end{pmatrix}\begin{pmatrix} 5 \\ 4 \end{pmatrix} = \begin{pmatrix} 7 \\ -10 \end{pmatrix}$.

We have $\begin{pmatrix} k & 3 \\ -2 & m \end{pmatrix}\begin{pmatrix} 5 \\ 4 \end{pmatrix} = \begin{pmatrix} 7 \\ -10 \end{pmatrix}$, so $5k + 12 = 7$ and $-10 + 4m = -10$. Therefore, $k = -1$ and $m = 0$.

4. Find values of k and m so that $\begin{pmatrix} 1 & 2 \\ -2 & 5 \end{pmatrix}\begin{pmatrix} k \\ m \end{pmatrix} = \begin{pmatrix} 0 \\ -9 \end{pmatrix}$.

Since $\begin{pmatrix} 1 & 2 \\ -2 & 5 \end{pmatrix}\begin{pmatrix} k \\ m \end{pmatrix} = \begin{pmatrix} k + 2m \\ -2k + 5m \end{pmatrix}$, we need to find values of k and m so that $k + 2m = 0$ and $-2k + 5m = -9$. Solving this first equation for k gives $k = -2m$, and substituting this expression for k into the second equation gives $-9 = -2(-2m) + 5m = 9m$, so we have $m = -1$. Then, $k = -2m$ gives $k = 2$. Therefore, $k = 2$ and $m = -1$.

5. Write the following transformations using matrix multiplication.

a. $L(x, y) = (3x - 2y, 4x - 5y)$

$L\begin{pmatrix} x \\ y \end{pmatrix} = \begin{pmatrix} 3 & -2 \\ 4 & -5 \end{pmatrix}\begin{pmatrix} x \\ y \end{pmatrix}$

b. $L(x, y) = (6x + 10y, -2x + y)$

$L\begin{pmatrix} x \\ y \end{pmatrix} = \begin{pmatrix} 6 & 10 \\ -2 & 1 \end{pmatrix}\begin{pmatrix} x \\ y \end{pmatrix}$

c. $L(x, y) = (25x + 10y, 8x - 64y)$

$L\begin{pmatrix} x \\ y \end{pmatrix} = \begin{pmatrix} 25 & 10 \\ 8 & -64 \end{pmatrix}\begin{pmatrix} x \\ y \end{pmatrix}$

d. $L(x, y) = (\pi x - y, -2x + 3y)$

$L\begin{pmatrix} x \\ y \end{pmatrix} = \begin{pmatrix} \pi & -1 \\ -2 & 3 \end{pmatrix}\begin{pmatrix} x \\ y \end{pmatrix}$

e. $L(x, y) = (10x, 100x)$

$L\begin{pmatrix} x \\ y \end{pmatrix} = \begin{pmatrix} 10 & 0 \\ 100 & 0 \end{pmatrix}\begin{pmatrix} x \\ y \end{pmatrix}$

f. $L(x, y) = (2y, 7x)$

$L\begin{pmatrix} x \\ y \end{pmatrix} = \begin{pmatrix} 0 & 2 \\ 7 & 0 \end{pmatrix}\begin{pmatrix} x \\ y \end{pmatrix}$

6. **Identify whether or not the following transformations have the geometric effect of rotation only, dilation only, rotation and dilation only, or none of these.**

a. $L\begin{pmatrix} x \\ y \end{pmatrix} = \begin{pmatrix} 3 & -2 \\ 4 & -5 \end{pmatrix}\begin{pmatrix} x \\ y \end{pmatrix}$

The matrix $\begin{pmatrix} 3 & -2 \\ 4 & -5 \end{pmatrix}$ cannot be written in the form $\begin{pmatrix} a & -b \\ b & a \end{pmatrix}$, because $3 \neq -5$, so this is neither a rotation nor a dilation. The transformation L is not one of the specified types of transformations.

b. $L\begin{pmatrix} x \\ y \end{pmatrix} = \begin{pmatrix} 42 & 0 \\ 0 & 42 \end{pmatrix}\begin{pmatrix} x \\ y \end{pmatrix}$

This transformation has the geometric effect of dilation by a scale factor of 42.

c. $L\begin{pmatrix} x \\ y \end{pmatrix} = \begin{pmatrix} -4 & -2 \\ 2 & -4 \end{pmatrix}\begin{pmatrix} x \\ y \end{pmatrix}$

The matrix $\begin{pmatrix} -4 & -2 \\ 2 & -4 \end{pmatrix}$ has the form $\begin{pmatrix} a & -b \\ b & a \end{pmatrix}$ with $a = -4$ and $b = 2$. Therefore, this transformation has the geometric effect of rotation and dilation.

d. $L\begin{pmatrix} x \\ y \end{pmatrix} = \begin{pmatrix} 5 & -1 \\ -1 & 5 \end{pmatrix}\begin{pmatrix} x \\ y \end{pmatrix}$

The matrix $\begin{pmatrix} 5 & -1 \\ -1 & 5 \end{pmatrix}$ cannot be written in the form $\begin{pmatrix} a & -b \\ b & a \end{pmatrix}$, because $-1 \neq -(-1)$, so this is neither a rotation nor a dilation. The transformation L is not one of the specified types of transformations.

e. $L\begin{pmatrix} x \\ y \end{pmatrix} = \begin{pmatrix} -7 & 1 \\ 1 & 7 \end{pmatrix}\begin{pmatrix} x \\ y \end{pmatrix}$

The matrix $\begin{pmatrix} -7 & 1 \\ 1 & 7 \end{pmatrix}$ cannot be written in the form $\begin{pmatrix} a & -b \\ b & a \end{pmatrix}$, because $-7 \neq 7$, so this is neither a rotation nor a dilation. The transformation L is not one of the specified types of transformations.

f. $L\begin{pmatrix} x \\ y \end{pmatrix} = \begin{pmatrix} 0 & -2 \\ 2 & 0 \end{pmatrix}\begin{pmatrix} x \\ y \end{pmatrix}$

We see that $\begin{pmatrix} 0 & -2 \\ 2 & 0 \end{pmatrix} = \begin{pmatrix} \cos\left(\frac{\pi}{2}\right) & -\sqrt{2}\sin\left(\frac{\pi}{2}\right) \\ \sqrt{2}\sin\left(\frac{\pi}{2}\right) & \cos\left(\frac{\pi}{2}\right) \end{pmatrix}$, so this transformation has the geometric effect of dilation by $\sqrt{2}$ and rotation by $\frac{\pi}{2}$.

7. **Create a matrix representation of a linear transformation that has the specified geometric effect.**

a. **Dilation by a factor of 4 and no rotation**

$L\begin{pmatrix} x \\ y \end{pmatrix} = \begin{pmatrix} 4 & 0 \\ 0 & 4 \end{pmatrix}\begin{pmatrix} x \\ y \end{pmatrix}$

b. Rotation by $180°$ and no dilation

$$L\begin{pmatrix}x\\y\end{pmatrix} = \begin{pmatrix}\cos(180°) & -\sin(180°)\\ \sin(180°) & \cos(180°)\end{pmatrix}\begin{pmatrix}x\\y\end{pmatrix}$$
$$= \begin{pmatrix}-1 & 0\\ 0 & -1\end{pmatrix}\begin{pmatrix}x\\y\end{pmatrix}$$

c. Rotation by $-\frac{\pi}{2}$ rad and dilation by a scale factor of 3

$$L\begin{pmatrix}x\\y\end{pmatrix} = \begin{pmatrix}3\cos\left(-\frac{\pi}{2}\right) & -3\sin\left(-\frac{\pi}{2}\right)\\ 3\sin\left(-\frac{\pi}{2}\right) & 3\cos\left(-\frac{\pi}{2}\right)\end{pmatrix}\begin{pmatrix}x\\y\end{pmatrix}$$
$$= \begin{pmatrix}0 & 3\\ -3 & 0\end{pmatrix}\begin{pmatrix}x\\y\end{pmatrix}$$

d. Rotation by $30°$ and dilation by a scale factor of 4

$$L\begin{pmatrix}x\\y\end{pmatrix} = \begin{pmatrix}4\cos(30°) & -4\sin(30°)\\ 4\sin(30°) & 4\cos(30°)\end{pmatrix}\begin{pmatrix}x\\y\end{pmatrix}$$
$$= \begin{pmatrix}2\sqrt{3} & -2\\ 2 & 2\sqrt{3}\end{pmatrix}\begin{pmatrix}x\\y\end{pmatrix}$$

8. Identify the geometric effect of the following transformations. Justify your answers.

a. $L\begin{pmatrix}x\\y\end{pmatrix} = \begin{pmatrix}-\frac{\sqrt{2}}{2} & -\frac{\sqrt{2}}{2}\\ \frac{\sqrt{2}}{2} & -\frac{\sqrt{2}}{2}\end{pmatrix}\begin{pmatrix}x\\y\end{pmatrix}$

Since $\cos\left(\frac{3\pi}{4}\right) = -\frac{\sqrt{2}}{2}$ *and* $\sin\left(\frac{3\pi}{4}\right) = \frac{\sqrt{2}}{2}$, *this transformation has the form*

$L\begin{pmatrix}x\\y\end{pmatrix} = \begin{pmatrix}\cos\left(\frac{3\pi}{4}\right) & -\sin\left(\frac{3\pi}{4}\right)\\ \sin\left(\frac{3\pi}{4}\right) & \cos\left(\frac{3\pi}{4}\right)\end{pmatrix}\begin{pmatrix}x\\y\end{pmatrix}$ *and, thus, represents counterclockwise rotation by* $\frac{3\pi}{4}$ *with no dilation.*

b. $L\begin{pmatrix}x\\y\end{pmatrix} = \begin{pmatrix}0 & -5\\ 5 & 0\end{pmatrix}\begin{pmatrix}x\\y\end{pmatrix}$

Since $\cos\left(\frac{\pi}{2}\right) = 0$ *and* $\sin\left(\frac{\pi}{2}\right) = 1$, *this transformation has the form* $L\begin{pmatrix}x\\y\end{pmatrix} = \begin{pmatrix}5\cos\left(\frac{\pi}{2}\right) & -5\sin\left(\frac{\pi}{2}\right)\\ 5\sin\left(\frac{\pi}{2}\right) & 5\cos\left(\frac{\pi}{2}\right)\end{pmatrix}\begin{pmatrix}x\\y\end{pmatrix}$ *and, thus, represents counterclockwise rotation by* $\frac{\pi}{2}$ *and dilation by a scale factor* 5.

c. $L\begin{pmatrix}x\\y\end{pmatrix} = \begin{pmatrix}-10 & 0\\ 0 & -10\end{pmatrix}\begin{pmatrix}x\\y\end{pmatrix}$

Since $\cos(\pi) = -1$ *and* $\sin(\pi) = 0$, *this transformation has the form*

$L\begin{pmatrix}x\\y\end{pmatrix} = \begin{pmatrix}10\cos(\pi) & -10\sin(\pi)\\ 10\sin(\pi) & 10\cos(\pi)\end{pmatrix}\begin{pmatrix}x\\y\end{pmatrix}$ *and, thus, represents counterclockwise rotation by* π *and dilation by a scale factor* 10.

d. $L\begin{pmatrix} x \\ y \end{pmatrix} = \begin{pmatrix} 6 & 6\sqrt{3} \\ -6\sqrt{3} & 6 \end{pmatrix}\begin{pmatrix} x \\ y \end{pmatrix}$

Since $\cos\left(\frac{5\pi}{3}\right) = \frac{1}{2}$ *and* $\sin\left(\frac{5\pi}{3}\right) = -\frac{\sqrt{3}}{2}$*, this transformation has the form*

$L\begin{pmatrix} x \\ y \end{pmatrix} = \begin{pmatrix} 12\cos\left(\frac{5\pi}{3}\right) & -12\sin\left(\frac{5\pi}{3}\right) \\ 12\sin\left(\frac{5\pi}{3}\right) & 12\cos\left(\frac{5\pi}{3}\right) \end{pmatrix}\begin{pmatrix} x \\ y \end{pmatrix}$ *and, thus, represents counterclockwise rotation by* $\frac{5\pi}{3}$ *and dilation with scale factor* 12.

Lesson 22: Modeling Video Game Motion with Matrices

Student Outcomes

- Students use matrix transformations to represent motion along a straight line.

Lesson Notes

This is the first of a two-day lesson where students use their knowledge of 2×2 matrices and their transformations to program video game motion. Lesson 22 focuses on straight-line motion. In Lesson 23, students extend that motion to include rotations. In programming, students multiply matrices and vectors (**N-VM.C.11**) and use matrices to perform transformations in the plane (**N-VM.C.12**). This lesson focuses on MP.4 as students use mathematics (matrices and transformations) to model a real-world situation (video game programming).

Classwork

Opening Exercise (2 minutes)

Opening Exercise

Let $D\begin{pmatrix} x \\ y \end{pmatrix} = \begin{pmatrix} 3 & 0 \\ 0 & 3 \end{pmatrix}\begin{pmatrix} x \\ y \end{pmatrix}$.

a. **Plot the point $\begin{pmatrix} 2 \\ 1 \end{pmatrix}$.**

b. **Find $D\begin{pmatrix} 2 \\ 1 \end{pmatrix}$ and plot it.**

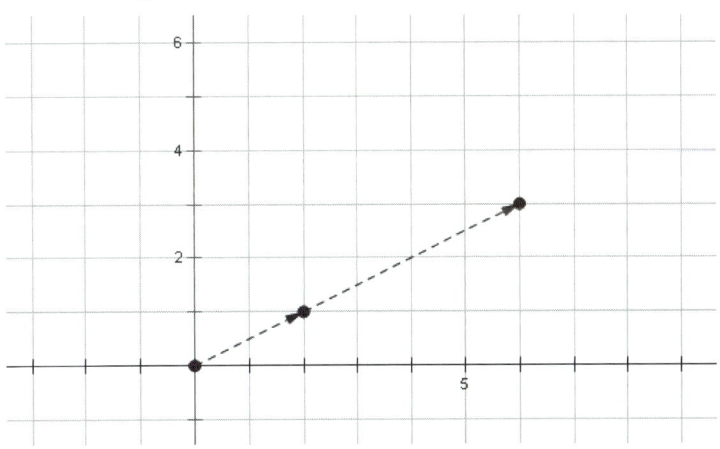

c. **Describe the geometric effect of performing the transformation $\begin{pmatrix} x \\ y \end{pmatrix} \to D\begin{pmatrix} x \\ y \end{pmatrix}$.**

MP.2

Each point in the plane gets dilated by a factor of 3. In other words, a point P gets moved to a new location that is on the line through P and the origin, but its distance from the origin increases by a factor of 3.

Discussion (9 minutes): Motion Along a Line

Let $f(t) = \begin{pmatrix} t & 0 \\ 0 & t \end{pmatrix}\begin{pmatrix} 3 \\ 1 \end{pmatrix}$.

- Find $f(0)$, $f(1)$, $f(2)$, $f(3)$, and $f(4)$.
  - $f(0) = \begin{pmatrix} 0 & 0 \\ 0 & 0 \end{pmatrix}\begin{pmatrix} 3 \\ 1 \end{pmatrix} = \begin{pmatrix} 0 \\ 0 \end{pmatrix}$
  - $f(1) = \begin{pmatrix} 1 & 0 \\ 0 & 1 \end{pmatrix}\begin{pmatrix} 3 \\ 1 \end{pmatrix} = \begin{pmatrix} 3 \\ 1 \end{pmatrix}$
  - $f(2) = \begin{pmatrix} 2 & 0 \\ 0 & 2 \end{pmatrix}\begin{pmatrix} 3 \\ 1 \end{pmatrix} = \begin{pmatrix} 6 \\ 2 \end{pmatrix}$
  - $f(3) = \begin{pmatrix} 3 & 0 \\ 0 & 3 \end{pmatrix}\begin{pmatrix} 3 \\ 1 \end{pmatrix} = \begin{pmatrix} 9 \\ 3 \end{pmatrix}$
  - $f(4) = \begin{pmatrix} 4 & 0 \\ 0 & 4 \end{pmatrix}\begin{pmatrix} 3 \\ 1 \end{pmatrix} = \begin{pmatrix} 12 \\ 4 \end{pmatrix}$

MP.7 & MP.8

- Plot each of these points on a graph. What do you notice?
  - *Each point appears to lie on a straight line through the origin.*

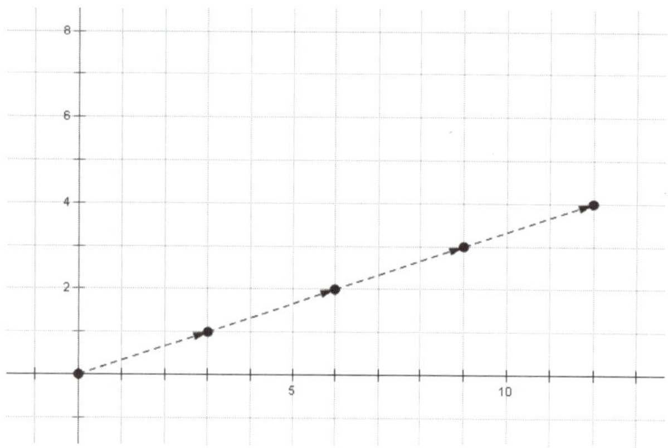

> **Scaffolding:**
> - Support students in understanding the functions defined by matrices using simpler examples.
> - Let $f(t) = 3t$. Find $f(0), f(1), f(2), f(3), f(4)$.
> - Explain how you determined your answer.
> - Advanced learners can be given the following prompt without supporting questions.
> - Suppose you were designing a computer game. You want an object to travel along a line from the origin to the point $(30, 10)$ in 20 seconds. Can you design a function $h(t)$ that does this?

- Let $P = f(t)$. How can we be sure that P actually does trace out a straight line as t varies?
  - *We could check to see if every pair of points forms a segment with the same slope.*
- What is the slope of the segment that joins $f(1)$ and $f(4)$?
  - *Since $f(1) = (3,1)$ and $f(4) = (12,4)$, the slope is $\frac{4-1}{12-3} = \frac{3}{9} = \frac{1}{3}$.*
- Now let's check to see if every pair of points forms a segment with slope $\frac{1}{3}$ also. Let t_1 and t_2 be two arbitrary times in the domain of f. What is the slope of the segment that joins $f(t_1)$ and $f(t_2)$?
  - *Since $f(t_1) = (3t_1, t_1)$ and $f(t_2) = (3t_2, t_2)$, the slope of the segment is $\frac{t_2-t_1}{3t_2-3t_1} = \frac{t_2-t_1}{3(t_2-t_1)} = \frac{1}{3}$. Since the slope of every segment is constant, we can conclude that the path traced out by P is indeed a straight line.*
- Now suppose that t represents time, measured in seconds, and $f(t)$ represents the location of an object at time t. How long would it take the object to travel from the origin to the point $(30, 10)$?
  - *We need to find a value of t such that $\begin{pmatrix} t & 0 \\ 0 & t \end{pmatrix}\begin{pmatrix} 3 \\ 1 \end{pmatrix} = \begin{pmatrix} 30 \\ 10 \end{pmatrix}$. Apparently, $t = 10$ works, which means it would take 10 seconds for the object to reach this point.*

- Now let $g(t) = \begin{pmatrix} 2t & 0 \\ 0 & 2t \end{pmatrix}\begin{pmatrix} 3 \\ 1 \end{pmatrix}$. Do you think the object will reach $(30,10)$ faster or slower? Go ahead and find out.
  - *If we choose t so that $2t = 10$, then we'd have $\begin{pmatrix} 10 & 0 \\ 0 & 10 \end{pmatrix}\begin{pmatrix} 3 \\ 1 \end{pmatrix} = \begin{pmatrix} 30 \\ 10 \end{pmatrix}$, so t must be 5. Therefore, the object reaches the desired location in 5 seconds.*
- Suppose you were designing a computer game. You want an object to travel along a line from the origin to the point $(30, 10)$ in 20 seconds. Can you design a function $h(t)$ that does this?
  - *We need to find a scale factor k such that $h(20) = \begin{pmatrix} k \cdot 20 & 0 \\ 0 & k \cdot 20 \end{pmatrix}\begin{pmatrix} 3 \\ 1 \end{pmatrix} = \begin{pmatrix} 30 \\ 10 \end{pmatrix}$. Seeing that we need to have $20k = 10$ for this to work, we must have $k = \frac{1}{2}$. Thus, $h(t) = \begin{pmatrix} \frac{1}{2}t & 0 \\ 0 & \frac{1}{2}t \end{pmatrix}\begin{pmatrix} 3 \\ 1 \end{pmatrix}$.*
- Before we move on, try to make sense of the relationship between $f(t)$, $g(t)$, and $h(t)$. Take about half a minute to think about this for yourself, then share with a partner, and then we'll discuss your responses as a whole class.
  - *$f(t)$, $g(t)$, and $h(t)$ use scale factors of t, $2t$, and $\frac{1}{2}t$, respectively.*
  - *Since g doubles the t-value, it makes sense that the object is moving twice as fast. For instance, to make the scale factor equal 10, we can use $t = 5$, since $2(5) = 10$. So, it takes 5 seconds instead of 10 to reach the desired point.*
  - *On the other hand, h cuts the t-value in half, so it would make sense to say that the object should move only half as fast. In particular, to make the scale factor 10, we have to use $t = 20$ because $\frac{1}{2}(20) = 10$. Thus, it took 20 seconds instead of 10 to reach the desired point, which is twice as much time as it took originally.*

Exercises 1–2 (3 minutes)

Give students time to perform the following exercises, and then instruct students to compare their responses with a partner. Select students to share their responses with the whole class.

Exercises 1–2

1. **Let $f(t) = \begin{pmatrix} t & 0 \\ 0 & t \end{pmatrix}\begin{pmatrix} 2 \\ 4 \end{pmatrix}$, where t represents time, measured in seconds. $P = f(t)$ represents the position of a moving object at time t. If the object starts at the origin, how long would it take to reach $(12, 24)$?**

   $\begin{pmatrix} t & 0 \\ 0 & t \end{pmatrix}\begin{pmatrix} 2 \\ 4 \end{pmatrix} = \begin{pmatrix} 12 \\ 24 \end{pmatrix}$, $t \times 2 + 0 \times 4 = 12$, $2t = 12$, $t = 6$, *or*

   $\begin{pmatrix} t & 0 \\ 0 & t \end{pmatrix}\begin{pmatrix} 2 \\ 4 \end{pmatrix} = \begin{pmatrix} 12 \\ 24 \end{pmatrix}$, $0 \times 2 + t \times 4 = 24$, $4t = 24$, $t = 6$

2. **Let $g(t) = \begin{pmatrix} kt & 0 \\ 0 & kt \end{pmatrix}\begin{pmatrix} 2 \\ 4 \end{pmatrix}$.**

   a. **Find the value of k that moves an object from the origin to $(12, 24)$ in just 2 seconds.**

   $t = 2$, $\begin{pmatrix} 2k & 0 \\ 0 & 2k \end{pmatrix}\begin{pmatrix} 2 \\ 4 \end{pmatrix} = \begin{pmatrix} 12 \\ 24 \end{pmatrix}$, $2k \times 2 + 0 \times 4 = 12$, $k = \frac{12}{4} = 3$, *or*

   $t = 2$, $\begin{pmatrix} 2k & 0 \\ 0 & 2k \end{pmatrix}\begin{pmatrix} 2 \\ 4 \end{pmatrix} = \begin{pmatrix} 12 \\ 24 \end{pmatrix}$, $0 \times 2k + 2k \times 4 = 24$, $k = \frac{24}{8} = 3$

b. Find the value of k that moves an object from the origin to $(12, 24)$ in 30 seconds.

$t = 30, \begin{pmatrix} 30k & 0 \\ 0 & 30k \end{pmatrix}\begin{pmatrix} 2 \\ 4 \end{pmatrix} = \begin{pmatrix} 12 \\ 24 \end{pmatrix}, 30k \times 2 + 0 \times 4 = 12, k = \frac{12}{60} = \frac{1}{5}$, *or*

$t = 30, \begin{pmatrix} 30k & 0 \\ 0 & 30k \end{pmatrix}\begin{pmatrix} 2 \\ 4 \end{pmatrix} = \begin{pmatrix} 12 \\ 24 \end{pmatrix}, 0 \times 30k + 30k \times 4 = 24, k = \frac{24}{120} = \frac{1}{5}$

Example 1 (3 minutes)

- Let's continue our exploration of the function $f(t) = \begin{pmatrix} t & 0 \\ 0 & t \end{pmatrix}\begin{pmatrix} 3 \\ 1 \end{pmatrix}$. To get some practice with different ways of representing transformations, let's write $f(t)$ in the form $\begin{pmatrix} x(t) \\ y(t) \end{pmatrix}$.
  - $f(t) = \begin{pmatrix} t & 0 \\ 0 & t \end{pmatrix}\begin{pmatrix} 3 \\ 1 \end{pmatrix} = \begin{pmatrix} 3t + 0 \\ 0 + t \end{pmatrix} = \begin{pmatrix} 3t \\ t \end{pmatrix}$
- Let's suppose that an object is moving in a straight line, with the x-coordinate increasing at 3 units per second and the y-coordinate increasing at 1 unit per second, as with $f(t)$ above. If the object starts at $(12, 4)$, how long would it take to reach $(30, 10)$?
  - *The x-coordinate is increasing at 3 units per second, so we have $12 + 3t = 30$, which gives $t = 6$ seconds.*
  - *The y-coordinate is increasing at 1 unit per second, so we have $4 + t = 10$, which also gives $t = 6$ seconds. So, our results corroborate each other.*
- Can you write a new function $g(t)$ that gives the position of the object above after t seconds?
  - *We have $x(t) = 12 + 3t$ and $y(t) = 4 + t$, so that gives $g(t) = \begin{pmatrix} 12 + 3t \\ 4 + t \end{pmatrix}$.*
- Can you find a way to write $g(t)$ as a matrix transformation?
  - $g(t) = \begin{pmatrix} 12 + 3t \\ 4 + t \end{pmatrix} = \begin{pmatrix} 3(4 + t) \\ 1(4 + t) \end{pmatrix}$ *This looks like a dilation of $(3, 1)$ with scale factor $4 + t$, so the matrix representation of this transformation is $g(t) = \begin{pmatrix} 4 + t & 0 \\ 0 & 4 + t \end{pmatrix}\begin{pmatrix} 3 \\ 1 \end{pmatrix}$.*

Exercises 3–4 (3 minutes)

Give students a minute to perform the following exercises; monitor their responses. Then, present the solutions, and ask students to check their answers.

3. Let $f(t) = \begin{pmatrix} 2 + t & 0 \\ 0 & 2 + t \end{pmatrix}\begin{pmatrix} 5 \\ 7 \end{pmatrix}$, where t represents time, measured in seconds, and $f(t)$ represents the position of a moving object at time t.

a. Find the position of the object at $t = 0, t = 1$, and $t = 2$.

$t = 0, \begin{pmatrix} 2 & 0 \\ 0 & 2 \end{pmatrix}\begin{pmatrix} 5 \\ 7 \end{pmatrix} = \begin{pmatrix} 2 \times 5 + 0 \times 7 \\ 0 \times 5 + 2 \times 7 \end{pmatrix} = \begin{pmatrix} 10 \\ 14 \end{pmatrix}$ *or* $10 + 14i$

$t = 1, \begin{pmatrix} 3 & 0 \\ 0 & 3 \end{pmatrix}\begin{pmatrix} 5 \\ 7 \end{pmatrix} = \begin{pmatrix} 3 \times 5 + 0 \times 7 \\ 0 \times 5 + 3 \times 7 \end{pmatrix} = \begin{pmatrix} 15 \\ 21 \end{pmatrix}$, *or* $15 + 21i$

$t = 2, \begin{pmatrix} 4 & 0 \\ 0 & 4 \end{pmatrix}\begin{pmatrix} 5 \\ 7 \end{pmatrix} = \begin{pmatrix} 4 \times 5 + 0 \times 7 \\ 0 \times 5 + 4 \times 7 \end{pmatrix} = \begin{pmatrix} 20 \\ 28 \end{pmatrix}$, *or* $20 + 28i$

b. **Write $f(t)$ in the form $\begin{pmatrix} x(t) \\ y(t) \end{pmatrix}$.**

$$\begin{pmatrix} 2+t & 0 \\ 0 & 2+t \end{pmatrix}\begin{pmatrix} 5 \\ 7 \end{pmatrix} = \begin{pmatrix} 10+5t \\ 14+7t \end{pmatrix}$$

4. **Write the transformation $g(t) = \begin{pmatrix} 15+5t \\ -6-2t \end{pmatrix}$ as a matrix transformation.**

Answers vary based on factoring of factors. However, they start at different points that are all from the line, and they all end up having the same result: $g(t) = \begin{pmatrix} 15+5t \\ -6-2t \end{pmatrix}$.

$$\begin{pmatrix} 15+5t \\ -6-2t \end{pmatrix} = \begin{pmatrix} 5(3+t) \\ -2(3+t) \end{pmatrix} = \begin{pmatrix} 3+t & 0 \\ 0 & 3+t \end{pmatrix}\begin{pmatrix} 5 \\ -2 \end{pmatrix} \text{ or } \begin{pmatrix} 15+5t \\ -6-2t \end{pmatrix} = \begin{pmatrix} -5(-3-t) \\ 2(-3-t) \end{pmatrix} = \begin{pmatrix} -3-t & 0 \\ 0 & -3-t \end{pmatrix}\begin{pmatrix} -5 \\ 2 \end{pmatrix}$$

Example 2 (4 minutes)

Let $f(t) = \begin{pmatrix} 1-t & 0 \\ 0 & 1-t \end{pmatrix}\begin{pmatrix} 3 \\ 1 \end{pmatrix}$.

- Graph the path traced out by $P = f(t)$ with $0 \le t \le 3$.
  - *We have $f(0) = (3,1)$, $f(1) = (0,0)$, $f(2) = (-3,-1)$, and $f(3) = (-6,-2)$.*

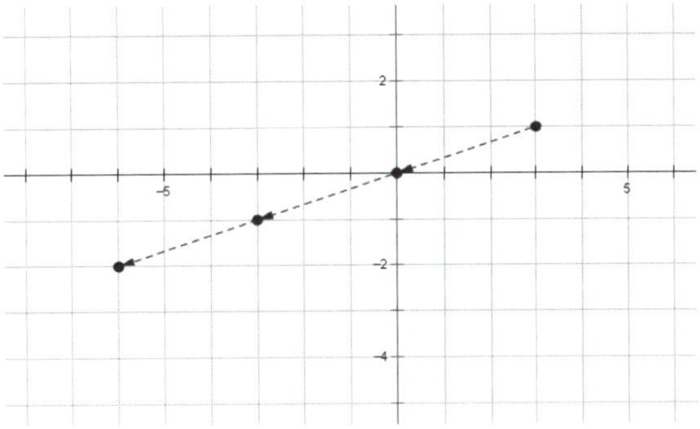

- Write $f(t)$ in the form $\begin{pmatrix} x(t) \\ y(t) \end{pmatrix}$.
  - $f(t) = \begin{pmatrix} 1-t & 0 \\ 0 & 1-t \end{pmatrix}\begin{pmatrix} 3 \\ 1 \end{pmatrix} = \begin{pmatrix} (1-t)(3)+0 \\ 0+(1-t)(1) \end{pmatrix} - \begin{pmatrix} 3-3t \\ 1-t \end{pmatrix}$
- Now suppose that an object starts at $(20, 16)$ and moves along a line, reaching the origin in 4 seconds. Write an equation $P = h(t)$ for the position of the object at time t.
  - *Looking at the x-coordinates, we see that $20 - k(4) = 0$, which means that $k = 5$. That is, the x-coordinate of the point is decreasing at 5 units per second. Thus, $x(t) = 20 - 5t$.*
  - *Looking at the y-coordinates, we see that $16 - m(4) = 0$, which means that $m = 4$. That is, the y-coordinate of the point is decreasing at 4 units per second. Thus, $y(t) = 16 - 4t$.*
  - *Putting these two results together, we get $h(t) = \begin{pmatrix} 20-5t \\ 16-4t \end{pmatrix}$.*
- Write $h(t)$ as a matrix transformation.
  - $h(t) = \begin{pmatrix} 20-5t \\ 16-4t \end{pmatrix} = \begin{pmatrix} 5(4-t) \\ 4(4-t) \end{pmatrix} = \begin{pmatrix} 4-t & 0 \\ 0 & 4-t \end{pmatrix}\begin{pmatrix} 5 \\ 4 \end{pmatrix}$

Exercise 5 (2 minutes)

Give students a minute to perform the following exercise, and monitor their responses. Have students compare their responses with a partner; then, select students to share their responses with the whole class.

5. **An object is moving in a straight line from $(18, 12)$ to the origin over a 6-second period of time. Find a function $f(t)$ that gives the position of the object after t seconds. Write your answer in the form $f(t) = \begin{pmatrix} x(t) \\ y(t) \end{pmatrix}$, and then express $f(t)$ as a matrix transformation.**

For the x-coordinates, we have $18 - 6k = 0$, $k = 3$. The x-coordinate of the point is decreasing at 3 units per second. Thus, $x(t) = 18 - 3t$.

For the y-coordinates, we have $12 - 6m = 0$, $m = 2$. The y-coordinate of the point is decreasing at 2 units per second. Thus, $x(t) = 12 - 2t$.

$$f(t) = \begin{pmatrix} 18 - 3t \\ 12 - 2t \end{pmatrix} = \begin{pmatrix} 3(6 - t) \\ 2(6 - t) \end{pmatrix} = \begin{pmatrix} 6 - t & 0 \\ 0 & 6 - t \end{pmatrix} \begin{pmatrix} 3 \\ 2 \end{pmatrix}$$

Discussion (7 minutes): Translations

- In a video game, the player controls a character named Steve. When Steve climbs a certain ladder, his vertical position on the screen increases by 5 units.

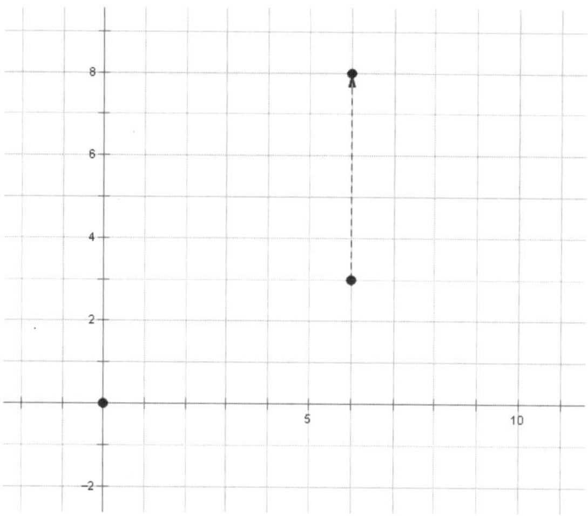

- Let $(x, y) \to V(x, y)$ represent the change in Steve's position when he climbs the ladder. The input represents his position before climbing the ladder, and the output represents his position after climbing the ladder. Find the outputs that correspond to each of the following inputs.
- $(3,4)$
  - $(3,4) \to (3,9)$
- $(10,12)$
  - $(10,12) \to (10,17)$
- $(7,20)$
  - $(7,20) \to (7,25)$

- Let's look more closely at that last input-output pair. Can you carefully explain the thinking that allowed you to produce the output here?
  - *Climbing a ladder does not affect Steve's horizontal position, so the x-coordinate is still 7. To get the new y-coordinate, we add 5 to 20, giving $20 + 5 = 25$.*
- To reveal the underlying structure of this transformation, let's write $(7, 20) \to (7, 20 + 5)$. Now let's generalize: What is the output that corresponds to a generic input (x, y)?
  - $(x, y) \to (x, y + 5)$
- Now let's write the transformation using the column notation that we have found useful for our work that involves matrices: $\binom{x}{y} \to \binom{x}{y+5}$.
- Next, let's analyze horizontal motion. When the player presses the control pad to the right, Steve moves to the right 3 units per second.

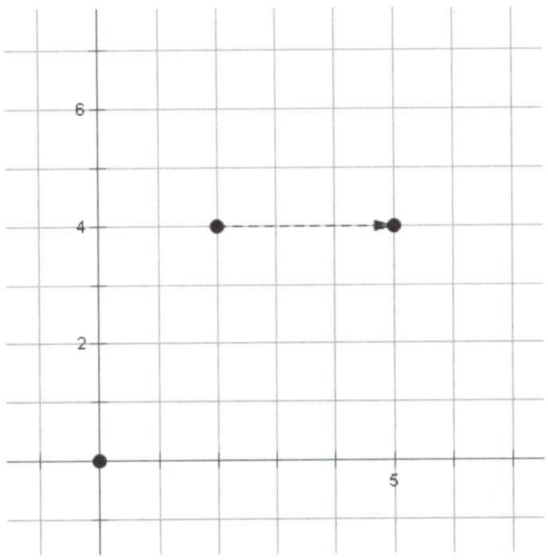

- Write a function rule that represents a translation that takes each point in the plane 3 units to the right. Practice using the column notation.
  - $\binom{x}{y} \to \binom{x+3}{y}$
- When Steve jumps while running at super speed, he moves to a new location that is 6 units to the right and 4 units above where he started the jump.

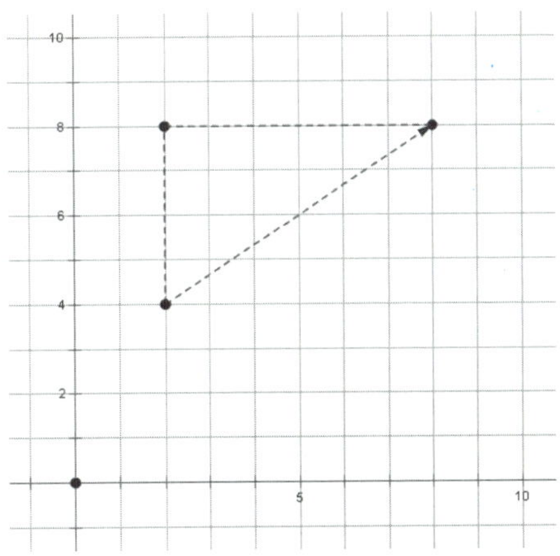

MP.4

- Write a function rule that represents the change in Steve's position when he does a jump while running at super speed. Use column notation.
  - $\begin{pmatrix} x \\ y \end{pmatrix} \to \begin{pmatrix} x+6 \\ y+4 \end{pmatrix}$
- Do you think there is a way we can represent a translation as a matrix transformation? In particular, can we encode the transformation $\begin{pmatrix} x \\ y \end{pmatrix} \to \begin{pmatrix} x+6 \\ y+4 \end{pmatrix}$ as a matrix mapping $\begin{pmatrix} x \\ y \end{pmatrix} \to \begin{pmatrix} a & b \\ c & d \end{pmatrix}\begin{pmatrix} x \\ y \end{pmatrix}$?

Let students consider this question for a few moments.

- Here's a hint: If we take the point $(0,0)$ as an input, what output is produced in each transformation above?
  - *The map* $\begin{pmatrix} x \\ y \end{pmatrix} \to \begin{pmatrix} x+6 \\ y+4 \end{pmatrix}$ *takes* $\begin{pmatrix} 0 \\ 0 \end{pmatrix} \to \begin{pmatrix} 0+6 \\ 0+4 \end{pmatrix} = \begin{pmatrix} 6 \\ 4 \end{pmatrix}$.
  - *On the other hand, the map* $\begin{pmatrix} x \\ y \end{pmatrix} \to \begin{pmatrix} a & b \\ c & d \end{pmatrix}\begin{pmatrix} x \\ y \end{pmatrix}$ *takes* $\begin{pmatrix} 0 \\ 0 \end{pmatrix} \to \begin{pmatrix} a & b \\ c & d \end{pmatrix}\begin{pmatrix} 0 \\ 0 \end{pmatrix} = \begin{pmatrix} 0 \\ 0 \end{pmatrix}$.

A matrix transformation always maps the origin to itself, whereas a translation shifts every point in the plane, including the origin. Thus, there is no way to encode a translation as a matrix transformation.

Exercises 6–9 (2 minutes)

Give students a minute to perform the following exercise, and monitor their responses. Have students compare their responses with a partner; then, select students to share their responses with the whole class.

6. Write a rule for the function that shifts every point in the plane 6 units to the left.

$\begin{pmatrix} x \\ y \end{pmatrix} \to \begin{pmatrix} x-6 \\ y \end{pmatrix}, f(x,y) = \begin{pmatrix} x-6 \\ y \end{pmatrix}$

7. Write a rule for the function that shifts every point in the plane 9 units upward.

$\begin{pmatrix} x \\ y \end{pmatrix} \to \begin{pmatrix} x \\ y+9 \end{pmatrix}, f(x,y) = \begin{pmatrix} x \\ y+9 \end{pmatrix}$

8. Write a rule for the function that shifts every point in the plane 10 units down and 4 units to the right.

$\begin{pmatrix} x \\ y \end{pmatrix} \to \begin{pmatrix} x+4 \\ y-10 \end{pmatrix}, f(x,y) = \begin{pmatrix} x+4 \\ y-10 \end{pmatrix}$

9. Consider the rule $\begin{pmatrix} x \\ y \end{pmatrix} \to \begin{pmatrix} x-7 \\ y+2 \end{pmatrix}$. Describe the effect this transformation has on the plane.

Every point in the plane is shifted 7 units to the left and 2 units upward.

Closing (2 minutes)

Have students take a minute to write a response to the following questions in their notebooks; then, ask them to share their responses with a partner. Select two students to share their responses with the whole class.

- What did you learn today about representing straight-line motion? Give an example of a function that represents this kind of motion.
  - *An example would be* $f(t) = \begin{pmatrix} 10 - 2t \\ 3 + 5t \end{pmatrix}$.
- What did you learn about representing translations? Give an example of a function that represents this kind of motion.
  - *An example would be* $f(x, y) = \begin{pmatrix} x \\ y \end{pmatrix} + \begin{pmatrix} 3 \\ 4 \end{pmatrix} = \begin{pmatrix} x + 3 \\ y + 4 \end{pmatrix}$.

Exit Ticket (8 minutes)

Name ______________________________ Date ______________

Lesson 22: Modeling Video Game Motion with Matrices

Exit Ticket

1. Consider the function $h(t) = \begin{pmatrix} t+5 \\ t-3 \end{pmatrix}$. Draw the path that the point $P = h(t)$ traces out as t varies within the interval $0 \le t \le 4$.

2. The position of an object is given by the function $f(t) = \begin{pmatrix} t & 0 \\ 0 & t \end{pmatrix}\begin{pmatrix} 5 \\ 2 \end{pmatrix}$, where t is measured in seconds.

   a. Write $f(t)$ in the form $\begin{pmatrix} x(t) \\ y(t) \end{pmatrix}$.

   b. Find how fast the object is moving in the horizontal direction and in the vertical direction.

3. Write a function $f(x, y)$, which will translate all points in the plane 2 units to the left and 5 units downward.

EUREKA MATH™

Exit Ticket Sample Solutions

1. Consider the function $h(t) = \begin{pmatrix} t+5 \\ t-3 \end{pmatrix}$. Draw the path that the point $P = h(t)$ traces out as t varies within the interval $0 \le t \le 4$.

   $h(0) = \begin{pmatrix} 5 \\ -3 \end{pmatrix}, h(1) = \begin{pmatrix} 6 \\ -2 \end{pmatrix}, h(2) = \begin{pmatrix} 7 \\ -1 \end{pmatrix}, h(3) = \begin{pmatrix} 8 \\ 0 \end{pmatrix}, h(4) = \begin{pmatrix} 9 \\ 1 \end{pmatrix}$

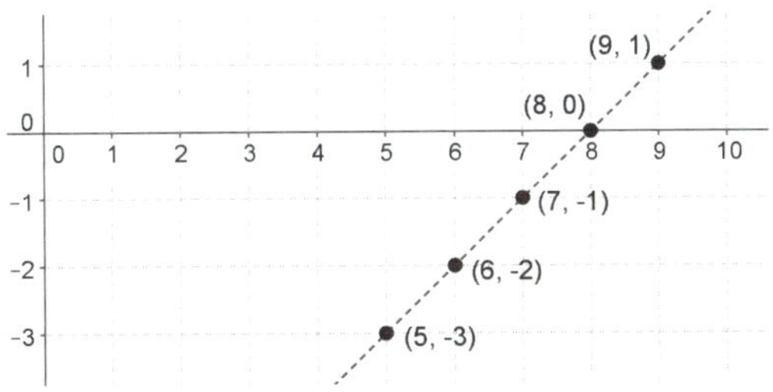

2. The position of an object is given by the function $f(t) = \begin{pmatrix} t & 0 \\ 0 & t \end{pmatrix}\begin{pmatrix} 5 \\ 2 \end{pmatrix}$, where t is measured in seconds.

   a. Write $f(t)$ in the form $\begin{pmatrix} x(t) \\ y(t) \end{pmatrix}$.

   $f(t) = \begin{pmatrix} t & 0 \\ 0 & t \end{pmatrix}\begin{pmatrix} 5 \\ 2 \end{pmatrix} = \begin{pmatrix} 5t + 0 \times 2 \\ 0 \times 5 + t \times 2 \end{pmatrix} = \begin{pmatrix} 5t \\ 2t \end{pmatrix}$. $f(0) = \begin{pmatrix} 0 \\ 0 \end{pmatrix}$, $f(1) = \begin{pmatrix} 5 \\ 2 \end{pmatrix}$, *and the slope of the line is* $m = \frac{2}{5}$.

   a. Find how fast the object is moving in the horizontal direction and in the vertical direction.

   The object is moving 2 *units upward vertically per second and* 5 *units to the right horizontally per second.*

3. Write a function $f(x, y)$, which will translate all points in the plane 2 units to the left and 5 units downward.

   $f(x, y) = \begin{pmatrix} x-2 \\ y-5 \end{pmatrix}$

Problem Set Sample Solutions

1. Let $D\begin{pmatrix} x \\ y \end{pmatrix} = \begin{pmatrix} 2 & 0 \\ 0 & 2 \end{pmatrix}\begin{pmatrix} x \\ y \end{pmatrix}$. Find and plot the following.

   a. Plot the point $\begin{pmatrix} -1 \\ 2 \end{pmatrix}$, and find $D\begin{pmatrix} -1 \\ 2 \end{pmatrix}$ and plot it.

   $$D\begin{pmatrix} -1 \\ 2 \end{pmatrix} = \begin{pmatrix} 2 & 0 \\ 0 & 2 \end{pmatrix}\begin{pmatrix} -1 \\ 2 \end{pmatrix} = \begin{pmatrix} -2 \\ 4 \end{pmatrix}$$

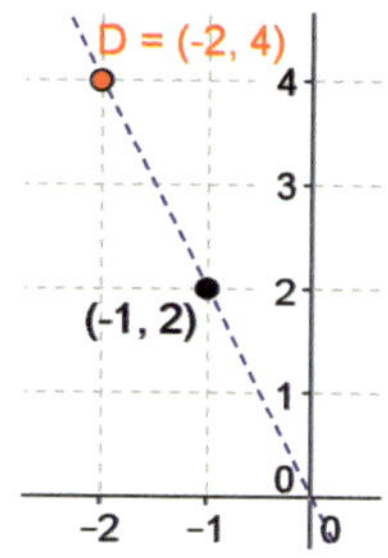

   b. Plot the point $\begin{pmatrix} 3 \\ 4 \end{pmatrix}$, and find $D\begin{pmatrix} 3 \\ 4 \end{pmatrix}$ and plot it.

   $$D\begin{pmatrix} 3 \\ 4 \end{pmatrix} = \begin{pmatrix} 2 & 0 \\ 0 & 2 \end{pmatrix}\begin{pmatrix} 3 \\ 4 \end{pmatrix} = \begin{pmatrix} 6 \\ 8 \end{pmatrix}$$

   c. Plot the point $\begin{pmatrix} 5 \\ 2 \end{pmatrix}$, and find $D\begin{pmatrix} 5 \\ 2 \end{pmatrix}$ and plot it.

   $$D\begin{pmatrix} 5 \\ 2 \end{pmatrix} = \begin{pmatrix} 2 & 0 \\ 0 & 2 \end{pmatrix}\begin{pmatrix} 5 \\ 2 \end{pmatrix} = \begin{pmatrix} 10 \\ 4 \end{pmatrix}$$

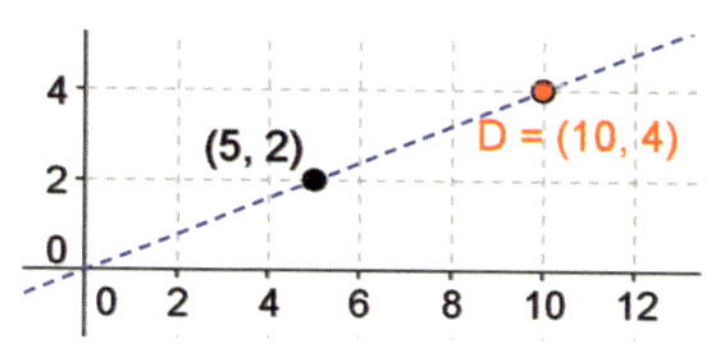

2. Let $f(t) = \begin{pmatrix} t & 0 \\ 0 & t \end{pmatrix}\begin{pmatrix} -1 \\ 2 \end{pmatrix}$. Find $f(0), f(1), f(2), f(3)$, and plot them on the same graph.

   $$f(0) = \begin{pmatrix} 0 & 0 \\ 0 & 0 \end{pmatrix}\begin{pmatrix} -1 \\ 2 \end{pmatrix} = \begin{pmatrix} 0 \\ 0 \end{pmatrix},$$

   $$f(1) = \begin{pmatrix} 1 & 0 \\ 0 & 1 \end{pmatrix}\begin{pmatrix} -1 \\ 2 \end{pmatrix} = \begin{pmatrix} -1 \\ 2 \end{pmatrix},$$

   $$f(2) = \begin{pmatrix} 2 & 0 \\ 0 & 2 \end{pmatrix}\begin{pmatrix} -1 \\ 2 \end{pmatrix} = \begin{pmatrix} -2 \\ 4 \end{pmatrix},$$

   $$f(3) = \begin{pmatrix} 3 & 0 \\ 0 & 3 \end{pmatrix}\begin{pmatrix} -1 \\ 2 \end{pmatrix} = \begin{pmatrix} -3 \\ 6 \end{pmatrix}$$

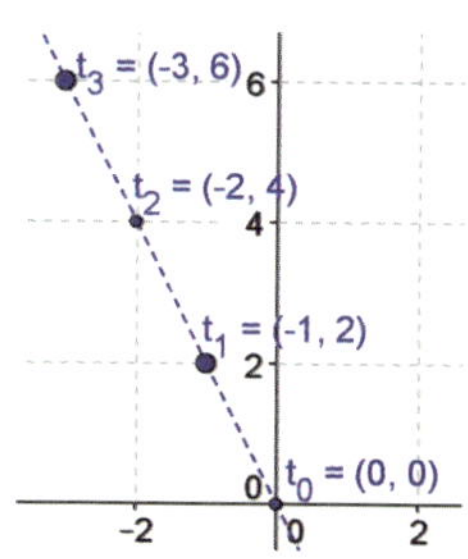

3. Let $f(t) = \begin{pmatrix} t & 0 \\ 0 & t \end{pmatrix}\begin{pmatrix} 3 \\ 2 \end{pmatrix}$ represent the location of an object at time t that is measured in seconds.

   a. How long does it take the object to travel from the origin to the point $\begin{pmatrix} 12 \\ 8 \end{pmatrix}$?

   $3t + 0 \times 2 = 12, t = 4$ *or* $0 \times 3 + 2t = 8, t = 4$

   b. Find the speed of the object in the horizontal direction and in the vertical direction.

   $f(t) = \begin{pmatrix} 3t \\ 2t \end{pmatrix}$ *The object is moving* 2 *units upward per second and* 3 *units to the right per second.*

4. Let $f(t) = \begin{pmatrix} 0.2t & 0 \\ 0 & 0.2t \end{pmatrix}\begin{pmatrix} 3 \\ 2 \end{pmatrix}, h(t) = \begin{pmatrix} 2t & 0 \\ 0 & 2t \end{pmatrix}\begin{pmatrix} 3 \\ 2 \end{pmatrix}$. Which one will reach the point $\begin{pmatrix} 12 \\ 8 \end{pmatrix}$ first? The time t is measured in seconds.

   For $f(t), 0.2t \times 3 + 0 \times 2 = 12, t = \frac{12}{0.6} = 20$ *seconds.*

   For $h(t), 2t \times 3 + 0 \times 2 = 12, t = \frac{12}{6} = 2$ *seconds; therefore,* $h(t)$ *will reach the point* $\begin{pmatrix} 12 \\ 8 \end{pmatrix}$ *first.*

5. Let $f(t) = \begin{pmatrix} kt & 0 \\ 0 & kt \end{pmatrix}\begin{pmatrix} 3 \\ 2 \end{pmatrix}$. Find the value of k that moves the object from the origin to $\begin{pmatrix} -45 \\ -30 \end{pmatrix}$ in 5 seconds.

   $f(5) = \begin{pmatrix} 5k & 0 \\ 0 & 5k \end{pmatrix}\begin{pmatrix} 3 \\ 2 \end{pmatrix} = \begin{pmatrix} -45 \\ -30 \end{pmatrix}, 5k \times 3 + 0 \times 2 = -45, k = -3$ *or* $0 \times 3 + 5k \times 2 = -30, k = -3$

6. Write $f(t)$ in the form $\begin{pmatrix} x(t) \\ y(t) \end{pmatrix}$ if

   a. $f(t) = \begin{pmatrix} t & 0 \\ 0 & t \end{pmatrix}\begin{pmatrix} 2 \\ 5 \end{pmatrix}$

   $f(t) = \begin{pmatrix} 2t \\ 5t \end{pmatrix}$

   b. $f(t) = \begin{pmatrix} 2t+1 & 0 \\ 0 & 2t+1 \end{pmatrix}\begin{pmatrix} 3 \\ 2 \end{pmatrix}$

   $f(t) = \begin{pmatrix} 6t+3 \\ 4t+2 \end{pmatrix}$

   c. $f(t) = \begin{pmatrix} \frac{t}{2} - 3 & 0 \\ 0 & \frac{t}{2} - 3 \end{pmatrix}\begin{pmatrix} 4 \\ -6 \end{pmatrix}$

   $f(t) = \begin{pmatrix} 2t - 12 \\ 3t - 18 \end{pmatrix}$

7. Let $f(t) = \begin{pmatrix} t & 0 \\ 0 & t \end{pmatrix}\begin{pmatrix} 2 \\ 5 \end{pmatrix}$ represent the location of an object after t seconds.

   a. If the object starts at $\begin{pmatrix} 6 \\ 15 \end{pmatrix}$, how long would it take to reach $\begin{pmatrix} 34 \\ 85 \end{pmatrix}$?

   $f(t) = \begin{pmatrix} 2t \\ 5t \end{pmatrix}$*; it starts at* $\begin{pmatrix} 6 \\ 15 \end{pmatrix}$*; therefore,* $f(t) = \begin{pmatrix} 2t+6 \\ 5t+15 \end{pmatrix}$.

   $2t + 6 = 34, t = 14$ *or* $5t + 15 = 85, t = 14$

b. Write the new function $f(t)$ that gives the position of the object after t seconds.

$$f(t) = \begin{pmatrix} 2t + 6 \\ 5t + 15 \end{pmatrix}$$

c. Write $f(t)$ as a matrix transformation.

$$f(t) = \begin{pmatrix} 2t + 6 \\ 5t + 15 \end{pmatrix} = \begin{pmatrix} (t+3)2 \\ (t+3)5 \end{pmatrix} = \begin{pmatrix} t+3 & 0 \\ 0 & t+3 \end{pmatrix}\begin{pmatrix} 2 \\ 5 \end{pmatrix}$$

$$or\ f(t) = \begin{pmatrix} 2t + 6 \\ 5t + 15 \end{pmatrix} = \begin{pmatrix} (-t-3)(-2) \\ (-t-3)(-5) \end{pmatrix} = \begin{pmatrix} -t-3 & 0 \\ 0 & -t-3 \end{pmatrix}\begin{pmatrix} -2 \\ -5 \end{pmatrix}$$

The answers vary; it depends on how the factoring is applied.

8. Write the following functions as a matrix transformation.

a. $f(t) = \begin{pmatrix} 10 + 2t \\ 15 + 3t \end{pmatrix}$

$$f(t) = \begin{pmatrix} 10 + 2t \\ 15 + 3t \end{pmatrix} = \begin{pmatrix} (5+t)2 \\ (5+t)3 \end{pmatrix} = \begin{pmatrix} 5+t & 0 \\ 0 & 5+t \end{pmatrix}\begin{pmatrix} 2 \\ 3 \end{pmatrix}$$

b. $f(t) = \begin{pmatrix} -6t + 15 \\ 8t - 20 \end{pmatrix}$

$$f(t) = \begin{pmatrix} -6t + 15 \\ 8t - 20 \end{pmatrix} = \begin{pmatrix} (2t-5)(-3) \\ (2t-5)4 \end{pmatrix} = \begin{pmatrix} 2t-5 & 0 \\ 0 & 2t-5 \end{pmatrix}\begin{pmatrix} -3 \\ 4 \end{pmatrix}$$

$$or\ f(t) = \begin{pmatrix} -6t + 15 \\ 8t - 20 \end{pmatrix} = \begin{pmatrix} (-2t+5)3 \\ (-2t+5)(-4) \end{pmatrix} = \begin{pmatrix} -2t+5 & 0 \\ 0 & -2t+5 \end{pmatrix}\begin{pmatrix} 3 \\ -4 \end{pmatrix}$$

9. Write a function rule that represents the change in position of the point $\begin{pmatrix} x \\ y \end{pmatrix}$ for the following.

a. 5 units to the right and 3 units downward

$$f(x, y) = \begin{pmatrix} x + 5 \\ y - 3 \end{pmatrix}$$

b. 2 units downward and 3 units to the left

$$f(x, y) = \begin{pmatrix} x - 3 \\ y - 2 \end{pmatrix}$$

c. 3 units upward, 5 units to the left, and then it dilates by 2.

$$f(x, y) = \begin{pmatrix} x - 5 \\ y + 3 \end{pmatrix}, f(x, y) = \begin{pmatrix} 2 & 0 \\ 0 & 2 \end{pmatrix}\begin{pmatrix} x - 5 \\ y + 3 \end{pmatrix}$$

d. 3 units upward, 5 units to the left, and then it rotates by $\frac{\pi}{2}$ counterclockwise.

$$f(x, y) = \begin{pmatrix} x - 5 \\ y + 3 \end{pmatrix}, f(x, y) = \begin{pmatrix} \cos\left(\frac{\pi}{2}\right) & -\sin\left(\frac{\pi}{2}\right) \\ \sin\left(\frac{\pi}{2}\right) & \cos\left(\frac{\pi}{2}\right) \end{pmatrix}\begin{pmatrix} x - 5 \\ y + 3 \end{pmatrix}$$

MP.4

10. Annie is designing a video game and wants her main character to be able to move from any given point $\begin{pmatrix} x \\ y \end{pmatrix}$ in the following ways: right 1 unit, jump up 1 unit, and both jump up and move right 1 unit each.

a. What function rules can she use to represent each time the character moves?

$f(x, y) = \begin{pmatrix} x+1 \\ y \end{pmatrix}$; $g(x, y) = \begin{pmatrix} x \\ y+1 \end{pmatrix}$; $h(x, y) = \begin{pmatrix} x+1 \\ y+1 \end{pmatrix}$

b. Annie is also developing a ski slope stage for her game and wants to model her character's position using matrix transformations. Annie wants the player to start at $\begin{pmatrix} -20 \\ 10 \end{pmatrix}$ and eventually pass through the origin moving 5 units per second down. How fast does the player need to move to the right in order to pass through the origin? What matrix transformation can Annie use to describe the movement of the character? If the far right of the screen is at $x = 20$, how long until the player moves off the screen traveling this path?

If the player is moving 5 units per second down, then she will reach $y = 0$ in $t = 2$ seconds. Thus, the player needs to move 10 units per second to the right.

$$\begin{aligned} f(t) &= \begin{pmatrix} -20+10t \\ 10-5t \end{pmatrix} \\ &= \begin{pmatrix} -10(2-t) \\ 5(2-t) \end{pmatrix} \\ &= \begin{pmatrix} 2-t & 0 \\ 0 & 2-t \end{pmatrix}\begin{pmatrix} -10 \\ 5 \end{pmatrix} \end{aligned}$$

The player will leave the screen in 4 seconds.

MP.3

11. Remy thinks that he has developed matrix transformations to model the movements of Annie's characters in Problem 10 from any given point $\begin{pmatrix} x \\ y \end{pmatrix}$, and he has tested them on the point $\begin{pmatrix} 1 \\ 1 \end{pmatrix}$. This is the work Remy did on the transformations:

$\begin{pmatrix} 2 & 0 \\ 0 & 1 \end{pmatrix}\begin{pmatrix} 1 \\ 1 \end{pmatrix} = \begin{pmatrix} 2 \\ 1 \end{pmatrix}$ $\quad\begin{pmatrix} 1 & 0 \\ 1 & 1 \end{pmatrix}\begin{pmatrix} 1 \\ 1 \end{pmatrix} = \begin{pmatrix} 1 \\ 2 \end{pmatrix}$ $\quad\begin{pmatrix} 2 & 0 \\ 1 & 1 \end{pmatrix}\begin{pmatrix} 1 \\ 1 \end{pmatrix} = \begin{pmatrix} 2 \\ 2 \end{pmatrix}$.

Do these matrix transformations accomplish the movements that Annie wants to program into the game? Explain why or why not.

These do not accomplish the movements. If we apply the transformations to any other point in the plane, then they will not produce the same results of moving one unit to the right, one unit up, and one unit up and right.

As a counterexample, any of the three matrix transformations applied to the origin do nothing.

12. Nolan has been working on how to know when the path of a point can be described with matrix transformations and how to know when it requires translations and cannot be described with matrix transformations. So far, he has been focusing on the following two functions, which both pass through the point $(2, 5)$:

$f(t) = \begin{pmatrix} 2t+6 \\ 5t+15 \end{pmatrix}$ and $g(t) = \begin{pmatrix} t+2 \\ t+5 \end{pmatrix}$.

a. If we simplify these functions algebraically, how does the rule for f differ from the rule for g? What does this say about which function can be expressed with matrix transformations?

$f(t) = \begin{pmatrix} 2(t+3) \\ 5(t+3) \end{pmatrix}$ *Thus, there is a common factor in both the x- and y-coordinate. Because there is a common factor, we can pull the factor out as a scalar and rewrite the scalar as a matrix multiplication. $g(t)$ does not have a common factor (other than 1) between the x- and y-coordinate.*

b. **Nolan has noticed that functions that can be expressed with matrix transformations always pass through the origin; does either f or g pass through the origin, and does this support or contradict Nolan's reasoning?**

At $t = -3$, the graph of f passes through the origin. On the other hand, the graph of g crosses the x-axis at $t = -2$ and the y-axis at $t = -5$, so it does not pass through the origin. This seems to support Nolan's reasoning. This agrees with our response to part (a), since the common factor has the same zero and causes the function to cross the origin.

c. **Summarize the results of parts (a) and (b) to describe how we can tell from the equation for a function or from the graph of a function that it can be expressed with matrix transformations.**

If a function has a common factor involving t that can be pulled out of both the x- and y-coordinates, then the function can be represented as a matrix transformation. If the graph of the function passes through the origin, then the function can be represented as a matrix transformation.

Lesson 23: Modeling Video Game Motion with Matrices

Student Outcomes

- Students use matrix transformations to model circular motion.
- Students use coordinate transformations to represent a combination of motions.

Lesson Notes

Students have recently learned how to represent rotations as matrix transformations. In this lesson, they apply that knowledge to represent dynamic motion, as seen in video games. Students analyze circular motion that involves a time parameter such as $G(t) = \begin{pmatrix} \cos\left(\frac{\pi}{2} \cdot t\right) & -\sin\left(\frac{\pi}{2} \cdot t\right) \\ \sin\left(\frac{\pi}{2} \cdot t\right) & \cos\left(\frac{\pi}{2} \cdot t\right) \end{pmatrix}$. The second part of the lesson involves modeling a combination of motions. For instance, students model motion along a circle followed by a translation or motion along a line followed by a translation.

Classwork

The Opening Exercise allows students to practice matrix transformations and plot the results. This prepares students for skills needed in this lesson. Work through this as a whole class, asking questions to assess student understanding. Use this as a way to clear up misconceptions.

Opening Exercise (5 minutes)

Opening Exercise

Let $R\begin{pmatrix} x \\ y \end{pmatrix} = \begin{pmatrix} \cos\left(\frac{\pi}{3}\right) & -\sin\left(\frac{\pi}{3}\right) \\ \sin\left(\frac{\pi}{3}\right) & \cos\left(\frac{\pi}{3}\right) \end{pmatrix}\begin{pmatrix} x \\ y \end{pmatrix}$.

a. **Describe the geometric effect of performing the transformation $\begin{pmatrix} x \\ y \end{pmatrix} \to R\begin{pmatrix} x \\ y \end{pmatrix}$.**

Applying R rotates each point in the plane about the origin through $\frac{\pi}{3}$ radians in a counterclockwise direction.

b. **Plot the point $\begin{pmatrix} 1 \\ 0 \end{pmatrix}$, and then find $R\begin{pmatrix} 1 \\ 0 \end{pmatrix}$ and plot it.**

$$R\begin{pmatrix} 1 \\ 0 \end{pmatrix} = \begin{pmatrix} \cos\left(\frac{\pi}{3}\right) \\ \sin\left(\frac{\pi}{3}\right) \end{pmatrix}$$

c. If we want to show that R has been applied twice to $(1,0)$, we can write $R^2\begin{pmatrix}1\\0\end{pmatrix}$, which represents $R\left(R\begin{pmatrix}1\\0\end{pmatrix}\right)$. Find $R^2\begin{pmatrix}1\\0\end{pmatrix}$ and plot it. Then, find $R^3\begin{pmatrix}1\\0\end{pmatrix} = R\left(R\left(R\begin{pmatrix}1\\0\end{pmatrix}\right)\right)$, and plot it.

$$R^2\begin{pmatrix}1\\0\end{pmatrix} = \begin{pmatrix}\cos\left(\frac{2\pi}{3}\right)\\ \sin\left(\frac{2\pi}{3}\right)\end{pmatrix};\ R^3\begin{pmatrix}1\\0\end{pmatrix} = \begin{pmatrix}\cos(\pi)\\ \sin(\pi)\end{pmatrix}$$

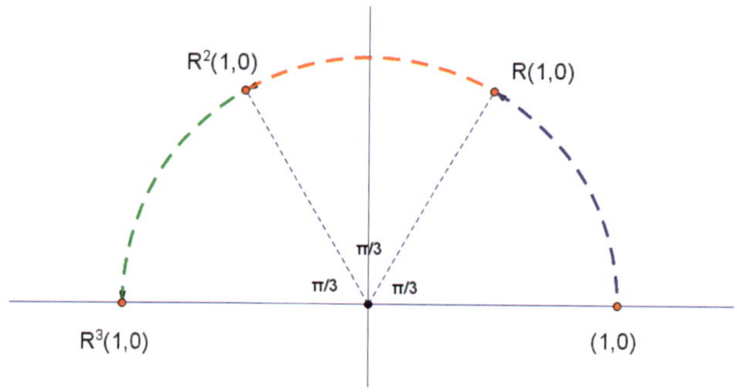

MP.7

d. Describe the matrix transformation $\begin{pmatrix}x\\y\end{pmatrix} \to R^2\begin{pmatrix}x\\y\end{pmatrix}$ using a single matrix.

$R^2\begin{pmatrix}x\\y\end{pmatrix}$ is the transformation that rotates points through $2\cdot\frac{\pi}{3}$ radians, so a formula for $R^2\begin{pmatrix}x\\y\end{pmatrix}$ is $\begin{pmatrix}\cos\left(\frac{2\pi}{3}\right) & -\sin\left(\frac{2\pi}{3}\right)\\ \sin\left(\frac{2\pi}{3}\right) & \cos\left(\frac{2\pi}{3}\right)\end{pmatrix}\begin{pmatrix}x\\y\end{pmatrix}$.

Discussion (3 minutes): Circular Motion over Time

Let $R(t) = \begin{pmatrix}\cos(t) & -\sin(t)\\ \sin(t) & \cos(t)\end{pmatrix}\begin{pmatrix}1\\0\end{pmatrix}$.

Suppose that t is measured in degrees. Let's place several input-output pairs for this function on a graph:

- $R(30) = (0.87, 0.50)$ $R(45) = (0.71, 0.71)$ $R(60) = (0.50, 0.87)$
- $R(90) = (0,1)$ $R(180) = (-1,0)$ $R(270) = (0,-1)$ $R(360) = (1,0)$

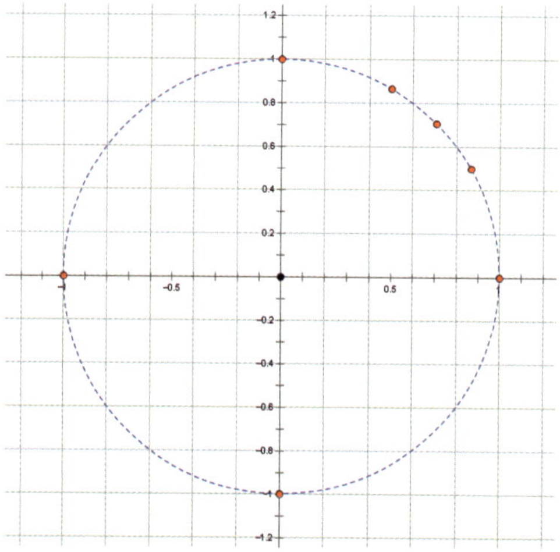

- What do you notice about the points on the graph?
    - *The points appear to lie on a circle.*
- How could we be sure the points are actually on a circle?
    - *If each point is the same distance from the origin, then the points form a circle.*
- Now check and see if this is, in fact, the case. Have different students find the distance to the origin from given points.
    - *Students check the distance from a given point to the origin and confirm using the distance formula. For example, $R(30) = (0.87, 0.50)$. Its distance from the origin is $\sqrt{(0.87-0)^2 + (0.5-0)^2} = 1$.*
- What is the result of R(t)?
    - $R(t) = \begin{pmatrix} \cos(t) & -\sin(t) \\ \sin(t) & \cos(t) \end{pmatrix} \begin{pmatrix} 1 \\ 0 \end{pmatrix} = \begin{pmatrix} \cos(t) \\ \sin(t) \end{pmatrix}$
- Does this result ensure points lie on a circle?
    - *Yes, this would confirm that the points lie on the unit circle because the x-value corresponds to cosine of an angle t and the y-value corresponds to sine of the same angle on the unit circle.*

Exercise 1 (4 minutes)

This exercise provides students more practice with matrices representing rotations. This time, the angle is different in each function, allowing them to compare the results. Give students time to work on the following problems independently; then, call on students to share their responses with the class.

Exercises

1. Let $f(t) = \begin{pmatrix} \cos(2t) & -\sin(2t) \\ \sin(2t) & \cos(2t) \end{pmatrix} \begin{pmatrix} 1 \\ 0 \end{pmatrix}$, and let $g(t) = \begin{pmatrix} \cos\left(\frac{t}{2}\right) & -\sin\left(\frac{t}{2}\right) \\ \sin\left(\frac{t}{2}\right) & \cos\left(\frac{t}{2}\right) \end{pmatrix} \begin{pmatrix} 1 \\ 0 \end{pmatrix}$.

a. Suppose $f(t)$ represents the position of a moving object that starts at $(1, 0)$. How long does it take for this object to return to its starting point? When the argument of the trigonometric function changes from t to $2t$, what effect does this have?

The object will return to $(1, 0)$ when $2t = 2\pi$. Thus, it will take $t = \pi$ seconds for this to happen. Changing the argument from t to $2t$ causes the object to move twice as fast.

b. If the position is given instead by $g(t)$, how long would it take the object to return to its starting point? When the argument of the trigonometric function changes from t to $\frac{t}{2}$, what effect does this have?

The object will return to $(1, 0)$ when $\frac{t}{2} = 2\pi$. Thus, it will take $t = 4\pi$ seconds for this to happen.

Changing the argument from t to $\frac{t}{2}$ causes the object to move half as fast.

Example 1 (4 minutes)

Let $F(t) = \begin{pmatrix} \cos(t) & -\sin(t) \\ \sin(t) & \cos(t) \end{pmatrix} \begin{pmatrix} 3 \\ 2 \end{pmatrix}$.

- This time, we'll measure t in radians. Find $F\left(\frac{\pi}{2}\right)$, $F(\pi)$, $F\left(\frac{3\pi}{2}\right)$, and $F(2\pi)$.
  - $F\left(\frac{\pi}{2}\right) = \begin{pmatrix} \cos\left(\frac{\pi}{2}\right) & -\sin\left(\frac{\pi}{2}\right) \\ \sin\left(\frac{\pi}{2}\right) & \cos\left(\frac{\pi}{2}\right) \end{pmatrix} \begin{pmatrix} 3 \\ 2 \end{pmatrix} = \begin{pmatrix} 0 & -1 \\ 1 & 0 \end{pmatrix} \begin{pmatrix} 3 \\ 2 \end{pmatrix} = \begin{pmatrix} 0-2 \\ 3+0 \end{pmatrix} = \begin{pmatrix} -2 \\ 3 \end{pmatrix}$
  - $F(\pi) = \begin{pmatrix} \cos(\pi) & -\sin(\pi) \\ \sin(\pi) & \cos(\pi) \end{pmatrix} \begin{pmatrix} 3 \\ 2 \end{pmatrix} = \begin{pmatrix} -1 & 0 \\ 0 & -1 \end{pmatrix} \begin{pmatrix} 3 \\ 2 \end{pmatrix} = \begin{pmatrix} -3+0 \\ 0-2 \end{pmatrix} = \begin{pmatrix} -3 \\ -2 \end{pmatrix}$
  - $F\left(\frac{3\pi}{2}\right) = \begin{pmatrix} \cos\left(\frac{3\pi}{2}\right) & -\sin\left(\frac{3\pi}{2}\right) \\ \sin\left(\frac{3\pi}{2}\right) & \cos\left(\frac{3\pi}{2}\right) \end{pmatrix} \begin{pmatrix} 3 \\ 2 \end{pmatrix} = \begin{pmatrix} 0 & 1 \\ -1 & 0 \end{pmatrix} \begin{pmatrix} 3 \\ 2 \end{pmatrix} = \begin{pmatrix} 0+2 \\ -3+0 \end{pmatrix} = \begin{pmatrix} 2 \\ -3 \end{pmatrix}$
  - $F(2\pi) = \begin{pmatrix} \cos(2\pi) & -\sin(2\pi) \\ \sin(2\pi) & \cos(2\pi) \end{pmatrix} \begin{pmatrix} 3 \\ 2 \end{pmatrix} = \begin{pmatrix} 1 & 0 \\ 0 & 1 \end{pmatrix} \begin{pmatrix} 3 \\ 2 \end{pmatrix} = \begin{pmatrix} 3+0 \\ 0+2 \end{pmatrix} = \begin{pmatrix} 3 \\ 2 \end{pmatrix}$
- When we plot points, we see once again that they appear to lie on a circle. Make sure this is really true.
  - $F(t) = \begin{pmatrix} \cos(t) & -\sin(t) \\ \sin(t) & \cos(t) \end{pmatrix} \begin{pmatrix} 3 \\ 2 \end{pmatrix} = \begin{pmatrix} 3\cos(t) - 2\sin(t) \\ 3\sin(t) + 2\cos(t) \end{pmatrix}$

    $(3\cos(t) - 2\sin(t))^2 + (3\sin(t) + 2\cos(t))^2$

    $9\cos^2(t) - 12\cos(t)\sin(t) + 4\sin^2(t) + 9\sin^2(t) + 12\cos(t)\sin(t) + 4\cos^2(t)$

    $9(\cos^2(t) + \sin^2(t)) + 4(\sin^2(t) + \cos^2(t))$

    $9(1) + 4(1) = 9 + 4 = 13$

    Thus, each point is $\sqrt{13}$ units from the origin, which confirms that the outputs lie on a circle.

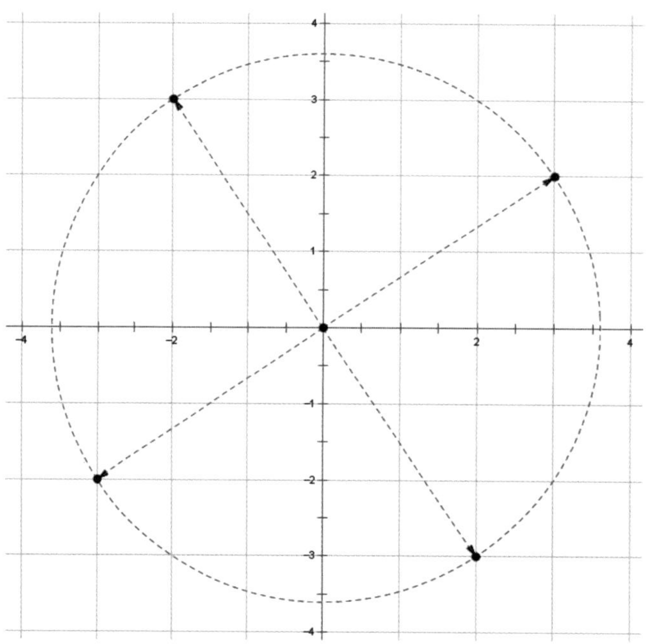

Discussion (4 minutes): Rotations That Use a Time Parameter

Let $F(t) = \begin{pmatrix} \cos\left(\frac{\pi}{2}\cdot t\right) & -\sin\left(\frac{\pi}{2}\cdot t\right) \\ \sin\left(\frac{\pi}{2}\cdot t\right) & \cos\left(\frac{\pi}{2}\cdot t\right) \end{pmatrix}\begin{pmatrix}1\\0\end{pmatrix}$.

- Draw the path that $P = F(t)$ traces out as t varies within each of the following intervals:
  - $0 \leq t \leq 1$
  - $1 \leq t \leq 2$
  - $2 \leq t \leq 3$
  - $3 \leq t \leq 4$

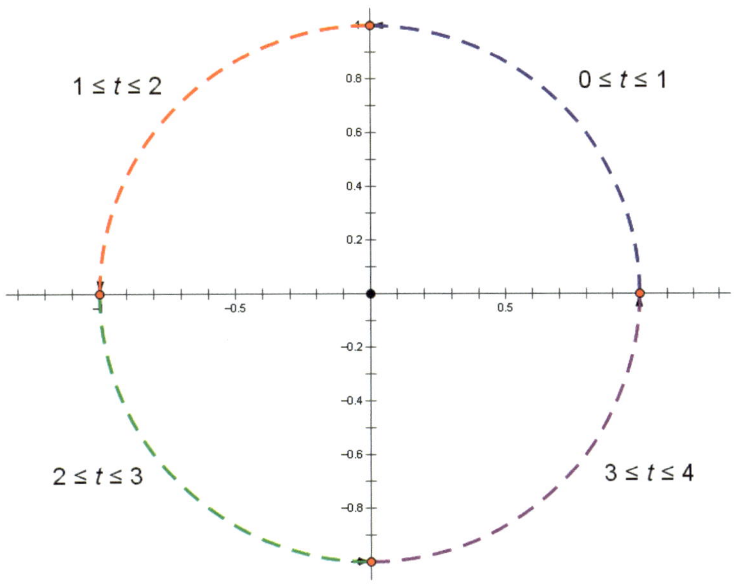

- Where will the object be located at $t = 0.5$ second?
  - $f(0.5) = \left(\cos\left(\frac{\pi}{4}\right), \sin\left(\frac{\pi}{4}\right)\right) \approx (0.71, 0.71)$
- How long will it take the object to reach $(0.71, -0.71)$?
  - *These coordinates represent* $\left(\cos\left(\frac{7\pi}{4}\right), \sin\left(\frac{7\pi}{4}\right)\right)$, *so* $\left(\cos\left(\frac{\pi}{2}\cdot\frac{7}{2}\right), \sin\left(\frac{\pi}{2}\cdot\frac{7}{2}\right)\right)$. *The object reaches this location at* $t = \frac{7}{2} = 3.5$ *seconds.*

Exercises 2–3 (5 minutes)

Give students time to complete the following exercises; then, ask them to compare their responses with a partner. Call on students to share their responses with the class.

2. Let $G(t) = \begin{pmatrix} \cos\left(\frac{\pi}{2}\cdot t\right) & -\sin\left(\frac{\pi}{2}\cdot t\right) \\ \sin\left(\frac{\pi}{2}\cdot t\right) & \cos\left(\frac{\pi}{2}\cdot t\right) \end{pmatrix}\begin{pmatrix} 0 \\ 1 \end{pmatrix}$.

a. **Draw the path that $P = G(t)$ traces out as t varies within the interval $0 \le t \le 1$.**

b. **Where will the object be at $t = 3$ seconds?**

$G(3) = (1, 0)$

c. **How long will it take the object to reach $(0, -1)$?**

These coordinates represent $(\cos(\pi), \sin(\pi))$*, so* $\left(\cos\left(\frac{\pi}{2}\cdot 2\right), \sin\left(\frac{\pi}{2}\cdot 2\right)\right)$*; the object reaches this location at* $t = 2$ *seconds.* $G(2) = (0, -1)$*, so it will take* 2 *seconds to reach that location.*

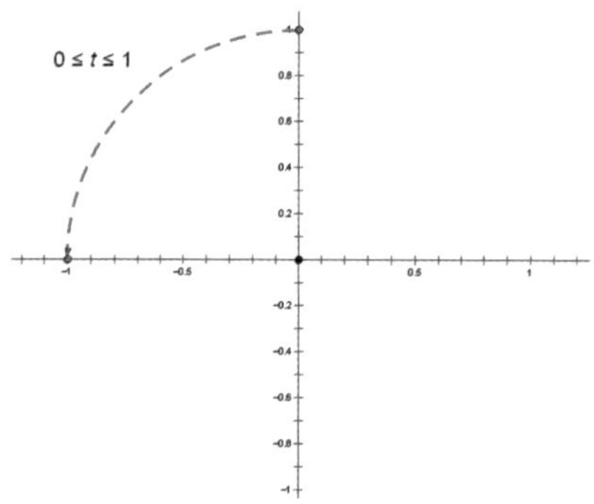

3. Let $H(t) = \begin{pmatrix} \cos\left(\frac{\pi}{2}\cdot t\right) & -\sin\left(\frac{\pi}{2}\cdot t\right) \\ \sin\left(\frac{\pi}{2}\cdot t\right) & \cos\left(\frac{\pi}{2}\cdot t\right) \end{pmatrix}\begin{pmatrix} 1 \\ 4 \end{pmatrix}$.

a. **Draw the path that $P = H(t)$ traces out as t varies within the interval $0 \le t \le 2$.**

b. **Where will the object be at $t = 1$ second?**

$H(1) = (-4, 1)$

c. **How long will it take the object to return to its starting point?**

$H(4) = (1, 4)$*, so it will take* 4 *seconds to return to its starting point.*

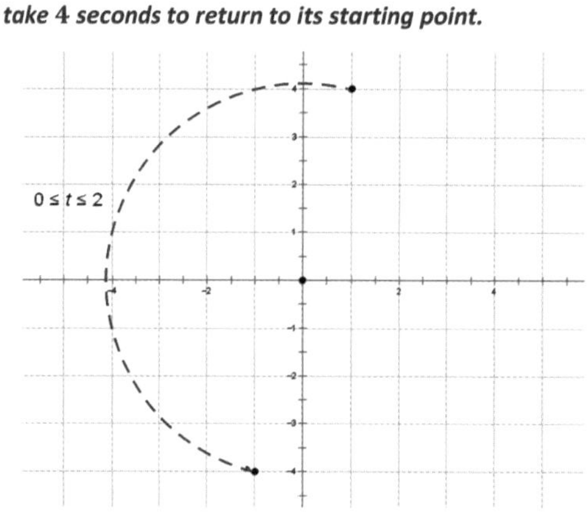

Example 2 (4 minutes)

Let $f(t) = \begin{pmatrix} \cos\left(\frac{\pi}{6}\cdot t\right) & -\sin\left(\frac{\pi}{6}\cdot t\right) \\ \sin\left(\frac{\pi}{6}\cdot t\right) & \cos\left(\frac{\pi}{6}\cdot t\right) \end{pmatrix}\begin{pmatrix} 6 \\ 2 \end{pmatrix}$.

- Draw the path that $P = f(t)$ traces out as t varies within each of the following intervals:
  - $0 \le t \le 1$ $1 \le t \le 2$ $2 \le t \le 3$
  - $3 \le t \le 4$ $4 \le t \le 5$ $5 \le t \le 6$
- As an example, can you describe what happens to the object as t varies within the interval $0 \le t \le 1$?
  - *Since $f(0) = (6,2)$, the object starts its trajectory there. When $t = 1$, the object will have moved through $\frac{\pi}{6}$ radians. So, in the time interval $0 \le t \le 1$, the object moves along a circular arc as shown below.*

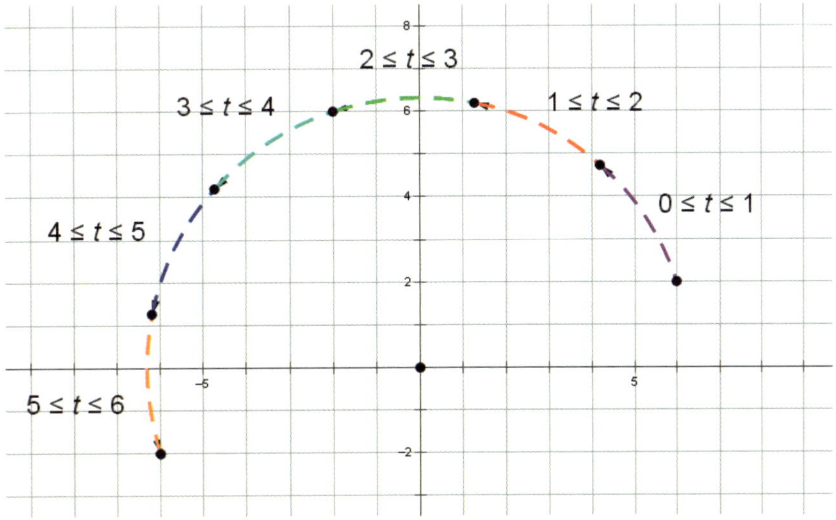

Exercises 4–5 (3 minutes)

Give students time to complete the following exercises; then ask them to compare their responses with a partner. Call on students to share their responses with the class, and use this as an opportunity to check for understanding.

4. Suppose you want to write a program that takes the point $(3,5)$ and rotates it about the origin to the point $(-3,-5)$ over a 1-second interval. Write a function $P = f(t)$ that encodes this rotation.

Let $f(t) = \begin{pmatrix} \cos(\pi\cdot t) & -\sin(\pi\cdot t) \\ \sin(\pi\cdot t) & \cos(\pi\cdot t) \end{pmatrix}\begin{pmatrix} 3 \\ 5 \end{pmatrix}$. We have $f(0) = (3,5)$ and $f(1) = (-3,-5)$, as required.

5. If instead you wanted the rotation to take place over a 1.5-second interval, how would your function change?

Let $f(t) = \begin{pmatrix} \cos\left(\pi\cdot\frac{t}{1}\cdot 5\right) & -\sin\left(\pi\cdot\frac{t}{1}\cdot 5\right) \\ \sin\left(\pi\cdot\frac{t}{1}\cdot 5\right) & \cos\left(\pi\cdot\frac{t}{1}\cdot 5\right) \end{pmatrix}\begin{pmatrix} 3 \\ 5 \end{pmatrix}$. We have $f(0) = (3,5)$ and $f(1.5) = (-3,-5)$, as required.

Example 3 (4 minutes)

Let's analyze the transformation $g(t) = \begin{pmatrix} 3\cos\left(\frac{\pi}{2}\cdot t\right) & -3\sin\left(\frac{\pi}{2}\cdot t\right) \\ 3\sin\left(\frac{\pi}{2}\cdot t\right) & 3\cos\left(\frac{\pi}{2}\cdot t\right) \end{pmatrix}\begin{pmatrix} 1 \\ 2 \end{pmatrix}$. In particular, we will compare $g(0)$ and $g(1)$.

- What is $g(0)$? What geometric effect does $g(t)$ have on $(1, 2)$ initially?
  - *We have* $g(0) = (3, 6)$*, which is a dilation of* $(1, 2)$ *using scale factor* 3.
- What is $g(1)$? Describe what is going on.
  - *We have* $g(1) = (-6, 3)$*, which represents a quarter turn of the point* $(3, 6)$ *about the origin in a counterclockwise direction.*
- Can you summarize the geometric effect of applying $g(t)$ to the point $(1, 2)$ during the time interval $0 \leq t \leq 1$?
  - *This transformation combines a quarter turn about the origin with a scaling by a factor of* 3.

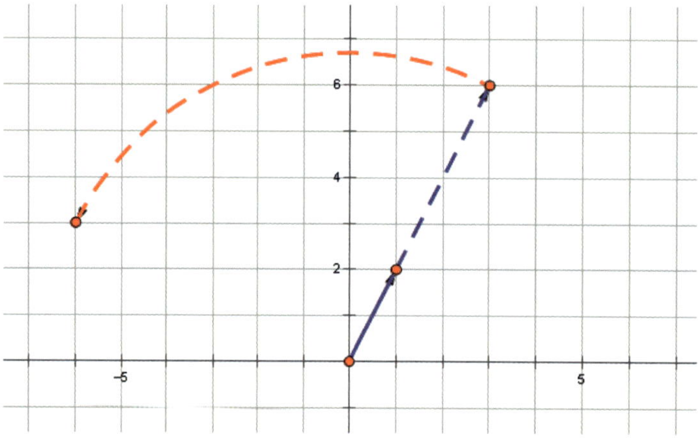

- What is $g(2)$? Describe what is going on.
  - *We have* $g(2) = (-3, -6)$*, which represents a quarter turn of the point* $(-6, 3)$ *about the origin in a counterclockwise direction.*
- What is $g(3)$? Describe what is going on.
  - *We have* $g(3) = (6, -3)$*, which represents a quarter turn of the point* $(-3, -6)$ *about the origin in a counterclockwise direction.*
- What is $g(4)$? Describe what is going on.
  - *We have* $g(1) = (3, 6)$*, which represents a quarter turn of the point* $(6, -3)$ *about the origin in a counterclockwise direction.*
- Compare $g(0)$ and $g(4)$. Does this make sense?
  - $g(0) = g(4)$ *This makes sense because 4 quarter turns would be a full rotation, so this would bring you back to the starting point.*

Closing (4 minutes)

- Write one to two sentences in your notebook describing what you learned in today's lesson; then, share your response with a partner.
  - *We learned how to use matrices to describe rotations that happen over a specific time interval. We also discussed how to model multiple transformations, such as a rotation followed by a translation.*

Exit Ticket (5 minutes)

Name ______________________________ Date ______________

Lesson 23: Modeling Video Game Motion with Matrices

Exit Ticket

Write a function $f(t)$ that incorporates the following actions. Make a drawing of the path the point follows during the time interval $0 \leq t \leq 3$.

a. During the time interval $0 \leq t \leq 1$, move the point $(8, 6)$ through $\frac{\pi}{4}$ radians about the origin in a counterclockwise direction.

b. During the time interval $1 < t \leq 3$, move the image along a straight line to $(6, -8)$.

Exit Ticket Sample Solutions

Write a function $f(t)$ that incorporates the following actions. Make a drawing of the path the point follows during the time interval $0 \le t \le 3$.

a. During the time interval $0 \le t \le 1$, move the point $(8, 6)$ through $\frac{\pi}{4}$ radians about the origin in a counterclockwise direction.

$$f(t) = \begin{pmatrix} \cos\left(\frac{\pi t}{4}\right) & -\sin\left(\frac{\pi t}{4}\right) \\ \sin\left(\frac{\pi t}{4}\right) & \cos\left(\frac{\pi t}{4}\right) \end{pmatrix}\begin{pmatrix} 8 \\ 6 \end{pmatrix}, \quad 0 \le t \le 1$$

$$f(0) = \begin{pmatrix} \cos(0) & -\sin(0) \\ \sin(0) & \cos(0) \end{pmatrix}\begin{pmatrix} 8 \\ 6 \end{pmatrix} = \begin{pmatrix} 1 & 0 \\ 0 & 1 \end{pmatrix}\begin{pmatrix} 8 \\ 6 \end{pmatrix} = \begin{pmatrix} 8 \\ 6 \end{pmatrix}$$

$$f(1) = \begin{pmatrix} \cos\left(\frac{\pi}{4}\right) & -\sin\left(\frac{\pi}{4}\right) \\ \sin\left(\frac{\pi}{4}\right) & \cos\left(\frac{\pi}{4}\right) \end{pmatrix}\begin{pmatrix} 8 \\ 6 \end{pmatrix} = \begin{pmatrix} \frac{\sqrt{2}}{2} & \frac{-\sqrt{2}}{2} \\ \frac{\sqrt{2}}{2} & \frac{\sqrt{2}}{2} \end{pmatrix}\begin{pmatrix} 8 \\ 6 \end{pmatrix} = \begin{pmatrix} \sqrt{2} \\ 7\sqrt{2} \end{pmatrix} \approx \begin{pmatrix} 1.41 \\ 9.90 \end{pmatrix}$$

b. During the time interval $1 < t \le 3$, move the image along a straight line to $(6, -8)$.

The image is $\begin{pmatrix} \sqrt{2} \\ 7\sqrt{2} \end{pmatrix} \to \begin{pmatrix} 6 \\ -8 \end{pmatrix}$ *in* 2 *seconds from* $1 < t \le 3$.

$$\sqrt{2} - kt = 6$$
$$\sqrt{2} - 2k = 6$$
$$k = \frac{\sqrt{2} - 6}{2}$$

$$\sqrt{72} - mt = -8$$
$$7\sqrt{2} - 2m = -8$$
$$m = \frac{7\sqrt{2} + 8}{2}$$

$$h(t) = \begin{pmatrix} \sqrt{2} - \frac{(\sqrt{2} - 6)t}{3} \\ 7\sqrt{2} - \frac{(7\sqrt{2} + 8)t}{3} \end{pmatrix}$$

Problem Set Sample Solutions

1. Let $R\begin{pmatrix} x \\ y \end{pmatrix} = \begin{pmatrix} \cos\left(\frac{\pi}{4}\right) & -\sin\left(\frac{\pi}{4}\right) \\ \sin\left(\frac{\pi}{4}\right) & \cos\left(\frac{\pi}{4}\right) \end{pmatrix}\begin{pmatrix} x \\ y \end{pmatrix}$. Find the following.

a. $R^2\begin{pmatrix} \sqrt{2} \\ \sqrt{2} \end{pmatrix}$

$$R^2\begin{pmatrix} \sqrt{2} \\ \sqrt{2} \end{pmatrix} = R\left(R\begin{pmatrix} \sqrt{2} \\ \sqrt{2} \end{pmatrix}\right) = R\left(\begin{pmatrix} \cos\left(\frac{\pi}{4}\right) & -\sin\left(\frac{\pi}{4}\right) \\ \sin\left(\frac{\pi}{4}\right) & \cos\left(\frac{\pi}{4}\right) \end{pmatrix}\begin{pmatrix} \sqrt{2} \\ \sqrt{2} \end{pmatrix}\right) = R\left(\begin{pmatrix} \frac{\sqrt{2}}{2} & -\frac{\sqrt{2}}{2} \\ \frac{\sqrt{2}}{2} & \frac{\sqrt{2}}{2} \end{pmatrix}\begin{pmatrix} \sqrt{2} \\ \sqrt{2} \end{pmatrix}\right)$$

$$= R\begin{pmatrix} 0 \\ 2 \end{pmatrix} = \begin{pmatrix} \cos\left(\frac{\pi}{4}\right) & -\sin\left(\frac{\pi}{4}\right) \\ \sin\left(\frac{\pi}{4}\right) & \cos\left(\frac{\pi}{4}\right) \end{pmatrix}\begin{pmatrix} 0 \\ 2 \end{pmatrix} = \begin{pmatrix} \frac{\sqrt{2}}{2} & -\frac{\sqrt{2}}{2} \\ \frac{\sqrt{2}}{2} & \frac{\sqrt{2}}{2} \end{pmatrix}\begin{pmatrix} 0 \\ 2 \end{pmatrix} = \begin{pmatrix} -\sqrt{2} \\ \sqrt{2} \end{pmatrix}$$

b. How many transformations do you need to take so that the image returns to where it started?

It rotates by $\frac{\pi}{4}$ radians for each transformation; therefore, it takes 8 times to get to 2π, which is where it started.

c. Describe the matrix transformation $\begin{pmatrix} x \\ y \end{pmatrix} \to R^2\begin{pmatrix} x \\ y \end{pmatrix}$ and $R^n\begin{pmatrix} x \\ y \end{pmatrix}$ using a single matrix.

$R^2 = \begin{pmatrix} x \\ y \end{pmatrix}$ is the transformation that rotates the point through $2 \times \frac{\pi}{4}$ radian, so a formula for $R^2\begin{pmatrix} x \\ y \end{pmatrix}$ is

$$\begin{pmatrix} \cos\left(\frac{2\pi}{4}\right) & -\sin\left(\frac{2\pi}{4}\right) \\ \sin\left(\frac{2\pi}{4}\right) & \cos\left(\frac{2\pi}{4}\right) \end{pmatrix}\begin{pmatrix} x \\ y \end{pmatrix} = \begin{pmatrix} \cos\left(\frac{\pi}{2}\right) & -\sin\left(\frac{\pi}{2}\right) \\ \sin\left(\frac{\pi}{2}\right) & \cos\left(\frac{\pi}{2}\right) \end{pmatrix}\begin{pmatrix} x \\ y \end{pmatrix} = \begin{pmatrix} x \cdot \cos\left(\frac{\pi}{2}\right) - y \cdot \sin\left(\frac{\pi}{2}\right) \\ x \cdot \sin\left(\frac{\pi}{2}\right) + y \cdot \cos\left(\frac{\pi}{2}\right) \end{pmatrix}.$$

$R^n = \begin{pmatrix} x \\ y \end{pmatrix}$ is $\begin{pmatrix} x \cdot \cos\left(\frac{n\pi}{4}\right) - y \cdot \sin\left(\frac{n\pi}{4}\right) \\ x \cdot \sin\left(\frac{n\pi}{4}\right) + y \cdot \cos\left(\frac{n\pi}{4}\right) \end{pmatrix}$.

2. For $f(t) = \begin{pmatrix} \cos(t) & -\sin(t) \\ \sin(t) & \cos(t) \end{pmatrix}\begin{pmatrix} 1 \\ 1 \end{pmatrix}$, it takes 2π to transform the object at $\begin{pmatrix} 1 \\ 1 \end{pmatrix}$ back to where it starts. How long does it take the following functions to return to their starting point?

a. $f(t) = \begin{pmatrix} \cos(3t) & -\sin(3t) \\ \sin(3t) & \cos(3t) \end{pmatrix}\begin{pmatrix} 1 \\ 1 \end{pmatrix}$

$3t = 2\pi, \quad t = \frac{2\pi}{3}$

b. $f(t) = \begin{pmatrix} \cos\left(\frac{t}{3}\right) & -\sin\left(\frac{t}{3}\right) \\ \sin\left(\frac{t}{3}\right) & \cos\left(\frac{t}{3}\right) \end{pmatrix}\begin{pmatrix} 1 \\ 1 \end{pmatrix}$

$\frac{t}{3} = 2\pi, \quad t = 6\pi$

c. $f(t) = \begin{pmatrix} \cos\left(\frac{2t}{5}\right) & -\sin\left(\frac{2t}{5}\right) \\ \sin\left(\frac{2t}{5}\right) & \cos\left(\frac{2t}{5}\right) \end{pmatrix}\begin{pmatrix} 1 \\ 1 \end{pmatrix}$

$\frac{2t}{5} = 2\pi, \quad t = 5\pi$

3. Let $F(t) = \begin{pmatrix} \cos(t) & -\sin(t) \\ \sin(t) & \cos(t) \end{pmatrix} \begin{pmatrix} 2 \\ 1 \end{pmatrix}$, where t is measured in radians. Find the following:

a. $F\left(\frac{3\pi}{2}\right)$, $F\left(\frac{7\pi}{6}\right)$, and the radius of the path

$$F\left(\frac{3\pi}{2}\right) = \begin{pmatrix} \cos\left(\frac{3\pi}{2}\right) & -\sin\left(\frac{3\pi}{2}\right) \\ \sin\left(\frac{3\pi}{2}\right) & \cos\left(\frac{3\pi}{2}\right) \end{pmatrix} \begin{pmatrix} 2 \\ 1 \end{pmatrix} = \begin{pmatrix} 0 & 1 \\ -1 & 0 \end{pmatrix} \begin{pmatrix} 2 \\ 1 \end{pmatrix} = \begin{pmatrix} 1 \\ -2 \end{pmatrix}$$

$$F\left(\frac{7\pi}{6}\right) = \begin{pmatrix} \cos\left(\frac{7\pi}{6}\right) & -\sin\left(\frac{7\pi}{6}\right) \\ \sin\left(\frac{7\pi}{6}\right) & \cos\left(\frac{7\pi}{6}\right) \end{pmatrix} \begin{pmatrix} 2 \\ 1 \end{pmatrix} = \begin{pmatrix} -\frac{\sqrt{3}}{2} & \frac{1}{2} \\ -\frac{1}{2} & -\frac{\sqrt{3}}{2} \end{pmatrix} \begin{pmatrix} 2 \\ 1 \end{pmatrix} = \begin{pmatrix} -\sqrt{3} + \frac{1}{2} \\ -1 - \frac{\sqrt{3}}{2} \end{pmatrix}$$

The path of the point from $0 \le t \le 2\pi$ is a circle with a center at $(0, 0)$.

Thus, the radius $= \sqrt{x^2 + y^2} = \sqrt{(1)^2 + (2)^2} = \sqrt{5}$ *or*

the radius $= \sqrt{x^2 + y^2} = \sqrt{\left(-\sqrt{3} + \frac{1}{2}\right)^2 + \left(-1 - \frac{\sqrt{3}}{2}\right)^2} = \sqrt{5}.$

b. Show that the radius is always $\sqrt{x^2 + y^2}$ for the path of this transformation $(t) = \begin{pmatrix} \cos(t) & -\sin(t) \\ \sin(t) & \cos(t) \end{pmatrix} \begin{pmatrix} x \\ y \end{pmatrix}$.

$$F(t) = \begin{pmatrix} \cos(t) & -\sin(t) \\ \sin(t) & \cos(t) \end{pmatrix} \begin{pmatrix} x \\ y \end{pmatrix} = \begin{pmatrix} x\cos(t) - y\sin(t) \\ x\sin(t) + y\cos(t) \end{pmatrix}$$

The radius $= \sqrt{(x\cos(t) - y\sin(t))^2 + (x\sin(t) + y\cos(t))^2}$

$= \sqrt{x^2\cos^2(t) - 2xy\cos(t)\sin(t) + y^2\sin^2(t) + x^2\sin^2(t) + 2xy\sin(t)\cos(t) + y^2\cos^2(t)}$

$= \sqrt{x^2\cos^2(t) + x^2\sin^2(t) + y^2\sin^2(t) + y^2\cos^2(t)}$

$= \sqrt{x^2(\cos^2(t) + \sin^2(t)) + y^2(\cos^2(t) + \sin^2(t))}$

$= \sqrt{x^2 + y^2}.$

4. Let $F(t) = \begin{pmatrix} \cos\left(\frac{\pi t}{2}\right) & -\sin\left(\frac{\pi t}{2}\right) \\ \sin\left(\frac{\pi t}{2}\right) & \cos\left(\frac{\pi t}{2}\right) \end{pmatrix} \begin{pmatrix} 4 \\ 4 \end{pmatrix}$, where t is a real number.

a. Draw the path that $P = F(t)$ traces out as t varies within each of the following intervals:

i. $0 \le t \le 1$

ii. $1 \le t \le 2$

iii. $2 \le t \le 3$

iv. $3 \le t \le 4$

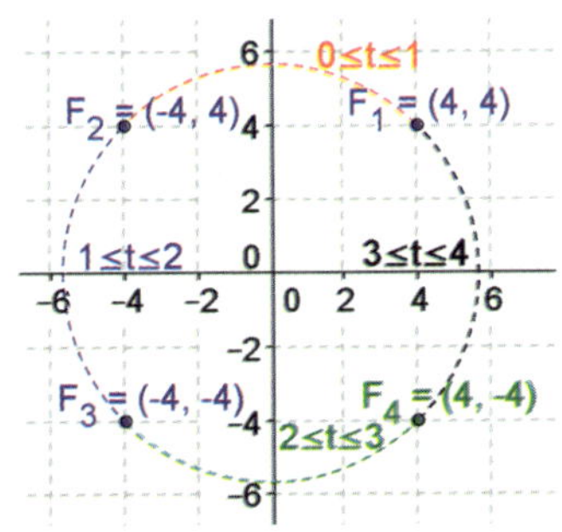

b. Where will the object be located at $t = 2.5$ seconds?

$$F(2.5) = \begin{pmatrix} \cos\left(\frac{5\pi}{4}\right) & -\sin\left(\frac{5\pi}{4}\right) \\ \sin\left(\frac{5\pi}{4}\right) & \cos\left(\frac{5\pi}{4}\right) \end{pmatrix}\begin{pmatrix} 4 \\ 4 \end{pmatrix} = \begin{pmatrix} \frac{-\sqrt{2}}{2} & \frac{\sqrt{2}}{2} \\ \frac{-\sqrt{2}}{2} & \frac{-\sqrt{2}}{2} \end{pmatrix}\begin{pmatrix} 4 \\ 4 \end{pmatrix} = \begin{pmatrix} 0 \\ -4\sqrt{2} \end{pmatrix}$$

c. How long does it take the object to reach $\begin{pmatrix} -8\sqrt{6} \\ 8\sqrt{2} \end{pmatrix}$?

The point $\begin{pmatrix} -8\sqrt{6} \\ 8\sqrt{2} \end{pmatrix}$ *is in Quadrant II; the reference angle is* $\frac{\pi t}{2} = \arctan\left(\frac{8\sqrt{2}}{8\sqrt{6}}\right) = \frac{\pi}{6}, \frac{\pi t}{2} = \frac{\pi}{6}, t = \frac{1}{3}$ *second.*

It takes 1.5 *seconds to rotate the point to* π*; therefore,* $1.5 - \frac{1}{3} = \frac{7}{6}$ *second.*

5. *Let* $F(t) = \begin{pmatrix} \cos\left(\frac{\pi t}{3}\right) & -\sin\left(\frac{\pi t}{3}\right) \\ \sin\left(\frac{\pi t}{3}\right) & \cos\left(\frac{\pi t}{3}\right) \end{pmatrix}\begin{pmatrix} -1 \\ -\sqrt{3} \end{pmatrix}$.

a. Draw the path that $P = F(t)$ traces out as t varies within the interval $0 \le t \le 1$.

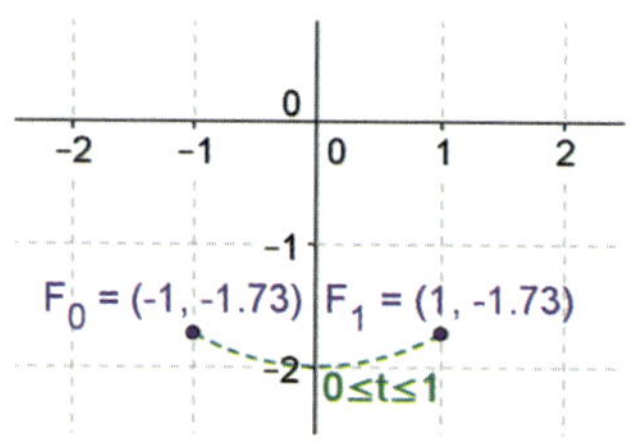

b. How long does it take the object to reach $(\sqrt{3}, 0)$?

The point $(\sqrt{3}, 0)$ *lies on the* x*-axis. Therefore,* $t = 2$ *seconds to rotate to the point* $(-1, -\sqrt{3})$.

c. How long does it take the object to return to its starting point?

It takes 6 *seconds.*

6. Find the function that will rotate the point $(4, 2)$ about the origin to the point $(-4, -2)$ over the following time intervals.

a. Over a 1-second interval

$$f(t) = \begin{pmatrix} \cos(\pi t) & -\sin(\pi t) \\ \sin(\pi t) & \cos(\pi t) \end{pmatrix}\begin{pmatrix} 4 \\ 2 \end{pmatrix}$$

b. Over a 2-second interval

$$f(t) = \begin{pmatrix} \cos\left(\frac{\pi t}{2}\right) & -\sin\left(\frac{\pi t}{2}\right) \\ \sin\left(\frac{\pi t}{2}\right) & \cos\left(\frac{\pi t}{2}\right) \end{pmatrix}\begin{pmatrix} 4 \\ 2 \end{pmatrix}$$

c. Over a $\frac{1}{3}$-second interval

$$f(t) = \begin{pmatrix} \cos(3\pi t) & -\sin(3\pi t) \\ \sin(3\pi t) & \cos(3\pi t) \end{pmatrix} \begin{pmatrix} 4 \\ 2 \end{pmatrix}$$

d. How about rotating it back to where it starts over a $\frac{4}{5}$-second interval?

$$f(t) = \begin{pmatrix} \cos\left(\frac{5\pi t}{2}\right) & -\sin\left(\frac{5\pi t}{2}\right) \\ \sin\left(\frac{5\pi t}{2}\right) & \cos\left(\frac{5\pi t}{2}\right) \end{pmatrix} \begin{pmatrix} 4 \\ 2 \end{pmatrix}$$

7. Summarize the geometric effect of the following function at the given point and the time interval.

a. $F(t) = \begin{pmatrix} 5\cos\left(\frac{\pi t}{4}\right) & -5\sin\left(\frac{\pi t}{4}\right) \\ 5\sin\left(\frac{\pi t}{4}\right) & 5\cos\left(\frac{\pi t}{4}\right) \end{pmatrix} \begin{pmatrix} 4 \\ 3 \end{pmatrix}, 0 \le t \le 1$

At $t = 0$, the point $\begin{pmatrix} 4 \\ 3 \end{pmatrix}$ is dilated by a factor of 5 to $\begin{pmatrix} 20 \\ 15 \end{pmatrix}$.

At $t = 1$, the image $\begin{pmatrix} 20 \\ 15 \end{pmatrix}$ then is rotated by $\frac{\pi}{4}$ radians counterclockwise about the origin.

b. $F(t) = \begin{pmatrix} \frac{1}{2}\cos\left(\frac{\pi t}{6}\right) & -\frac{1}{2}\sin\left(\frac{\pi t}{6}\right) \\ \frac{1}{2}\sin\left(\frac{\pi t}{6}\right) & \frac{1}{2}\cos\left(\frac{\pi t}{6}\right) \end{pmatrix} \begin{pmatrix} 6 \\ 2 \end{pmatrix}, 0 \le t \le 1$

At $t = 0$, the point $\begin{pmatrix} 6 \\ 2 \end{pmatrix}$ is dilated by a factor of $\frac{1}{2}$ to $\begin{pmatrix} 3 \\ 1 \end{pmatrix}$.

At $t = 1$, the image $\begin{pmatrix} 3 \\ 1 \end{pmatrix}$ then is rotated by $\frac{\pi}{6}$ radians counterclockwise about the origin.

8. In programming a computer video game, Grace coded the changing location of a rocket as follows:

At the time t second between $t = 0$ seconds and $t = 4$ seconds, the location $\begin{pmatrix} x \\ y \end{pmatrix}$ of the rocket is given by

$$\begin{pmatrix} \cos\left(\frac{\pi}{4}t\right) & -\sin\left(\frac{\pi}{4}t\right) \\ \sin\left(\frac{\pi}{4}t\right) & \cos\left(\frac{\pi}{4}t\right) \end{pmatrix} \begin{pmatrix} \sqrt{2} \\ \sqrt{2} \end{pmatrix}.$$

At a time of t seconds between $t = 4$ and $t = 8$ seconds, the location of the rocket is given by

$$\begin{pmatrix} -\sqrt{2} + \frac{\sqrt{2}}{2}(t-4) \\ -\sqrt{2} + \frac{\sqrt{2}}{2}(t-4) \end{pmatrix}.$$

a. What is the location of the rocket at time $t = 0$? What is its location at time $t = 8$?

At $t = 0$, $\begin{pmatrix} \cos(0) & -\sin(0) \\ \sin(0) & \cos(0) \end{pmatrix} \begin{pmatrix} \sqrt{2} \\ \sqrt{2} \end{pmatrix} = \begin{pmatrix} 1 & 0 \\ 0 & 1 \end{pmatrix} \begin{pmatrix} \sqrt{2} \\ \sqrt{2} \end{pmatrix} = \begin{pmatrix} \sqrt{2} \\ \sqrt{2} \end{pmatrix}.$

At $t = 4$, $\begin{pmatrix} -\sqrt{2} + \frac{\sqrt{2}}{2}(8-4) \\ -\sqrt{2} + \frac{\sqrt{2}}{2}(8-4) \end{pmatrix} = \begin{pmatrix} -\sqrt{2} + 2\sqrt{2} \\ -\sqrt{2} + 2\sqrt{2} \end{pmatrix} = \begin{pmatrix} \sqrt{2} \\ \sqrt{2} \end{pmatrix}.$

b. **Mason is worried that Grace may have made a mistake and the location of the rocket is unclear at time $t = 4$ seconds. Explain why there is no inconsistency in the location of the rocket at this time.**

At $t = 4$, $\begin{pmatrix} \cos(\pi) & -\sin(\pi) \\ \sin(\pi) & \cos(\pi) \end{pmatrix}\begin{pmatrix} 1 \\ 1 \end{pmatrix} = \begin{pmatrix} -1 & 0 \\ 0 & -1 \end{pmatrix}\begin{pmatrix} \sqrt{2} \\ \sqrt{2} \end{pmatrix} = \begin{pmatrix} -\sqrt{2} \\ -\sqrt{2} \end{pmatrix}.$

At $t = 4$, $\begin{pmatrix} -\sqrt{2} + \frac{\sqrt{2}}{2}(4-4) \\ -\sqrt{2} + \frac{\sqrt{2}}{2}(4-4) \end{pmatrix} = \begin{pmatrix} -\sqrt{2} \\ -\sqrt{2} \end{pmatrix}.$

These are consistent.

c. **What is the area of the region enclosed by the path of the rocket from time $t = 0$ to $t = 8$?**

The path traversed is a semicircle with a radius of 2; the area enclosed is $A = \frac{\pi r^2}{2} = \frac{4\pi}{2} = 2\pi$ square units.

Lesson 24: Matrix Notation Encompasses New Transformations!

Student Outcomes

- Students work with 2×2 matrices as transformations of the plane.
- Students understand the role of the multiplicative identity matrix.

Lesson Notes

In the preceding lessons, students learned how to represent rotations and dilations in matrix notation. They saw that not all transformations can be represented this way (translations, for example). In Lessons 24 and 25, they continue to explore the power of this new notation for finding new transformations, some that they may not have even conceived of. This lesson begins as students study the multiplicative identity matrix and discover it is similar to multiplying by 1 within the real number system. Students experiment with various matrices to discover the transformation each produces. The use of a transparency and a dry erase marker assists students in conducting their exploration.

Classwork

Opening (5 minutes)

Review the types of matrices that, when used to transform a point or vector, produce rotations, dilations, or both.

MP.2

Show each matrix, and ask students to describe the effect of the matrix in words.

- $\begin{pmatrix} \cos(\theta) & -\sin(\theta) \\ \sin(\theta) & \cos(\theta) \end{pmatrix}$
  - *Produces a counterclockwise rotation of $\theta°$*
- $\begin{pmatrix} k & 0 \\ 0 & k \end{pmatrix}$
  - *Produces a dilation with a scale factor of k*
- $\begin{pmatrix} a & -b \\ b & a \end{pmatrix}$
  - *Produces a counterclockwise rotation and a dilation*
- $\begin{pmatrix} 1 & 0 \\ 0 & 1 \end{pmatrix}$
  - *No effect; this is the multiplicative identity matrix.*

Scaffolding:

- These more specific matrices can be used to help students visualize the results of matrix multiplication.
  $\begin{pmatrix} \cos(90°) & -\sin(90°) \\ \sin(90°) & \cos(90°) \end{pmatrix}$ is a $90°$ rotation counterclockwise.
  $\begin{pmatrix} 4 & 0 \\ 0 & 4 \end{pmatrix}$ is a dilation with a scale factor of 4.
  $\begin{pmatrix} 4 & -3 \\ 3 & 4 \end{pmatrix}$ is a dilation and a rotation.
- For advanced students, instead of Example 1, pose the following: Multiply several matrices by the matrix $\begin{pmatrix} 1 & 0 \\ 0 & 1 \end{pmatrix}$. What effect does this matrix have on other matrices? Can you predict the name of this matrix and what number it is similar to in the real number system?

Example 1 (8 minutes)

Students should work on this example individually; after they have finished, pull the class together to debrief.

Example 1

Determine the following:

a. $\begin{pmatrix} 1 & 0 \\ 0 & 1 \end{pmatrix}\begin{pmatrix} 3 \\ -2 \end{pmatrix}$ $\qquad \begin{pmatrix} 3 \\ -2 \end{pmatrix}$

b. $\begin{pmatrix} 1 & 0 \\ 0 & 1 \end{pmatrix}\begin{pmatrix} -7 \\ 12 \end{pmatrix}$ $\qquad \begin{pmatrix} -7 \\ 12 \end{pmatrix}$

c. $\begin{pmatrix} 1 & 0 \\ 0 & 1 \end{pmatrix}\begin{pmatrix} 3 & 5 \\ -2 & 1 \end{pmatrix}$ $\qquad \begin{pmatrix} 3 & 5 \\ -2 & 1 \end{pmatrix}$

d. $\begin{pmatrix} 1 & 0 \\ 0 & 1 \end{pmatrix}\begin{pmatrix} -1 & -3 \\ -7 & 6 \end{pmatrix}$ $\qquad \begin{pmatrix} -1 & -3 \\ -7 & 6 \end{pmatrix}$

e. $\begin{pmatrix} 9 & 12 \\ -3 & -1 \end{pmatrix}\begin{pmatrix} 1 & 0 \\ 0 & 1 \end{pmatrix}$ $\qquad \begin{pmatrix} 9 & 12 \\ -3 & -1 \end{pmatrix}$

f. $\begin{pmatrix} 1 & 0 \\ 0 & 1 \end{pmatrix}\begin{pmatrix} a & b \\ c & d \end{pmatrix}$ $\qquad \begin{pmatrix} a & b \\ c & d \end{pmatrix}$

g. $\begin{pmatrix} x & y \\ z & w \end{pmatrix}\begin{pmatrix} 1 & 0 \\ 0 & 1 \end{pmatrix}$ $\qquad \begin{pmatrix} x & y \\ z & w \end{pmatrix}$

- What did you notice about the result of each matrix multiplication problem?
  - *The result was always one of the matrices.*
- Which matrix?
  - *The one that was not* $\begin{pmatrix} 1 & 0 \\ 0 & 1 \end{pmatrix}$
- What is this similar to in the real number system?
  - *When we multiply by the number* 1
- What do we call the number 1 when multiplying real numbers? Explain.
  - *The multiplicative identity because you always get the number that was multiplied by* 1 *as the product.*
- Can you predict what we call the matrix $\begin{pmatrix} 1 & 0 \\ 0 & 1 \end{pmatrix}$?
  - *The multiplicative identity matrix*
- Explain to your neighbor what the multiplicative identity matrix is and why it is called the multiplicative identity.
  - *Students explain.*

Example 2 (8 minutes)

Give students time to think about the question individually and then pair up to discuss with a partner. After each pair has come up with an answer, discuss as a class.

> **Example 2**
>
> **Can the reflection about the real axis $L(z) = \bar{z}$ be expressed in matrix notation?**
>
> *Yes, using the matrix* $\begin{pmatrix} 1 & 0 \\ 0 & -1 \end{pmatrix}$ $\begin{pmatrix} x \\ y \end{pmatrix} \rightarrow \begin{pmatrix} 1 & 0 \\ 0 & -1 \end{pmatrix}\begin{pmatrix} x \\ y \end{pmatrix} = \begin{pmatrix} x \\ -y \end{pmatrix}$

- Is the matrix you found in the same form as those studied in the previous lessons, $\begin{pmatrix} a & -b \\ b & a \end{pmatrix}$?
  - *No*

MP.3

- How did you verify that the matrix you found did in fact produce a reflection about the real axis?
  - *I multiplied it by a sample point to see if it reflected the point about the real axis. Teacher note: Allow students to share various responses, even incorrect ones, until the class comes to a consensus on the correct answer.*
- Do you think all matrices in the form $\begin{pmatrix} a_{11} & a_{12} \\ a_{21} & a_{22} \end{pmatrix}$ correspond to a transformation of some kind?
  - *Yes, multiplying* $\begin{pmatrix} a_{11} & a_{12} \\ a_{21} & a_{22} \end{pmatrix}$ *times some point* $\begin{pmatrix} x \\ y \end{pmatrix}$ *will affect the point in some way, thereby producing a transformation of some sort.*

Exercises 1–3 (8 minutes)

Allow students time to work on Exercises 1–3 either individually or in pairs, and then debrief.

> **Exercises**
>
> **1. Express a reflection about the vertical axis in matrix notation. Prove that it produces the desired reflection by using matrix multiplication.**
>
> $\begin{pmatrix} -1 & 0 \\ 0 & 1 \end{pmatrix}$
>
> $\begin{pmatrix} x \\ y \end{pmatrix} \rightarrow \begin{pmatrix} -1 & 0 \\ 0 & 1 \end{pmatrix}\begin{pmatrix} x \\ y \end{pmatrix} = \begin{pmatrix} -x \\ y \end{pmatrix}$
>
> **2. Express a reflection about both the horizontal and vertical axes in matrix notation. Prove that it produces the desired reflection by using matrix multiplication.**
>
> $\begin{pmatrix} -1 & 0 \\ 0 & -1 \end{pmatrix}$
>
> $\begin{pmatrix} x \\ y \end{pmatrix} \rightarrow \begin{pmatrix} -1 & 0 \\ 0 & -1 \end{pmatrix}\begin{pmatrix} x \\ y \end{pmatrix} = \begin{pmatrix} -x \\ -y \end{pmatrix}$

Scaffolding:

- If students are struggling, have them work with a sample point before generalizing.

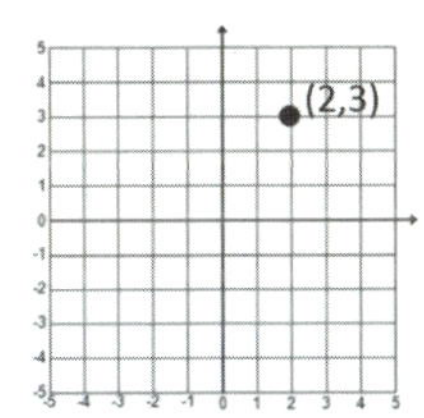

- What are the coordinates of $(2, 3)$ after a reflection in the x-axis?
- What are the coordinates of $(2, 3)$ after a reflection in the y-axis?
- What happens to the coordinates of a point after a reflection in the x-axis? y-axis?

3. Express a reflection about the vertical axis and a dilation with a scale factor of 6 in matrix notation. Prove that it produces the desired reflection by using matrix multiplication.

$\begin{pmatrix} -6 & 0 \\ 0 & 6 \end{pmatrix}$

$\begin{pmatrix} x \\ y \end{pmatrix} \rightarrow \begin{pmatrix} -6 & 0 \\ 0 & 6 \end{pmatrix}\begin{pmatrix} x \\ y \end{pmatrix} = \begin{pmatrix} -6x \\ 6y \end{pmatrix}$

- For Exercise 2, could this transformation also be viewed as a rotation?
  - *Yes. Reflecting about both axes is the same as a $180°$ rotation.*
- Is it possible to have a dilation where the x-coordinate is increased by a different scale factor than the y-coordinate?
  - *Yes. For example, $\begin{pmatrix} 2 & 0 \\ 0 & 3 \end{pmatrix}$ would dilate the x by a factor of 2 and the y by a factor of 3.*

Exercises 4–8 (15 minutes)

Provide each student with a transparency and a dry erase marker. Instruct them to use the graph provided to experiment with the effect of each matrix. While circulating around the room providing assistance as needed, allow students time to work either individually or in pairs.

Explore the transformation given by each matrix below. Use the graph of the rectangle provided to assist in the exploration. Describe the effect on the graph of the rectangle, and then show the general effect of the transformation by using matrix multiplication.

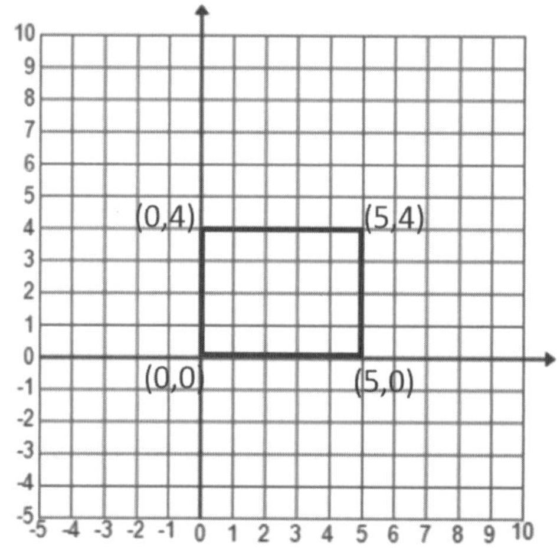

	Matrix	Transformation of the Rectangle	General Effect of the Matrix
4.	$\begin{pmatrix} 1 & 0 \\ 0 & 1 \end{pmatrix}$	*None; the points do not change.*	$\begin{pmatrix} x \\ y \end{pmatrix} \rightarrow \begin{pmatrix} 1 & 0 \\ 0 & 1 \end{pmatrix}\begin{pmatrix} x \\ y \end{pmatrix} = \begin{pmatrix} x \\ y \end{pmatrix}$
5.	$\begin{pmatrix} 0 & 1 \\ 1 & 0 \end{pmatrix}$	*The x- and y-coordinates switched. The rectangle flipped sideways.*	$\begin{pmatrix} x \\ y \end{pmatrix} \rightarrow \begin{pmatrix} 0 & 1 \\ 1 & 0 \end{pmatrix}\begin{pmatrix} x \\ y \end{pmatrix} = \begin{pmatrix} y \\ x \end{pmatrix}$
6.	$\begin{pmatrix} 0 & 0 \\ 0 & 0 \end{pmatrix}$	*It takes all the points to the origin.*	$\begin{pmatrix} x \\ y \end{pmatrix} \rightarrow \begin{pmatrix} 0 & 0 \\ 0 & 0 \end{pmatrix}\begin{pmatrix} x \\ y \end{pmatrix} = \begin{pmatrix} 0 \\ 0 \end{pmatrix}$

7. $\begin{pmatrix} 1 & 1 \\ 0 & 1 \end{pmatrix}$	*The points on the x-axis remained fixed, but all other points shifted horizontally by y units. The rectangle became a parallelogram.*	$\begin{pmatrix} x \\ y \end{pmatrix} \rightarrow \begin{pmatrix} 1 & 1 \\ 0 & 1 \end{pmatrix}\begin{pmatrix} x \\ y \end{pmatrix} = \begin{pmatrix} x+y \\ y \end{pmatrix}$
8. $\begin{pmatrix} 1 & 0 \\ 1 & 1 \end{pmatrix}$	*The points on the y-axis remained fixed, but all other points shifted vertically by x units. The rectangle became a parallelogram.*	$\begin{pmatrix} x \\ y \end{pmatrix} \rightarrow \begin{pmatrix} 1 & 0 \\ 1 & 1 \end{pmatrix}\begin{pmatrix} x \\ y \end{pmatrix} = \begin{pmatrix} x \\ y+x \end{pmatrix}$

- What effect did the matrix $\begin{pmatrix} 1 & 0 \\ 0 & 1 \end{pmatrix}$ have on the graph?
  - *None*
- This is called an identity matrix because it is equivalent to multiplying by 1.
- Is the matrix $\begin{pmatrix} 0 & 1 \\ 1 & 0 \end{pmatrix}$ also an identity matrix?
  - *No. It did change the coordinates. Teacher note: The teacher may demonstrate or lead students to the fact that the graph was reflected about the line $y = x$, but it is not the focal point of this lesson.*
- Can we view the zero matrix $\begin{pmatrix} 0 & 0 \\ 0 & 0 \end{pmatrix}$ as a dilation?
  - *Yes, with a scale factor of* 0
- How would you classify the last two transformations?
  - *It was sort of like a translation because some of the points moved to the right, but some of them remained fixed. It does not exactly fit into any of our transformations. Teacher note: These are both examples of shears, but the main point is that all of these matrices produce some sort of transformation but not necessarily ones that we know about at this point.*

Closing (4 minutes)

Ask students to summarize the types of transformation matrices seen in this lesson. Add these to the list started at the beginning of class.

Lesson Summary

All matrices in the form $\begin{pmatrix} a_{11} & a_{12} \\ a_{21} & a_{22} \end{pmatrix}$ correspond to a transformation of some kind.

- **The matrix $\begin{pmatrix} 1 & 0 \\ 0 & -1 \end{pmatrix}$ reflects all coordinates about the horizontal axis.**
- **The matrix $\begin{pmatrix} -1 & 0 \\ 0 & 1 \end{pmatrix}$ reflects all coordinates about the vertical axis.**
- **The matrix $\begin{pmatrix} 1 & 0 \\ 0 & 1 \end{pmatrix}$ is the identity matrix and corresponds to a transformation that leaves points alone.**
- **The matrix $\begin{pmatrix} 0 & 0 \\ 0 & 0 \end{pmatrix}$ is the zero matrix and corresponds to a dilation of scale factor 0.**

Exit Ticket (5 minutes)

Name ______________________________ Date ______________

Lesson 24: Matrix Notation Encompasses New Transformations!

Exit Ticket

What type of transformation is shown in the following examples? What is the resulting matrix?

1. $\begin{pmatrix} 0 & -1 \\ 1 & 0 \end{pmatrix}\begin{pmatrix} 3 \\ 2 \end{pmatrix}$

2. $\begin{pmatrix} 3 & 0 \\ 0 & 3 \end{pmatrix}\begin{pmatrix} 3 \\ 2 \end{pmatrix}$

3. $\begin{pmatrix} -1 & 0 \\ 0 & 1 \end{pmatrix}\begin{pmatrix} 3 \\ 2 \end{pmatrix}$

4. $\begin{pmatrix} 1 & 0 \\ 0 & -1 \end{pmatrix}\begin{pmatrix} 3 \\ 2 \end{pmatrix}$

5. What is the multiplicative identity matrix? What is it similar to in the set of real numbers? Explain your answer.

Exit Ticket Sample Solutions

What type of transformation is shown in the following examples? What is the resulting matrix?

1. $\begin{pmatrix} 0 & -1 \\ 1 & 0 \end{pmatrix}\begin{pmatrix} 3 \\ 2 \end{pmatrix}$

   It is a pure rotation. The point $\begin{pmatrix} 3 \\ 2 \end{pmatrix}$ *is rotated* $\frac{\pi}{2}$ *radians, and the image is* $\begin{pmatrix} -2 \\ 3 \end{pmatrix}$.

2. $\begin{pmatrix} 3 & 0 \\ 0 & 3 \end{pmatrix}\begin{pmatrix} 3 \\ 2 \end{pmatrix}$

   It is dilation with a factor of 3. *The point* $\begin{pmatrix} 3 \\ 2 \end{pmatrix}$ *is dilated by a factor of* 3, *and the image is* $\begin{pmatrix} 9 \\ 6 \end{pmatrix}$.

3. $\begin{pmatrix} -1 & 0 \\ 0 & 1 \end{pmatrix}\begin{pmatrix} 3 \\ 2 \end{pmatrix}$

   It is a reflection about the y*-axis. The point* $\begin{pmatrix} 3 \\ 2 \end{pmatrix}$ *is reflected about the* y*-axis, and the image is* $\begin{pmatrix} -3 \\ 2 \end{pmatrix}$.

4. $\begin{pmatrix} 1 & 0 \\ 0 & -1 \end{pmatrix}\begin{pmatrix} 3 \\ 2 \end{pmatrix}$

   It is a reflection about the x*-axis. The point* $\begin{pmatrix} 3 \\ 2 \end{pmatrix}$ *is reflected about the* x*-axis, and the image is* $\begin{pmatrix} 3 \\ -2 \end{pmatrix}$.

5. What is the multiplicative identity matrix? What is it similar to in the set of real numbers? Explain your answer.

   The multiplicative identity matrix is $\begin{bmatrix} 1 & 0 \\ 0 & 1 \end{bmatrix}$. *It is similar to the number* 1 *in the real number system because any matrix multiplied by* $\begin{bmatrix} 1 & 0 \\ 0 & 1 \end{bmatrix}$ *produces the original matrix.*

Problem Set Sample Solutions

1. What matrix do you need to use to reflect the following points about the y-axis? What is the resulting matrix when this is done? Show all work and sketch it.

   a. $\begin{pmatrix} 3 \\ 0 \end{pmatrix}$

   $\begin{pmatrix} -1 & 0 \\ 0 & 1 \end{pmatrix}\begin{pmatrix} 3 \\ 0 \end{pmatrix} = \begin{pmatrix} -3 \\ 0 \end{pmatrix}$

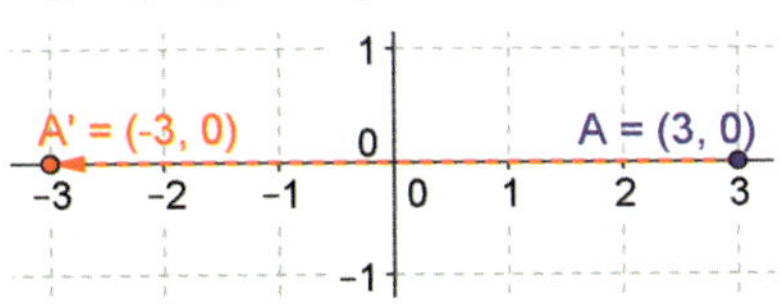

b. $\begin{pmatrix} 3 \\ 2 \end{pmatrix}$

$\begin{pmatrix} -1 & 0 \\ 0 & 1 \end{pmatrix}\begin{pmatrix} 3 \\ 2 \end{pmatrix} = \begin{pmatrix} -3 \\ 2 \end{pmatrix}$

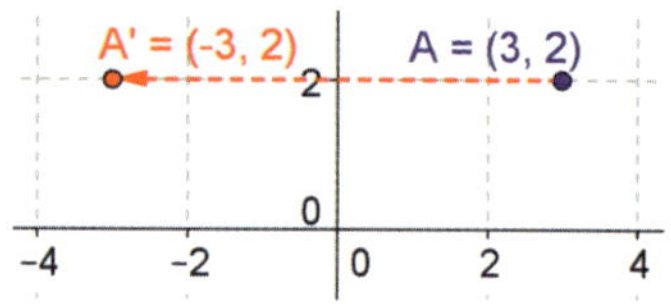

c. $\begin{pmatrix} -4 \\ 3 \end{pmatrix}$

$\begin{pmatrix} -1 & 0 \\ 0 & 1 \end{pmatrix}\begin{pmatrix} -4 \\ 3 \end{pmatrix} = \begin{pmatrix} 4 \\ 3 \end{pmatrix}$

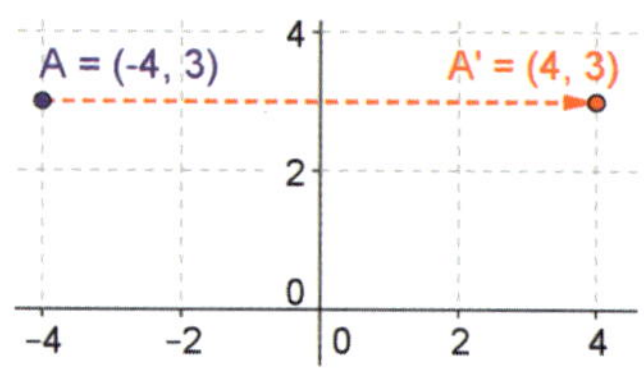

d. $\begin{pmatrix} -3 \\ -2 \end{pmatrix}$

$\begin{pmatrix} -1 & 0 \\ 0 & 1 \end{pmatrix}\begin{pmatrix} -3 \\ -2 \end{pmatrix} = \begin{pmatrix} 3 \\ -2 \end{pmatrix}$

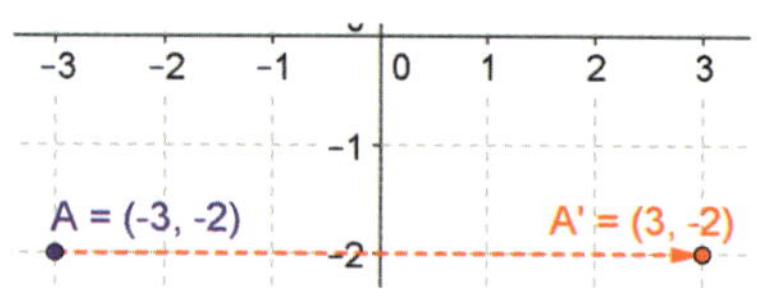

e. $\begin{pmatrix} 3 \\ -3 \end{pmatrix}$

$\begin{pmatrix} -1 & 0 \\ 0 & 1 \end{pmatrix}\begin{pmatrix} 3 \\ -3 \end{pmatrix} = \begin{pmatrix} -3 \\ -3 \end{pmatrix}$

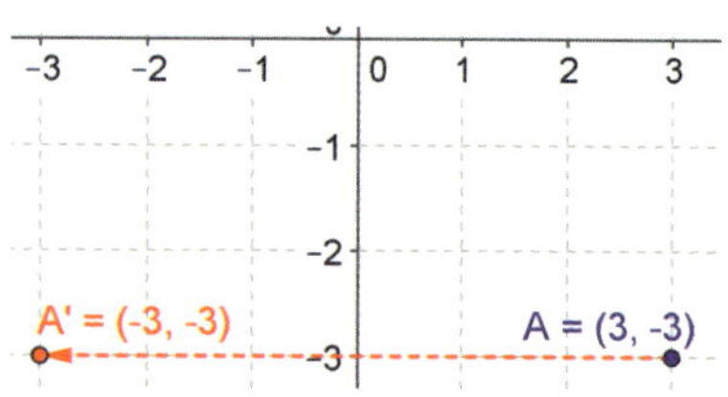

f. $\begin{pmatrix} 5 \\ 4 \end{pmatrix}$

$\begin{pmatrix} -1 & 0 \\ 0 & 1 \end{pmatrix}\begin{pmatrix} 5 \\ 4 \end{pmatrix} = \begin{pmatrix} -5 \\ 4 \end{pmatrix}$

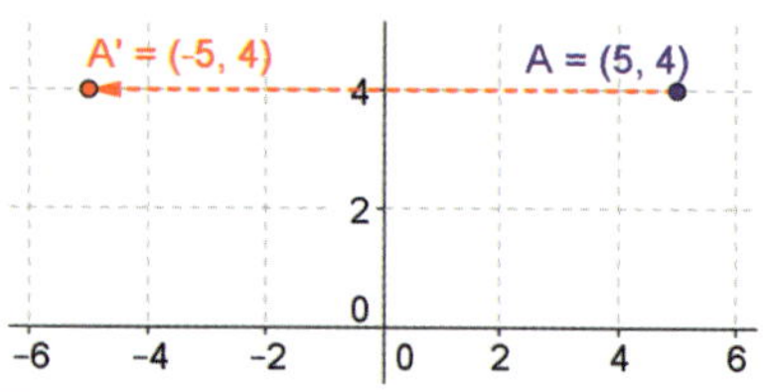

2. What matrix do you need to use to reflect the following points about the x-axis? What is the resulting matrix when this is done? Show all work and sketch it.

a. $\begin{pmatrix} 0 \\ 3 \end{pmatrix}$

$\begin{pmatrix} 1 & 0 \\ 0 & -1 \end{pmatrix}\begin{pmatrix} 0 \\ 3 \end{pmatrix} = \begin{pmatrix} 0 \\ -3 \end{pmatrix}$

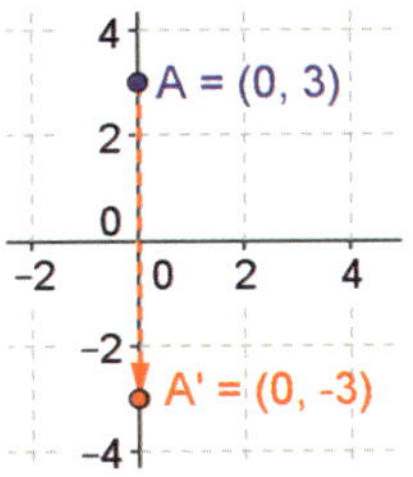

b. $\begin{pmatrix} 2 \\ 3 \end{pmatrix}$

$\begin{pmatrix} 1 & 0 \\ 0 & -1 \end{pmatrix}\begin{pmatrix} 2 \\ 3 \end{pmatrix} = \begin{pmatrix} 2 \\ -3 \end{pmatrix}$

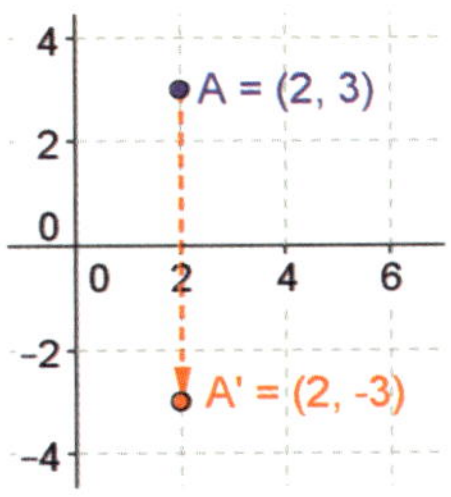

c. $\begin{pmatrix} -2 \\ 3 \end{pmatrix}$

$\begin{pmatrix} 1 & 0 \\ 0 & -1 \end{pmatrix}\begin{pmatrix} -2 \\ 3 \end{pmatrix} = \begin{pmatrix} -2 \\ -3 \end{pmatrix}$

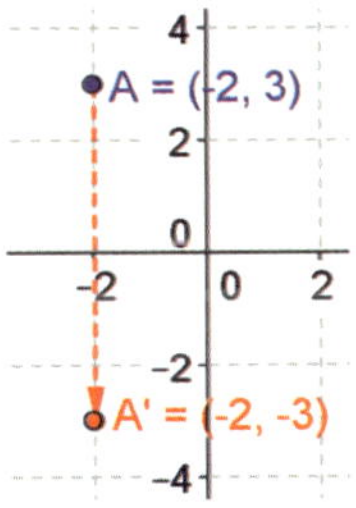

d. $\begin{pmatrix} -3 \\ -3 \end{pmatrix}$

$\begin{pmatrix} 1 & 0 \\ 0 & -1 \end{pmatrix}\begin{pmatrix} -3 \\ -3 \end{pmatrix} = \begin{pmatrix} -3 \\ 3 \end{pmatrix}$

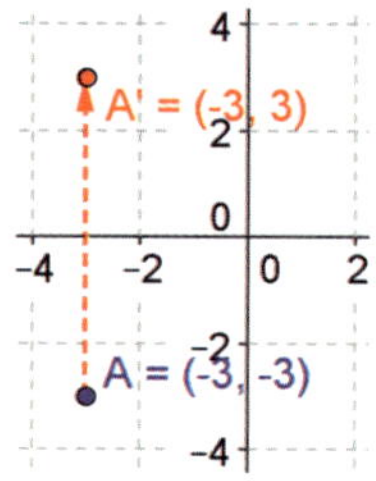

e. $\begin{pmatrix} 3 \\ -3 \end{pmatrix}$

$\begin{pmatrix} 1 & 0 \\ 0 & -1 \end{pmatrix}\begin{pmatrix} 3 \\ -3 \end{pmatrix} = \begin{pmatrix} 3 \\ 3 \end{pmatrix}$

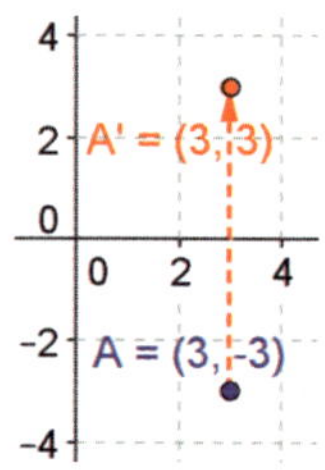

f. $\begin{pmatrix} -3 \\ 4 \end{pmatrix}$

$\begin{pmatrix} 1 & 0 \\ 0 & -1 \end{pmatrix}\begin{pmatrix} -3 \\ -4 \end{pmatrix} = \begin{pmatrix} -3 \\ 4 \end{pmatrix}$

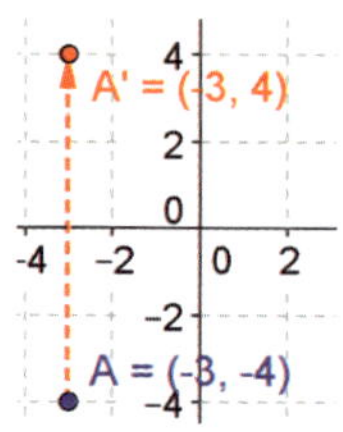

3. **What matrix do you need to use to dilate the following points by a given factor? What is the resulting matrix when this is done? Show all work and sketch it.**

a. $\begin{pmatrix} 1 \\ 0 \end{pmatrix}$**, a factor of 3**

$\begin{pmatrix} 3 & 0 \\ 0 & 3 \end{pmatrix}\begin{pmatrix} 1 \\ 0 \end{pmatrix} = \begin{pmatrix} 3 \\ 0 \end{pmatrix}$

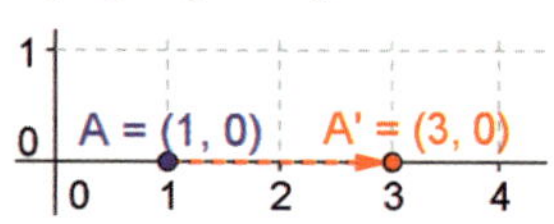

b. $\begin{pmatrix}3\\2\end{pmatrix}$, a factor of 2

$\begin{pmatrix}2 & 0\\0 & 2\end{pmatrix}\begin{pmatrix}3\\2\end{pmatrix}=\begin{pmatrix}6\\4\end{pmatrix}$

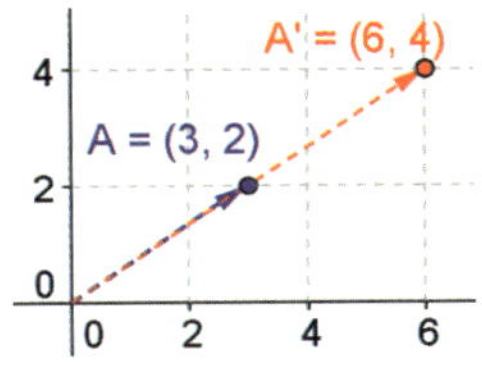

c. $\begin{pmatrix}1\\-2\end{pmatrix}$, a factor of 1

$\begin{pmatrix}1 & 0\\0 & 1\end{pmatrix}\begin{pmatrix}1\\-2\end{pmatrix}=\begin{pmatrix}1\\-2\end{pmatrix}$

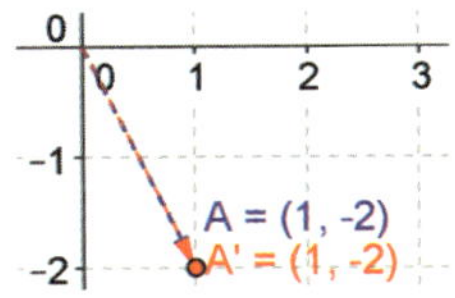

d. $\begin{pmatrix}-4\\-6\end{pmatrix}$, a factor of $\frac{1}{2}$

$\begin{pmatrix}\frac{1}{2} & 0\\0 & \frac{1}{2}\end{pmatrix}\begin{pmatrix}-4\\-6\end{pmatrix}=\begin{pmatrix}-2\\-3\end{pmatrix}$

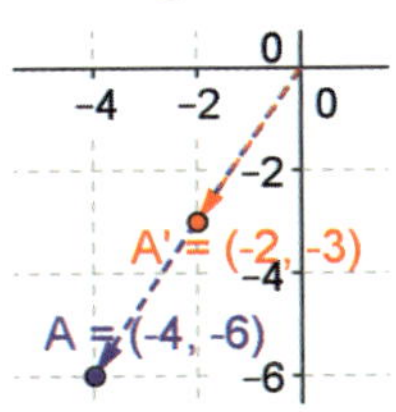

e. $\begin{pmatrix}9\\3\end{pmatrix}$, a factor of $\frac{1}{3}$

$\begin{pmatrix}\frac{1}{3} & 0\\0 & \frac{1}{3}\end{pmatrix}\begin{pmatrix}9\\3\end{pmatrix}=\begin{pmatrix}3\\1\end{pmatrix}$

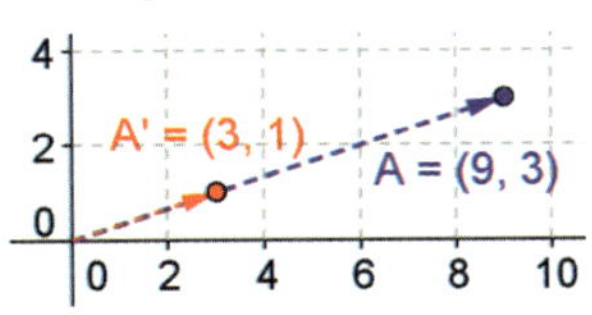

f. $\begin{pmatrix}\sqrt{3}\\\sqrt{11}\end{pmatrix}$, a factor of $\sqrt{2}$

$$\begin{pmatrix}\sqrt{2} & 0\\0 & \sqrt{2}\end{pmatrix}\begin{pmatrix}\sqrt{3}\\\sqrt{11}\end{pmatrix}=\begin{pmatrix}\sqrt{6}\\\sqrt{22}\end{pmatrix}$$

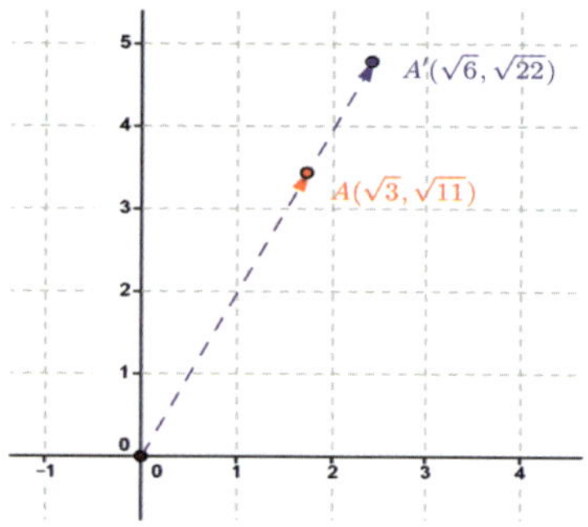

4. What matrix will rotate the given point by the angle? What is the resulting matrix when this is done? Show all work and sketch it.

a. $\begin{pmatrix}1\\0\end{pmatrix}$, $\frac{\pi}{2}$ radians

$$\begin{pmatrix}\cos\left(\frac{\pi}{2}\right) & -\sin\left(\frac{\pi}{2}\right)\\ \sin\left(\frac{\pi}{2}\right) & \cos\left(\frac{\pi}{2}\right)\end{pmatrix}\begin{pmatrix}1\\0\end{pmatrix}=\begin{pmatrix}0 & -1\\1 & 0\end{pmatrix}\begin{pmatrix}1\\0\end{pmatrix}=\begin{pmatrix}0\\1\end{pmatrix}$$

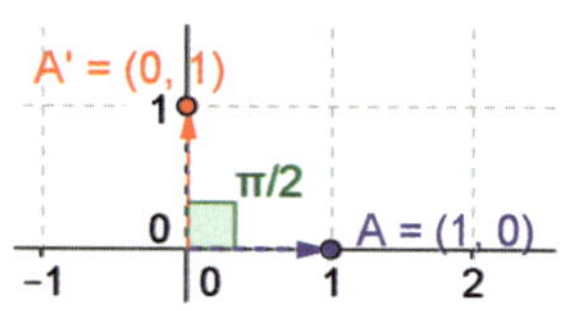

b. $\begin{pmatrix}1\\0\end{pmatrix}$, $\frac{\pi}{3}$ radians

$$\begin{pmatrix}\cos\left(\frac{\pi}{3}\right) & -\sin\left(\frac{\pi}{3}\right)\\ \sin\left(\frac{\pi}{3}\right) & \cos\left(\frac{\pi}{3}\right)\end{pmatrix}\begin{pmatrix}1\\0\end{pmatrix}=\begin{pmatrix}\frac{1}{2} & -\frac{\sqrt{3}}{2}\\ \frac{\sqrt{3}}{2} & \frac{1}{2}\end{pmatrix}\begin{pmatrix}1\\0\end{pmatrix}=\begin{pmatrix}\frac{1}{2}\\ \frac{\sqrt{3}}{2}\end{pmatrix}$$

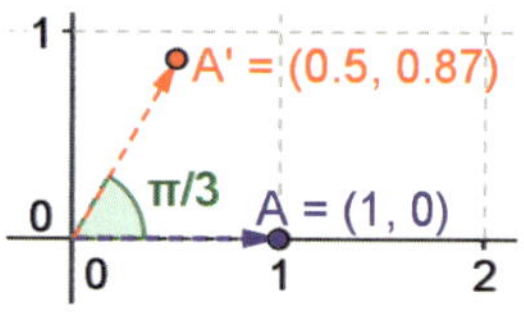

c. $\begin{pmatrix} 1 \\ 0 \end{pmatrix}$, $\frac{\pi}{6}$ radians

$$\begin{pmatrix} \cos\left(\frac{\pi}{6}\right) & -\sin\left(\frac{\pi}{6}\right) \\ \sin\left(\frac{\pi}{6}\right) & \cos\left(\frac{\pi}{6}\right) \end{pmatrix} \begin{pmatrix} 1 \\ 0 \end{pmatrix} = \begin{pmatrix} \frac{\sqrt{3}}{2} & -\frac{1}{2} \\ \frac{1}{2} & \frac{\sqrt{3}}{2} \end{pmatrix} \begin{pmatrix} 1 \\ 0 \end{pmatrix} = \begin{pmatrix} \frac{\sqrt{3}}{2} \\ \frac{1}{2} \end{pmatrix}$$

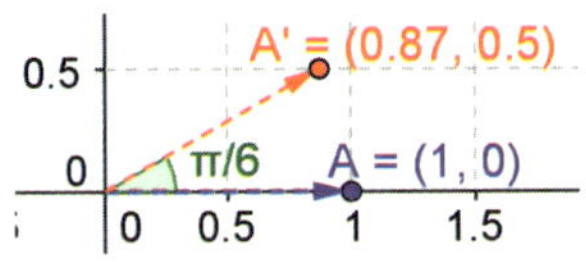

d. $\begin{pmatrix} 1 \\ 0 \end{pmatrix}$, $\frac{\pi}{4}$ radians

$$\begin{pmatrix} \cos\left(\frac{\pi}{4}\right) & -\sin\left(\frac{\pi}{4}\right) \\ \sin\left(\frac{\pi}{4}\right) & \cos\left(\frac{\pi}{4}\right) \end{pmatrix} \begin{pmatrix} 1 \\ 0 \end{pmatrix} = \begin{pmatrix} \frac{\sqrt{2}}{2} & -\frac{\sqrt{2}}{2} \\ \frac{\sqrt{2}}{2} & \frac{\sqrt{2}}{2} \end{pmatrix} \begin{pmatrix} 1 \\ 0 \end{pmatrix} = \begin{pmatrix} \frac{\sqrt{2}}{2} \\ \frac{\sqrt{2}}{2} \end{pmatrix}$$

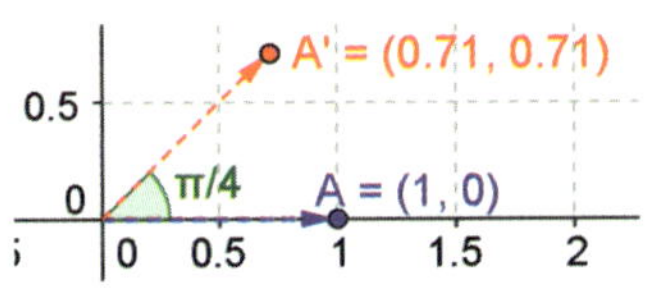

e. $\begin{pmatrix} \frac{\sqrt{3}}{2} \\ \frac{1}{2} \end{pmatrix}$, $\frac{\pi}{6}$ radians

$$\begin{pmatrix} \cos\left(\frac{\pi}{6}\right) & -\sin\left(\frac{\pi}{6}\right) \\ \sin\left(\frac{\pi}{6}\right) & \cos\left(\frac{\pi}{6}\right) \end{pmatrix} \begin{pmatrix} \frac{\sqrt{3}}{2} \\ \frac{1}{2} \end{pmatrix} = \begin{pmatrix} \frac{\sqrt{3}}{2} & -\frac{1}{2} \\ \frac{1}{2} & \frac{\sqrt{3}}{2} \end{pmatrix} \begin{pmatrix} \frac{\sqrt{3}}{2} \\ \frac{1}{2} \end{pmatrix} = \begin{pmatrix} \frac{1}{2} \\ \frac{\sqrt{3}}{2} \end{pmatrix}$$

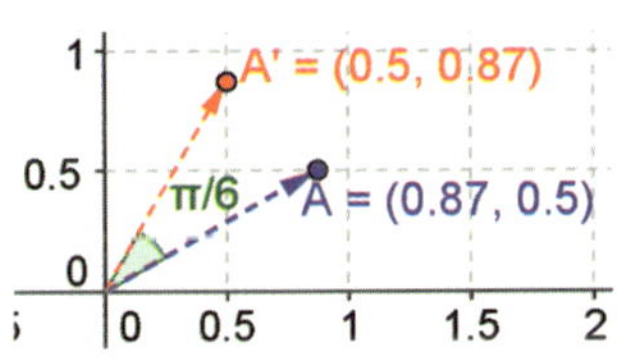

f. $\begin{pmatrix} \frac{\sqrt{2}}{2} \\ \frac{\sqrt{2}}{2} \end{pmatrix}$, $\frac{\pi}{4}$ radians

$$\begin{pmatrix} \cos\left(\frac{\pi}{4}\right) & -\sin\left(\frac{\pi}{4}\right) \\ \sin\left(\frac{\pi}{4}\right) & \cos\left(\frac{\pi}{4}\right) \end{pmatrix}\begin{pmatrix} \frac{\sqrt{2}}{2} \\ \frac{\sqrt{2}}{2} \end{pmatrix} = \begin{pmatrix} \frac{\sqrt{2}}{2} & -\frac{\sqrt{2}}{2} \\ \frac{\sqrt{2}}{2} & \frac{\sqrt{2}}{2} \end{pmatrix}\begin{pmatrix} \frac{\sqrt{2}}{2} \\ \frac{\sqrt{2}}{2} \end{pmatrix} = \begin{pmatrix} 0 \\ 1 \end{pmatrix}$$

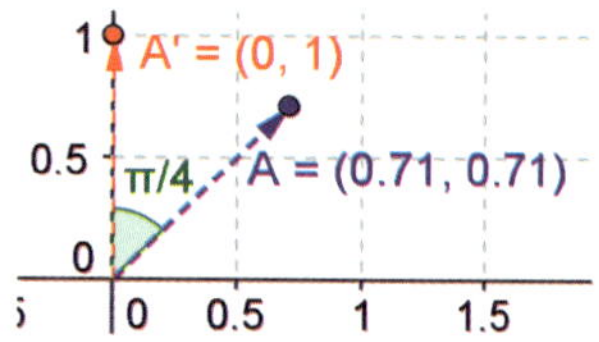

g. $\begin{pmatrix} \frac{\sqrt{3}}{2} \\ \frac{1}{2} \end{pmatrix}$, π radians

$$\begin{pmatrix} \cos(\pi) & -\sin(\pi) \\ \sin(\pi) & \cos(\pi) \end{pmatrix}\begin{pmatrix} \frac{\sqrt{3}}{2} \\ \frac{1}{2} \end{pmatrix} = \begin{pmatrix} -1 & 0 \\ 0 & -1 \end{pmatrix}\begin{pmatrix} \frac{\sqrt{3}}{2} \\ \frac{1}{2} \end{pmatrix} = \begin{pmatrix} -\frac{\sqrt{3}}{2} \\ -\frac{1}{2} \end{pmatrix}$$

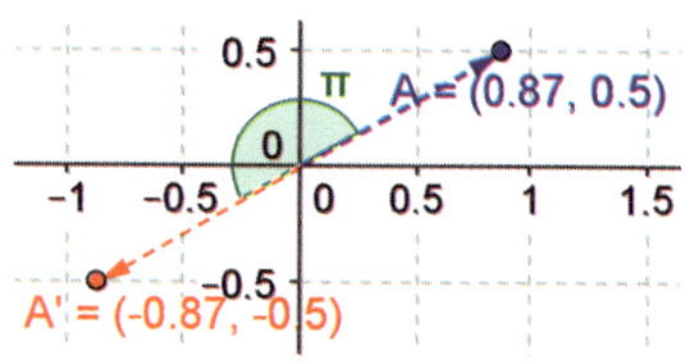

h. $\begin{pmatrix} 1 \\ 0 \end{pmatrix}$, $-\frac{\pi}{6}$ radians

$$\begin{pmatrix} \cos\left(-\frac{\pi}{6}\right) & -\sin\left(-\frac{\pi}{6}\right) \\ \sin\left(-\frac{\pi}{6}\right) & \cos\left(-\frac{\pi}{6}\right) \end{pmatrix}\begin{pmatrix} 1 \\ 0 \end{pmatrix} = \begin{pmatrix} \frac{\sqrt{3}}{2} & \frac{1}{2} \\ -\frac{1}{2} & \frac{\sqrt{3}}{2} \end{pmatrix}\begin{pmatrix} 1 \\ 0 \end{pmatrix} = \begin{pmatrix} \frac{\sqrt{3}}{2} \\ -\frac{1}{2} \end{pmatrix}$$

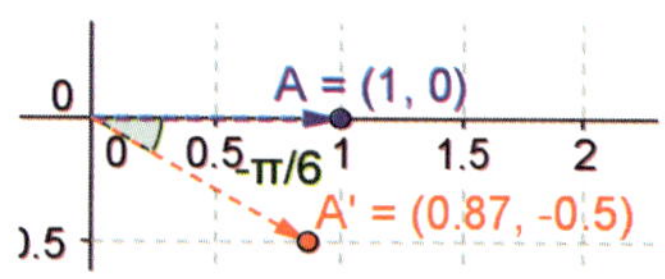

5. **For the transformation shown below, find the matrix that will transform point A to A', and verify your answer.**

$A = (2, 1)$, $OA = \sqrt{(2)^2 + (1)^2} = \sqrt{5}$

$A' = (-2, 4)$, $OA = \sqrt{(-2)^2 + (4)^2} = \sqrt{20} = 2\sqrt{5}$; therefore, it has a dilation of 2.

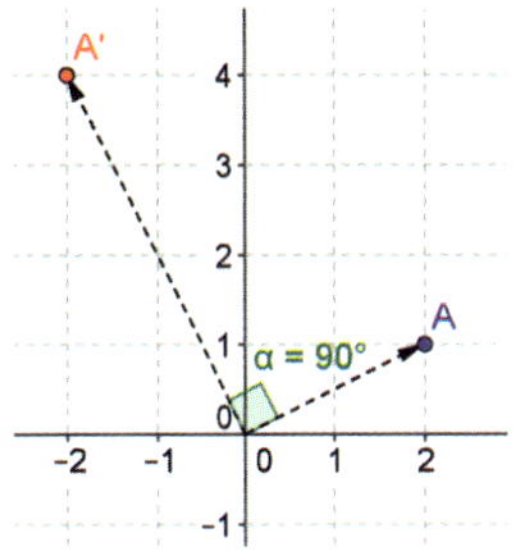

$$2\begin{pmatrix} \cos\left(\frac{\pi}{2}\right) & -\sin\left(\frac{\pi}{2}\right) \\ \sin\left(\frac{\pi}{2}\right) & \cos\left(\frac{\pi}{2}\right) \end{pmatrix} = 2\begin{pmatrix} 0 & -1 \\ 1 & 0 \end{pmatrix} = \begin{pmatrix} 0 & -2 \\ 2 & 0 \end{pmatrix}$$

$$\begin{pmatrix} 0 & -2 \\ 2 & 0 \end{pmatrix}\begin{pmatrix} 2 \\ 1 \end{pmatrix} = \begin{pmatrix} -2 \\ 4 \end{pmatrix}$$

6. **In this lesson, we learned $\begin{bmatrix} 0 & 1 \\ 1 & 0 \end{bmatrix}$ will produce a reflection about the line $y = x$. What matrix will produce a reflection about the line $y = -x$? Verify your answers by testing the given point $\begin{pmatrix} 3 \\ 1 \end{pmatrix}$ and graphing them on the coordinate plane.**

$$\begin{pmatrix} 0 & -1 \\ -1 & 0 \end{pmatrix}$$

$$\begin{pmatrix} 0 & -1 \\ -1 & 0 \end{pmatrix}\begin{pmatrix} 3 \\ 1 \end{pmatrix} = \begin{pmatrix} -1 \\ -3 \end{pmatrix}$$

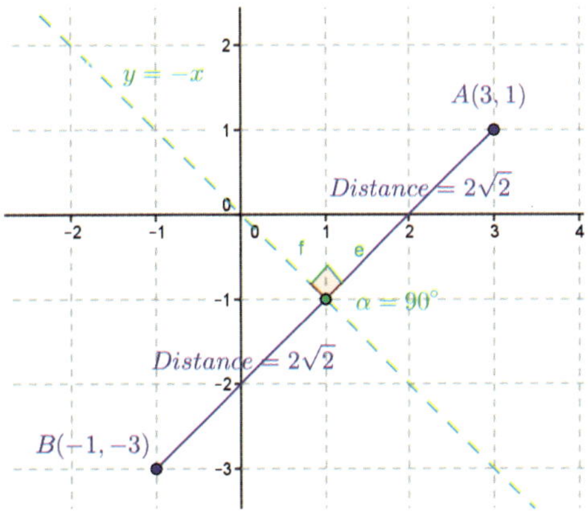

7. **Describe the transformation and the translations in the diagram below. Write the matrices that will perform the tasks. What is the area that these transformations and translations have enclosed?**

First, A is rotated a π radian counterclockwise to A'; next, it is translated to A''; finally, it is translated back to A.

From A to A': $\begin{pmatrix} \cos(\pi) & -\sin(\pi) \\ \sin(\pi) & \cos(\pi) \end{pmatrix} = \begin{pmatrix} -1 & 0 \\ 0 & -1 \end{pmatrix}$

From A' to A'': $\begin{pmatrix} x_1 - x_1 \\ y_1 - y_2 \end{pmatrix}$

From A'' to A: $\begin{pmatrix} x_2 + x \\ y_2 - y_2 \end{pmatrix}$

Area: Semicircle + triangle, which is $\dfrac{\pi|x - x_1|^2}{2} + \dfrac{|x - x_1| \cdot |y_2|}{2}$

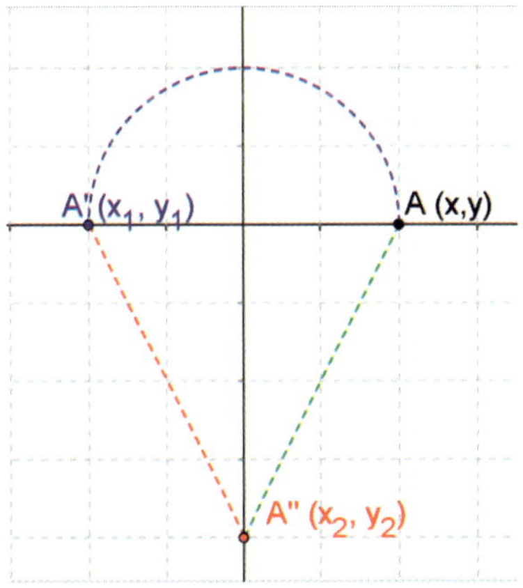

8. Given the kite figure $ABCD$ below, answer the following questions.

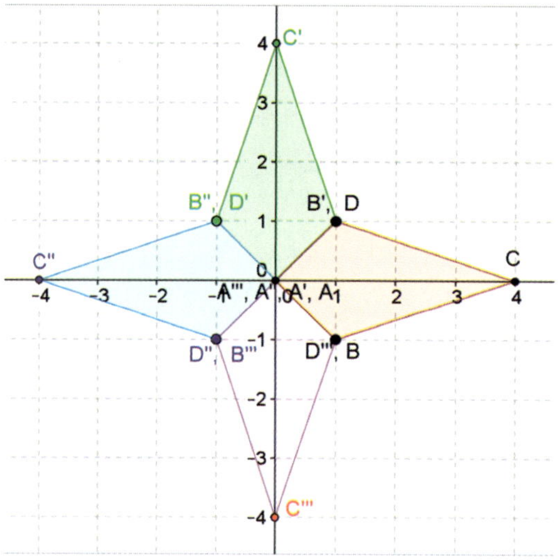

a. Explain how you would create the star figure above using only rotations.

By rotating the kite figure $ABCD$ $\frac{\pi}{2}$ radians counterclockwise or clockwise four times

b. Explain how to create the star figure above using reflections and rotation.

Answers will vary. One explanation is to rotate the kite figure $ABCD$ $\frac{\pi}{2}$ radians counterclockwise to get to $A'B'C'D'$ and then reflect both figures about the line $y = -x$, which is $\begin{pmatrix} 0 & -1 \\ -1 & 0 \end{pmatrix}$.

c. Explain how to create the star figure above using only reflections. Explain your answer.

First, reflect the kite figure $ABCD$ about the line $y = x$, which is $\begin{pmatrix} 0 & 1 \\ 1 & 0 \end{pmatrix}$; then, reflect $A'B'C'D'$ about the line $y = -x$, which is $\begin{pmatrix} 0 & -1 \\ -1 & 0 \end{pmatrix}$; next, reflect $A''B''C''D''$ about the line $y = x$, which is $\begin{pmatrix} 0 & 1 \\ 1 & 0 \end{pmatrix}$; finally, reflect $A'''B'''C'''D'''$ about the line $y = -x$, which is $\begin{pmatrix} 0 & -1 \\ -1 & 0 \end{pmatrix}$.

9. Given the rectangle $ABCD$ below, answer the following questions.

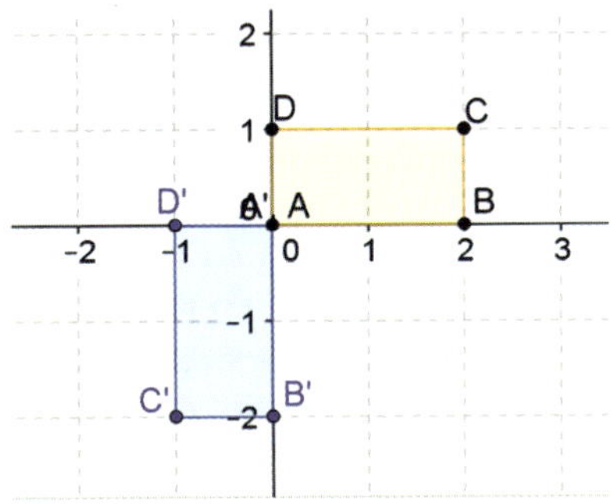

a. Can you transform the rectangle $A'B'C'D'$ above using only rotations? Explain your answer.

No, because no matter how it is rotated, the vertices of the rectangle will not stay the same with respect to each other.

b. Describe a way to create the rectangle $A'B'C'D'$.

Answers will vary. One way is to reflect the rectangle $ABCD$ about the y-axis first, which is $\begin{pmatrix} -1 & 0 \\ 0 & 1 \end{pmatrix}$, and then rotate $\frac{\pi}{2}$ radians counterclockwise, which is $\begin{pmatrix} 0 & -1 \\ 1 & 0 \end{pmatrix}$.

c. Can you make the rectangle $A'B'C'D'$ above using only reflections? Explain your answer.

Yes, reflect the rectangle $ABCD$ about the line $y = -x$, which is $\begin{pmatrix} 0 & -1 \\ -1 & 0 \end{pmatrix}$.

Lesson 25: Matrix Multiplication and Addition

Student Outcomes

- Students work with 2×2 matrices as transformations of the plane.
- Students combine matrices using matrix multiplication and addition.
- Students understand the role of the zero matrix in matrix addition.

Lesson Notes

In Lesson 24, students continued to explore matrices and their connection to transformations. In this lesson, students work with the zero matrix and discover that it is the additive identity matrix with a role similar to 0 in the real number system. Students focus on the result of performing one transformation followed by another and discover:

If L is given by $\begin{pmatrix} a & c \\ b & d \end{pmatrix}$ and M is given by $\begin{pmatrix} p & r \\ q & s \end{pmatrix}$, then $ML\begin{pmatrix} x \\ y \end{pmatrix}$ is the same as applying the matrix $\begin{pmatrix} pa+rb & pc+rd \\ qa+sb & qc+sd \end{pmatrix}$ to $\begin{pmatrix} x \\ y \end{pmatrix}$.

This motivates the definition of matrix multiplication. **N-VM.C.8** is introduced in this lesson but treated more fully in Module 2.

Classwork

Opening Exercise (8 minutes)

Allow students time to complete the Opening Exercise independently. Encourage students to think/write independently, chat with a partner, and then share as a class.

Opening Exercise

Consider the point $\begin{pmatrix} 4 \\ 1 \end{pmatrix}$ that undergoes a series of two transformations: a dilation of scale factor 4 followed by a reflection about the horizontal axis.

a. **What matrix produces the dilation of scale factor 4? What is the coordinate of the point after the dilation?**

The dilation matrix is $\begin{pmatrix} 4 & 0 \\ 0 & 4 \end{pmatrix}$.

$\begin{pmatrix} 4 & 0 \\ 0 & 4 \end{pmatrix}\begin{pmatrix} 4 \\ 1 \end{pmatrix} = \begin{pmatrix} 16 \\ 4 \end{pmatrix}$

The coordinate is now $\begin{pmatrix} 16 \\ 4 \end{pmatrix}$.

b. **What matrix produces the reflection about the horizontal axis? What is the coordinate of the point after the reflection?**

The reflection matrix is $\begin{pmatrix} 1 & 0 \\ 0 & -1 \end{pmatrix}$.

$\begin{pmatrix} 1 & 0 \\ 0 & -1 \end{pmatrix}\begin{pmatrix} 16 \\ 4 \end{pmatrix} = \begin{pmatrix} 16 \\ -4 \end{pmatrix}$

The coordinate is now $\begin{pmatrix} 16 \\ -4 \end{pmatrix}$.

MP.3

c. **Could we have produced both the dilation and the reflection using a single matrix? If so, what matrix would both dilate by a scale factor of 4 and produce a reflection about the horizontal axis? Show that the matrix you came up with combines these two matrices.**

Yes, by using the matrix $\begin{pmatrix} 4 & 0 \\ 0 & -4 \end{pmatrix}$

$\begin{pmatrix} 4 & 0 \\ 0 & -4 \end{pmatrix}\begin{pmatrix} 4 \\ 1 \end{pmatrix} = \begin{pmatrix} 16 \\ -4 \end{pmatrix}$

The dilation matrix was $\begin{pmatrix} 4 & 0 \\ 0 & 4 \end{pmatrix}$. *The rotation matrix was* $\begin{pmatrix} 1 & 0 \\ 0 & -1 \end{pmatrix}$. *The product of these matrices gives the matrix that produces a dilation and then a rotation.* $\begin{pmatrix} 4 & 0 \\ 0 & 4 \end{pmatrix}\begin{pmatrix} 1 & 0 \\ 0 & -1 \end{pmatrix} = \begin{pmatrix} 4 & 0 \\ 0 & -4 \end{pmatrix}$

- How did you come up with the dilation matrix?
  - *I know that a dilation matrix is in the form* $\begin{pmatrix} k & 0 \\ 0 & k \end{pmatrix}$ *where* k *the scale factor is.*
- How did you come up with the reflection matrix?
  - *I know that the matrix* $\begin{pmatrix} 1 & 0 \\ 0 & -1 \end{pmatrix}$ *reflects coordinates about the horizontal axis.*
- How did you come up with a matrix that was both a dilation and a reflection?
  - *I knew that I wanted to multiply both the* x *and* y *by a factor of* 4 *and that I also wanted to multiply the* y *by* -1*; from that, I combined the two matrices to get* $\begin{pmatrix} 4 & 0 \\ 0 & -4 \end{pmatrix}$.
- In what sense did we combine the two matrices?
  - *We multiplied* 4×-1.

MP.7

- We know that transformations are produced through matrix multiplication. What if we have more than one transformation? Could we multiply the two transformation matrices together first instead of completing the transformations in two separate steps?

Write this on the board:

$$\begin{pmatrix} 4 & 0 \\ 0 & 4 \end{pmatrix}\begin{pmatrix} 1 & 0 \\ 0 & -1 \end{pmatrix} =$$

- This should be equivalent to applying the dilation and then the reflection. So, what should the product equal?
  - $\begin{pmatrix} 4 & 0 \\ 0 & -4 \end{pmatrix}$

MP.7

- Ask students to think about how we multiply a 2×2 matrix and a 2×1 matrix. Based on the fact that this product should be $\begin{pmatrix} 4 & 0 \\ 0 & -4 \end{pmatrix}$ and what you know about multiplying a matrix by a vector, develop an explanation for how to multiply these two matrices together.
  - $\begin{pmatrix} 4 & 0 \\ 0 & 4 \end{pmatrix}\begin{pmatrix} 1 & 0 \\ 0 & -1 \end{pmatrix} = \begin{pmatrix} 4 \times 1 + 0 \times 0 & 4 \times 0 + 0 \times -1 \\ 0 \times 1 + 0 \times 0 & 4 \times 0 + 4 \times -1 \end{pmatrix} = \begin{pmatrix} 4 & 0 \\ 0 & -4 \end{pmatrix}$
- Does this technique align with our earlier definition of multiplying a 2×2 times a 2×1?
  - *Yes, we follow the same process to get the numbers in column 2 that we did to get the numbers in column 1—multiplying each row by the numbers in the column and then adding.*
- Can matrices of any size be multiplied together? For example, can you multiply $\begin{pmatrix} 3 \\ 2 \end{pmatrix}\begin{pmatrix} 1 & -1 \\ 4 & 5 \end{pmatrix}$? Why or why not?
  - *No, the number of rows and columns do not match up.*
- What must be true about the dimensions of matrices in order for them to be able to be multiplied?
  - *The number of columns of the first matrix must equal the number of rows of the second matrix.*

Example 1 (7 minutes): Is Matrix Multiplication Commutative?

Conduct each part of the example as a think-pair-share. Allow students time to think of the answer independently, pair with a partner to discuss, and then share as a class.

Example 1: Is Matrix Multiplication Commutative?

a. **Take the point $\begin{pmatrix} 2 \\ 1 \end{pmatrix}$ through the following transformations: a rotation of $\frac{\pi}{2}$ and a reflection across the y-axis.**

$\begin{pmatrix} 1 \\ 2 \end{pmatrix}$

b. **Will the resulting point be the same if the order of the transformations is reversed?**

No, if the reflection is applied first, followed by the rotation, the resulting point is $\begin{pmatrix} -1 \\ -2 \end{pmatrix}$.

c. **Are transformations commutative?**

Not necessarily; the order in which the transformations are applied can affect the results in some cases.

d. **Let $A = \begin{pmatrix} 0 & -1 \\ 1 & 0 \end{pmatrix}$ and $B = \begin{pmatrix} -1 & 0 \\ 0 & 1 \end{pmatrix}$. Find AB and then BA.**

$AB = \begin{pmatrix} 0 & -1 \\ 1 & 0 \end{pmatrix}\begin{pmatrix} -1 & 0 \\ 0 & 1 \end{pmatrix} = \begin{pmatrix} 0 & -1 \\ -1 & 0 \end{pmatrix}$

$BA = \begin{pmatrix} -1 & 0 \\ 0 & 1 \end{pmatrix}\begin{pmatrix} 0 & -1 \\ 1 & 0 \end{pmatrix} = \begin{pmatrix} 0 & 1 \\ 1 & 0 \end{pmatrix}$

Scaffolding:

- For students who are struggling, start with a review of the commutative property of multiplication.

  If a and b are real numbers, then $a \times b = b \times a$.
- Have them highlight or circle each row and column as they multiply.
- What does it mean for multiplication of real numbers to be commutative? Explain with an example.
- Can you think of an operation that is not commutative? Explain with an example.

e. **Is matrix multiplication commutative?**

No, $AB \neq BA$

f. **If we apply matrix AB to the point $\begin{pmatrix} 2 \\ 1 \end{pmatrix}$, in what order are the transformations applied?**

$\begin{pmatrix} 0 & -1 \\ -1 & 0 \end{pmatrix}\begin{pmatrix} 2 \\ 1 \end{pmatrix} = \begin{pmatrix} -1 \\ -2 \end{pmatrix}$

The reflection is applied first, followed by the dilation.

g. **If we apply matrix BA to the point $\begin{pmatrix} 2 \\ 1 \end{pmatrix}$, in what order are the transformations applied?**

$\begin{pmatrix} 0 & 1 \\ 1 & 0 \end{pmatrix}\begin{pmatrix} 2 \\ 1 \end{pmatrix} = \begin{pmatrix} 1 \\ 2 \end{pmatrix}$

The dilation is applied first, followed by the reflection.

h. **Can we apply $\begin{pmatrix} 2 \\ 1 \end{pmatrix}$ to matrix BA?**

$\begin{pmatrix} 2 \\ 1 \end{pmatrix}\begin{pmatrix} 0 & 1 \\ 1 & 0 \end{pmatrix}$

No, $\begin{pmatrix} 2 \\ 1 \end{pmatrix}$ has two rows and one column. It would have to have two columns to be able to multiply by the 2×2 matrix BA.

Exercises 1–3 (10 minutes)

Allow students time to work on the exercises either independently or in groups. Circulate around the room providing assistance as needed, particularly watching for students who are struggling with matrix multiplication.

Exercises 1–3

1. **Let $I = \begin{pmatrix} 1 & 0 \\ 0 & 1 \end{pmatrix}$ and $M = \begin{pmatrix} 4 & -6 \\ 3 & -2 \end{pmatrix}$.**

   a. **Find IM.**

   $\begin{pmatrix} 4 & -6 \\ 3 & -2 \end{pmatrix}$

   b. **Find MI.**

   $\begin{pmatrix} 4 & -6 \\ 3 & -2 \end{pmatrix}$

   c. **Do these results make sense based on what you know about the matrix $\begin{pmatrix} 1 & 0 \\ 0 & 1 \end{pmatrix}$?**

   Yes, the matrix $\begin{pmatrix} 1 & 0 \\ 0 & 1 \end{pmatrix}$ is the identity matrix. It corresponds to a transformation that leaves points alone. Therefore, geometrically we must have $IM = MI = M$.

2. Calculate AB and then BA. Is matrix multiplication commutative?

a. $A = \begin{pmatrix} 2 & 0 \\ -2 & 3 \end{pmatrix}$, $B = \begin{pmatrix} 1 & 5 \\ 0 & 1 \end{pmatrix}$

$AB = \begin{pmatrix} 2\times 1 + 0\times 0 & 2\times 5 + 0\times 1 \\ -2\times 1 + 3\times 0 & -2\times 5 + 3\times 1 \end{pmatrix} = \begin{pmatrix} 2 & 10 \\ -2 & -7 \end{pmatrix}$

$BA = \begin{pmatrix} 1\times 2 + 5\times -2 & 1\times 0 + 5\times 3 \\ 0\times 2 + 1\times -2 & 0\times 0 + 1\times 3 \end{pmatrix} = \begin{pmatrix} -8 & 15 \\ -2 & 3 \end{pmatrix}$

$AB \neq BA$; matrix multiplication is not commutative.

b. $A = \begin{pmatrix} -10 & 1 \\ 3 & 7 \end{pmatrix}$, $B = \begin{pmatrix} -3 & 2 \\ 4 & -1 \end{pmatrix}$

$AB = \begin{pmatrix} -10\times -3 + 1\times 4 & -10\times 2 + 1\times -1 \\ 3\times -3 + 7\times 4 & 3\times 2 + 7\times -1 \end{pmatrix} = \begin{pmatrix} 34 & -21 \\ 19 & -1 \end{pmatrix}$

$BA = \begin{pmatrix} -3\times -10 + 2\times 3 & -3\times 1 + 2\times 7 \\ 4\times -10 + -1\times 3 & 4\times 1 + -1\times 7 \end{pmatrix} = \begin{pmatrix} 36 & 11 \\ -43 & -3 \end{pmatrix}$

$AB \neq BA$; matrix multiplication is not commutative.

3. Write a matrix that would perform the following transformations in this order: a rotation of $180°$, a dilation by a scale factor of 4, and a reflection across the horizontal axis. Use the point $\begin{pmatrix} 2 \\ 1 \end{pmatrix}$ to illustrate that your matrix is correct.

$\begin{pmatrix} 1 & 0 \\ 0 & -1 \end{pmatrix}\begin{pmatrix} -4 & 0 \\ 0 & -4 \end{pmatrix} = \begin{pmatrix} -4 & 0 \\ 0 & 4 \end{pmatrix}$

$\begin{pmatrix} -4 & 0 \\ 0 & 4 \end{pmatrix}\begin{pmatrix} 2 \\ 1 \end{pmatrix} = \begin{pmatrix} -8 \\ 4 \end{pmatrix}$

Scaffolding:

For students who like a challenge, pose one of the following:

- Multiply two 3×3 matrices.
- Could we use our definition for multiplying matrices to multiply a 2×1 times a 2×2?

Example 2 (5 minutes): More Operations on Matrices

Discuss this question as a class before completing the example.

- We know that there is matrix multiplication. Does it seem logical that there would be matrix addition?
- If $\begin{pmatrix} 2 & 0 \\ -2 & 3 \end{pmatrix} + \begin{pmatrix} 1 & 5 \\ 0 & 1 \end{pmatrix} = \begin{pmatrix} 3 & 5 \\ -2 & 4 \end{pmatrix}$, explain how to add matrices.
  - *Add the numbers that are in the same position in the corresponding matrices.*
- Each number within a matrix is called an element. In order to add matrices, the elements in the same row and same column are added. We refer to elements in the same row and same column as *corresponding elements*. So, to recap, in order to add matrices, we add corresponding elements.
- How would you subtract matrices?
  - *Subtract corresponding elements.*
- Find $\begin{pmatrix} 2 & 0 \\ -2 & 3 \end{pmatrix} - \begin{pmatrix} 1 & 5 \\ 0 & 1 \end{pmatrix}$.
  - $\begin{pmatrix} 1 & -5 \\ -2 & 2 \end{pmatrix}$
- Can you add the following matrices: $\begin{pmatrix} 3 & 5 \\ -2 & 4 \end{pmatrix} + \begin{pmatrix} 1 \\ -3 \end{pmatrix}$? Explain your answer.
  - *No, the matrix is not the same size, so there are not corresponding elements.*

- When can matrices be added and subtracted?
  - *When they are the same size*
- In Lesson 24, we studied the multiplicative identity matrix. In what ways is the identity matrix similar to the number 1 within the set of real numbers? Why?
  - *The identity matrix is similar to the number 1 in the real number system because any number times 1 is itself, and any matrix times the identity matrix is itself.*
- What is the additive identity in the real number system? Why?
  - 0*, because any number plus 0 is itself.*
- What matrix would you hypothesize has the same effect in matrix addition? Write the matrix.
  - $\begin{pmatrix} 0 & 0 \\ 0 & 0 \end{pmatrix}$ *is the additive identity matrix because any matrix plus this matrix is itself.*
- We call this the *zero matrix*. What would be the impact of multiplying a matrix by the zero matrix? How does the impact of multiplying a matrix by the zero matrix compare to multiplying by zero in the real number system?
  - *Any number multiplied by 0 is 0, and any matrix multiplied by the zero matrix has all terms of 0.*
- Note that it is difficult to know what matrix addition means geometrically in terms of transformations, but we will see a natural interpretation of matrix addition in Module 3.

Example 2: More Operations on Matrices

Find the sum. $\begin{pmatrix} 2 & 0 \\ -2 & 3 \end{pmatrix} + \begin{pmatrix} 1 & 5 \\ 0 & 1 \end{pmatrix}$

$$\begin{pmatrix} 2 & 0 \\ -2 & 3 \end{pmatrix} + \begin{pmatrix} 1 & 5 \\ 0 & 1 \end{pmatrix} = \begin{pmatrix} 2+1 & 0+5 \\ -2+0 & 3+1 \end{pmatrix} = \begin{pmatrix} 3 & 5 \\ -2 & 4 \end{pmatrix}$$

Find the difference. $\begin{pmatrix} 2 & 0 \\ -2 & 3 \end{pmatrix} - \begin{pmatrix} 1 & 5 \\ 0 & 1 \end{pmatrix}$

$$\begin{pmatrix} 2 & 0 \\ -2 & 3 \end{pmatrix} - \begin{pmatrix} 1 & 5 \\ 0 & 1 \end{pmatrix} = \begin{pmatrix} 2-1 & 0-5 \\ -2-0 & 3-1 \end{pmatrix} = \begin{pmatrix} 1 & -5 \\ -2 & 2 \end{pmatrix}$$

Find the sum. $\begin{pmatrix} 2 & 0 \\ -2 & 3 \end{pmatrix} + \begin{pmatrix} 0 & 0 \\ 0 & 0 \end{pmatrix}$

$$\begin{pmatrix} 2 & 0 \\ -2 & 3 \end{pmatrix} + \begin{pmatrix} 0 & 0 \\ 0 & 0 \end{pmatrix} = \begin{pmatrix} 2+0 & 0+0 \\ -2+0 & 3+0 \end{pmatrix} = \begin{pmatrix} 2 & 0 \\ -2 & 3 \end{pmatrix}$$

Exercises 4–5 (5 minutes)

Allow students time to work on the exercises either independently or in groups. Circulate around the room providing assistance as needed.

Exercises 4–5

4. **Express each of the following as a single matrix.**

   a. $\begin{pmatrix} 6 & -3 \\ 10 & -1 \end{pmatrix} + \begin{pmatrix} -2 & 8 \\ 3 & -12 \end{pmatrix}$

   $\begin{pmatrix} 4 & 5 \\ 13 & -13 \end{pmatrix}$

b. $\begin{pmatrix} -2 & 7 \\ -3 & 1 \end{pmatrix}\begin{pmatrix} 3 \\ 4 \end{pmatrix} + \begin{pmatrix} 1 \\ 5 \end{pmatrix}$

$\begin{pmatrix} 23 \\ 0 \end{pmatrix}$

c. $\begin{pmatrix} 8 & 5 \\ 0 & 15 \end{pmatrix} - \begin{pmatrix} 4 & -6 \\ -3 & 18 \end{pmatrix}$

$\begin{pmatrix} 4 & 11 \\ 3 & -3 \end{pmatrix}$

5. **In arithmetic, the additive identity says that for some number a, $a + 0 = 0 + a = 0$. What would be an additive identity in matrix arithmetic?**

We would use the zero matrix. $\begin{pmatrix} a & c \\ b & d \end{pmatrix} + \begin{pmatrix} 0 & 0 \\ 0 & 0 \end{pmatrix} = \begin{pmatrix} a & c \\ b & d \end{pmatrix}$

Closing (5 minutes)

Ask students to summarize what they have learned about matrix multiplication and addition either in writing or orally.

- When we multiply two matrices, what is the geometric interpretation?
  - *It is a series of transformations.*
- Can all matrices be multiplied? Why or why not?
  - *Matrices can be multiplied if the number of columns of the first matrix is equal to the number of rows of the second matrix.*

MP.6

- Can two matrices be combined through addition? If so, explain how.
  - *Yes, if the matrices are the same size, they can be added by adding corresponding elements.*

Lesson Summary

- **If L is given by $\begin{pmatrix} a & c \\ b & d \end{pmatrix}$ and M is given by $\begin{pmatrix} p & r \\ q & s \end{pmatrix}$, then $ML\begin{pmatrix} x \\ y \end{pmatrix}$ is the same as applying the matrix $\begin{pmatrix} pa + rb & pc + rd \\ qa + sb & qc + sd \end{pmatrix}$ to $\begin{pmatrix} x \\ y \end{pmatrix}$.**
- **If L is given by $\begin{pmatrix} a & c \\ b & d \end{pmatrix}$ and I is given by $\begin{pmatrix} 1 & 0 \\ 0 & 1 \end{pmatrix}$, then I acts as a multiplicative identity, and $IL = LI = L$.**
- **If L is given by $\begin{pmatrix} a & c \\ b & d \end{pmatrix}$ and O is given by $\begin{pmatrix} 0 & 0 \\ 0 & 0 \end{pmatrix}$, then O acts as an additive identity, and $O + L = L + O = L$.**

Exit Ticket (5 minutes)

EUREKA MATH

Name ______________________________ Date ______________

Lesson 25: Matrix Multiplication and Addition

Exit Ticket

1. Carmine has never seen matrices before but must quickly understand how to add, subtract, and multiply matrices. Explain the following problems to Carmine.

   a. $\begin{pmatrix} 2 & 3 \\ -1 & 5 \end{pmatrix} + \begin{pmatrix} 0 & 1 \\ 4 & -3 \end{pmatrix} = \begin{pmatrix} 2 & 4 \\ 3 & 2 \end{pmatrix}$

   b. $\begin{pmatrix} 2 & 3 \\ -1 & 5 \end{pmatrix} - \begin{pmatrix} 0 & 1 \\ 4 & -3 \end{pmatrix} = \begin{pmatrix} 2 & 2 \\ -5 & 8 \end{pmatrix}$

   c. $\begin{pmatrix} 2 & 3 \\ -1 & 5 \end{pmatrix} \begin{pmatrix} 0 & 1 \\ 4 & -3 \end{pmatrix} = \begin{pmatrix} 12 & -7 \\ 20 & -16 \end{pmatrix}$

2. Explain to Carmine the significance of the zero matrix and the multiplicative identity matrix.

Exit Ticket Sample Solutions

1. **Carmine has never seen matrices before but must quickly understand how to add, subtract, and multiply matrices. Explain the following problems to Carmine.**

   a. $\begin{pmatrix} 2 & 3 \\ -1 & 5 \end{pmatrix} + \begin{pmatrix} 0 & 1 \\ 4 & -3 \end{pmatrix} = \begin{pmatrix} 2 & 4 \\ 3 & 2 \end{pmatrix}$

   To add matrices, add the corresponding elements. So, add the 2 and the 0 because they are both in the first row, first column.

   b. $\begin{pmatrix} 2 & 3 \\ -1 & 5 \end{pmatrix} - \begin{pmatrix} 0 & 1 \\ 4 & -3 \end{pmatrix} = \begin{pmatrix} 2 & 2 \\ -5 & 8 \end{pmatrix}$

   To subtract matrices, subtract the corresponding elements. So, subtract the 0 from the 2 because they are both in the first row, first column.

   c. $\begin{pmatrix} 2 & 3 \\ -1 & 5 \end{pmatrix}\begin{pmatrix} 0 & 1 \\ 4 & -3 \end{pmatrix} = \begin{pmatrix} 12 & -7 \\ 20 & -16 \end{pmatrix}$

   To multiply matrices, multiply the elements in the first row by the elements in the first column, and then add the products together.

   $$\begin{pmatrix} 2 \times 0 + 3 \times 4 & 2 \times 1 + 3 \times -3 \\ -1 \times 0 + 5 \times 4 & -1 \times 1 + 5 \times -3 \end{pmatrix} = \begin{pmatrix} 12 & -7 \\ 20 & -16 \end{pmatrix}$$

2. **Explain to Carmine the significance of the zero matrix and the multiplicative identity matrix.**

   The zero matrix is $\begin{pmatrix} 0 & 0 \\ 0 & 0 \end{pmatrix}$ and is similar to 0 in the real number system. Any matrix added to the zero matrix is itself, and any matrix multiplied by the zero matrix has all terms of 0.

   The multiplicative identity matrix is $\begin{pmatrix} 1 & 0 \\ 0 & 1 \end{pmatrix}$ and is similar to 1 in the real number system. Any matrix times the multiplicative identity matrix has a product of itself.

Problem Set Sample Solutions

1. **What type of transformation is shown in the following examples? What is the resulting matrix?**

   a. $\begin{pmatrix} \cos(\pi) & -\sin(\pi) \\ \sin(\pi) & \cos(\pi) \end{pmatrix}\begin{pmatrix} 3 \\ 2 \end{pmatrix}$

   It is a pure rotation. The point $\begin{pmatrix} 3 \\ 2 \end{pmatrix}$ is rotated π radians, and the image is $\begin{pmatrix} -3 \\ -2 \end{pmatrix}$.

   b. $\begin{pmatrix} 0 & -1 \\ 1 & 0 \end{pmatrix}\begin{pmatrix} 3 \\ 2 \end{pmatrix}$

   It is a pure rotation. The point $\begin{pmatrix} 3 \\ 2 \end{pmatrix}$ is rotated $\frac{\pi}{2}$ radians, and the image is $\begin{pmatrix} -2 \\ 3 \end{pmatrix}$.

   c. $\begin{pmatrix} 3 & 0 \\ 0 & 3 \end{pmatrix}\begin{pmatrix} 3 \\ 2 \end{pmatrix}$

   It is dilation with a factor of 3. The point $\begin{pmatrix} 3 \\ 2 \end{pmatrix}$ is dilated by a factor of 3, and the image is $\begin{pmatrix} 9 \\ 6 \end{pmatrix}$.

d. $\begin{pmatrix} -1 & 0 \\ 0 & 1 \end{pmatrix}\begin{pmatrix} 3 \\ 2 \end{pmatrix}$

It is a reflection about the y-axis. The point $\begin{pmatrix} 3 \\ 2 \end{pmatrix}$ is reflected about the y-axis, and the image is $\begin{pmatrix} -3 \\ 2 \end{pmatrix}$.

e. $\begin{pmatrix} 1 & 0 \\ 0 & -1 \end{pmatrix}\begin{pmatrix} 3 \\ 2 \end{pmatrix}$

It is a reflection about the x-axis. The point $\begin{pmatrix} 3 \\ 2 \end{pmatrix}$ is reflected about the x-axis, and the image is $\begin{pmatrix} 3 \\ -2 \end{pmatrix}$.

f. $\begin{pmatrix} \cos(2\pi) & -\sin(2\pi) \\ \sin(2\pi) & \cos(2\pi) \end{pmatrix}\begin{pmatrix} 3 \\ 2 \end{pmatrix}$

It is a pure rotation. The point $\begin{pmatrix} 3 \\ 2 \end{pmatrix}$ is rotated 2π radians, and the image is $\begin{pmatrix} 3 \\ 2 \end{pmatrix}$.

g. $\begin{pmatrix} \frac{\sqrt{2}}{2} & -\frac{\sqrt{2}}{2} \\ \frac{\sqrt{2}}{2} & \frac{\sqrt{2}}{2} \end{pmatrix}\begin{pmatrix} 3 \\ 2 \end{pmatrix}$

It is a pure rotation. The point $\begin{pmatrix} 3 \\ 2 \end{pmatrix}$ is rotated $\frac{\pi}{4}$ radians, and the image is $\begin{pmatrix} \frac{\sqrt{2}}{2} \\ \frac{5\sqrt{2}}{2} \end{pmatrix}$.

h. $\begin{pmatrix} \frac{\sqrt{3}}{2} & -\frac{1}{2} \\ \frac{1}{2} & \frac{\sqrt{3}}{2} \end{pmatrix}\begin{pmatrix} 3 \\ 2 \end{pmatrix}$

It is a pure rotation. The point $\begin{pmatrix} 3 \\ 2 \end{pmatrix}$ is rotated $\frac{\pi}{6}$ radians, and the image is $\begin{pmatrix} \frac{3\sqrt{2}}{2} - 1 \\ \frac{3}{2} - \sqrt{3} \end{pmatrix}$.

2. Calculate each of the following products.

a. $\begin{pmatrix} 2 & 3 \\ 5 & 4 \end{pmatrix}\begin{pmatrix} 2 \\ 3 \end{pmatrix}$

$\begin{pmatrix} 13 \\ 22 \end{pmatrix}$

b. $\begin{pmatrix} 3 & 2 \\ 4 & 5 \end{pmatrix}\begin{pmatrix} 3 & 2 \\ 1 & 0 \end{pmatrix}$

$\begin{pmatrix} 11 & 6 \\ 17 & 8 \end{pmatrix}$

c. $\begin{pmatrix} -1 & -3 \\ -2 & -4 \end{pmatrix}\begin{pmatrix} -3 \\ -1 \end{pmatrix}$

$\begin{pmatrix} 6 \\ 10 \end{pmatrix}$

d. $\begin{pmatrix} -3 & -1 \\ -4 & -2 \end{pmatrix}\begin{pmatrix} -2 & -1 \\ 0 & -3 \end{pmatrix}$

$\begin{pmatrix} 6 & 6 \\ 8 & 10 \end{pmatrix}$

e. $\begin{pmatrix} 5 & 0 \\ -1 & 2 \end{pmatrix}\begin{pmatrix} 0 \\ 3 \end{pmatrix}$

$\begin{pmatrix} 0 \\ 6 \end{pmatrix}$

f. $\begin{pmatrix} 3 & -1 \\ 4 & -2 \end{pmatrix}\begin{pmatrix} 1 & 0 \\ -2 & -3 \end{pmatrix}$

$\begin{pmatrix} 5 & 3 \\ 8 & 6 \end{pmatrix}$

3. Calculate each sum or difference.

a. $\begin{pmatrix} 1 & 3 \\ 2 & 4 \end{pmatrix} + \begin{pmatrix} 4 & 2 \\ 3 & 1 \end{pmatrix}$

$\begin{pmatrix} 5 & 5 \\ 5 & 5 \end{pmatrix}$

b. $\begin{pmatrix} -4 & -5 \\ -6 & -7 \end{pmatrix} + \begin{pmatrix} -2 & 3 \\ -1 & 4 \end{pmatrix}$

$\begin{pmatrix} -6 & -2 \\ -7 & -3 \end{pmatrix}$

c. $\begin{pmatrix} -5 \\ 4 \end{pmatrix} + \begin{pmatrix} 3 \\ 2 \end{pmatrix}$

$\begin{pmatrix} -2 \\ 6 \end{pmatrix}$

d. $\begin{pmatrix} 3 \\ -5 \end{pmatrix} - \begin{pmatrix} 7 \\ 9 \end{pmatrix}$

$\begin{pmatrix} -4 \\ -14 \end{pmatrix}$

e. $\begin{pmatrix} -4 & -5 \\ -6 & -7 \end{pmatrix} - \begin{pmatrix} -2 & 3 \\ -1 & 4 \end{pmatrix}$

$\begin{pmatrix} -2 & -8 \\ -5 & -11 \end{pmatrix}$

4. In video game programming, Fahad translates a car, whose coordinate is $\begin{pmatrix} 1 \\ 1 \end{pmatrix}$, 2 units up and 4 units to the right, rotates it $\frac{\pi}{2}$ radians counterclockwise, reflects it about the x-axis, reflects it about the y-axis, rotates it $\frac{\pi}{2}$ radians counterclockwise, and finally translates it 4 units down and 2 units to the left. What point represents the final location of the car?

$\begin{pmatrix} 1 \\ 1 \end{pmatrix} \to \begin{pmatrix} 1+4 \\ 1+2 \end{pmatrix} = \begin{pmatrix} 5 \\ 3 \end{pmatrix}$

$\begin{pmatrix} 0 & -1 \\ 1 & 0 \end{pmatrix}\begin{pmatrix} 5 \\ 3 \end{pmatrix} = \begin{pmatrix} -3 \\ 5 \end{pmatrix}$

$\begin{pmatrix} 1 & 0 \\ 0 & -1 \end{pmatrix}\begin{pmatrix} -3 \\ 5 \end{pmatrix} = \begin{pmatrix} -3 \\ -5 \end{pmatrix}$

$\begin{pmatrix} -1 & 0 \\ 0 & 1 \end{pmatrix}\begin{pmatrix} -3 \\ -5 \end{pmatrix} = \begin{pmatrix} 3 \\ -5 \end{pmatrix}$

$\begin{pmatrix} 0 & -1 \\ 1 & 0 \end{pmatrix}\begin{pmatrix} 3 \\ -5 \end{pmatrix} = \begin{pmatrix} 5 \\ 3 \end{pmatrix}$

$\begin{pmatrix} 5 \\ 3 \end{pmatrix} \to \begin{pmatrix} 5-4 \\ 3-2 \end{pmatrix} = \begin{pmatrix} 1 \\ 1 \end{pmatrix}$

The final location of the car is its initial starting point.

Lesson 26: Getting a Handle on New Transformations

Student Outcomes

- Students understand that the absolute value of the determinant of a 2×2 matrix is the area of the image of the unit square.

Lesson Notes

This is day one of a two-day lesson on transformations using matrix notation. Students begin with the unit square and look at the geometric results of simple transformations on the unit square. Students then calculate the area of the transformed figure and understand that it is the absolute value of the determinant of the 2×2 matrix representing the transformation.

Classwork

Have students work on the Opening Exercise individually, and then check solutions as a class. This exercise allows students to practice the matrix operations of addition and subtraction and prepares them for concepts they need in Lessons 26 and 27.

In the next few exercises, matrices are represented with square brackets. Discuss with students that matrices can be represented with soft or square brackets. $\begin{bmatrix} 1 & 2 \\ 3 & 4 \end{bmatrix}$ and $\begin{pmatrix} 1 & 2 \\ 3 & 4 \end{pmatrix}$ represent the same matrix.

Opening Exercise (8 minutes)

Opening Exercise

Perform the following matrix operations:

a. $\begin{bmatrix} 3 & -2 \\ 1 & 5 \end{bmatrix}\begin{bmatrix} -1 \\ 3 \end{bmatrix}$

$\begin{bmatrix} -9 \\ 14 \end{bmatrix}$

b. $\begin{bmatrix} 3 & -2 \\ 1 & 5 \end{bmatrix}\begin{bmatrix} 1 & 0 \\ 0 & 1 \end{bmatrix}$

$\begin{bmatrix} 3 & -2 \\ 1 & 5 \end{bmatrix}$

c. $\begin{bmatrix} 3 & -2 \\ 1 & 5 \end{bmatrix}\begin{bmatrix} 1 & -3 \\ 2 & 4 \end{bmatrix}$

$\begin{bmatrix} -1 & -17 \\ 11 & 17 \end{bmatrix}$

Scaffolding:

- To aid students in matrix operations, ask questions, such as "How do we multiply matrices?" or "How do we add matrices?" The images below can be displayed in the classroom as guides.
- To multiply matrices:

$$\begin{bmatrix} a & b \\ c & d \end{bmatrix}\begin{bmatrix} x \\ y \end{bmatrix} = \begin{bmatrix} ax + by \\ cx + dy \end{bmatrix}$$

$$\begin{bmatrix} a & b \\ c & d \end{bmatrix}\begin{bmatrix} x & y \\ z & w \end{bmatrix} = \begin{bmatrix} ax + bz & ay + bw \\ cx + dz & cy + dw \end{bmatrix}$$

- To add matrices:

$$\begin{pmatrix} a & b \\ c & d \end{pmatrix} + \begin{pmatrix} x & y \\ z & w \end{pmatrix} = \begin{pmatrix} a + x & b + y \\ c + z & d + w \end{pmatrix}$$

d. $\begin{bmatrix} 3 & -2 \\ 1 & 5 \end{bmatrix}\begin{bmatrix} a & b \\ c & d \end{bmatrix}$

$\begin{bmatrix} 3a-2c & 3b-2d \\ 1a+5c & 1b+5d \end{bmatrix}$

e. $\begin{bmatrix} 3 & -2 \\ 1 & 5 \end{bmatrix}+\begin{bmatrix} 1 & 0 \\ 0 & 1 \end{bmatrix}$

$\begin{bmatrix} 4 & -2 \\ 1 & 6 \end{bmatrix}$

f. $\begin{bmatrix} 3 & -2 \\ 1 & 5 \end{bmatrix}+\begin{bmatrix} 1 & -3 \\ 2 & 4 \end{bmatrix}$

$\begin{bmatrix} 4 & -5 \\ 3 & 9 \end{bmatrix}$

g. $\begin{bmatrix} 3 & -2 \\ 1 & 5 \end{bmatrix}+\begin{bmatrix} a & b \\ c & d \end{bmatrix}$

$\begin{bmatrix} 3+a & -2+b \\ 1+c & 5+d \end{bmatrix}$

h. **Can you add the two matrices in part (a)? Why or why not?**

No, the matrices do not have the same dimensions, so they cannot be added.

Exploratory Challenge (20 minutes)

In this Exploratory Challenge, students discover what matrix transformations do to a unit square geometrically. Students calculate the area of the image of the unit square and discover that the area is the absolute value of the determinant of the resulting 2×2 matrix. In this challenge, let students work in pairs, but lead the class together from step to step. Students should have graph paper and a ruler.

- We have seen that every matrix $\begin{bmatrix} a & b \\ c & d \end{bmatrix}$ corresponds to some kind of transformation of the plane, but it can be hard to see what the transformation actually does. For example, what does the matrix $\begin{bmatrix} 109 & 3 \\ 1 & -2 \end{bmatrix}$ do to points, shapes, and lines in the plane?
  - *Allow students to share ideas. We answer this question later after looking at more basic matrices.*
- Let's draw the unit square in the coordinate plane with each side 1 inch long.
  - *Students draw the unit square. (Check to make sure the squares are* 1 *inch* $\times$ 1 *inch.)*
- Now label the vertices of the square.
  - *Students label the vertices as shown.*
- Write a set of matrices that represents the vertices of the unit square.
  - $\begin{bmatrix} 0 \\ 0 \end{bmatrix}, \begin{bmatrix} 1 \\ 0 \end{bmatrix}, \begin{bmatrix} 0 \\ 1 \end{bmatrix}$, *and* $\begin{bmatrix} 1 \\ 1 \end{bmatrix}$

Scaffolding:

- Some student pairs may need targeted one-to-one guidance on this challenge.
- For advanced students, give them the challenge without guiding questions, and allow them to work in pairs on their own, checking their steps periodically.
- Provide unlabeled graphs for students who have difficulties with eye-hand coordination or fine motor skills.

- As we learned in previous lessons, any matrix of the form $\begin{bmatrix} a & -b \\ b & a \end{bmatrix}$ is a rotation and a dilation. Perform this transformation on the vertices of the unit square if $a > 0$ and $b > 0$. Show your work.
    - $\begin{bmatrix} a & -b \\ b & a \end{bmatrix}\begin{bmatrix} 0 \\ 0 \end{bmatrix} = \begin{bmatrix} 0 \\ 0 \end{bmatrix}$, $\begin{bmatrix} a & -b \\ b & a \end{bmatrix}\begin{bmatrix} 1 \\ 0 \end{bmatrix} = \begin{bmatrix} a \\ b \end{bmatrix}$, $\begin{bmatrix} a & -b \\ b & a \end{bmatrix}\begin{bmatrix} 0 \\ 1 \end{bmatrix} = \begin{bmatrix} -b \\ a \end{bmatrix}$, *and* $\begin{bmatrix} a & -b \\ b & a \end{bmatrix}\begin{bmatrix} 1 \\ 1 \end{bmatrix} = \begin{bmatrix} a-b \\ b+a \end{bmatrix}$

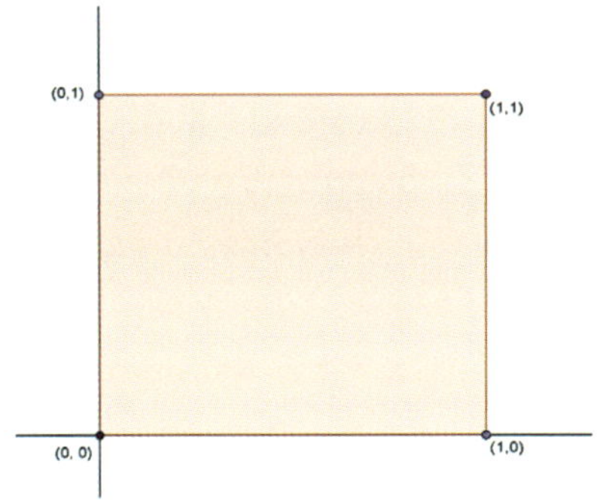

- What are the coordinates of the image of $(1, 0)$ and $(0, 1)$?
    - $(1,0) \to (a,b)$
    - $(0,1) \to (-b,a)$
- Graph the image on the same graph as the original unit square in a different color.
    - *See the diagram to the right.*
- Label the coordinates of the vertices.

  $(0,0), (a,b), (-b,a)$, *and* $(a-b, b+a)$

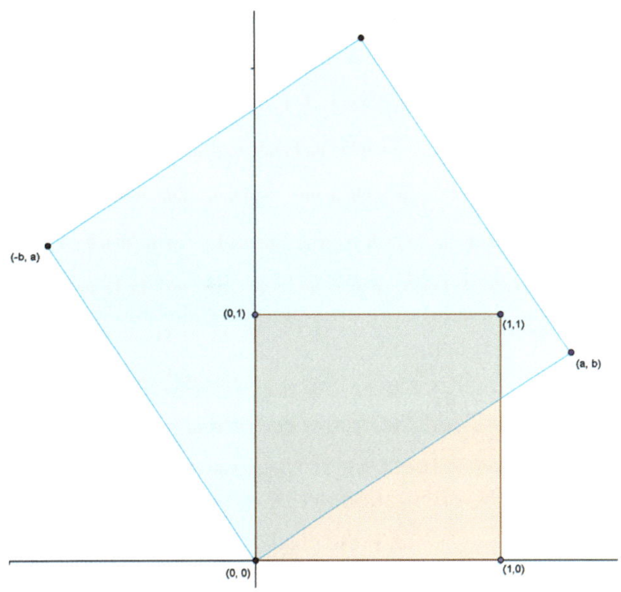

- This picture allows you to see the rotation and dilation that took place on the unit square. Let's try another transformation.
- Draw another unit square with side lengths of 1 inch.
    - *Students draw a second unit square.*
- Perform the general transformation $\begin{bmatrix} a & c \\ b & d \end{bmatrix}$ on the vertices of the unit square.
    - $\begin{bmatrix} a & c \\ b & d \end{bmatrix}\begin{bmatrix} 0 \\ 0 \end{bmatrix} = \begin{bmatrix} 0 \\ 0 \end{bmatrix}$
    - $\begin{bmatrix} a & c \\ b & d \end{bmatrix}\begin{bmatrix} 1 \\ 0 \end{bmatrix} = \begin{bmatrix} a \\ b \end{bmatrix}$
    - $\begin{bmatrix} a & c \\ b & d \end{bmatrix}\begin{bmatrix} 0 \\ 1 \end{bmatrix} = \begin{bmatrix} c \\ d \end{bmatrix}$
    - $\begin{bmatrix} a & c \\ b & d \end{bmatrix}\begin{bmatrix} 1 \\ 1 \end{bmatrix} = \begin{bmatrix} a+c \\ b+d \end{bmatrix}$
- What are the coordinates of the image of $(1, 0)$ and $(1, 1)$?
    - $(1,0) \to (a,b)$
    - $(1,1) \to (a+c, b+d)$
- Graph the image on the same graph as the second unit square in a different color, and label the vertices.
    - *See the diagram to the right.*

      The vertices are $(0,0), (a,b), (c,d)$, *and* $(a+c, b+d)$.

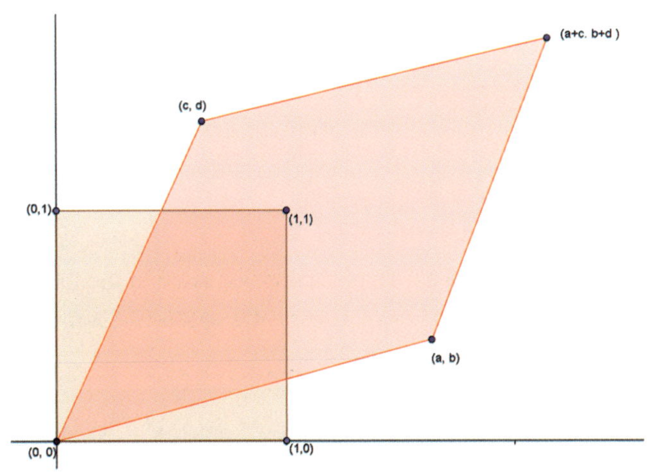

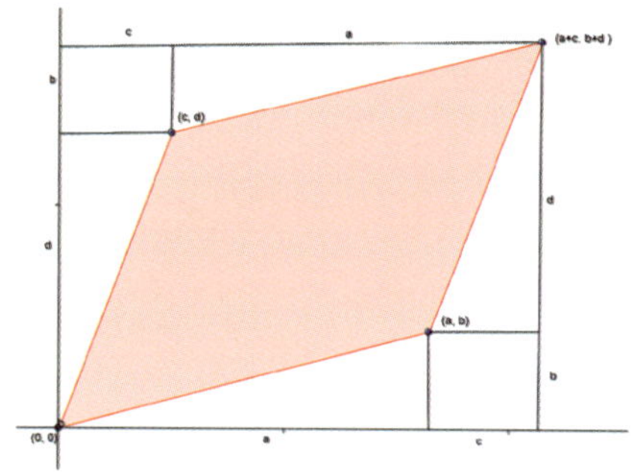

- Look at the two diagrams that we have created. The original unit square had four straight sides. After the transformations, were the straight segments mapped to straight segments? Was the square mapped to a square? Explain.
    - *Straight segments were mapped to straight segments.*
    - *The square was mapped to a parallelogram.*
- Does this transformation change the area of the unit square?
    - *It seems to—yes.*
- Let's find the area of the image from the general transformation. Allow students to work in pairs to find the area of the image by enclosing the parallelogram in a rectangle and subtracting the areas of the right triangles and rectangles surrounding the parallelogram. The area of the first image may be slightly easier to find.

$$\text{Area} = (a+c)(b+d) - 2\left(\frac{1}{2}ab + \frac{1}{2}cd + bc\right)$$

$$\text{Area} = ab + ad + bc + cd - ab - cd - 2bc$$

$$\text{Area} = ad - bc$$

- When we drew the image, we kept the orientation of the vertices; in other words, we mapped $(1, 0)$ to (a, b) and $(1, 1)$ to $(a + c, b + d)$. We could have switched the order of vertices (a, b) and (c, d). Redraw the picture and calculate the area of the parallelogram image. Do you get the same area? Explain.
    - *The area is the opposite of what we calculated before. The area is $bc - ad$.*
- What could we do to ensure this formula always works for the area regardless of the orientation of the vertices?
    - *Take the absolute value.*

MP.2

- Write the general formula for the area of the parallelogram that is the image of the transformation of the unit square.
    - $\text{Area} = |ad - bc|$
- The *determinant* of a 2×2 matrix $\begin{bmatrix} a & c \\ b & d \end{bmatrix}$ is $|ad - bc|$. Explain this geometrically to your neighbor.
    - *The determinant of a 2×2 matrix is the area of the image of the unit square that has undergone the transformation $\begin{bmatrix} a & c \\ b & d \end{bmatrix}$.*
- **DETERMINANT**: The area of the image of the unit square under the linear transformation represented by a 2×2 matrix is called the *determinant* of that matrix.

Scaffolding:

- If students are struggling to understand the need for absolute value with variables, have them perform this activity using the following matrices.
- $\begin{bmatrix} 3 & 1 \\ 2 & 4 \end{bmatrix}$ has a determinant of 10.
- $\begin{bmatrix} 1 & 3 \\ 4 & 2 \end{bmatrix}$ has a determinant of -10.
- Both give the same transformed figure, so to get the area, we must take the absolute value of the determinant (MP.2).

Exercises 1–3 (10 minutes)

In Exercises 1 and 2, two problems discussed in the Exploratory Challenge are revisited, and questions are answered using what students have discovered. Exercise 3 revisits the pure dilation and rotation matrices to see their effect on area. Students should work on these exercises in pairs and then debrief as a class. Any problems not completed can be assigned for homework.

Exercises

1. Perform the transformation $\begin{bmatrix} 109 & 3 \\ 1 & -2 \end{bmatrix}$ on the unit square.

   a. Sketch the image. What is the shape of the image?

   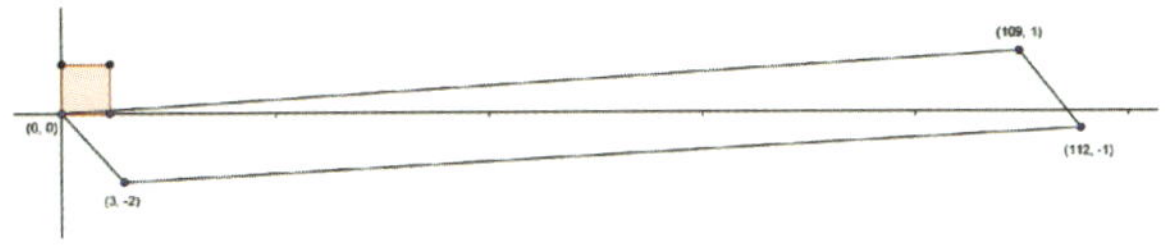

   The image is a parallelogram.

   b. What are the coordinates of the vertices of the image?

   $(0,0), (3,-2), (109,1)$, *and* $(112,-1)$

   c. What is the area of the image? Show your work.

   $\text{Area} = |(109)(-2) - (3)(1)| = |-221| = 221$

2. In the Exploratory Challenge, we drew the image of a general rotation/dilation of the unit square $\begin{bmatrix} a & -b \\ b & a \end{bmatrix}$.

   a. Calculate the area of the image by enclosing the image in a rectangle and subtracting the area of the surrounding right triangles. Show your work.

   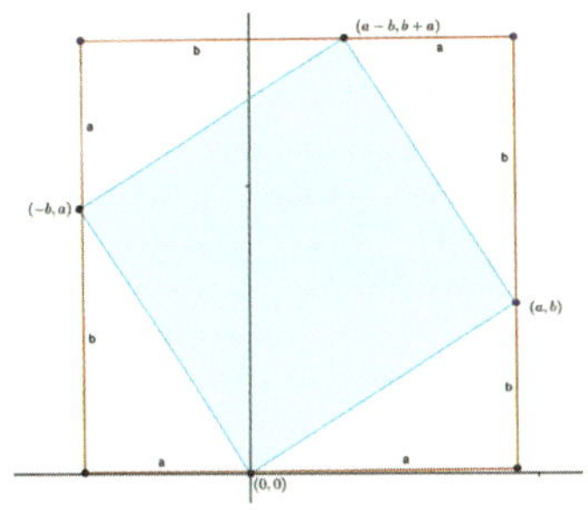

   $\text{Area} = (a+b)^2 - 4\left(\frac{1}{2}ab\right)$

   $\text{Area} = a^2 + 2ab + b^2 - 2ab$

   $\text{Area} = a^2 + b^2$

   b. Confirm the area using the determinant of the resulting matrix.

   $\begin{bmatrix} a & -b \\ b & a \end{bmatrix}$

   $\text{Area} = |(a)(a) - (b)(-b)| = a^2 + b^2$

3. We have looked at several general matrix transformations in Module 1. Answer the questions below about these familiar matrices, and explain your answers.

   a. What effect does the identity transformation have on the unit square? What is the area of the image? Confirm your answer using the determinant.

   The identity transformation does nothing to the unit square. The area is 1*, as is the determinant of the unit matrix.*

   b. How does a dilation with a scale factor of k change the area of the unit square? Calculate the determinant of a matrix representing a pure dilation of k.

   The dilation changes all areas by k^2*. The pure dilation matrix is* $\begin{bmatrix} k & 0 \\ 0 & k \end{bmatrix}$*, which has a determinant of* k^2*.*

   c. Does a rotation with no dilation change the area of the unit square? Confirm your answer by calculating the determinant of a pure rotation matrix, and explain it.

   A pure rotation does not change the area. The pure rotation matrix is $\begin{bmatrix} \cos(\theta) & -\sin(\theta) \\ \sin(\theta) & \cos(\theta) \end{bmatrix}$*. Its determinant is* $(\cos(\theta))^2 - (-\sin(\theta))^2 = \cos^2(\theta) + \sin^2(\theta) = 1$*, which confirms that the area does not change.*

Closing (2 minutes)

Have students do a 30-second quick write on the following question, and then debrief as a class.

- What effect does the general transformation $\begin{bmatrix} a & c \\ b & d \end{bmatrix}$ have on the unit square?
  - *The image of this transformation is a parallelogram with vertices* $(0,0), (a,b), (c,d)$, *and* $(a+c, b+d)$.
- What is the easiest way to calculate the area of the image of this transformation?
  - *Calculate the determinant of the resulting matrix* $|ad - bc|$.

> **Lesson Summary**
>
> **Definition**
>
> - **The area of the image of the unit square under the linear transformation represented by a 2×2 matrix is called the *determinant* of that matrix.**

Exit Ticket (5 minutes)

Name ______________________________ Date ______________

Lesson 26: Getting a Handle on New Transformations

Exit Ticket

Perform the transformation $\begin{bmatrix} -2 & 5 \\ 4 & -1 \end{bmatrix}$ on the unit square.

a. Draw the unit square and the image after this transformation.

b. Label the vertices. Explain the effect of this transformation on the unit square.

c. Calculate the area of the image. Show your work.

Exit Ticket Sample Solutions

Perform the transformation $\begin{bmatrix} -2 & 5 \\ 4 & -1 \end{bmatrix}$ on the unit square.

a. **Draw the unit square and the image after this transformation.**

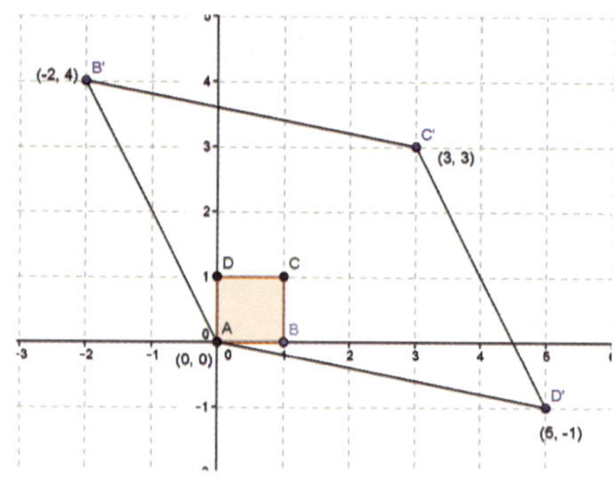

b. **Label the vertices. Explain the effect of this transformation on the unit square.**

$B(1,0) \rightarrow B'(-2,4)$

$D(0,1) \rightarrow D'(5,-1)$

$C(1,1) \rightarrow C'(3,3)$

$A(0,0) \rightarrow A(0,0)$

c. **Calculate the area of the image. Show your work.**

$|(-2)(-1) - (5)(4)| = |2 - 20| = |-18| = 18$

Problem Set Sample Solutions

1. **Perform the following transformation on the unit square: Sketch and state the area of the image.**

a. $\begin{bmatrix} 3 & -1 \\ 1 & 3 \end{bmatrix}$

$\begin{bmatrix} 3 & -1 \\ 1 & 3 \end{bmatrix}\begin{bmatrix} 0 \\ 0 \end{bmatrix} = \begin{bmatrix} 0 \\ 0 \end{bmatrix}$, $\begin{bmatrix} 3 & -1 \\ 1 & 3 \end{bmatrix}\begin{bmatrix} 1 \\ 0 \end{bmatrix} = \begin{bmatrix} 3 \\ 1 \end{bmatrix}$, $\begin{bmatrix} 3 & -1 \\ 1 & 3 \end{bmatrix}\begin{bmatrix} 1 \\ 1 \end{bmatrix} = \begin{bmatrix} 2 \\ 4 \end{bmatrix}$, *and* $\begin{bmatrix} 3 & -1 \\ 1 & 3 \end{bmatrix}\begin{bmatrix} 0 \\ 1 \end{bmatrix} = \begin{bmatrix} -1 \\ 3 \end{bmatrix}$

Area $= a^2 + b^2 = 3^2 + 1^2 = 10$

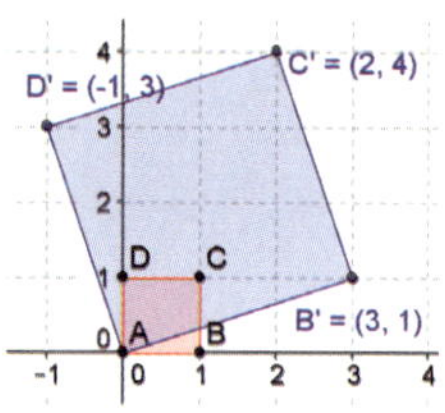

The area is 10 *units*^{2}.

b. $\begin{bmatrix} 1 & -3 \\ 3 & 1 \end{bmatrix}$

$\begin{bmatrix} 1 & -3 \\ 3 & 1 \end{bmatrix}\begin{bmatrix} 0 \\ 0 \end{bmatrix} = \begin{bmatrix} 0 \\ 0 \end{bmatrix}$, $\begin{bmatrix} 1 & -3 \\ 3 & 1 \end{bmatrix}\begin{bmatrix} 1 \\ 0 \end{bmatrix} = \begin{bmatrix} 1 \\ 3 \end{bmatrix}$, $\begin{bmatrix} 1 & -3 \\ 3 & 1 \end{bmatrix}\begin{bmatrix} 1 \\ 1 \end{bmatrix} = \begin{bmatrix} -2 \\ 4 \end{bmatrix}$, *and* $\begin{bmatrix} 1 & -3 \\ 3 & 1 \end{bmatrix}\begin{bmatrix} 0 \\ 1 \end{bmatrix} = \begin{bmatrix} -3 \\ 1 \end{bmatrix}$

Area $= a^2 + b^2 = 1^2 + 3^2 = 10$

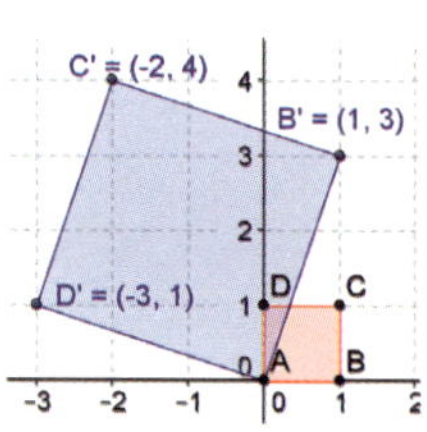

The area is 10 *units*^{2}.

c. $\begin{bmatrix}4 & -2\\2 & 4\end{bmatrix}$

$\begin{bmatrix}4 & -2\\2 & 4\end{bmatrix}\begin{bmatrix}0\\0\end{bmatrix}=\begin{bmatrix}0\\0\end{bmatrix}$, $\begin{bmatrix}4 & -2\\2 & 4\end{bmatrix}\begin{bmatrix}1\\0\end{bmatrix}=[2]$, $\begin{bmatrix}4 & -2\\2 & 4\end{bmatrix}\begin{bmatrix}1\\1\end{bmatrix}=\begin{bmatrix}2\\6\end{bmatrix}$, *and* $\begin{bmatrix}4 & -2\\2 & 4\end{bmatrix}\begin{bmatrix}0\\1\end{bmatrix}=\begin{bmatrix}-2\\4\end{bmatrix}$

Area $= a^2 + b^2 = 4^2 + 2^2 = 20$

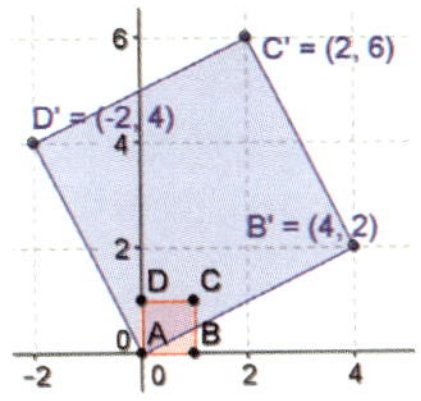

The area is 20 *units*^{2}.

d. $\begin{bmatrix}2 & -4\\4 & 2\end{bmatrix}$

$\begin{bmatrix}2 & -4\\4 & 2\end{bmatrix}\begin{bmatrix}0\\0\end{bmatrix}=\begin{bmatrix}0\\0\end{bmatrix}$, $\begin{bmatrix}2 & -4\\4 & 2\end{bmatrix}\begin{bmatrix}1\\0\end{bmatrix}=\begin{bmatrix}2\\4\end{bmatrix}$, $\begin{bmatrix}2 & -4\\4 & 2\end{bmatrix}\begin{bmatrix}1\\1\end{bmatrix}=\begin{bmatrix}-2\\6\end{bmatrix}$, *and* $\begin{bmatrix}2 & -4\\4 & 2\end{bmatrix}\begin{bmatrix}0\\1\end{bmatrix}=\begin{bmatrix}-4\\2\end{bmatrix}$

Area $= a^2 + b^2 = 2^2 + 4^2 = 20$

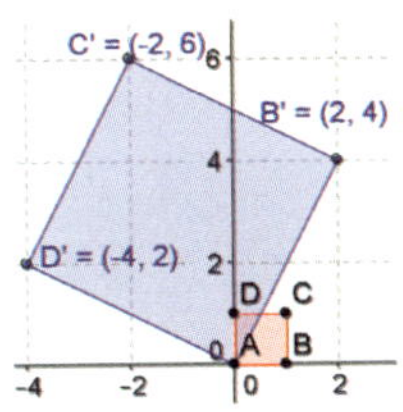

The area is 20 *units*^{2}.

2. Perform the following transformation on the unit square: Sketch the image, find the determinant of the given matrix, and find the area of the image.

a. $\begin{bmatrix}1 & 3\\2 & 4\end{bmatrix}$

$\begin{bmatrix}1 & 3\\2 & 4\end{bmatrix}\begin{bmatrix}0\\0\end{bmatrix}=\begin{bmatrix}0\\0\end{bmatrix}$, $\begin{bmatrix}1 & 3\\2 & 4\end{bmatrix}\begin{bmatrix}1\\0\end{bmatrix}=\begin{bmatrix}1\\2\end{bmatrix}$, $\begin{bmatrix}1 & 3\\2 & 4\end{bmatrix}\begin{bmatrix}1\\1\end{bmatrix}=\begin{bmatrix}4\\6\end{bmatrix}$, *and* $\begin{bmatrix}1 & 3\\2 & 4\end{bmatrix}\begin{bmatrix}0\\1\end{bmatrix}=\begin{bmatrix}3\\4\end{bmatrix}$

Determinant: $ad - bc = 4 - 6 = -2$ Area $= |ad - bc| = |4 - 6| = 2$

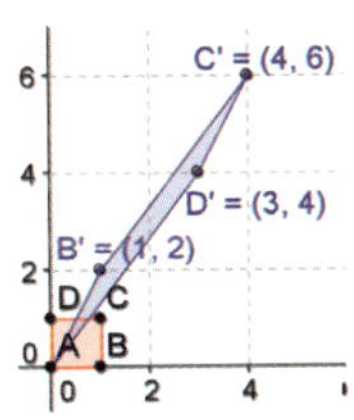

The area is 2 *units*^{2}.

b. $\begin{bmatrix} 1 & 2 \\ 3 & 4 \end{bmatrix}$

$\begin{bmatrix} 1 & 2 \\ 3 & 4 \end{bmatrix}\begin{bmatrix} 0 \\ 0 \end{bmatrix} = \begin{bmatrix} 0 \\ 0 \end{bmatrix}$, $\begin{bmatrix} 1 & 2 \\ 3 & 4 \end{bmatrix}\begin{bmatrix} 1 \\ 0 \end{bmatrix} = \begin{bmatrix} 1 \\ 3 \end{bmatrix}$, $\begin{bmatrix} 1 & 2 \\ 3 & 4 \end{bmatrix}\begin{bmatrix} 1 \\ 1 \end{bmatrix} = \begin{bmatrix} 3 \\ 7 \end{bmatrix}$, *and* $\begin{bmatrix} 1 & 2 \\ 3 & 4 \end{bmatrix}\begin{bmatrix} 0 \\ 1 \end{bmatrix} = \begin{bmatrix} 2 \\ 4 \end{bmatrix}$

Determinant: $ad - bc = 4 - 6 = -2$ Area $= |ad - bc| = |4 - 6| = 2$

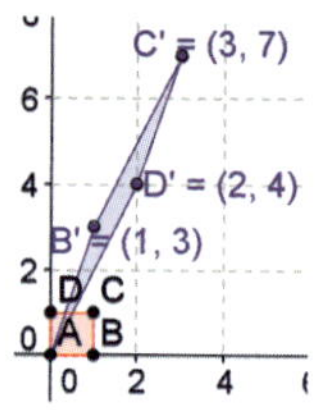

The area is 2 *units*^{2}.

c. $\begin{bmatrix} 3 & 1 \\ 2 & 4 \end{bmatrix}$

$\begin{bmatrix} 3 & 1 \\ 2 & 4 \end{bmatrix}\begin{bmatrix} 0 \\ 0 \end{bmatrix} = \begin{bmatrix} 0 \\ 0 \end{bmatrix}$, $\begin{bmatrix} 3 & 1 \\ 2 & 4 \end{bmatrix}\begin{bmatrix} 1 \\ 0 \end{bmatrix} = \begin{bmatrix} 3 \\ 2 \end{bmatrix}$, $\begin{bmatrix} 3 & 1 \\ 2 & 4 \end{bmatrix}\begin{bmatrix} 1 \\ 1 \end{bmatrix} = \begin{bmatrix} 4 \\ 6 \end{bmatrix}$, *and* $\begin{bmatrix} 3 & 1 \\ 2 & 4 \end{bmatrix}\begin{bmatrix} 0 \\ 1 \end{bmatrix} = \begin{bmatrix} 1 \\ 4 \end{bmatrix}$

Determinant: $ad - bc = 12 - 2 = 10$ Area $= |ad - bc| = |12 - 2| = 10$

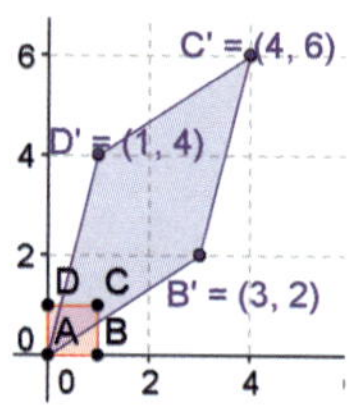

The area is 10 *units*^{2}.

d. $\begin{bmatrix} 4 & 2 \\ 3 & 1 \end{bmatrix}$

$\begin{bmatrix} 4 & 2 \\ 3 & 1 \end{bmatrix}\begin{bmatrix} 0 \\ 0 \end{bmatrix} = \begin{bmatrix} 0 \\ 0 \end{bmatrix}$, $\begin{bmatrix} 4 & 2 \\ 3 & 1 \end{bmatrix}\begin{bmatrix} 1 \\ 0 \end{bmatrix} = \begin{bmatrix} 4 \\ 3 \end{bmatrix}$, $\begin{bmatrix} 4 & 2 \\ 3 & 1 \end{bmatrix}\begin{bmatrix} 1 \\ 1 \end{bmatrix} = \begin{bmatrix} 6 \\ 4 \end{bmatrix}$, *and* $\begin{bmatrix} 4 & 2 \\ 3 & 1 \end{bmatrix}\begin{bmatrix} 0 \\ 1 \end{bmatrix} = \begin{bmatrix} 2 \\ 1 \end{bmatrix}$

Determinant: $ad - bc = 4 - 6 = -2$ Area $= |ad - bc| = |4 - 6| = 2$

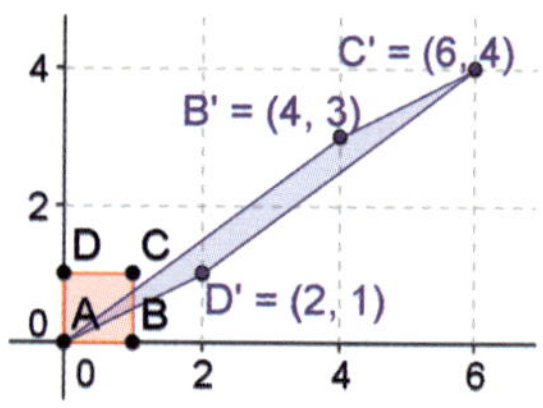

The area is 2 *units*^{2}.

e. **The determinants in parts (a), (b), (c), and (d) have positive or negative values. What is the value of the determinants if the vertices (b, c) and (c, d) are switched?**

The value is negative.

3. **Perform the following transformation on the unit square: Sketch the image, find the determinant of the given matrix, and find the area of the image.**

a. $\begin{bmatrix} -1 & -3 \\ -2 & -4 \end{bmatrix}$

$\begin{bmatrix} -1 & -2 \\ -3 & -4 \end{bmatrix}\begin{bmatrix} 0 \\ 0 \end{bmatrix} = \begin{bmatrix} 0 \\ 0 \end{bmatrix}$, $\begin{bmatrix} -1 & -2 \\ -3 & -4 \end{bmatrix}\begin{bmatrix} 1 \\ 0 \end{bmatrix} = \begin{bmatrix} -1 \\ -3 \end{bmatrix}$, $\begin{bmatrix} -1 & -2 \\ -3 & -4 \end{bmatrix}\begin{bmatrix} 1 \\ 1 \end{bmatrix} = \begin{bmatrix} -3 \\ -7 \end{bmatrix}$, *and* $\begin{bmatrix} -1 & -2 \\ -3 & -4 \end{bmatrix}\begin{bmatrix} 0 \\ 1 \end{bmatrix} = \begin{bmatrix} -2 \\ -4 \end{bmatrix}$

Determinant: $ad - bc = 4 - 6 = -2$ Area $= |ad - bc| = |4 - 6| = 2$

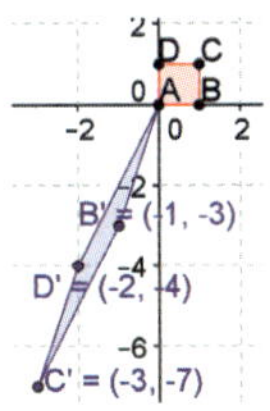

The area is 2 *units*^{2}.

b. $\begin{bmatrix} -1 & -3 \\ 2 & 4 \end{bmatrix}$

$\begin{bmatrix} -1 & -3 \\ 2 & 4 \end{bmatrix}\begin{bmatrix} 0 \\ 0 \end{bmatrix} = \begin{bmatrix} 0 \\ 0 \end{bmatrix}$, $\begin{bmatrix} -1 & -3 \\ 2 & 4 \end{bmatrix}\begin{bmatrix} 1 \\ 0 \end{bmatrix} = \begin{bmatrix} -1 \\ 2 \end{bmatrix}$, $\begin{bmatrix} -1 & -3 \\ 2 & 4 \end{bmatrix}\begin{bmatrix} 1 \\ 1 \end{bmatrix} = \begin{bmatrix} -4 \\ 6 \end{bmatrix}$, *and* $\begin{bmatrix} -1 & -3 \\ 2 & 4 \end{bmatrix}\begin{bmatrix} 0 \\ 1 \end{bmatrix} = \begin{bmatrix} -3 \\ 4 \end{bmatrix}$

Determinant: $ad - bc = -4 + 6 = 2$ Area $= |ad - bc| = |-4 + 6| = 2$

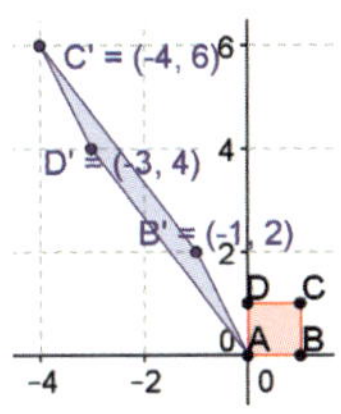

The area is 2 *units*^{2}.

c. $\begin{bmatrix} 1 & 3 \\ -2 & -4 \end{bmatrix}$

$\begin{bmatrix} 1 & 3 \\ -2 & -4 \end{bmatrix}\begin{bmatrix} 0 \\ 0 \end{bmatrix} = \begin{bmatrix} 0 \\ 0 \end{bmatrix}$, $\begin{bmatrix} 1 & 3 \\ -2 & -4 \end{bmatrix}\begin{bmatrix} 1 \\ 0 \end{bmatrix} = \begin{bmatrix} 1 \\ -2 \end{bmatrix}$, $\begin{bmatrix} 1 & 3 \\ -2 & -4 \end{bmatrix}\begin{bmatrix} 1 \\ 1 \end{bmatrix} = \begin{bmatrix} 4 \\ -6 \end{bmatrix}$, *and* $\begin{bmatrix} 1 & 3 \\ -2 & -4 \end{bmatrix}\begin{bmatrix} 0 \\ 1 \end{bmatrix} = \begin{bmatrix} 3 \\ -4 \end{bmatrix}$

Determinant: $ad - bc = -4 + 6 = 2$ Area $= |ad - bc| = |-4 + 6| = 2$

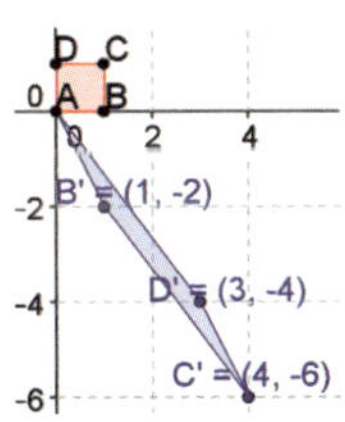

The area is 2 *units*^{2}.

d. $\begin{bmatrix} -1 & 3 \\ -2 & 4 \end{bmatrix}$

$\begin{bmatrix} -1 & 3 \\ -2 & 4 \end{bmatrix}\begin{bmatrix} 0 \\ 0 \end{bmatrix} = \begin{bmatrix} 0 \\ 0 \end{bmatrix}$, $\begin{bmatrix} -1 & 3 \\ -2 & 4 \end{bmatrix}\begin{bmatrix} 1 \\ 0 \end{bmatrix} = \begin{bmatrix} -1 \\ -2 \end{bmatrix}$, $\begin{bmatrix} -1 & 3 \\ -2 & 4 \end{bmatrix}\begin{bmatrix} 1 \\ 1 \end{bmatrix} = \begin{bmatrix} 2 \\ 2 \end{bmatrix}$, *and* $\begin{bmatrix} -1 & 3 \\ -2 & 4 \end{bmatrix}\begin{bmatrix} 0 \\ 1 \end{bmatrix} = \begin{bmatrix} 3 \\ 4 \end{bmatrix}$

Determinant: $ad - bc = -4 + 6 = 2$ Area $= |ad - bc| = |-4 + 6| = 2$

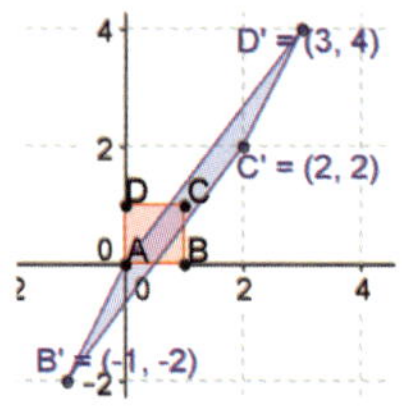

The area is 2 ***units***^{2}.

e. $\begin{bmatrix} 1 & -3 \\ 2 & -4 \end{bmatrix}$

$\begin{bmatrix} 1 & -3 \\ 2 & -4 \end{bmatrix}\begin{bmatrix} 0 \\ 0 \end{bmatrix} = \begin{bmatrix} 0 \\ 0 \end{bmatrix}$, $\begin{bmatrix} 1 & -3 \\ 2 & -4 \end{bmatrix}\begin{bmatrix} 1 \\ 0 \end{bmatrix} = \begin{bmatrix} 1 \\ 2 \end{bmatrix}$, $\begin{bmatrix} 1 & -3 \\ 2 & -4 \end{bmatrix}\begin{bmatrix} 1 \\ 1 \end{bmatrix} = \begin{bmatrix} -2 \\ -2 \end{bmatrix}$, *and* $\begin{bmatrix} 1 & -3 \\ 2 & -4 \end{bmatrix}\begin{bmatrix} 0 \\ 1 \end{bmatrix} = \begin{bmatrix} -3 \\ -4 \end{bmatrix}$

Determinant: $ad - bc = -4 + 6 = -2$ Area $= |ad - bc| = |-4 + 6| = 2$

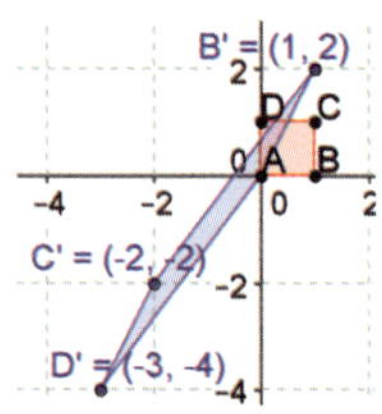

The area is 2 ***units***^{2}.

f. $\begin{bmatrix} -1 & 3 \\ 2 & -4 \end{bmatrix}$

$\begin{bmatrix} -1 & 3 \\ 2 & -4 \end{bmatrix}\begin{bmatrix} 0 \\ 0 \end{bmatrix} = \begin{bmatrix} 0 \\ 0 \end{bmatrix}$, $\begin{bmatrix} -1 & 3 \\ 2 & -4 \end{bmatrix}\begin{bmatrix} 1 \\ 0 \end{bmatrix} = \begin{bmatrix} -1 \\ 2 \end{bmatrix}$, $\begin{bmatrix} -1 & 3 \\ 2 & -4 \end{bmatrix}\begin{bmatrix} 1 \\ 1 \end{bmatrix} = \begin{bmatrix} 2 \\ -2 \end{bmatrix}$, *and* $\begin{bmatrix} -1 & 3 \\ 2 & -4 \end{bmatrix}\begin{bmatrix} 0 \\ 1 \end{bmatrix} = \begin{bmatrix} 3 \\ -4 \end{bmatrix}$

Determinant: $ad - bc = 4 - 6 = -2$ Area $= |ad - bc| = |4 - 6| = 2$

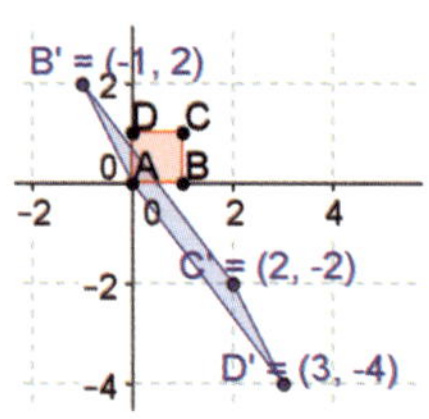

The area is 2 ***units***^{2}.

g. $\begin{bmatrix} 1 & -3 \\ -2 & 4 \end{bmatrix}$

$\begin{bmatrix} 1 & -3 \\ -2 & 4 \end{bmatrix}\begin{bmatrix} 0 \\ 0 \end{bmatrix} = \begin{bmatrix} 0 \\ 0 \end{bmatrix}$, $\begin{bmatrix} 1 & -3 \\ -2 & 4 \end{bmatrix}\begin{bmatrix} 1 \\ 0 \end{bmatrix} = \begin{bmatrix} 1 \\ -2 \end{bmatrix}$, $\begin{bmatrix} 1 & -3 \\ -2 & 4 \end{bmatrix}\begin{bmatrix} 1 \\ 1 \end{bmatrix} = \begin{bmatrix} -2 \\ 2 \end{bmatrix}$, *and* $\begin{bmatrix} 1 & -3 \\ -2 & 4 \end{bmatrix}\begin{bmatrix} 0 \\ 1 \end{bmatrix} = \begin{bmatrix} -3 \\ 4 \end{bmatrix}$

Determinant: $ad - bc = 4 - 6 = 2$ Area $= |ad - bc| = |4 - 6| = 2$

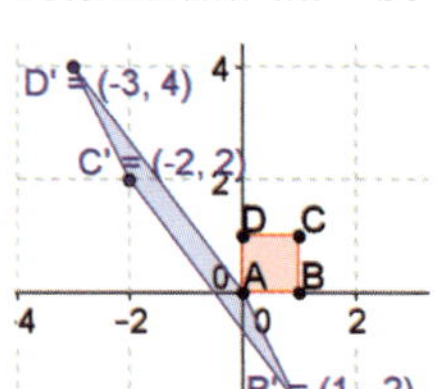

The area is 2 ***units***^{2}.

Lesson 27: Getting a Handle on New Transformations

Student Outcomes

- Students understand that 2×2 matrix transformations are linear transformations taking straight lines to straight lines.
- Students understand that the absolute value of the determinant of a 2×2 matrix is the area of the image of the unit square.

Lesson Notes

This is day two of a two-day lesson on transformations using matrix notation. In Lesson 26, students looked at general matrix transformations on the unit square and discovered that the area of the image was the determinant of the resulting matrix. In Lesson 27, students get more practice with this concept and connect it to the study of linearity.

Classwork

This Opening Exercise reminds students of general matrices studied in prior lessons and their geometric effect. Show one matrix at a time to the class, and discuss the geometric significance of each matrix.

Opening Exercise (8 minutes)

MP.2

Opening Exercise

Explain the geometric effect of each matrix.

a. $\begin{bmatrix} a & -b \\ b & a \end{bmatrix}$

A rotation of $\arctan\left(\frac{b}{a}\right)$ ***and a dilation with scale factor*** $\sqrt{a^2+b^2}$

b. $\begin{bmatrix} \cos(\theta) & -\sin(\theta) \\ \sin(\theta) & \cos(\theta) \end{bmatrix}$

A pure rotation of θ

c. $\begin{bmatrix} k & 0 \\ 0 & k \end{bmatrix}$

A pure dilation of scale factor k

d. $\begin{bmatrix} 1 & 0 \\ 0 & 1 \end{bmatrix}$

The multiplicative identify matrix has no geometric effect.

Scaffolding:

- Have students create an example of each transformation and show it graphically in a graphic organizer (see sample below).
- Ask advanced learners to create a matrix that produces a dilation of 3 and a rotation of 30° counterclockwise.

Matrix	Transformation	Picture
$\begin{bmatrix} k & 0 \\ 0 & k \end{bmatrix}$	Pure dilation of scale factor k	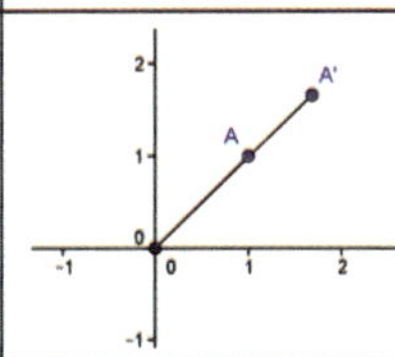

MP.2

e. $\begin{bmatrix} 0 & 0 \\ 0 & 0 \end{bmatrix}$

The additive identify matrix maps all points to the origin.

f. $\begin{bmatrix} a & c \\ b & d \end{bmatrix}$

Transforms the unit square to a parallelogram with vertices $(0, 0)$, (a, b), (c, d), and $(a + c, b + d)$ with area of $|ad - bc|$.

Example 1 (10 minutes)

In Example 1, students perform a transformation on the unit square, calculate and confirm the area of the image, and then solve a system of equations that would map the transformation to a given point. Students should complete this example in groups with guiding questions from the teacher as needed.

Example 1

Given the transformation $\begin{bmatrix} k & 0 \\ k & 1 \end{bmatrix}$ with $k > 0$:

a. **Perform this transformation on the vertices of the unit square. Sketch the image, and label the vertices.**

$\begin{bmatrix} k & 0 \\ k & 1 \end{bmatrix}\begin{bmatrix} 0 \\ 0 \end{bmatrix} = \begin{bmatrix} 0 \\ 0 \end{bmatrix}$

$\begin{bmatrix} k & 0 \\ k & 1 \end{bmatrix}\begin{bmatrix} 1 \\ 0 \end{bmatrix} = \begin{bmatrix} k \\ k \end{bmatrix}$

$\begin{bmatrix} k & 0 \\ k & 1 \end{bmatrix}\begin{bmatrix} 0 \\ 1 \end{bmatrix} = \begin{bmatrix} 0 \\ 1 \end{bmatrix}$

$\begin{bmatrix} k & 0 \\ k & 1 \end{bmatrix}\begin{bmatrix} 1 \\ 1 \end{bmatrix} = \begin{bmatrix} k \\ k + 1 \end{bmatrix}$

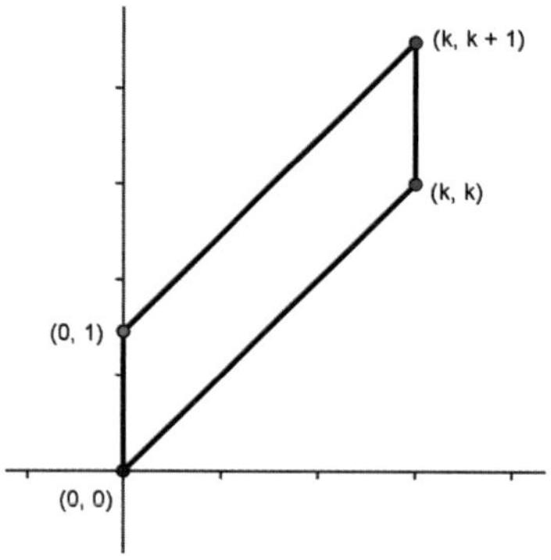

Scaffolding:

- Give advanced students a single task: "Write a formula for the application of this transformation n times." Ask them to develop an answer without the questions shown.
- Provide labeled graphs for students who have difficulties with eye-hand coordination or fine motor skills.

b. **Calculate the area of the image using the dimensions of the image parallelogram.**

The parallelogram is 1 unit high, and the perpendicular distance between parallel bases is k units wide, so the area is $1 \cdot k = k$ square units.

c. **Confirm the area of the image using the determinant.**

The area of the unit square is 1, and the determinant of the transformation matrix is $|(k)(1) - (0)(k)| = k$. The area of the parallelogram is $1 \cdot k = k$ square units. The area is confirmed.

d. **Perform the transformation on $\begin{bmatrix} x \\ y \end{bmatrix}$.**

$\begin{bmatrix} k & 0 \\ k & 1 \end{bmatrix}\begin{bmatrix} x \\ y \end{bmatrix} = \begin{bmatrix} kx \\ kx + y \end{bmatrix}$

EUREKA MATH™

e. **In order for two matrices to be equivalent, each of the corresponding elements must be equivalent. Given that, if the image of this transformation is $\begin{bmatrix}5\\4\end{bmatrix}$, find $\begin{bmatrix}x\\y\end{bmatrix}$.**

$$\begin{bmatrix}kx\\kx+y\end{bmatrix}=\begin{bmatrix}5\\4\end{bmatrix}$$

$$kx=5 \qquad x=\frac{5}{k}$$

$$kx+y=4$$
$$k\left(\frac{5}{k}\right)+y=4$$
$$5+y=4$$
$$y=-1$$

$$\begin{bmatrix}x\\y\end{bmatrix}=\begin{bmatrix}\frac{5}{k}\\-1\end{bmatrix}$$

f. **Perform the transformation on $\begin{bmatrix}1\\1\end{bmatrix}$. Write the image matrix.**

$$\begin{bmatrix}k&0\\k&1\end{bmatrix}\begin{bmatrix}1\\1\end{bmatrix}=\begin{bmatrix}k\\k+1\end{bmatrix}$$

g. **Perform the transformation on the image again, and then repeat until the transformation has been performed four times on the image of the preceding matrix.**

$$\begin{bmatrix}k&0\\k&1\end{bmatrix}\begin{bmatrix}k\\k+1\end{bmatrix}=\begin{bmatrix}k^2\\k^2+k+1\end{bmatrix}$$
$$\begin{bmatrix}k&0\\k&1\end{bmatrix}\begin{bmatrix}k^2\\k^2+k+1\end{bmatrix}=\begin{bmatrix}k^3\\k^3+k^2+k+1\end{bmatrix}$$
$$\begin{bmatrix}k&0\\k&1\end{bmatrix}\begin{bmatrix}k^3\\k^3+k^2+k+1\end{bmatrix}=\begin{bmatrix}k^4\\k^4+k^3+k^2+k+1\end{bmatrix}$$

- What are the vertices of the image?
  - $(0,0)$, (k,k), $(0,1)$, *and* $(k,k+1)$
- What is the formula for the area of a parallelogram?
  - *The area of a parallelogram is* base × height.
- What is the base and height of the parallelogram that is the image of this transformation? How do you know?
  - *The base is the length of one of the parallel sides, which is* 1 *unit. The height is the perpendicular distance between parallel sides, and that is* k *units.*
- Using the formula, calculate the area of the parallelogram.
  - *The area is* k *square units.*
- Now find the area using the determinant. Is the area confirmed?
  - Area $=|(k)(1)-(0)k|=k$ *square units. This is the same area.*
- Now perform the transformation on a point (x,y). What is the matrix that results?
  - $\begin{bmatrix}kx\\kx+y\end{bmatrix}$
- If we want the image of this transformation on (x,y) to map to $(5,4)$, how could we find (x,y)?
  - *We could write a system of equations and solve for* (x,y).

- Write the system of equations.
  - $kx = 5$
  - $kx + y = 4$
- Solve for x and y in terms of k. Which variable is easiest to solve for? Explain and solve for it.
  - *It is easiest to solve for x because the first equation only has x, not y.* $x = \frac{5}{k}$
- Now solve for the other variable.
  - $y = 4 - k\left(\frac{5}{k}\right) = 4 - 5 = -1$
- So, $\begin{bmatrix} x \\ y \end{bmatrix}$ is equal to what?
  - $\begin{bmatrix} \frac{5}{k} \\ -1 \end{bmatrix}$

Exercise 1 (8 minutes)

This exercise should be completed in pairs and gives students practice solving for $\begin{bmatrix} x \\ y \end{bmatrix}$ and writing a general formula to represent n transformations. Some groups may need the leading questions presented in Example 1 to help them.

Exercise 1

1. **Perform the transformation $\begin{bmatrix} k & 0 \\ 1 & k \end{bmatrix}$ with $k > 1$ on the vertices of the unit square.**

   a. **What are the vertices of the image?**

   $(0, 0)$, $(k, 1)$, $(0, k)$, *and* $(k, k+1)$

   b. **Calculate the area of the image.**

   k^2

   c. **If the image of the transformation on $\begin{bmatrix} x \\ y \end{bmatrix}$ is $\begin{bmatrix} -2 \\ -1 \end{bmatrix}$, find $\begin{bmatrix} x \\ y \end{bmatrix}$ in terms of k.**

   $$\begin{bmatrix} k & 0 \\ 1 & k \end{bmatrix}\begin{bmatrix} x \\ y \end{bmatrix} = \begin{bmatrix} kx \\ x + ky \end{bmatrix} = \begin{bmatrix} -2 \\ -1 \end{bmatrix}$$

   $kx = -2$

   $x = \frac{-2}{k}$

   $x + ky = -1$

   $\frac{-2}{k} + ky = -1$

   $ky = -1 + \frac{2}{k}$

   $y = \frac{-1}{k} + \frac{2}{k^2}$

   $$\begin{bmatrix} x \\ y \end{bmatrix} = \begin{bmatrix} \frac{-2}{k} \\ \frac{-1}{k} + \frac{2}{k^2} \end{bmatrix}$$

Example 2 (10 minutes)

In Lesson 26, the claim was made that a matrix transformation takes straight lines to straight lines. This example explores that claim, because students discover that matrix transformations are indeed linear, and sets the groundwork for the work of Lessons 27–30. Students should work in pairs with the teacher leading the discussion.

Example 2

Consider the matrix $L = \begin{bmatrix} 2 & 5 \\ -1 & 3 \end{bmatrix}$. For each real number $0 \le t \le 1$, consider the point $(3 + t, 10 + 2t)$.

a. **Find point A when $t = 0$.**

$A(3, 10)$

b. **Find point B when $t = 1$.**

$B(4, 12)$

c. **Show that for $t = \frac{1}{2}$, $(3 + t, 10 + 2t)$ is the midpoint of $\overline{AB}$.**

When $t = \frac{1}{2}$, point M is $\left(3 + \frac{1}{2}, 10 + 2\left(\frac{1}{2}\right)\right)$, or $(3.5, 11)$. The midpoint of $\overline{AB}$ is $\left(\frac{3+4}{2}, \frac{10+12}{2}\right)$, or $\left(\frac{7}{2}, 11\right)$.

The midpoint is at $t = \frac{1}{2}$.

d. **Show that for each value of t, $(3 + t, 10 + 2t)$ is a point on the line through A and B.**

The equation of the line through A and B is $y - 10 = \frac{12-10}{4-3}(x - 3)$, or $y - 10 = 2(x - 3)$, or $y = 2x + 4$. If we substitute $(3 + t, 10 + 2t)$ into the equation, we get $10 + 2t = 2(3 + t) + 4$ or $10 + 2t = 2t + 10$, which is a statement that is true for all real values of t. Therefore, the point $(3 + t, 10 + 2t)$ lies on the line through A and B for all values of t.

e. **Find LA and LB.**

$LA = \begin{bmatrix} 2 & 5 \\ -1 & 3 \end{bmatrix}\begin{bmatrix} 3 \\ 10 \end{bmatrix} = \begin{bmatrix} 56 \\ 27 \end{bmatrix}$

$LB = \begin{bmatrix} 2 & 5 \\ -1 & 3 \end{bmatrix}\begin{bmatrix} 4 \\ 12 \end{bmatrix} = \begin{bmatrix} 68 \\ 32 \end{bmatrix}$

f. **What is the equation of the line through LA and LB?**

The line through LA and LB is $y - 27 = \frac{32-27}{68-56}(x - 56)$ or $y - 27 = \frac{5}{12}(x - 56)$.

g. **Show that $L\begin{bmatrix} 3+t \\ 10+2t \end{bmatrix}$ lies on the line through LA and LB.**

$L\begin{bmatrix} 3+t \\ 10+2t \end{bmatrix} = \begin{bmatrix} 2 & 5 \\ -1 & 3 \end{bmatrix}\begin{bmatrix} 3+t \\ 10+2t \end{bmatrix} = \begin{bmatrix} 6+2t+50+10t \\ -3-t+30+6t \end{bmatrix} = \begin{bmatrix} 12t+56 \\ 5t+27 \end{bmatrix}$

$(5t + 27) - 27 = \frac{5}{12}((12t + 56) - 56)$

$5t = 5t$, which is true for all real values of t, so $L\begin{bmatrix} 3+t \\ 10+2t \end{bmatrix}$ and lies on the line through LA and LB.

- Will the midpoint always occur at $t = \frac{1}{2}$? Explain.
  - *It will always occur at the* $\frac{1}{2}(t_1 + t_2)$. *Since* $t_1 + t_2 = 1$ *in this problem, the midpoint occurred at* $t = \frac{1}{2}$.
- Write an equation for the line through A and B. Explain your work.
  - $A(3, 10)$ *and* $B(4, 12)$, *so the slope is* $m = \frac{12 - 10}{4 - 3} = 2$. *In point slope form, the equation is* $y - 10 = 2(x - 3)$ *or* $y - 12 = 2(x - 4)$. *In slope-intercept form, the equation is* $y = 2x + 4$.
- Substitute $x = 3 + t$ and $y = 10 + 2t$ into this equation. What do you discover?
  - $10 + 2t = 2(3 + t) + 4$

    $10 + 2t = 6 + 2t + 4$

    $10 + 2t = 10 + 2t$

    We get a statement that is true for all real values of t.
- What does this mean?
  - *The point* $(3 + t, 10 + 2t)$ *lies on the line through* A *and* B *for all values of* t.
- Write an equation for the line through LA and LB.
  - $y - 27 = \frac{5}{12}(x - 56)$ *or* $y - 32 = \frac{5}{12}(x - 68)$
- Does every point on $L\begin{bmatrix} 3 + t \\ 10 + 2t \end{bmatrix}$ lie on the line through LA and LB? Explain.
  - *Yes* $(5t + 27) - 27 = \frac{5}{12}\left((12t + 56) - 56\right)$
  - $5t = 5t$, *which is true for all real values of t; therefore,* $L\begin{bmatrix} 3 + t \\ 10 + 2t \end{bmatrix}$ *lies on the line through LA and LB.*

Closing (4 minutes)

Have students explain to a neighbor everything that they learned about matrix transformations in Lessons 26 and 27; then, pull the class together to debrief.

- Explain to your neighbor everything that you learned about matrix transformations in Lessons 26 and 27.
  - *The image of this transformation is a parallelogram with vertices.*
  - $(0, 0)$, (a, b), (c, d), and $(a + c, b + d)$
  - *The determinant of the* 2×2 *transformation matrix is the area of the image of the unit square after the transformation.*
  - *A* 2×2 *transformation can rotate, dilate, and/or change the shape of the unit square.*
  - *A* 2×2 *transformation takes straight lines and maps them to straight lines.*

Exit Ticket (5 minutes)

Name ______________________________ Date ______________

Lesson 27: Getting a Handle on New Transformations

Exit Ticket

Given the transformation $\begin{bmatrix} 0 & k \\ 1 & k \end{bmatrix}$ with $k > 0$:

a. Find the area of the image of the transformation performed on the unit matrix.

b. The image of the transformation on $\begin{bmatrix} x \\ y \end{bmatrix}$ is $\begin{bmatrix} 1 \\ 5 \end{bmatrix}$; find $\begin{bmatrix} x \\ y \end{bmatrix}$ in terms of k. Show your work.

Exit Ticket Sample Solutions

Given the transformation $\begin{bmatrix} 0 & k \\ 1 & k \end{bmatrix}$ with $k > 0$:

a. **Find the area of the image of the transformation performed on the unit matrix.**

$|(0)(k) - (k)(1)| = |-k| = k$

b. **The image of the transformation on $\begin{bmatrix} x \\ y \end{bmatrix}$ is $\begin{bmatrix} 1 \\ 5 \end{bmatrix}$; find $\begin{bmatrix} x \\ y \end{bmatrix}$ in terms of k. Show your work.**

$$\begin{bmatrix} 0 & k \\ 1 & k \end{bmatrix}\begin{bmatrix} x \\ y \end{bmatrix} = \begin{bmatrix} ky \\ x + ky \end{bmatrix} = \begin{bmatrix} 1 \\ 5 \end{bmatrix}$$

$$ky = 1$$
$$y = \frac{1}{k}$$

$$x + ky = 5$$
$$x + k\left(\frac{1}{k}\right) = 5$$
$$x + 1 = 5$$
$$x = 4$$

$$\begin{bmatrix} x \\ y \end{bmatrix} = \begin{bmatrix} 4 \\ \frac{1}{k} \end{bmatrix}$$

Problem Set Sample Solutions

1. **Perform the following transformation on the vertices of the unit square. Sketch the image, label the vertices, and find the area of the image parallelogram.**

a. $\begin{bmatrix} 1 & 0 \\ 1 & 1 \end{bmatrix}$

$\begin{bmatrix} 1 & 0 \\ 1 & 1 \end{bmatrix}\begin{bmatrix} 0 \\ 0 \end{bmatrix} = \begin{bmatrix} 0 \\ 0 \end{bmatrix}$, $\begin{bmatrix} 1 & 0 \\ 1 & 1 \end{bmatrix}\begin{bmatrix} 1 \\ 0 \end{bmatrix} = \begin{bmatrix} 1 \\ 1 \end{bmatrix}$, $\begin{bmatrix} 1 & 0 \\ 1 & 1 \end{bmatrix}\begin{bmatrix} 1 \\ 1 \end{bmatrix} = \begin{bmatrix} 1 \\ 2 \end{bmatrix}$, *and* $\begin{bmatrix} 1 & 0 \\ 1 & 1 \end{bmatrix}\begin{bmatrix} 0 \\ 1 \end{bmatrix} = \begin{bmatrix} 0 \\ 1 \end{bmatrix}$

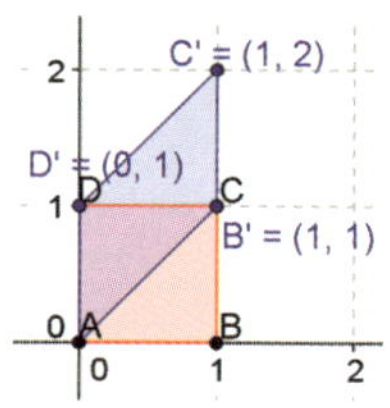

$\text{Area} = |1 \times 1 - 1 \times 0| = 1$

The area is 1 *units*^{2}

b. $\begin{bmatrix} 2 & 0 \\ 2 & 1 \end{bmatrix}$

$\begin{bmatrix} 2 & 0 \\ 2 & 1 \end{bmatrix}\begin{bmatrix} 0 \\ 0 \end{bmatrix} = \begin{bmatrix} 0 \\ 0 \end{bmatrix}$, $\begin{bmatrix} 2 & 0 \\ 2 & 1 \end{bmatrix}\begin{bmatrix} 1 \\ 0 \end{bmatrix} = \begin{bmatrix} 2 \\ 2 \end{bmatrix}$, $\begin{bmatrix} 2 & 0 \\ 2 & 1 \end{bmatrix}\begin{bmatrix} 1 \\ 1 \end{bmatrix} = \begin{bmatrix} 2 \\ 3 \end{bmatrix}$, *and* $\begin{bmatrix} 2 & 0 \\ 2 & 1 \end{bmatrix}\begin{bmatrix} 0 \\ 1 \end{bmatrix} = \begin{bmatrix} 0 \\ 1 \end{bmatrix}$

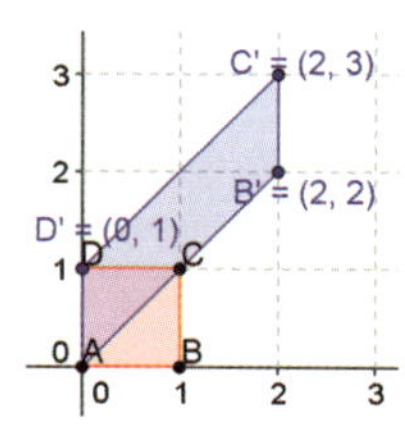

$\text{Area} = |2 \times 1 - 2 \times 0| = 2$

The area is 2 *units*^{2}.

c. $\begin{bmatrix} 3 & 0 \\ 3 & 1 \end{bmatrix}$

$\begin{bmatrix} 3 & 0 \\ 3 & 1 \end{bmatrix}\begin{bmatrix} 0 \\ 0 \end{bmatrix} = \begin{bmatrix} 0 \\ 0 \end{bmatrix}$, $\begin{bmatrix} 3 & 0 \\ 3 & 1 \end{bmatrix}\begin{bmatrix} 1 \\ 0 \end{bmatrix} = \begin{bmatrix} 3 \\ 3 \end{bmatrix}$, $\begin{bmatrix} 3 & 0 \\ 3 & 1 \end{bmatrix}\begin{bmatrix} 1 \\ 1 \end{bmatrix} = \begin{bmatrix} 3 \\ 4 \end{bmatrix}$, *and* $\begin{bmatrix} 3 & 0 \\ 3 & 1 \end{bmatrix}\begin{bmatrix} 0 \\ 1 \end{bmatrix} = \begin{bmatrix} 0 \\ 1 \end{bmatrix}$

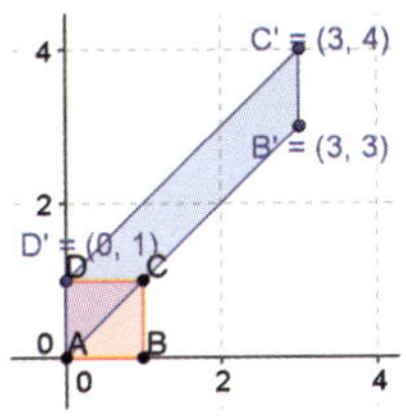

$\text{Area} = |3 \times 1 - 3 \times 0| = 3$

The area is 3 *units*^{2}.

d. $\begin{bmatrix} 2 & 0 \\ 2 & 2 \end{bmatrix}$

$\begin{bmatrix} 2 & 0 \\ 2 & 2 \end{bmatrix}\begin{bmatrix} 0 \\ 0 \end{bmatrix} = \begin{bmatrix} 0 \\ 0 \end{bmatrix}$, $\begin{bmatrix} 2 & 0 \\ 2 & 2 \end{bmatrix}\begin{bmatrix} 1 \\ 0 \end{bmatrix} = \begin{bmatrix} 2 \\ 2 \end{bmatrix}$, $\begin{bmatrix} 2 & 0 \\ 2 & 2 \end{bmatrix}\begin{bmatrix} 1 \\ 1 \end{bmatrix} = \begin{bmatrix} 2 \\ 4 \end{bmatrix}$, *and* $\begin{bmatrix} 2 & 0 \\ 2 & 2 \end{bmatrix}\begin{bmatrix} 0 \\ 1 \end{bmatrix} = \begin{bmatrix} 0 \\ 2 \end{bmatrix}$

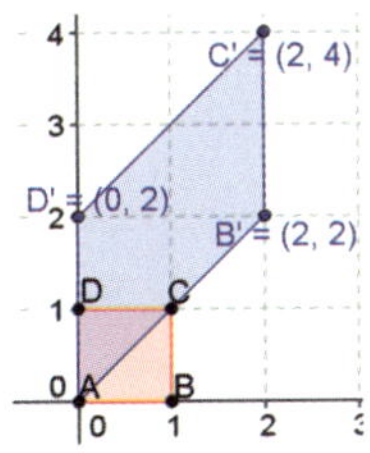

$\text{Area} = |2 \times 2 - 2 \times 0| = 4$

The area is 4 *units*^{2}.

e. $\begin{bmatrix} 3 & 0 \\ 3 & 3 \end{bmatrix}$

$\begin{bmatrix} 3 & 0 \\ 3 & 3 \end{bmatrix}\begin{bmatrix} 0 \\ 0 \end{bmatrix} = \begin{bmatrix} 0 \\ 0 \end{bmatrix}$, $\begin{bmatrix} 3 & 0 \\ 3 & 3 \end{bmatrix}\begin{bmatrix} 1 \\ 0 \end{bmatrix} = \begin{bmatrix} 3 \\ 3 \end{bmatrix}$, $\begin{bmatrix} 3 & 0 \\ 3 & 3 \end{bmatrix}\begin{bmatrix} 1 \\ 1 \end{bmatrix} = \begin{bmatrix} 3 \\ 6 \end{bmatrix}$, *and* $\begin{bmatrix} 3 & 0 \\ 3 & 3 \end{bmatrix}\begin{bmatrix} 0 \\ 1 \end{bmatrix} = \begin{bmatrix} 0 \\ 3 \end{bmatrix}$

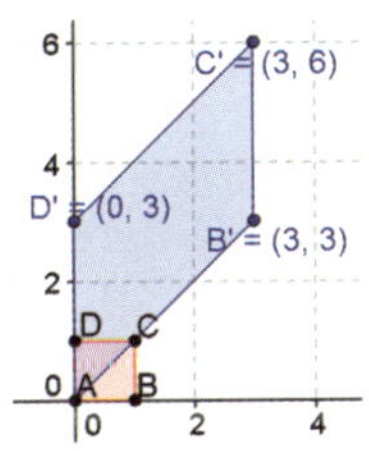

$\text{Area} = |3 \times 3 - 3 \times 0| = 9$

The area is 9 *units*^{2}.

f. $\begin{bmatrix} 1 & 0 \\ 1 & 2 \end{bmatrix}$

$\begin{bmatrix} 1 & 0 \\ 1 & 2 \end{bmatrix}\begin{bmatrix} 0 \\ 0 \end{bmatrix} = \begin{bmatrix} 0 \\ 0 \end{bmatrix}$, $\begin{bmatrix} 1 & 0 \\ 1 & 2 \end{bmatrix}\begin{bmatrix} 1 \\ 0 \end{bmatrix} = \begin{bmatrix} 1 \\ 1 \end{bmatrix}$, $\begin{bmatrix} 1 & 0 \\ 1 & 2 \end{bmatrix}\begin{bmatrix} 1 \\ 1 \end{bmatrix} = \begin{bmatrix} 1 \\ 3 \end{bmatrix}$, *and* $\begin{bmatrix} 1 & 0 \\ 1 & 2 \end{bmatrix}\begin{bmatrix} 0 \\ 1 \end{bmatrix} = \begin{bmatrix} 0 \\ 2 \end{bmatrix}$

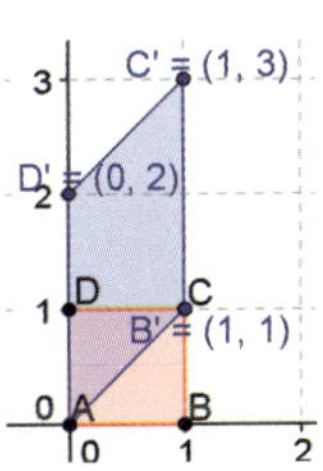

$\text{Area} = |1 \times 2 - 1 \times 0| = 2$

The area is 2 *units*^{2}.

g. $\begin{bmatrix} 1 & 0 \\ 1 & 3 \end{bmatrix}$

$\begin{bmatrix} 1 & 0 \\ 1 & 3 \end{bmatrix}\begin{bmatrix} 0 \\ 0 \end{bmatrix} = \begin{bmatrix} 0 \\ 0 \end{bmatrix}$, $\begin{bmatrix} 1 & 0 \\ 1 & 3 \end{bmatrix}\begin{bmatrix} 1 \\ 0 \end{bmatrix} = \begin{bmatrix} 1 \\ 1 \end{bmatrix}$, $\begin{bmatrix} 1 & 0 \\ 1 & 3 \end{bmatrix}\begin{bmatrix} 1 \\ 1 \end{bmatrix} = \begin{bmatrix} 1 \\ 4 \end{bmatrix}$, *and* $\begin{bmatrix} 1 & 0 \\ 1 & 3 \end{bmatrix}\begin{bmatrix} 0 \\ 1 \end{bmatrix} = \begin{bmatrix} 0 \\ 3 \end{bmatrix}$

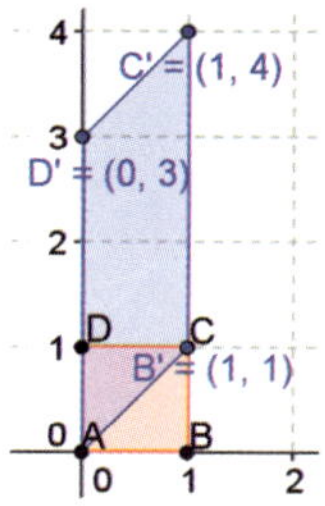

$\textbf{Area} = |1 \times 3 - 1 \times 0| = 3$

The area is 3 *units*^{2}.

h. $\begin{bmatrix} 2 & 0 \\ 2 & 3 \end{bmatrix}$

$\begin{bmatrix} 2 & 0 \\ 2 & 3 \end{bmatrix}\begin{bmatrix} 0 \\ 0 \end{bmatrix} = \begin{bmatrix} 0 \\ 0 \end{bmatrix}$, $\begin{bmatrix} 2 & 0 \\ 2 & 3 \end{bmatrix}\begin{bmatrix} 1 \\ 0 \end{bmatrix} = \begin{bmatrix} 2 \\ 2 \end{bmatrix}$, $\begin{bmatrix} 2 & 0 \\ 2 & 3 \end{bmatrix}\begin{bmatrix} 1 \\ 1 \end{bmatrix} = \begin{bmatrix} 2 \\ 5 \end{bmatrix}$, *and* $\begin{bmatrix} 2 & 0 \\ 2 & 3 \end{bmatrix}\begin{bmatrix} 0 \\ 1 \end{bmatrix} = \begin{bmatrix} 0 \\ 3 \end{bmatrix}$

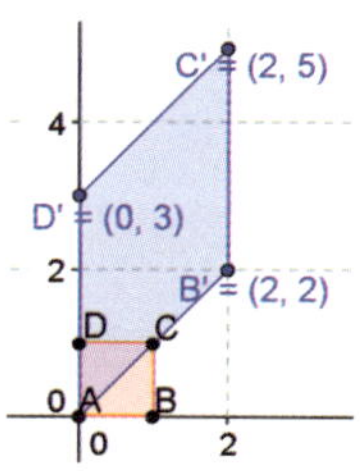

$\textbf{Area} = |2 \times 3 - 2 \times 0| = 6$

The area is 6 *units*^{2}.

i. $\begin{bmatrix} 3 & 0 \\ 3 & 5 \end{bmatrix}$

$\begin{bmatrix} 3 & 0 \\ 3 & 5 \end{bmatrix}\begin{bmatrix} 0 \\ 0 \end{bmatrix} = \begin{bmatrix} 0 \\ 0 \end{bmatrix}$, $\begin{bmatrix} 3 & 0 \\ 3 & 5 \end{bmatrix}\begin{bmatrix} 1 \\ 0 \end{bmatrix} = \begin{bmatrix} 3 \\ 3 \end{bmatrix}$, $\begin{bmatrix} 3 & 0 \\ 3 & 5 \end{bmatrix}\begin{bmatrix} 1 \\ 1 \end{bmatrix} = \begin{bmatrix} 3 \\ 8 \end{bmatrix}$, *and* $\begin{bmatrix} 3 & 0 \\ 3 & 5 \end{bmatrix}\begin{bmatrix} 0 \\ 1 \end{bmatrix} = \begin{bmatrix} 0 \\ 5 \end{bmatrix}$

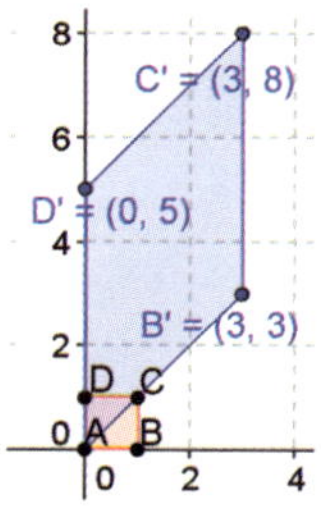

$\textbf{Area} = |3 \times 5 - 3 \times 0| = 15$

The area is 15 *units*^{2}.

2. Given $\begin{bmatrix} k & 0 \\ k & 1 \end{bmatrix}\begin{bmatrix} x \\ y \end{bmatrix} = \begin{bmatrix} kx \\ kx+y \end{bmatrix}$. Find $\begin{bmatrix} x \\ y \end{bmatrix}$ if the image of the transformation is the following:

a. $\begin{bmatrix} 4 \\ 5 \end{bmatrix}$

$$\begin{bmatrix} kx \\ kx+y \end{bmatrix} = \begin{bmatrix} 4 \\ 5 \end{bmatrix}$$

$kx = 4$

$x = \frac{4}{k}$

$kx + y = 5$

$k\frac{4}{k} + y = 5$

$4 + y = 5$

$y = 1$

$$\begin{bmatrix} x \\ y \end{bmatrix} = \begin{bmatrix} \frac{4}{k} \\ 1 \end{bmatrix}$$

b. $\begin{bmatrix} -3 \\ 2 \end{bmatrix}$

$$\begin{bmatrix} kx \\ kx+y \end{bmatrix} = \begin{bmatrix} -3 \\ 2 \end{bmatrix}$$

$kx = -3$

$x = -\frac{3}{k}$

$kx + y = 2$

$k\frac{-3}{k} + y = 2$

$-3 + y = 2$

$y = 5$

$$\begin{bmatrix} x \\ y \end{bmatrix} = \begin{bmatrix} -\frac{3}{k} \\ 5 \end{bmatrix}$$

c. $\begin{bmatrix} 5 \\ -6 \end{bmatrix}$

$$\begin{bmatrix} kx \\ kx+y \end{bmatrix} = \begin{bmatrix} 5 \\ -6 \end{bmatrix}$$

$kx = 5$

$x = \frac{5}{k}$

$kx + y = -6$

$k\frac{5}{k} + y = -6$

$5 + y = -6$

$y = -11$

$$\begin{bmatrix} x \\ y \end{bmatrix} = \begin{bmatrix} \frac{5}{k} \\ -11 \end{bmatrix}$$

3. Given $\begin{bmatrix} k & 0 \\ k & 1 \end{bmatrix}\begin{bmatrix} x \\ y \end{bmatrix} = \begin{bmatrix} kx \\ kx+y \end{bmatrix}$. Find the value of k so that:

a. $\begin{bmatrix} x \\ y \end{bmatrix} = \begin{bmatrix} 3 \\ -2 \end{bmatrix}$ and the image is $\begin{bmatrix} 24 \\ 22 \end{bmatrix}$.

$$\begin{bmatrix} k & 0 \\ k & 1 \end{bmatrix}\begin{bmatrix} 3 \\ -2 \end{bmatrix} = \begin{bmatrix} 24 \\ 22 \end{bmatrix}$$

$3k + 0 = 24$ *OR* $3k - 2 = 22$

$k = 8$ $\qquad$ $k = 8$

b. $\begin{bmatrix} x \\ y \end{bmatrix} = \begin{bmatrix} 27 \\ 3 \end{bmatrix}$ and the image is $\begin{bmatrix} 18 \\ 21 \end{bmatrix}$.

$$\begin{bmatrix} k & 0 \\ k & 1 \end{bmatrix}\begin{bmatrix} 27 \\ 3 \end{bmatrix} = \begin{bmatrix} 18 \\ 21 \end{bmatrix}$$

$27k + 0 = 18$ *OR* $27k + 3 = 21$

$k = \frac{18}{27} = \frac{2}{3}$ $\qquad$ $k = \frac{18}{27} = \frac{2}{3}$

4. Given $\begin{bmatrix} k & 0 \\ 1 & k \end{bmatrix}\begin{bmatrix} x \\ y \end{bmatrix} = \begin{bmatrix} kx \\ x+ky \end{bmatrix}$. Find the value of k so that:

a. $\begin{bmatrix} x \\ y \end{bmatrix} = \begin{bmatrix} -4 \\ 5 \end{bmatrix}$ and the image is $\begin{bmatrix} -12 \\ 11 \end{bmatrix}$.

$$\begin{bmatrix} k & 0 \\ 1 & k \end{bmatrix}\begin{bmatrix} -4 \\ 5 \end{bmatrix} = \begin{bmatrix} -12 \\ 11 \end{bmatrix}$$

$-4k + 0 = -12$ *OR* $-4 + 5k = 11$

$k = 3$ $\qquad$ $k = 3$

b. $\begin{bmatrix} x \\ y \end{bmatrix} = \begin{bmatrix} \frac{5}{3} \\ \frac{2}{9} \end{bmatrix}$ and the image is $\begin{bmatrix} -15 \\ -\frac{1}{3} \end{bmatrix}$.

$$\begin{bmatrix} k & 0 \\ 1 & k \end{bmatrix}\begin{bmatrix} \frac{5}{3} \\ \frac{2}{9} \end{bmatrix} = \begin{bmatrix} -15 \\ -\frac{1}{3} \end{bmatrix}$$

$\frac{5}{3}k + 0 = -15$ *OR* $\frac{5}{3} + \frac{2}{9}k = -\frac{1}{3}$

$k = -9$ $\qquad$ $k = -9$

5. Perform the following transformation on the vertices of the unit square. Sketch the image, label the vertices, and find the area of the image parallelogram.

a. $\begin{bmatrix} 2 & 0 \\ 1 & 2 \end{bmatrix}$

$\begin{bmatrix} 2 & 0 \\ 1 & 2 \end{bmatrix}\begin{bmatrix} 0 \\ 0 \end{bmatrix} = \begin{bmatrix} 0 \\ 0 \end{bmatrix}$, $\begin{bmatrix} 2 & 0 \\ 1 & 2 \end{bmatrix}\begin{bmatrix} 1 \\ 0 \end{bmatrix} = \begin{bmatrix} 2 \\ 1 \end{bmatrix}$, $\begin{bmatrix} 2 & 0 \\ 1 & 2 \end{bmatrix}\begin{bmatrix} 1 \\ 1 \end{bmatrix} = \begin{bmatrix} 2 \\ 3 \end{bmatrix}$, *and* $\begin{bmatrix} 2 & 0 \\ 1 & 2 \end{bmatrix}\begin{bmatrix} 0 \\ 1 \end{bmatrix} = \begin{bmatrix} 0 \\ 2 \end{bmatrix}$

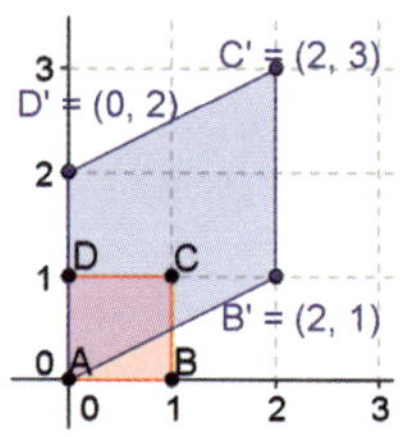

Area $= |k \times k - 1 \times 0| = k^2 = 4$

The area is 4 units^2.

b. $\begin{bmatrix} 3 & 0 \\ 1 & 3 \end{bmatrix}$

$\begin{bmatrix} 3 & 0 \\ 1 & 3 \end{bmatrix}\begin{bmatrix} 0 \\ 0 \end{bmatrix} = \begin{bmatrix} 0 \\ 0 \end{bmatrix}, \begin{bmatrix} 3 & 0 \\ 1 & 3 \end{bmatrix}\begin{bmatrix} 1 \\ 0 \end{bmatrix} = \begin{bmatrix} 3 \\ 1 \end{bmatrix}, \begin{bmatrix} 3 & 0 \\ 1 & 3 \end{bmatrix}\begin{bmatrix} 1 \\ 1 \end{bmatrix} = \begin{bmatrix} 3 \\ 4 \end{bmatrix}$, *and* $\begin{bmatrix} 3 & 0 \\ 1 & 3 \end{bmatrix}\begin{bmatrix} 0 \\ 1 \end{bmatrix} = \begin{bmatrix} 0 \\ 3 \end{bmatrix}$

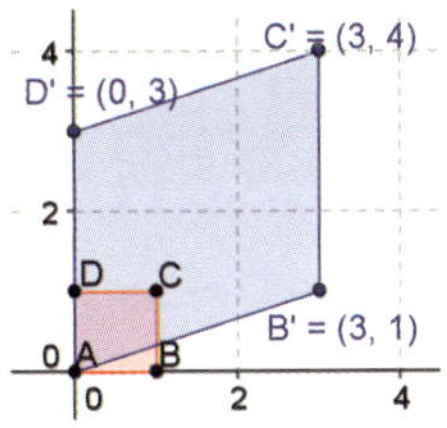

$\textbf{Area} = |k \times k - 1 \times 0| = k^2 = 9$

The area is 9 *units*^{2}.

c. $\begin{bmatrix} 2 & 0 \\ 3 & 2 \end{bmatrix}$

$\begin{bmatrix} 2 & 0 \\ 3 & 2 \end{bmatrix}\begin{bmatrix} 0 \\ 0 \end{bmatrix} = \begin{bmatrix} 0 \\ 0 \end{bmatrix}, \begin{bmatrix} 2 & 0 \\ 3 & 2 \end{bmatrix}\begin{bmatrix} 1 \\ 0 \end{bmatrix} = \begin{bmatrix} 2 \\ 3 \end{bmatrix}, \begin{bmatrix} 2 & 0 \\ 3 & 2 \end{bmatrix}\begin{bmatrix} 1 \\ 1 \end{bmatrix} = \begin{bmatrix} 2 \\ 5 \end{bmatrix}$, *and* $\begin{bmatrix} 2 & 0 \\ 3 & 2 \end{bmatrix}\begin{bmatrix} 0 \\ 1 \end{bmatrix} = \begin{bmatrix} 0 \\ 2 \end{bmatrix}$

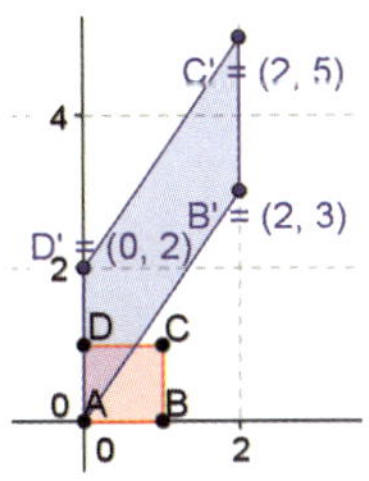

$\textbf{Area} = |k \times k - 1 \times 0| = k^2 = 4$

The area is 4 *units*^{2}.

d. $\begin{bmatrix} 2 & 0 \\ 4 & 2 \end{bmatrix}$

$\begin{bmatrix} 2 & 0 \\ 4 & 2 \end{bmatrix}\begin{bmatrix} 0 \\ 0 \end{bmatrix} = \begin{bmatrix} 0 \\ 0 \end{bmatrix}, \begin{bmatrix} 2 & 0 \\ 4 & 2 \end{bmatrix}\begin{bmatrix} 1 \\ 0 \end{bmatrix} = \begin{bmatrix} 2 \\ 4 \end{bmatrix}, \begin{bmatrix} 2 & 0 \\ 4 & 2 \end{bmatrix}\begin{bmatrix} 1 \\ 1 \end{bmatrix} = \begin{bmatrix} 2 \\ 6 \end{bmatrix}$, *and* $\begin{bmatrix} 2 & 0 \\ 4 & 2 \end{bmatrix}\begin{bmatrix} 0 \\ 1 \end{bmatrix} = \begin{bmatrix} 0 \\ 2 \end{bmatrix}$

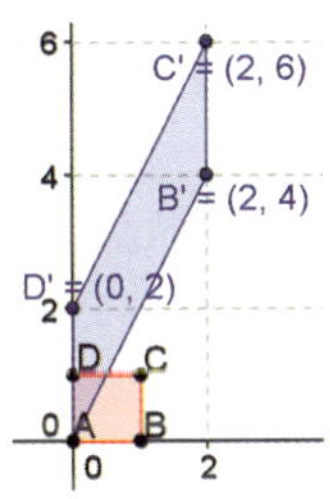

$\textbf{Area} = |k \times k - 1 \times 0| = k^2 = 4$

The area is 4 *units*^{2}.

e. $\begin{bmatrix} 4 & 0 \\ 2 & 4 \end{bmatrix}$

$\begin{bmatrix} 4 & 0 \\ 2 & 4 \end{bmatrix}\begin{bmatrix} 0 \\ 0 \end{bmatrix} = \begin{bmatrix} 0 \\ 0 \end{bmatrix}, \begin{bmatrix} 4 & 0 \\ 2 & 4 \end{bmatrix}\begin{bmatrix} 1 \\ 0 \end{bmatrix} = \begin{bmatrix} 4 \\ 2 \end{bmatrix}, \begin{bmatrix} 4 & 0 \\ 2 & 4 \end{bmatrix}\begin{bmatrix} 1 \\ 1 \end{bmatrix} = \begin{bmatrix} 4 \\ 6 \end{bmatrix}$, *and* $\begin{bmatrix} 4 & 0 \\ 2 & 4 \end{bmatrix}\begin{bmatrix} 0 \\ 1 \end{bmatrix} = \begin{bmatrix} 0 \\ 4 \end{bmatrix}$

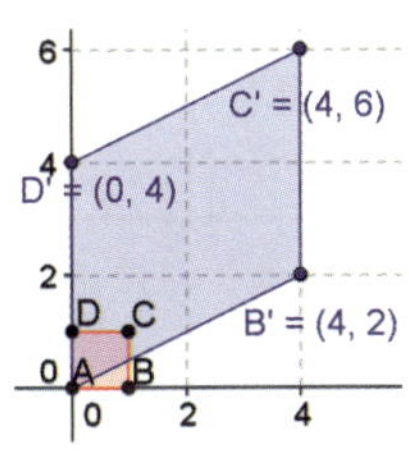

$\textbf{Area} = |k \times k - 1 \times 0| = k^2 = 16$

The area is 16 *units*^{2}.

f. $\begin{bmatrix} 3 & 0 \\ 5 & 3 \end{bmatrix}$

$\begin{bmatrix} 3 & 0 \\ 5 & 3 \end{bmatrix}\begin{bmatrix} 0 \\ 0 \end{bmatrix} = \begin{bmatrix} 0 \\ 0 \end{bmatrix}, \begin{bmatrix} 3 & 0 \\ 5 & 3 \end{bmatrix}\begin{bmatrix} 1 \\ 0 \end{bmatrix} = \begin{bmatrix} 3 \\ 5 \end{bmatrix}, \begin{bmatrix} 3 & 0 \\ 5 & 3 \end{bmatrix}\begin{bmatrix} 1 \\ 1 \end{bmatrix} = \begin{bmatrix} 3 \\ 8 \end{bmatrix}$, *and* $\begin{bmatrix} 3 & 0 \\ 5 & 3 \end{bmatrix}\begin{bmatrix} 0 \\ 1 \end{bmatrix} = \begin{bmatrix} 0 \\ 3 \end{bmatrix}$

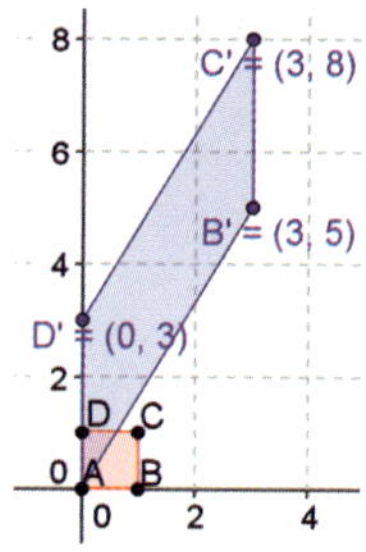

Area $= |k \times k - 1 \times 0| = k^2 = 9$

The area is 9 units^2.

6. Consider the matrix $L = \begin{bmatrix} 1 & 3 \\ 2 & 4 \end{bmatrix}$. For each real number $0 \le t \le 1$, consider the point $(3 + 2t, 12 + 2t)$.

a. Find the point A when $t = 0$.

$A(3, 12)$

b. Find the point B when $t = 1$.

$B(5, 14)$

c. Show that for $t = \frac{1}{2}$, $(3 + 2t, 12 + 2t)$ is the midpoint of $\overline{AB}$.

When $t = \frac{1}{2}$, *the point* M *is* $(3 + 1, 12 + 1)$, *or* $(4, 13)$.

And the midpoint of $\overline{AB}$ *is* $\left(\frac{3+5}{2}, \frac{12+14}{2}\right)$, *or* $(4, 13)$. *Thus, the midpoint is at* $t = \frac{1}{2}$.

d. Show that for each value of t, $(3 + 2t, 12 + 2t)$ is a point on the line through A and B.

The equation of the line through $\overline{AB}$:

$$y - 12 = \frac{12 - 14}{3 - 5}(x - 3)$$
$$y = x + 9$$

If we substitute $(3 + 2t, 12 + 2t)$ *into the equation, we get* $12 + 2t = 3 + 2t + 9$, *or* $12 + 2t = 12 + 2t$, *which is a statement that is true for all real values of* t. *Therefore, the point* $(3 + 2t, 12 + 2t)$ *lies on the line through* A *and* B *for all values of* t.

e. Find LA and LB.

$LA = \begin{bmatrix} 1 & 3 \\ 2 & 4 \end{bmatrix}\begin{bmatrix} 3 \\ 12 \end{bmatrix} = \begin{bmatrix} 39 \\ 54 \end{bmatrix}$

$LB = \begin{bmatrix} 1 & 3 \\ 2 & 4 \end{bmatrix}\begin{bmatrix} 5 \\ 14 \end{bmatrix} = \begin{bmatrix} 47 \\ 66 \end{bmatrix}$

f. **What is the equation of the line through LA and LB?**

$$y - 54 = \frac{66 - 54}{47 - 39}(x - 39)$$

$$y - 54 = \frac{3}{2}(x - 39)$$

g. **Show that $L\begin{bmatrix} 3 + 2t \\ 12 + 2t \end{bmatrix}$ lies on the line through LA and LB.**

$$L\begin{bmatrix} 3 + 2t \\ 12 + 2t \end{bmatrix} = \begin{bmatrix} 1 & 3 \\ 2 & 4 \end{bmatrix}\begin{bmatrix} 3 + 2t \\ 12 + 2t \end{bmatrix} = \begin{bmatrix} 3 + 2t + 36 + 6t \\ 6 + 4t + 48 + 8t \end{bmatrix} = \begin{bmatrix} 39 + 8t \\ 54 + 12t \end{bmatrix}$$

We substitute it into the equation in part (f):

$$54 + 12t - 54 = \frac{3}{2}(39 + 8t - 39)$$

$$12t = \frac{3}{2}(8t)$$

$$12t = 12t,$$

which is true for all real values of t, so $L\begin{bmatrix} 3 + 2t \\ 12 + 2t \end{bmatrix}$ lies on the line through LA and LB.

Lesson 28: When Can We Reverse a Transformation?

Student Outcomes

- Students determine inverse matrices using linear systems.

Lesson Notes

In the final three lessons of this module, students discover how to reverse a transformation by discovering the inverse matrix. In Lesson 28, students are introduced to inverse matrices and find inverses of matrices with a determinant of 1 by solving a system of equations. Lesson 29 expands this idea to include inverses of matrices with a determinant other than 1 and finding a general formula for an inverse matrix. In Lesson 30, students discover that matrices with determinants of zero do not have an inverse.

Classwork

The Opening Exercise can be done individually or in pairs. It allows students to practice a 2×2 matrix transformation on a unit square. Students need graph paper.

Opening Exercise (8 minutes)

Opening Exercise

Perform the operation $\begin{bmatrix} 3 & -2 \\ 1 & 1 \end{bmatrix}$ on the unit square.

a. **State the vertices of the transformation.**

$(0, 0)$, $(3, 1)$, $(-2, 1)$, and $(1, 2)$

b. **Explain the transformation in words.**

$(0, 0)$ stays at the origin, the vertex $(1, 0)$ moves to $(3, 1)$, $(0, 1)$ moves to $(-2, 1)$, and $(1, 1)$ moves to $(1, 2)$.

c. **Find the area of the transformed figure.**

$|(3)(1) - (-2)(1)| = 5$

The area is 5 units^2.

d. **If the original square was 2×2 instead of a unit square, how would the transformation change?**

The coordinates of the vertices of the image would all double. The vertices would be $(0, 0)$, $(6, 2)$, $(-4, 2)$, and $(2, 4)$.

e. **What is the area of the image? Explain how you know.**

The area of the image is 20 square units. The area of the original square was 4 square units; multiply that by the determinant, which is 5, and the area of the new figure is $4 \times 5 = 20$ square units.

Discussion (10 minutes)

This Discussion is a whole-class discussion that wraps up the Opening Exercise and gets students to think about reversing transformations.

Scaffolding:

- Advanced learners can do the Discussion in small homogenous groups and do Example 1 with no guiding questions.
- Remind students how to multiply 2×2 matrices by displaying this graphic:
  $\begin{bmatrix} a & b \\ c & d \end{bmatrix}\begin{bmatrix} r & s \\ t & u \end{bmatrix} = \begin{bmatrix} ab + bt & as + bu \\ cr + dt & cs + du \end{bmatrix}$.
- Remind students of the trigonometry Pythagorean identities by displaying the following:
  $\sin^2(\theta) + \cos^2(\theta) = 1$.

- What are some differences between the unit square and a 2×2 square?
  - *The coordinates of the vertices of the 2×2 square were double the coordinates of the vertices of the unit square.*
  - *The area of the 2×2 square is four times the area of the unit square.*
- Was the same thing true when the matrix transformation was applied?
  - *Yes*
- How can the determinant of the transformation matrix be used to find the area of a transformed image if the original image was not a unit square?
  - *Find the area of the original square, and then multiply that area by the value of the determinant.*
- What matrix have we studied that produces only a counterclockwise rotation through an angle θ about the origin? Call it R_θ.
  - $R_\theta = \begin{bmatrix} \cos(\theta) & -\sin(\theta) \\ \sin(\theta) & \cos(\theta) \end{bmatrix}$

MP.3

- What transformation would "undo" this transformation? Describe it in words or symbols.
  - *We can undo this transformation by rotating in the opposite direction or through an angle of $-\theta$.*
  - $\sin(-\theta) = -\sin(\theta)$ *because it is an odd function and is symmetric about the origin.* $f(-x) = -f(x)$
  - $\cos(-\theta) = \cos(\theta)$ *because it is an even function and is symmetric about the y-axis.* $f(-x) = f(x)$
- Write the matrix that represents the rotation through $-\theta$. Call it $R_{-\theta}$.
  - $R_{-\theta} = \begin{bmatrix} \cos(-\theta) & -\sin(-\theta) \\ \sin(-\theta) & \cos(-\theta) \end{bmatrix} = \begin{bmatrix} \cos(\theta) & \sin(\theta) \\ -\sin(\theta) & \cos(\theta) \end{bmatrix}$
- What do you think will happen if we apply R_θ and then $R_{-\theta}$?
  - *We should end up with what we started with.*
- Let's confirm this. What matrix do you expect to see when you compute the product $R_{-\theta}R_\theta$?
  - *The identity matrix* $\begin{bmatrix} 1 & 0 \\ 0 & 1 \end{bmatrix}$

- Perform this operation. Were you correct?
  - $\begin{bmatrix} \cos(\theta) & \sin(\theta) \\ -\sin(\theta) & \cos(\theta) \end{bmatrix} \begin{bmatrix} \cos(\theta) & -\sin(\theta) \\ \sin(\theta) & \cos(\theta) \end{bmatrix} =$

    $\begin{bmatrix} \cos^2(\theta) + \sin^2(\theta) & -\cos(\theta)\sin(\theta) + \sin(\theta)\cos(\theta) \\ -\sin(\theta)\cos(\theta) + \cos(\theta)\sin(\theta) & \sin^2(\theta) + \cos^2(\theta) \end{bmatrix} = \begin{bmatrix} 1 & 0 \\ 0 & 1 \end{bmatrix}$
  - *Yes, we got the identity matrix.*
- What about $R_\theta R_{-\theta}$?
  - $\begin{bmatrix} \cos(\theta) & -\sin(\theta) \\ \sin(\theta) & \cos(\theta) \end{bmatrix} \begin{bmatrix} \cos(\theta) & \sin(\theta) \\ -\sin(\theta) & \cos(\theta) \end{bmatrix} =$

    $\begin{bmatrix} \cos^2(\theta) + \sin^2(\theta) & -\sin(\theta)\cos(\theta) + \cos(\theta)\sin(\theta) \\ -\cos(\theta)\sin(\theta) + \sin(\theta)\cos(\theta) & \sin^2(\theta) + \cos^2(\theta) \end{bmatrix} = \begin{bmatrix} 1 & 0 \\ 0 & 1 \end{bmatrix}$
  - *Yes, we get the identify matrix again.*
- Explain to your neighbor what we have just discovered.
  - $R_{-\theta}R_\theta = R_\theta R_{-\theta} = \begin{bmatrix} 1 & 0 \\ 0 & 1 \end{bmatrix}$, *the identify matrix*
  - *If the transformation R_θ is performed, it can be reversed by performing the transformation $R_{-\theta}$.*

Example 1 (10 minutes)

In this example, students solve a system of equations to find the transformation that reverses the pure dilation matrix with a scale factor of k. This example concludes with students writing their own definition of an inverse matrix and then comparing it to the formal definition. Students should work in small homogenous groups or pairs. Some groups can work through without guided questions while others may need targeted teacher support.

Example

What transformation reverses a pure dilation from the origin with a scale factor of k?

a. Write the pure dilation matrix, and multiply it by $\begin{bmatrix} a & c \\ b & d \end{bmatrix}$.

$\begin{bmatrix} a & c \\ b & d \end{bmatrix} \begin{bmatrix} k & 0 \\ 0 & k \end{bmatrix} = \begin{bmatrix} ak & ck \\ bk & dk \end{bmatrix}$

b. What values of a, b, c, and d would produce the identity matrix? (Hint: Write and solve a system of equations.)

$\begin{bmatrix} ak & ck \\ bk & dk \end{bmatrix} = \begin{bmatrix} 1 & 0 \\ 0 & 1 \end{bmatrix}$

$ak = 1, \quad ck = 0, \quad bk = 0, \quad dk = 1$

$a = \frac{1}{k}, \quad c = 0, \quad b = 0, \quad d = \frac{1}{k}$

c. Write the matrix, and confirm that it reverses the pure dilation with a scale factor of k.

$\begin{bmatrix} \frac{1}{k} & 0 \\ 0 & \frac{1}{k} \end{bmatrix} \begin{bmatrix} k & 0 \\ 0 & k \end{bmatrix} = \begin{bmatrix} 1 & 0 \\ 0 & 1 \end{bmatrix}$

- What is the pure dilation matrix with a scale factor of k?
  - $\begin{bmatrix} k & 0 \\ 0 & k \end{bmatrix}$
- Multiply by a general matrix $\begin{bmatrix} a & c \\ b & d \end{bmatrix}$. What is the resulting matrix?
  - $\begin{bmatrix} ak & ck \\ bk & dk \end{bmatrix}$
- What matrix would this have to be equal to if the transformation had been reversed? Write that matrix.
  - *The identity matrix,* $\begin{bmatrix} 1 & 0 \\ 0 & 1 \end{bmatrix}$
- Equate the two matrices, and write a system of equations that would have to be true for the matrices to be equal.
  - $\begin{bmatrix} ak & ck \\ bk & dk \end{bmatrix} = \begin{bmatrix} 1 & 0 \\ 0 & 1 \end{bmatrix}$
  - $ak = 1, bk = 0, ck = 0,$ *and* $dk = 1$
- Solve this system for a, b, c, and d in terms of k.
  - $a = \frac{1}{k}, b = 0, c = 0,$ *and* $d = \frac{1}{k}$
- Write the matrix that reverses the pure dilation transformation with a scale factor of k.
  - $\begin{bmatrix} \frac{1}{k} & 0 \\ 0 & \frac{1}{k} \end{bmatrix}$
- Confirm that this is the matrix that reverses the transformation. Explain.
  - $\begin{bmatrix} \frac{1}{k} & 0 \\ 0 & \frac{1}{k} \end{bmatrix} \begin{bmatrix} k & 0 \\ 0 & k \end{bmatrix} = \begin{bmatrix} 1 & 0 \\ 0 & 1 \end{bmatrix}$
  - *When the two matrices are multiplied, you get the identity matrix, which means that the transformation has been reversed.*
- If the transformations were done in the reverse order, would they still "undo" each other? Show your work.
  - *Yes* $\begin{bmatrix} k & 0 \\ 0 & k \end{bmatrix} \begin{bmatrix} \frac{1}{k} & 0 \\ 0 & \frac{1}{k} \end{bmatrix} = \begin{bmatrix} 1 & 0 \\ 0 & 1 \end{bmatrix}$
- Let's call original matrix A, the matrix that reverses the transformation B, and the identity matrix I. Write a statement that is true that relates the three matrices.
  - $AB = I, BA = I$
- We call matrix B an *inverse matrix* of matrix A. Write a definition of the inverse matrix.
  - *Matrix B is an inverse matrix to matrix A if $AB = I$ and $BA = I$.*

Exercises 1–3 (10 minutes)

Students find the inverse matrices of each matrix given. Exercises 1 and 2 require students to solve a system of four equations and four variables. This is not as difficult as it may seem, since two of the equations are equal to zero. In Lesson 29, a general formula is developed for the inverse of any matrix; in this exercise, students start seeing patterns relating the inverse matrix and the original matrix.

These problems were all chosen because their determinant is zero, so students can focus on the movement of terms and changing of signs. All students should do Exercises 1 and 2. Early finishers can also do Exercise 3. The results of this exercise are used in the Opening Exercise of Lesson 29, asking if students see a pattern.

Exercises

Find the inverse matrix and verify.

1. $\begin{bmatrix} 1 & 0 \\ 1 & 1 \end{bmatrix}$

$\begin{bmatrix} a & c \\ b & d \end{bmatrix}\begin{bmatrix} 1 & 0 \\ 1 & 1 \end{bmatrix} = \begin{bmatrix} 1 & 0 \\ 0 & 1 \end{bmatrix}$

$\begin{bmatrix} a+c & c \\ b+d & d \end{bmatrix} = \begin{bmatrix} 1 & 0 \\ 0 & 1 \end{bmatrix}$

$\begin{bmatrix} a & c \\ b & d \end{bmatrix} = \begin{bmatrix} 1 & 0 \\ -1 & 1 \end{bmatrix}$

2. $\begin{bmatrix} 3 & 1 \\ 5 & 2 \end{bmatrix}$

$\begin{bmatrix} a & c \\ b & d \end{bmatrix}\begin{bmatrix} 3 & 1 \\ 5 & 2 \end{bmatrix} = \begin{bmatrix} 1 & 0 \\ 0 & 1 \end{bmatrix}$

$\begin{bmatrix} 3a+5c & a+2c \\ 3b+5d & b+2d \end{bmatrix} = \begin{bmatrix} 1 & 0 \\ 0 & 1 \end{bmatrix}$

$\begin{bmatrix} a & c \\ b & d \end{bmatrix} = \begin{bmatrix} 2 & -1 \\ -5 & 3 \end{bmatrix}$

3. $\begin{bmatrix} -2 & -5 \\ 1 & 2 \end{bmatrix}$

$\begin{bmatrix} a & c \\ b & d \end{bmatrix}\begin{bmatrix} -2 & -5 \\ 1 & 2 \end{bmatrix} = \begin{bmatrix} 1 & 0 \\ 0 & 1 \end{bmatrix}$

$\begin{bmatrix} -2a+c & -5a+2c \\ -2b+d & -5b+2d \end{bmatrix} = \begin{bmatrix} 1 & 0 \\ 0 & 1 \end{bmatrix}$

$\begin{bmatrix} a & c \\ b & d \end{bmatrix} = \begin{bmatrix} 2 & 5 \\ -1 & -2 \end{bmatrix}$

Closing (2 minutes)

Students should do a 30-second quick write and then share with the class the answer to the following:

- What is an inverse matrix?
  - *An inverse matrix is a matrix that when multiplied by a given matrix, the product is the identity matrix.*
  - *An inverse matrix "undoes" a transformation.*
- Explain how to find an inverse matrix.
  - *Multiply a general matrix $\begin{bmatrix} a & c \\ b & d \end{bmatrix}$ by a given matrix, and set it equal to the identify matrix. Solve the system of equations for a, b, c, and d.*

Exit Ticket (5 minutes)

Name ______________________________ Date ______________

Lesson 28: When Can We Reverse a Transformation?

Exit Ticket

$$A = \begin{bmatrix} 4 & -2 \\ -1 & 3 \end{bmatrix} \qquad B = \begin{bmatrix} 3 & 2 \\ 1 & 4 \end{bmatrix}$$

1. Is matrix A the inverse of matrix B? Show your work, and explain your answer.

2. What is the determinant of matrix B? Of matrix A?

Exit Ticket Sample Solutions

$$A = \begin{bmatrix} 4 & -2 \\ -1 & 3 \end{bmatrix} \qquad B = \begin{bmatrix} 3 & 2 \\ 1 & 4 \end{bmatrix}$$

1. **Is matrix A the inverse of matrix B? Show your work, and explain your answer.**

   No, the product of the two matrices is not the identity matrix.

   $$\begin{bmatrix} 4 & -2 \\ -1 & 3 \end{bmatrix}\begin{bmatrix} 3 & 2 \\ 1 & 4 \end{bmatrix} = \begin{bmatrix} 10 & 0 \\ 0 & 10 \end{bmatrix}$$

2. **What is the determinant of matrix B? Of matrix A?**

   The determinant of matrix $A = |(4)(3) - (-2)(-1)| = 10$.

   The determinant of matrix $B = |(3)(4) - (2)(1)| = 10$.

Problem Set Sample Solutions

1. **In this lesson, we learned $R_\theta R_{-\theta} = \begin{bmatrix} 1 & 0 \\ 0 & 1 \end{bmatrix}$. Chad was saying that he found an easy way to find the inverse matrix, which is $R_{-\theta} = \dfrac{\begin{bmatrix} 1 & 0 \\ 0 & 1 \end{bmatrix}}{R_\theta}$. His argument is that if we have $2x = 1$, then $x = \frac{1}{2}$.**

   a. **Is Chad correct? Explain your reason.**

   Chad is not correct. Matrices cannot be divided.

   b. **If Chad is not correct, what is the correct way to find the inverse matrix?**

   To find the inverse of $R_{-\theta}$, calculate the determinant, switch the terms on the forward diagonal and change the signs on the back diagonal, and then divide all terms by the absolute value of the determinant.

2. **Find the inverse matrix and verify it.**

   a. $\begin{bmatrix} 3 & 2 \\ 7 & 5 \end{bmatrix}$

   $$\begin{bmatrix} 3 & 2 \\ 7 & 5 \end{bmatrix}\begin{bmatrix} a & c \\ b & d \end{bmatrix} = \begin{bmatrix} 1 & 0 \\ 0 & 1 \end{bmatrix}$$

   $$\begin{bmatrix} 3a+2b & 3c+2d \\ 7a+5b & 7c+5d \end{bmatrix} = \begin{bmatrix} 1 & 0 \\ 0 & 1 \end{bmatrix}, \quad 3a+2b=1, \quad 3c+2d=0, \quad 7a+5b=0, \quad 7c+5d=1$$

   Solve a, b, c, d: $\begin{bmatrix} a & c \\ b & d \end{bmatrix} = \begin{bmatrix} 5 & -2 \\ -7 & 3 \end{bmatrix}$

   Verify: $\begin{bmatrix} 3 & 2 \\ 7 & 5 \end{bmatrix}\begin{bmatrix} 5 & -2 \\ -7 & 3 \end{bmatrix} = \begin{bmatrix} 15-14 & -6+6 \\ 35-35 & -14+15 \end{bmatrix} = \begin{bmatrix} 1 & 0 \\ 0 & 1 \end{bmatrix}$

b. $\begin{bmatrix} -2 & -1 \\ 3 & 1 \end{bmatrix}$

$\begin{bmatrix} -3 & -1 \\ 3 & 1 \end{bmatrix}\begin{bmatrix} a & c \\ b & d \end{bmatrix} = \begin{bmatrix} 1 & 0 \\ 0 & 1 \end{bmatrix}$

$\begin{bmatrix} -3a-b & -3c-d \\ 3a+b & 3c+d \end{bmatrix} = \begin{bmatrix} 1 & 0 \\ 0 & 1 \end{bmatrix}$, $\quad -3a-b=1$, $\quad -3c-d=0$, $\quad 3a+b=0$, $\quad 3c+d=1$

Solve a, b, c, d: $\begin{bmatrix} a & c \\ b & d \end{bmatrix} = \begin{bmatrix} 1 & 1 \\ -3 & -2 \end{bmatrix}$

Verify: $\begin{bmatrix} -2 & -1 \\ 3 & 1 \end{bmatrix}\begin{bmatrix} 1 & 1 \\ -3 & -2 \end{bmatrix} = \begin{bmatrix} -2+3 & -2+2 \\ 3-3 & 3-2 \end{bmatrix} = \begin{bmatrix} 1 & 0 \\ 0 & 1 \end{bmatrix}$

c. $\begin{bmatrix} 3 & -3 \\ -2 & 2 \end{bmatrix}$

The determinant is 0; *therefore, there is no inverse matrix.*

d. $\begin{bmatrix} 0 & 1 \\ -1 & 3 \end{bmatrix}$

$\begin{bmatrix} 0 & 1 \\ -1 & 3 \end{bmatrix}\begin{bmatrix} a & c \\ b & d \end{bmatrix} = \begin{bmatrix} 1 & 0 \\ 0 & 1 \end{bmatrix}$

$\begin{bmatrix} b & d \\ -a+3b & -c+3d \end{bmatrix} = \begin{bmatrix} 1 & 0 \\ 0 & 1 \end{bmatrix}$, $\quad b=1$, $\quad d=0$, $\quad -a+3b=0$, $\quad -c+3d=1$

Solve a, b, c, d: $\begin{bmatrix} a & c \\ b & d \end{bmatrix} = \begin{bmatrix} 3 & -1 \\ 1 & 0 \end{bmatrix}$

Verify: $\begin{bmatrix} 0 & 1 \\ -1 & 3 \end{bmatrix}\begin{bmatrix} 3 & -1 \\ 1 & 0 \end{bmatrix} = \begin{bmatrix} 0+1 & 0 \\ -3+3 & 1+0 \end{bmatrix} = \begin{bmatrix} 1 & 0 \\ 0 & 1 \end{bmatrix}$

e. $\begin{bmatrix} 4 & 1 \\ 2 & 1 \end{bmatrix}$

$\begin{bmatrix} 4 & 1 \\ 2 & 1 \end{bmatrix}\begin{bmatrix} a & c \\ b & d \end{bmatrix} = \begin{bmatrix} 1 & 0 \\ 0 & 1 \end{bmatrix}$

$\begin{bmatrix} 4a+b & 4c+d \\ 2a+b & 2c+d \end{bmatrix} = \begin{bmatrix} 1 & 0 \\ 0 & 1 \end{bmatrix}$, $\quad 4a+b=1$, $\quad \begin{matrix} 4a+b & 4c+d \\ 2a+b & 2c+d \end{matrix} = 0$, $\quad 2a+b=0$, $\quad 2c+d=1$

Solve a, b, c, d: $\begin{bmatrix} a & c \\ b & d \end{bmatrix} = \begin{bmatrix} \frac{1}{2} & -\frac{1}{2} \\ -1 & 2 \end{bmatrix}$

Verify: $\begin{bmatrix} 4 & 1 \\ 2 & 1 \end{bmatrix}\begin{bmatrix} \frac{1}{2} & -\frac{1}{2} \\ -1 & 2 \end{bmatrix} = \begin{bmatrix} 2-1 & -2+2 \\ 1-1 & -1+2 \end{bmatrix} = \begin{bmatrix} 1 & 0 \\ 0 & 1 \end{bmatrix}$

3. Find the starting point $\begin{bmatrix} x \\ y \end{bmatrix}$ if:

a. The point $\begin{bmatrix} 4 \\ 2 \end{bmatrix}$ is the image of a pure dilation with a factor of 2.

$\begin{bmatrix} x \\ y \end{bmatrix} = \begin{bmatrix} \frac{4}{2} \\ \frac{2}{2} \end{bmatrix} = \begin{bmatrix} 2 \\ 1 \end{bmatrix}$

b. The point $\begin{bmatrix}4\\2\end{bmatrix}$ is the image of a pure dilation with a factor of $\frac{1}{2}$.

$$\begin{bmatrix}x\\y\end{bmatrix}=\begin{bmatrix}\frac{4}{\frac{1}{2}}\\ \frac{2}{\frac{1}{2}}\end{bmatrix}=\begin{bmatrix}8\\4\end{bmatrix}$$

c. The point $\begin{bmatrix}-10\\35\end{bmatrix}$ is the image of a pure dilation with a factor of 5.

$$\begin{bmatrix}x\\y\end{bmatrix}=\begin{bmatrix}\frac{-10}{5}\\ \frac{35}{5}\end{bmatrix}=\begin{bmatrix}-2\\7\end{bmatrix}$$

d. The point $\begin{bmatrix}\frac{4}{9}\\ \frac{16}{21}\end{bmatrix}$ is the image of a pure dilation with a factor of $\frac{2}{3}$.

$$\begin{bmatrix}x\\y\end{bmatrix}=\begin{bmatrix}\frac{\frac{4}{9}}{\frac{2}{3}}\\ \frac{\frac{16}{21}}{\frac{2}{3}}\end{bmatrix}=\begin{bmatrix}\frac{2}{3}\\ \frac{8}{7}\end{bmatrix}$$

4. Find the starting point if:

a. $3+2i$ is the image of a reflection about the real axis.

$\bar{z}=3-2i$

b. $3+2i$ is the image of a reflection about the imaginary axis.

$-\bar{z}=-(\overline{3+2i})=-(3-2i)=-3+2i$

c. $3+2i$ is the image of a reflection about the real axis and then the imaginary axis.

$-\bar{\bar{z}}=-(\overline{\overline{3+2i}})=-(\overline{3-2i})=-(3+2i)=-3-2i$

d. $-3-2i$ is the image of a π radians counterclockwise rotation.

$$\begin{bmatrix}x\\y\end{bmatrix}=\frac{3+2i}{i\cdot i}=3+2i$$

5. Let's call the pure counterclockwise rotation of the matrix $\begin{bmatrix}\cos(\theta) & -\sin(\theta)\\ \sin(\theta) & \cos(\theta)\end{bmatrix}$ as R_θ, and the "undo" of the pure rotation is $\begin{bmatrix}\cos(-\theta) & -\sin(-\theta)\\ \sin(-\theta) & \cos(-\theta)\end{bmatrix}$ as $R_{-\theta}$.

a. Simplify $\begin{bmatrix}\cos(-\theta) & -\sin(-\theta)\\ \sin(-\theta) & \cos(-\theta)\end{bmatrix}$.

$$\begin{bmatrix}\cos(-\theta) & -\sin(-\theta)\\ \sin(-\theta) & \cos(-\theta)\end{bmatrix}=\begin{bmatrix}\cos(\theta) & \sin(\theta)\\ -\sin(\theta) & \cos(\theta)\end{bmatrix}$$

b. What would you get if you multiply R_θ to $R_{-\theta}$?

$$R_\theta \times R_{-\theta} = \begin{bmatrix} \cos(\theta) & -\sin(\theta) \\ \sin(\theta) & \cos(\theta) \end{bmatrix} \begin{bmatrix} \cos(\theta) & \sin(\theta) \\ -\sin(\theta) & \cos(\theta) \end{bmatrix}$$
$$= \begin{bmatrix} \cos^2(\theta) + \sin^2(\theta) & \cos(\theta)\cdot\sin(\theta) - \sin(\theta)\cdot\cos(\theta) \\ \sin(\theta)\cdot\cos(\theta) - \cos(\theta)\cdot\sin(\theta) & \sin^2(\theta) + \cos^2(\theta) \end{bmatrix}$$
$$= \begin{bmatrix} 1 & 0 \\ 0 & 1 \end{bmatrix}$$

c. Write the matrix if you want to rotate $\frac{\pi}{2}$ radians counterclockwise.

$$\begin{bmatrix} \cos\left(\frac{\pi}{2}\right) & -\sin\left(\frac{\pi}{2}\right) \\ \sin\left(\frac{\pi}{2}\right) & \cos\left(\frac{\pi}{2}\right) \end{bmatrix} = \begin{bmatrix} 0 & -1 \\ 1 & 0 \end{bmatrix}$$

d. Write the matrix if you want to rotate $\frac{\pi}{2}$ radians clockwise.

$$\begin{bmatrix} \cos\left(-\frac{\pi}{2}\right) & -\sin\left(-\frac{\pi}{2}\right) \\ \sin\left(-\frac{\pi}{2}\right) & \cos\left(-\frac{\pi}{2}\right) \end{bmatrix} = \begin{bmatrix} 0 & 1 \\ -1 & 0 \end{bmatrix}$$

e. Write the matrix if you want to rotate $\frac{\pi}{6}$ radians counterclockwise.

$$\begin{bmatrix} \cos\left(\frac{\pi}{6}\right) & -\sin\left(\frac{\pi}{6}\right) \\ \sin\left(\frac{\pi}{6}\right) & \cos\left(\frac{\pi}{6}\right) \end{bmatrix} = \begin{bmatrix} \frac{\sqrt{3}}{2} & -\frac{1}{2} \\ \frac{1}{2} & \frac{\sqrt{3}}{2} \end{bmatrix}$$

f. Write the matrix if you want to rotate $\frac{\pi}{4}$ radians counterclockwise.

$$\begin{bmatrix} \cos\left(\frac{\pi}{4}\right) & -\sin\left(\frac{\pi}{4}\right) \\ \sin\left(\frac{\pi}{4}\right) & \cos\left(\frac{\pi}{4}\right) \end{bmatrix} = \begin{bmatrix} \frac{\sqrt{2}}{2} & -\frac{\sqrt{2}}{2} \\ \frac{\sqrt{2}}{2} & \frac{\sqrt{2}}{2} \end{bmatrix}$$

g. If the point $\begin{bmatrix} 0 \\ 2 \end{bmatrix}$ is the image of $\frac{\pi}{4}$ radians counterclockwise rotation, find the starting point $\begin{bmatrix} x \\ y \end{bmatrix}$.

$$\begin{bmatrix} x \\ y \end{bmatrix} = \begin{bmatrix} \cos\left(-\frac{\pi}{4}\right) & -\sin\left(-\frac{\pi}{4}\right) \\ \sin\left(-\frac{\pi}{4}\right) & \cos\left(-\frac{\pi}{4}\right) \end{bmatrix} \begin{bmatrix} 0 \\ 2 \end{bmatrix} = \begin{bmatrix} \frac{\sqrt{2}}{2} & \frac{\sqrt{2}}{2} \\ -\frac{\sqrt{2}}{2} & \frac{\sqrt{2}}{2} \end{bmatrix} \begin{bmatrix} 0 \\ 2 \end{bmatrix} = \begin{bmatrix} \sqrt{2} \\ \sqrt{2} \end{bmatrix}$$

h. If the point $\begin{bmatrix} \frac{1}{2} \\ \frac{\sqrt{3}}{2} \end{bmatrix}$ is the image of $\frac{\pi}{6}$ radians counterclockwise rotation, find the starting point $\begin{bmatrix} x \\ y \end{bmatrix}$.

$$\begin{bmatrix} x \\ y \end{bmatrix} = \begin{bmatrix} \cos\left(-\frac{\pi}{6}\right) & -\sin\left(-\frac{\pi}{6}\right) \\ \sin\left(-\frac{\pi}{6}\right) & \cos\left(-\frac{\pi}{6}\right) \end{bmatrix} \begin{bmatrix} \frac{1}{2} \\ \frac{\sqrt{3}}{2} \end{bmatrix} = \begin{bmatrix} \frac{\sqrt{3}}{2} & \frac{1}{2} \\ -\frac{1}{2} & \frac{\sqrt{3}}{2} \end{bmatrix} \begin{bmatrix} \frac{1}{2} \\ \frac{\sqrt{3}}{2} \end{bmatrix} = \begin{bmatrix} \frac{\sqrt{3}}{2} \\ \frac{1}{2} \end{bmatrix}$$

Lesson 29: When Can We Reverse a Transformation?

Student Outcomes

- Students understand that an inverse transformation, when represented by a 2×2 matrix, exists precisely when the determinant of that matrix is nonzero.

Lesson Notes

Lesson 29 is the second of a three-day lesson sequence. In Lesson 28, students were introduced to inverse matrices and asked to find inverses of matrices with a determinant of 1 by solving a system of equations. Lesson 29 has students finding the inverse of any matrix and understanding when a matrix does not have an inverse.

Classwork

The Opening Exercise can be done individually or in pairs. Students use the skills learned in Lesson 28 to find an inverse matrix and then compare that inverse to inverses of other matrices determined in Lesson 28. Students see a pattern. Then, they see that that pattern only works if the determinant is 1. This leads to a general formula for any matrix followed by the question, "Do all matrices have inverses?"

Opening Exercise (5 minutes)

Opening Exercise

Find the inverse of $\begin{bmatrix} -7 & -2 \\ 4 & 1 \end{bmatrix}$. Show your work. Confirm that the matrices are inverses.

$\begin{bmatrix} a & c \\ b & d \end{bmatrix}\begin{bmatrix} -7 & -2 \\ 4 & 1 \end{bmatrix} = \begin{bmatrix} 1 & 0 \\ 0 & 1 \end{bmatrix}$

$-7a + 4c = 1, -2a + c = 0, -7b + 4d = 0, -2b + d = 1$

$a = 1,\ b = -4,\ c = 2, \text{and } d = -7$

$\begin{bmatrix} 1 & 2 \\ -4 & -7 \end{bmatrix}\begin{bmatrix} -7 & -2 \\ 4 & 1 \end{bmatrix} = \begin{bmatrix} 1 & 0 \\ 0 & 1 \end{bmatrix}$

Exploratory Challenge (10 minutes)

In this Exploratory Challenge, students look at the patterns of matrices and their inverses that they found in Exercises 1–3 of Lesson 28 and the Opening Exercise of Lesson 29. This leads to the discovery of the general formula for the inverse of any matrix. Students should work in small groups.

MP.3

- Do you think all matrices have inverses? Explain why or why not.
  - *Answers will vary. Allow students to state their opinions and explain. Do not add to the discussion; students discover the correct answer in this Exploratory Challenge.*

Post or project the following:

Matrix	Inverse
$\begin{bmatrix} 1 & 0 \\ 1 & 1 \end{bmatrix}$	$\begin{bmatrix} 1 & 0 \\ -1 & 1 \end{bmatrix}$
$\begin{bmatrix} 3 & 1 \\ 5 & 2 \end{bmatrix}$	$\begin{bmatrix} 2 & -1 \\ -5 & 3 \end{bmatrix}$
$\begin{bmatrix} -2 & -5 \\ 1 & 2 \end{bmatrix}$	$\begin{bmatrix} 2 & 5 \\ -1 & -2 \end{bmatrix}$
$\begin{bmatrix} -7 & -2 \\ 4 & 1 \end{bmatrix}$	$\begin{bmatrix} 1 & 2 \\ -4 & -7 \end{bmatrix}$

- In Lesson 28, you found the inverses of the first three matrices, and in the Opening Exercise, you found the inverse of the last matrix. Do you see any patterns between the original matrix and its inverse?
  - *The numbers in the top left and bottom right corners seem to change places.*
  - *The numbers in the top right and bottom left corners change signs.*
- Do you think this is true for the inverse of all matrices?

Answers will vary, but most students will think that yes, this is true.

- Let's see if we are right. Find the inverse of the matrix in Exercise 1 using the pattern we discovered, and confirm that it is indeed the inverse.

Scaffolding:

- Some student pairs may need targeted one-to-one guidance on this challenge. Consider pairing groups and having a larger group that is teacher led.
- Give advanced students a single task: "Write a formula for an inverse matrix after studying the patterns, and verify your formula." Ask them to develop an answer without the questions shown.

Exercise 1 (3 minutes)

Exercises

1. **Find the inverse of $\begin{bmatrix} 5 & 3 \\ 2 & 4 \end{bmatrix}$. Confirm your answer.**

   $$\begin{bmatrix} 4 & -3 \\ -2 & 5 \end{bmatrix}\begin{bmatrix} 5 & 3 \\ 2 & 4 \end{bmatrix} = \begin{bmatrix} 14 & 0 \\ 0 & 14 \end{bmatrix}$$

- Was the matrix that you found using the pattern the inverse? What was missing?
  - *No. Where we needed* 1*'s, we had* 14*'s.*
- Let's look at this a little further. Look at the matrices in the table. Find the determinant of the matrices. (Assign different groups/pairs different matrices from above.)
  - *All of the determinants were* 1.
- Do you think that makes a difference? What was the determinant of the matrix in Exercise 1?
  - *The determinant was* 14.

- How does that compare to the matrix that resulted from multiplying the matrices in Exercise 1?
  - *That was the number that was in the position that should have been a* 1.
- How do you think this ties into the way we find an inverse matrix?
  - *We can still use our pattern, but we need to divide each term by the determinant. (Answers may vary, but let students try out their hypotheses to come up with the right answer.)*
- Try it on the inverse matrix in Exercise 1. Write the inverse matrix.
  - $\begin{bmatrix} \frac{4}{14} & -\frac{3}{14} \\ -\frac{2}{14} & \frac{5}{14} \end{bmatrix}$
- Verify that is the inverse. Were you correct?
  - $\begin{bmatrix} \frac{4}{14} & -\frac{3}{14} \\ -\frac{2}{14} & \frac{5}{14} \end{bmatrix}\begin{bmatrix} 5 & 3 \\ 2 & 4 \end{bmatrix} = \begin{bmatrix} 1 & 0 \\ 0 & 1 \end{bmatrix}$
  - *Yes. We get the identify matrix again.*
- Explain to your neighbor how to find the inverse of a matrix.
  - *Switch the numbers in the top left and bottom right. Change the signs of the numbers in the top right and bottom left. Divide all of the terms by the determinant of the original matrix.*

Exercises 2–4 (10 minutes)

In Exercises 2 and 3, students practice finding an inverse matrix and confirm their results. In Exercise 4, students find the inverse matrix of a general matrix. Choose exercises based on the needs of students; Exercises 2 and 3 are simpler while Exercise 4 is more complicated. Students should complete this exercise in small groups and then present their findings to the class.

Find the inverse matrix and verify.

2. $\begin{bmatrix} 3 & -3 \\ 1 & 4 \end{bmatrix}$

Determinant $= \mathbf{(3)(4) - (-3)(1) = 12 + 3 = 15}$

$\begin{bmatrix} \frac{4}{15} & \frac{3}{15} \\ \frac{-1}{15} & \frac{3}{15} \end{bmatrix}\begin{bmatrix} 3 & -3 \\ 1 & 4 \end{bmatrix} = \begin{bmatrix} 1 & 0 \\ 0 & 1 \end{bmatrix}$

3. $\begin{bmatrix} 5 & -2 \\ 4 & -3 \end{bmatrix}$

Determinant $= \mathbf{(5)(-3) - (-2)(4) = -15 + 8 = -7}$

$\begin{bmatrix} \frac{3}{7} & -\frac{2}{7} \\ \frac{4}{7} & -\frac{5}{7} \end{bmatrix}\begin{bmatrix} 5 & -2 \\ 4 & -3 \end{bmatrix} = \begin{bmatrix} 1 & 0 \\ 0 & 1 \end{bmatrix}$

4. $\begin{bmatrix} a & c \\ b & d \end{bmatrix}$

Determinant $= (a)(d) - (c)(b) = ad - cb$

$$\begin{bmatrix} \frac{d}{ad-cb} & -\frac{c}{ad-cb} \\ -\frac{b}{ad-cb} & \frac{a}{ad-cb} \end{bmatrix} \begin{bmatrix} a & c \\ b & d \end{bmatrix} = \begin{bmatrix} 1 & 0 \\ 0 & 1 \end{bmatrix}$$

Example 1 (10 minutes)

In this example, students calculate the determinant of a matrix and find that it is 0; they then try to find the inverse of the matrix. They discover that there is no inverse and then explore what that means about the resulting image. Students conclude that matrices with a determinant of 0 do not have inverses. Students need graph paper.

Example 1

Find the determinant of $\begin{bmatrix} 1 & 2 \\ 2 & 4 \end{bmatrix}$.

The determinant is 0.

- Now that we have calculated the determinant and found it to be 0, let's examine the inverse of $\begin{bmatrix} 1 & 2 \\ 2 & 4 \end{bmatrix}$.

Students may struggle, but they should see that they cannot divide by 0, so there will be an issue finding the inverse.

- Let's try to solve for the inverse with a system of equations.
  - $\begin{bmatrix} a & c \\ b & d \end{bmatrix}\begin{bmatrix} 1 & 2 \\ 2 & 4 \end{bmatrix} = \begin{bmatrix} a+2c & 2a+4c \\ b+2d & 2b+4d \end{bmatrix} = \begin{bmatrix} 1 & 0 \\ 0 & 1 \end{bmatrix}$
  - $a + 2c = 1,\ 2a + 4c = 0,\ b + 2d = 0,\ 2b + 4d = 1$
- What did you discover?
  - *We get a system of equations with no solutions.*
- What do you think this means about the inverse of this matrix?
  - *This matrix does not have an inverse.*
- Let's explore this further using what we know about matrix transformations of the unit square.
- Perform the operation $\begin{bmatrix} 1 & 2 \\ 2 & 4 \end{bmatrix}$ on the unit square. What are the coordinates of the vertices of the unit square on the image?
  - $\begin{bmatrix} 1 & 2 \\ 2 & 4 \end{bmatrix}\begin{bmatrix} 0 \\ 0 \end{bmatrix} = \begin{bmatrix} 0 \\ 0 \end{bmatrix}$
  - $\begin{bmatrix} 1 & 2 \\ 2 & 4 \end{bmatrix}\begin{bmatrix} 1 \\ 0 \end{bmatrix} = \begin{bmatrix} 1 \\ 2 \end{bmatrix}$
  - $\begin{bmatrix} 1 & 2 \\ 2 & 4 \end{bmatrix}\begin{bmatrix} 0 \\ 1 \end{bmatrix} = \begin{bmatrix} 2 \\ 4 \end{bmatrix}$
  - $\begin{bmatrix} 1 & 2 \\ 2 & 4 \end{bmatrix}\begin{bmatrix} 1 \\ 1 \end{bmatrix} = \begin{bmatrix} 3 \\ 6 \end{bmatrix}$

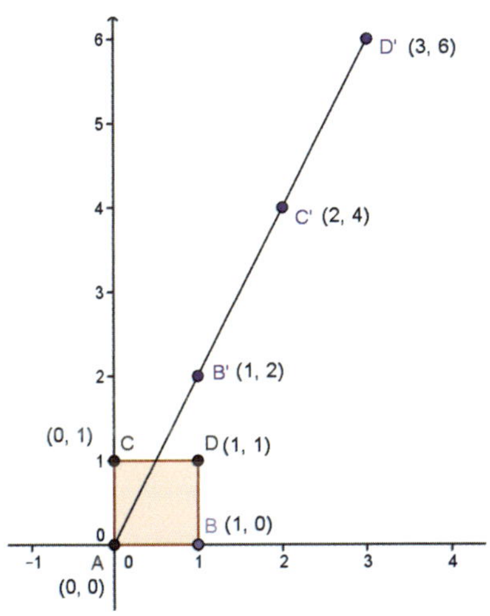

- Plot the unit square and the transformation. What do you notice?
  - *The image is a line.*
- What is the area of the image?
  - *The image is a line, not a parallelogram, so the area is* 0.
- What does the determinant of the transformation represent?
  - *It represents the area of the image of the unit square after the transformation.*
- Is the area confirmed?
  - *Yes. The determinant is* 0*, so the area of the transformation is* 0.
- The points $(1,0)$ and $\left(0,\frac{1}{2}\right)$ are both on the unit square. Perform this transformation on each of these points.
  - $\begin{bmatrix}1 & 2\\2 & 4\end{bmatrix}\begin{bmatrix}1\\0\end{bmatrix}=\begin{bmatrix}1\\2\end{bmatrix}$
  - $\begin{bmatrix}1 & 2\\2 & 4\end{bmatrix}\begin{bmatrix}0\\\frac{1}{2}\end{bmatrix}=\begin{bmatrix}1\\2\end{bmatrix}$
- What does this mean?
  - *Both points map to the same location.*
- When the unit square "collapses" to a straight line under a transformation, we will always have more than one point mapping to the same location. This means that we cannot "undo" this transformation, because there is no clear way to reverse the transformation. Would $(1,2)$ map back to $(1,0)$ or $\left(0,\frac{1}{2}\right)$? We are not sure.
- When does a matrix not have an inverse?
  - *When the image of the unit square "collapses" to a figure of* 0 *area, we have distinct points mapping to the same location, so there is no inverse.*
  - *When the determinant of the matrix is* 0

Closing (2 minutes)

Students should do a 30-second quick write and then share with the class the answer to the following:

- What is an inverse matrix?
  - *An inverse matrix is a matrix that when multiplied by a given matrix, the product is the identity matrix.*
  - *An inverse matrix "undoes" a transformation.*
- Explain how to find an inverse matrix.
  - *Multiply a general matrix* $\begin{bmatrix}a & c\\b & d\end{bmatrix}$ *by a given matrix, and set it equal to the identify matrix. Solve the system of equations for* a, b, c, *and* d.

Exit Ticket (5 minutes)

Name ______________________________ Date ____________

Lesson 29: When Can We Reverse a Transformation?

Exit Ticket

$$A = \begin{bmatrix} 4 & -2 \\ -1 & 3 \end{bmatrix}$$

1. Find the inverse of A. Show your work, and confirm your answer.

2. Explain why the matrix $\begin{bmatrix} 6 & 3 \\ 4 & 2 \end{bmatrix}$ has no inverse.

Exit Ticket Sample Solutions

$$A = \begin{bmatrix} 4 & -2 \\ -1 & 3 \end{bmatrix}$$

1. Find the inverse of A. Show your work, and confirm your answer.

   Determinant $= (4)(3) - (-2)(-1) = 12 - 2 = 10$

   $$\begin{bmatrix} \frac{3}{10} & \frac{2}{10} \\ \frac{1}{10} & \frac{4}{10} \end{bmatrix} \begin{bmatrix} 4 & -2 \\ -1 & 3 \end{bmatrix} = \begin{bmatrix} 1 & 0 \\ 0 & 1 \end{bmatrix}$$

2. Explain why the matrix $\begin{bmatrix} 6 & 3 \\ 4 & 2 \end{bmatrix}$ has no inverse.

   Determinant $= (6)(2) - (3)(4) = 0$

   This means the area of the image is 0 *because the image of the unit square maps to a straight line, which has no area. This also means that distinct points map to the same location, so the transformation cannot be reversed.*

Problem Set Sample Solutions

Find the inverse matrix of the following.

a. $\begin{bmatrix} 1 & 0 \\ 0 & 1 \end{bmatrix}$

   Determinant $= 1 - 0 = 1$ *Inverse matrix:* $\begin{bmatrix} 1 & 0 \\ 0 & 1 \end{bmatrix}$

   Verify:
   $\begin{bmatrix} 1 & 0 \\ 0 & 1 \end{bmatrix} \begin{bmatrix} 1 & 0 \\ 0 & 1 \end{bmatrix} = \begin{bmatrix} 1 & 0 \\ 0 & 1 \end{bmatrix}$

b. $\begin{bmatrix} 0 & 1 \\ 1 & 0 \end{bmatrix}$

   Determinant $= 0 - 1 = -1$ *Inverse matrix:* $\begin{bmatrix} 0 & 1 \\ 1 & 0 \end{bmatrix}$

   Verify:
   $\begin{bmatrix} 0 & 1 \\ 1 & 0 \end{bmatrix} \begin{bmatrix} 0 & 1 \\ 1 & 0 \end{bmatrix} = \begin{bmatrix} 1 & 0 \\ 0 & 1 \end{bmatrix}$

c. $\begin{bmatrix} 1 & 1 \\ 1 & 1 \end{bmatrix}$

   Determinant $= 1 - 1 = 0$ *No inverse matrix*

d. $\begin{bmatrix} 1 & 0 \\ 1 & 0 \end{bmatrix}$

   Determinant $= 0$ *No inverse matrix*

e. $\begin{bmatrix} 0 & 1 \\ 0 & 1 \end{bmatrix}$

   Determinant $= 0$ *No inverse matrix*

f. $\begin{bmatrix} -2 & 2 \\ -5 & 4 \end{bmatrix}$

Determinant $= -8 + 10 = 2$ *Inverse matrix:* $\begin{bmatrix} 2 & -1 \\ \frac{5}{2} & -1 \end{bmatrix}$

Verify:

$\begin{bmatrix} -2 & 2 \\ -5 & 4 \end{bmatrix}\begin{bmatrix} 2 & -1 \\ \frac{5}{2} & -1 \end{bmatrix} = \begin{bmatrix} 1 & 0 \\ 0 & 1 \end{bmatrix}$

g. $\begin{bmatrix} 4 & 6 \\ 5 & 8 \end{bmatrix}$

Determinant $= 32 - 30 = 2$ *Inverse matrix:* $\begin{bmatrix} 4 & -3 \\ \frac{5}{2} & 2 \end{bmatrix}$

Verify:

$\begin{bmatrix} 4 & 6 \\ 5 & 8 \end{bmatrix}\begin{bmatrix} 4 & -3 \\ \frac{5}{2} & 2 \end{bmatrix} = \begin{bmatrix} 1 & 0 \\ 0 & 1 \end{bmatrix}$

h. $\begin{bmatrix} 6 & -9 \\ 5 & -7 \end{bmatrix}$

Determinant $= -42 + 45 = 3$ *Inverse matrix:* $\begin{bmatrix} -\frac{7}{3} & 3 \\ -\frac{5}{3} & 2 \end{bmatrix}$

Verify:

$\begin{bmatrix} 6 & -9 \\ 5 & -7 \end{bmatrix}\begin{bmatrix} -\frac{7}{3} & 3 \\ -\frac{5}{3} & 2 \end{bmatrix} = \begin{bmatrix} 1 & 0 \\ 0 & 1 \end{bmatrix}$

i. $\begin{bmatrix} \frac{1}{2} & -\frac{2}{3} \\ -6 & 4 \end{bmatrix}$

Determinant $= 2 - 4 = -2$ *Inverse matrix:* $\begin{bmatrix} -2 & -\frac{1}{3} \\ -3 & \frac{1}{4} \end{bmatrix}$

Verify:

$\begin{bmatrix} \frac{1}{2} & -\frac{2}{3} \\ -6 & 4 \end{bmatrix}\begin{bmatrix} -2 & -\frac{1}{3} \\ -3 & \frac{1}{4} \end{bmatrix} = \begin{bmatrix} 1 & 0 \\ 0 & 1 \end{bmatrix}$

j. $\begin{bmatrix} 0.8 & 0.4 \\ -0.75 & -0.5 \end{bmatrix}$

Determinant $= -0.4 + 0.3 = -0.1$ *Inverse matrix:* $\begin{bmatrix} 5 & -4 \\ -7.5 & -8 \end{bmatrix}$

Verify:

$\begin{bmatrix} 0.8 & 0.4 \\ -0.75 & -0.5 \end{bmatrix}\begin{bmatrix} 5 & -4 \\ -7.5 & -8 \end{bmatrix} = \begin{bmatrix} 1 & 0 \\ 0 & 1 \end{bmatrix}$

Lesson 30: When Can We Reverse a Transformation?

Student Outcomes

- Students understand that an inverse transformation, when represented by a 2×2 matrix, exists precisely when the determinant of that matrix is nonzero.

Lesson Notes

Lesson 30 is the last of a three-day lesson sequence and the last lesson of Module 1. In Lessons 28 and 29, students studied inverse matrices and found that some matrices do not have inverses. Lesson 30 allows students to practice these concepts while revisiting rotations and dilations.

Classwork

The Opening Exercise serves as a review of concepts studied in the second half of Module 1. Conduct this exercise as a Rapid White Board Exchange, using it as a way to informally assess students. This allows teachers to assign homogeneous groups for the lesson.

Teachers should show one problem at a time either by projecting them or writing them on a personal white board. Allow students time to write answers on their personal white boards, and then signal students when to show their answers. Simple mistakes can be explained immediately. Students struggling can be assigned to groups that get more teacher attention during the lesson.

Opening Exercise (13 minutes)

Give students five minutes to complete the Opening Exercise individually, and then group students to compare answers. Have groups present their answers to each question and discuss.

MP.2

Opening Exercise

a. **What is the geometric effect of the following matrices?**

i. $\begin{bmatrix} k & 0 \\ 0 & k \end{bmatrix}$

A pure dilation with a scale factor of k

ii. $\begin{bmatrix} a & -b \\ b & a \end{bmatrix}$

A pure rotation

iii. $\begin{bmatrix} \cos(\theta) & -\sin(\theta) \\ \sin(\theta) & \cos(\theta) \end{bmatrix}$

A pure rotation $\theta°$ counterclockwise

MP.3

b. Jadavis says that the identity matrix is $\begin{bmatrix} 0 & 0 \\ 0 & 0 \end{bmatrix}$. Sophie disagrees and states that the identity matrix is $\begin{bmatrix} 1 & 0 \\ 0 & 1 \end{bmatrix}$.

i. Their teacher, Mr. Kuzy, says they are both correct and asks them to explain their thinking about matrices to each other but to also use a similar example in the real number system. Can you state each of their arguments?

Jadavis says that any matrix added to the matrix $\begin{bmatrix} 0 & 0 \\ 0 & 0 \end{bmatrix}$ does not change. This matrix is similar to 0 in the real number system, so it is the additive identity matrix.

Sophie explains that when a matrix is multiplied by $\begin{bmatrix} 1 & 0 \\ 0 & 1 \end{bmatrix}$, the matrix does not change, just like the number 1 in the real number system; so, this matrix is the multiplicative identity.

MP.2

ii. Mr. Kuzy then asks each of them to explain the geometric effect that their matrix would have on the unit square.

$\begin{bmatrix} 0 & 0 \\ 0 & 0 \end{bmatrix}$ *would collapse the entire square to $(0, 0)$.*

$\begin{bmatrix} 1 & 0 \\ 0 & 1 \end{bmatrix}$ *would have no effect on the unit square.*

c. Given the matrices below, answer the following:

$A = \begin{bmatrix} 2 & 3 \\ 1 & 4 \end{bmatrix}$ $B = \begin{bmatrix} 5 & 2 \\ 10 & 4 \end{bmatrix}$

i. Which matrix does not have an inverse? Explain algebraically and geometrically how you know.

Matrix B does not have an inverse. The determinant is 0, which means it would transform the unit square to a straight line with no area.

ii. If a matrix has an inverse, find it.

$$A^{-1} = \begin{bmatrix} \frac{4}{5} & -\frac{3}{5} \\ -\frac{1}{5} & \frac{2}{5} \end{bmatrix}$$

Scaffolding:

- Ask advanced learners, in place of Example 1, to determine and explain the meaning of the matrix that is the inverse of $\begin{bmatrix} \frac{1}{2} & -\frac{\sqrt{3}}{2} \\ \frac{\sqrt{3}}{2} & \frac{1}{2} \end{bmatrix}$
- After assessing students in the Opening Exercise, choose a small group for targeted instruction. This is an opportunity to solidify the concepts of this module with all students.

Example 1 (5 minutes)

Example 1 has students determine the transformation performed on the unit circle, determine the area of the image, determine the exact transformation, and then find the inverse matrix. This problem should be modeled with the entire class. Students need graph paper.

Example

Given $\begin{bmatrix} \frac{1}{2} & -\frac{\sqrt{3}}{2} \\ \frac{\sqrt{3}}{2} & \frac{1}{2} \end{bmatrix}$

a. Perform this transformation on the unit square, and sketch the results on graph paper. Label the vertices.

See diagram at right.

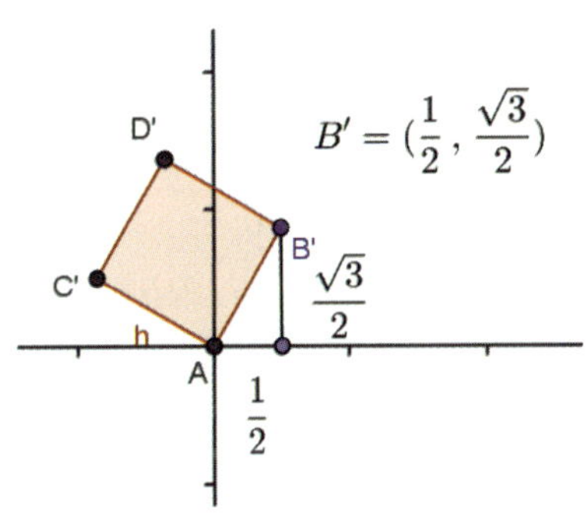

b. **Explain the transformation that occurred to the unit square.**

The transformation is a $60°$ *rotation counterclockwise about the origin.*

c. **Find the area of the image.**

$$\left|\left(\frac{1}{2}\right)\left(\frac{1}{2}\right)-\left(-\frac{\sqrt{3}}{2}\right)\left(\frac{\sqrt{3}}{2}\right)\right|=\frac{1}{4}+\frac{3}{4}=1$$

The area is 1 *units*^{2}.

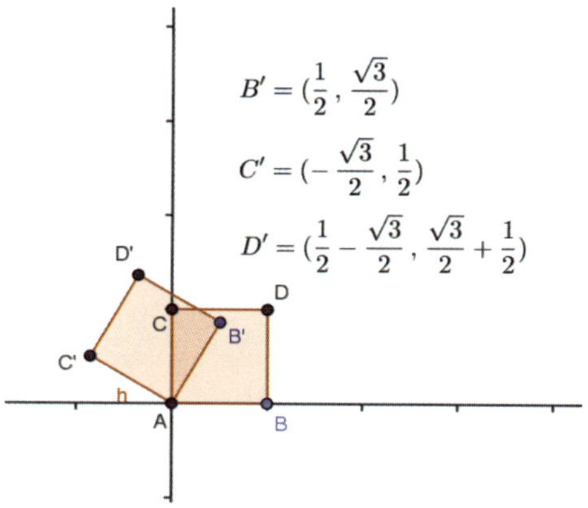

d. **Find the inverse of this transformation.**

$$\begin{bmatrix}\frac{1}{2} & \frac{\sqrt{3}}{2}\\ -\frac{\sqrt{3}}{2} & \frac{1}{2}\end{bmatrix}$$

MP.2

e. **Explain the meaning of the inverse transformation on the unit square.**

The inverse is a $-60°$ *rotation counterclockwise about the origin.*

- What are the vertices of the transformed image?
  - $(0,0)$, $\left(\frac{1}{2},\frac{\sqrt{3}}{2}\right)$, $\left(-\frac{\sqrt{3}}{2},\frac{1}{2}\right)$, *and* $\left(\frac{1}{2}-\frac{\sqrt{3}}{2},\frac{\sqrt{3}}{2}+\frac{1}{2}\right)$
- Look at the special triangle formed connecting the origin, the vertex of the image $\left(\frac{1}{2},\frac{\sqrt{3}}{2}\right)$, and the x-axis. What type of transformation occurred? Explain.
  - *This is a* $30°$-$60°$-$90°$ *triangle. The unit square was rotated* $60°$ *counterclockwise about the origin.*
- What is the area of the image? Explain.
  - *Only a rotation occurred, so the area did not change. The area is* 1 *square unit. This is confirmed because the determinant is* 1.
- What is the inverse matrix?
  - $\begin{bmatrix}\frac{1}{2} & \frac{\sqrt{3}}{2}\\ -\frac{\sqrt{3}}{2} & \frac{1}{2}\end{bmatrix}$
- Explain this transformation.
  - *If the original transformation was a* $60°$ *counterclockwise rotation about the origin, the inverse is a* $-60°$ *rotation, or a* $60°$ *clockwise rotation about the origin.*

Exercises 1–8 (20 minutes)

Students should work on these exercises in pairs or small groups. All students should complete Exercises 1 and 2. The other exercises can be assigned to specific groups or all groups. If problems are assigned to specific groups, have groups work their assigned problems and then post solutions and either present them to the class or have a gallery walk. Exercises 3 and 4 are similar, but Exercise 4 is more challenging. Exercises 5, 6, and 7 are also similar, with 6 and 7 being more challenging. Exercise 8 can be assigned to stronger students. Students need graph paper.

Exercises

1. Given $\begin{bmatrix} \frac{1}{2} & 0 \\ 0 & \frac{1}{2} \end{bmatrix}$

   a. Perform this transformation on the unit square, and sketch the results on graph paper. Label the vertices.

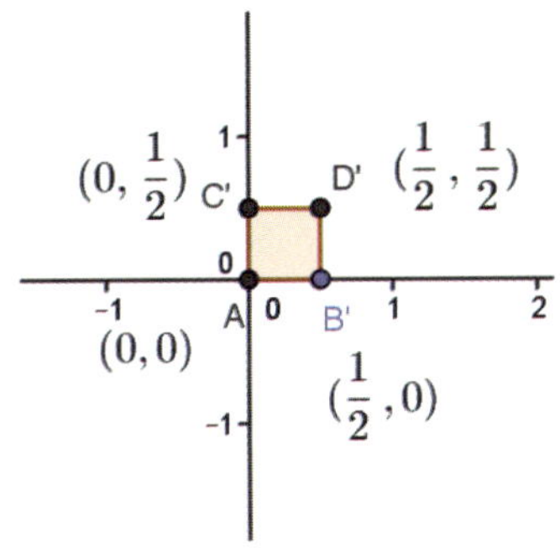

   b. Explain the transformation that occurred to the unit square.

   The transformation is a dilation with a scale factor of $\frac{1}{2}$.

   c. Find the area of the image.

   $\left|\left(\frac{1}{2}\right)\left(\frac{1}{2}\right) - (0)(0)\right| = \frac{1}{4}$

   The area is $\frac{1}{4}$ *units*^{2}.

   d. Find the inverse of this transformation.

   $\begin{bmatrix} 2 & 0 \\ 0 & 2 \end{bmatrix}$

   e. Explain the meaning of the inverse transformation on the unit square.

   The inverse is a dilation with a scale factor of 2.

   f. If any matrix produces a dilation with a scale factor of k, what would the inverse matrix produce?

   It would produce a dilation with a scale factor of $\frac{1}{k}$.

2. Given $\begin{bmatrix} \frac{1}{\sqrt{2}} & -\frac{1}{\sqrt{2}} \\ \frac{1}{\sqrt{2}} & \frac{1}{\sqrt{2}} \end{bmatrix}$

   a. Perform this transformation on the unit square, and sketch the results on graph paper. Label the vertices.

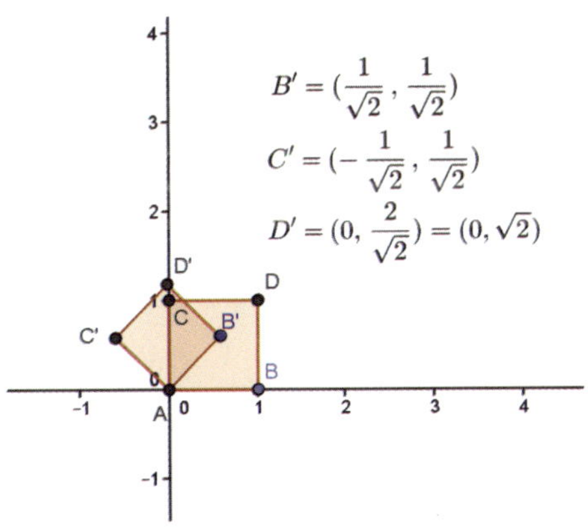

   b. Explain the transformation that occurred to the unit square.

   The transformation is a $45°$ *rotation counterclockwise about the origin.*

   c. Find the area of the image.

   $\left|\left(\frac{1}{\sqrt{2}}\right)\left(\frac{1}{\sqrt{2}}\right) - \left(-\frac{1}{\sqrt{2}}\right)\left(\frac{1}{\sqrt{2}}\right)\right| = \frac{1}{2} + \frac{1}{2} = 1$

   The area is 1 *units*^{2}.

d. **Find the inverse of the transformation.**

$$\begin{bmatrix} \frac{1}{\sqrt{2}} & \frac{1}{\sqrt{2}} \\ -\frac{1}{\sqrt{2}} & \frac{1}{\sqrt{2}} \end{bmatrix}$$

e. **Explain the meaning of the inverse transformation on the unit square.**

The inverse is a $-45°$ *rotation counterclockwise about the origin.*

f. **Rewrite the original matrix if it also included a dilation with a scale factor of** 2.

$$\begin{bmatrix} \frac{2}{\sqrt{2}} & -\frac{2}{\sqrt{2}} \\ \frac{2}{\sqrt{2}} & \frac{2}{\sqrt{2}} \end{bmatrix} \text{ or } \begin{bmatrix} \sqrt{2} & -\sqrt{2} \\ \sqrt{2} & \sqrt{2} \end{bmatrix}$$

g. **What is the inverse of this matrix?**

$$\begin{bmatrix} \frac{1}{2\sqrt{2}} & \frac{1}{2\sqrt{2}} \\ -\frac{1}{2\sqrt{2}} & \frac{1}{2\sqrt{2}} \end{bmatrix} \text{ or } \begin{bmatrix} \frac{\sqrt{2}}{4} & \frac{\sqrt{2}}{4} \\ -\frac{\sqrt{2}}{4} & \frac{\sqrt{2}}{4} \end{bmatrix}$$

3. **Find a transformation that would create a** $90°$ **counterclockwise rotation about the origin. Set up a system of equations, and solve to find the matrix.**

$$\begin{bmatrix} a & c \\ b & d \end{bmatrix}\begin{bmatrix} 1 \\ 0 \end{bmatrix} = \begin{bmatrix} 0 \\ 1 \end{bmatrix}, \begin{bmatrix} a & c \\ b & d \end{bmatrix}\begin{bmatrix} 0 \\ 1 \end{bmatrix} = \begin{bmatrix} -1 \\ 0 \end{bmatrix}$$

$$\begin{bmatrix} 0 & -1 \\ 1 & 0 \end{bmatrix}$$

4.

a. **Find a transformation that would create a** $180°$ **counterclockwise rotation about the origin. Set up a system of equations, and solve to find the matrix.**

$$\begin{bmatrix} a & c \\ b & d \end{bmatrix}\begin{bmatrix} 1 \\ 0 \end{bmatrix} = \begin{bmatrix} -1 \\ 0 \end{bmatrix}, \begin{bmatrix} a & c \\ b & d \end{bmatrix}\begin{bmatrix} 0 \\ 1 \end{bmatrix} = \begin{bmatrix} 0 \\ -1 \end{bmatrix}$$

$$\begin{bmatrix} -1 & 0 \\ 0 & -1 \end{bmatrix}$$

b. **Rewrite the matrix to also include a dilation with a scale factor of** 5.

$$\begin{bmatrix} -5 & 0 \\ 0 & -5 \end{bmatrix}$$

5. **For which values of** a **does** $\begin{bmatrix} 3 & -100 \\ 900 & a \end{bmatrix}$ **have an inverse matrix?**

$$3a - (-100)(900) \neq 0$$

$$3a + 90000 \neq 0$$

$$a \neq -30000$$

EUREKA MATH

6. **For which values of a does $\begin{bmatrix} a & a+4 \\ 2 & a \end{bmatrix}$ have an inverse matrix?**

$$(a)(a)-(a+4)(2)\neq 0$$
$$a^2-2a-8\neq 0$$
$$(a-4)(a+2)\neq 0$$
$$a\neq 4, a\neq -2$$

7. **For which values of a does $\begin{bmatrix} a+2 & a-4 \\ a-3 & a+3 \end{bmatrix}$ have an inverse matrix?**

$$(a+2)(a+3)-(a-4)(a-3)\neq 0$$
$$(a^2+5a+6)-(a^2-7a+12)\neq 0$$
$$12a-6\neq 0$$
$$a\neq \frac{1}{2}$$

8. **Chethan says that the matrix $\begin{bmatrix} \cos(\theta°) & -\sin(\theta°) \\ \sin(\theta°) & \cos(\theta°) \end{bmatrix}$ produces a rotation $\theta°$ counterclockwise. He justifies his work by showing that when $\theta = 60$, the rotation matrix is $\begin{bmatrix} \cos(60°) & -\sin(60°) \\ \sin(60°) & \cos(60°) \end{bmatrix} = \begin{bmatrix} \frac{1}{2} & -\frac{\sqrt{3}}{2} \\ \frac{\sqrt{3}}{2} & \frac{1}{2} \end{bmatrix}$. Shayla disagrees and says that the matrix $\begin{bmatrix} 1 & -\sqrt{3} \\ \sqrt{3} & 1 \end{bmatrix}$ produces a $60°$ rotation counterclockwise. Tyler says that he has found that the matrix $\begin{bmatrix} 2 & -2\sqrt{3} \\ 2\sqrt{3} & 2 \end{bmatrix}$ produces a $60°$ rotation counterclockwise, too.**

a. **Who is correct? Explain.**

They are all correct. All of the matrices produce a $60°$ rotation counterclockwise, but each has a different scale factor.

b. **Which matrix has the largest scale factor? Explain.**

$\begin{bmatrix} 2 & -2\sqrt{3} \\ 2\sqrt{3} & 2 \end{bmatrix}$ *has the largest scale factor of 4. The first matrix has a scale factor of 1, and the second matrix a scale factor of 2.*

c. **Create a matrix with a scale factor less than 1 that would produce the same rotation.**

Answers will vary. $\begin{bmatrix} \frac{1}{4} & -\frac{\sqrt{3}}{4} \\ \frac{\sqrt{3}}{4} & \frac{1}{4} \end{bmatrix}$ *would have a scale factor of $\frac{1}{2}$.*

Closing (2 minutes)

Have a whole-class discussion using the following questions.

- What effect does performing the transformation $\begin{bmatrix} 1 & 0 \\ 0 & 1 \end{bmatrix}$ have on the unit square?
  - *No effect—it is the multiplicative identity matrix.*
- What effect does performing the transformation $\begin{bmatrix} 4 & 0 \\ 0 & 4 \end{bmatrix}$ have on the unit square?
  - *It is a dilation with a scale factor of* 4.
- What effect does performing the transformation $\begin{bmatrix} a & -b \\ b & a \end{bmatrix}$ have on the unit square?
  - *It rotates the unit square in a counterclockwise direction about the origin.*
- What effect does performing the transformation $\begin{bmatrix} 2 & 4 \\ 4 & 8 \end{bmatrix}$ have on the unit square?
  - *The unit square collapses onto a line because the determinant is* 0.

Exit Ticket (5 minutes)

Name ______________________________ Date ______________

Lesson 30: When Can We Reverse a Transformation?

Exit Ticket

A and B are 2×2 matrices. I is the 2×2 multiplicative identity matrix.

1. If $AB = A$, name the matrix represented by B.

2. If $A + B = A$, name the matrix represented by B.

3. If $AB = I$, name the matrix represented by B.

4. Do the matrices have inverses? Justify your answer.

   a. $\begin{bmatrix} -2 & 6 \\ -3 & 9 \end{bmatrix}$

   b. $\begin{bmatrix} -2 & 6 \\ 3 & 9 \end{bmatrix}$

5. Find a value of a, such that the given matrix has an inverse.

   a. $\begin{bmatrix} -4 & 3a \\ 2 & 9 \end{bmatrix}$

   b. $\begin{bmatrix} 5 & a \\ -a & 5 \end{bmatrix}$

Exit Ticket Sample Solutions

A and B are 2×2 matrices. I is the 2×2 multiplicative identity matrix.

1. If $AB = A$, name the matrix represented by B.

   $B = \begin{bmatrix} 1 & 0 \\ 0 & 1 \end{bmatrix}$

2. If $A + B = A$, name the matrix represented by B.

   $B = \begin{bmatrix} 0 & 0 \\ 0 & 0 \end{bmatrix}$

3. If $AB = I$, name the matrix represented by B.

   B must be the inverse matrix of A.

4. Do the matrices have inverses? Justify your answer.

   a. $\begin{bmatrix} -2 & 6 \\ -3 & 9 \end{bmatrix}$

   No. The determinant $(-2)(9) - (6)(-3) = 0$.

   b. $\begin{bmatrix} -2 & 6 \\ 3 & 9 \end{bmatrix}$

   Yes. The determinant $(-2)(9) - (6)(3) \neq 0$.

5. Find a value of a, such that the given matrix has an inverse.

   a. $\begin{bmatrix} -4 & 3a \\ 2 & 9 \end{bmatrix}$

   $$(-4)(9) - (3a)(2) \neq 0$$
   $$-36 - 6a \neq 0$$
   $$a \neq -6$$

   b. $\begin{bmatrix} 5 & a \\ -a & 5 \end{bmatrix}$

   $$(5)(5) - (a)(-a) \neq 0$$
   $$25 + a^2 \neq 0$$
   For all real values of a

Problem Set Sample Solutions

The first seven problems are more procedural. Assign problems based on student abilities. All students do not have to complete all problems.

1. **Find a transformation that would create a $30°$ counterclockwise rotation about the origin and then its inverse.**

$$\begin{bmatrix} \frac{\sqrt{3}}{2} & -\frac{1}{2} \\ \frac{1}{2} & \frac{\sqrt{3}}{2} \end{bmatrix}, \begin{bmatrix} \frac{\sqrt{3}}{2} & \frac{1}{2} \\ -\frac{1}{2} & \frac{\sqrt{3}}{2} \end{bmatrix}$$

2. **Find a transformation that would create a $30°$ counterclockwise rotation about the origin, a dilation with a scale factor of 4, and then its inverse.**

$$\begin{bmatrix} 2\sqrt{3} & -2 \\ 2 & 2\sqrt{2} \end{bmatrix}, \begin{bmatrix} \frac{\sqrt{3}}{8} & \frac{1}{8} \\ -\frac{1}{8} & \frac{\sqrt{3}}{8} \end{bmatrix}$$

3. **Find a transformation that would create a $270°$ counterclockwise rotation about the origin. Set up a system of equations, and solve to find the matrix.**

$$\begin{bmatrix} a & c \\ b & d \end{bmatrix}\begin{bmatrix} 1 \\ 0 \end{bmatrix} = \begin{bmatrix} 0 \\ -1 \end{bmatrix}, \begin{bmatrix} a & c \\ b & d \end{bmatrix}\begin{bmatrix} 0 \\ 1 \end{bmatrix} = \begin{bmatrix} 1 \\ 0 \end{bmatrix}$$

$$\begin{bmatrix} 0 & 1 \\ -1 & 0 \end{bmatrix}$$

4. **Find a transformation that would create a $270°$ counterclockwise rotation about the origin, a dilation with a scale factor of 3, and its inverse.**

$$\begin{bmatrix} 0 & 3 \\ -3 & 0 \end{bmatrix}, \begin{bmatrix} 0 & -\frac{1}{3} \\ \frac{1}{3} & 0 \end{bmatrix}$$

5. **For which values of a does $\begin{bmatrix} 8 & a \\ a & 2 \end{bmatrix}$ have an inverse matrix?**

$$(8)(2) - (a)(a) \neq 0$$
$$16 - a^2 \neq 0$$
$$(4 - a)(4 + a) \neq 0$$
$$a \neq 4, a \neq -4$$

6. **For which values of a does $\begin{bmatrix} a & a-4 \\ a+4 & a \end{bmatrix}$ have an inverse matrix?**

$$(a)(a) - (a - 4)(a + 4) \neq 0$$
$$a^2 - (a^2 - 16) \neq 0$$
$$16 \neq 0$$

All real values of a will produce an inverse matrix.

7. For which values of a does $\begin{bmatrix} 3a & 2a-6 \\ 6a & 4a-12 \end{bmatrix}$ have an inverse matrix?

$$(3a)(4a-12)-(2a-6)(6a) \neq 0$$
$$(12a^2-36a)-(12a^2-36a) \neq 0$$
$$0 \neq 0$$

No real values of a will produce an inverse matrix.

8. In Lesson 27, we learned the effect of a transformation on a unit square by multiplying a matrix. For example, $A = \begin{bmatrix} 2 & 1 \\ 1 & 2 \end{bmatrix}$, $\begin{bmatrix} 2 & 1 \\ 1 & 2 \end{bmatrix}\begin{bmatrix} 1 \\ 0 \end{bmatrix} = \begin{bmatrix} 2 \\ 1 \end{bmatrix}$, $\begin{bmatrix} 2 & 1 \\ 1 & 2 \end{bmatrix}\begin{bmatrix} 1 \\ 1 \end{bmatrix} = \begin{bmatrix} 3 \\ 3 \end{bmatrix}$, and $\begin{bmatrix} 2 & 1 \\ 1 & 2 \end{bmatrix}\begin{bmatrix} 0 \\ 1 \end{bmatrix} = \begin{bmatrix} 1 \\ 2 \end{bmatrix}$.

a. Sasha says that we can multiply the inverse of A to those resultants of the square after the transformation to get back to the unit square. Is her conjecture correct? Justify your answer.

Yes, she is correct. $A^{-1} = \frac{1}{3}\begin{bmatrix} 2 & -1 \\ -1 & 2 \end{bmatrix} = \begin{bmatrix} \frac{2}{3} & -\frac{1}{3} \\ -\frac{1}{3} & \frac{2}{3} \end{bmatrix}$

$$\begin{bmatrix} \frac{2}{3} & -\frac{1}{3} \\ -\frac{1}{3} & \frac{2}{3} \end{bmatrix}\begin{bmatrix} 0 \\ 0 \end{bmatrix} = \begin{bmatrix} 0 \\ 0 \end{bmatrix}, \begin{bmatrix} \frac{2}{3} & -\frac{1}{3} \\ -\frac{1}{3} & \frac{2}{3} \end{bmatrix}\begin{bmatrix} 2 \\ 1 \end{bmatrix} = \begin{bmatrix} 1 \\ 0 \end{bmatrix}, \begin{bmatrix} \frac{2}{3} & -\frac{1}{3} \\ -\frac{1}{3} & \frac{2}{3} \end{bmatrix}\begin{bmatrix} 3 \\ 3 \end{bmatrix} = \begin{bmatrix} 1 \\ 1 \end{bmatrix}, \begin{bmatrix} \frac{2}{3} & -\frac{1}{3} \\ -\frac{1}{3} & \frac{2}{3} \end{bmatrix}\begin{bmatrix} 1 \\ 2 \end{bmatrix} = \begin{bmatrix} 0 \\ 1 \end{bmatrix}$$

b. From part (a), what would you say about the inverse matrix with regard to the geometric effect of transformations?

Multiplying the inverse matrix, A^{-1}, will "undo" the transformation that was done by multiplying matrix A.

c. A pure rotation matrix is $\begin{bmatrix} \cos(\theta) & -\sin(\theta) \\ \sin(\theta) & \cos(\theta) \end{bmatrix}$. Prove the inverse matrix for a pure rotation of $\frac{\pi}{4}$ radians counterclockwise is $\begin{bmatrix} \cos\left(-\frac{\pi}{4}\right) & -\sin\left(-\frac{\pi}{4}\right) \\ \sin\left(-\frac{\pi}{4}\right) & \cos\left(-\frac{\pi}{4}\right) \end{bmatrix}$, which is the same as $\begin{bmatrix} \frac{d}{ad-bc} & \frac{-c}{ad-bc} \\ \frac{-b}{ad-bc} & \frac{d}{ad-bc} \end{bmatrix}$.

The matrix for a pure rotation $\frac{\pi}{4}$ radians counterclockwise rotation is $\begin{bmatrix} \cos\frac{-\pi}{4} & -\sin\frac{-\pi}{4} \\ \sin\frac{-\pi}{4} & \cos\frac{-\pi}{4} \end{bmatrix} = \begin{bmatrix} \frac{\sqrt{2}}{2} & \frac{\sqrt{2}}{2} \\ \frac{-\sqrt{2}}{2} & \frac{\sqrt{2}}{2} \end{bmatrix}$,

$A = \begin{bmatrix} \cos\frac{\pi}{4} & -\sin\frac{\pi}{4} \\ \sin\frac{\pi}{4} & \cos\frac{\pi}{4} \end{bmatrix} = \begin{bmatrix} \frac{\sqrt{2}}{2} & \frac{-\sqrt{2}}{2} \\ \frac{\sqrt{2}}{2} & \frac{\sqrt{2}}{2} \end{bmatrix}$. *Det* $A = ad - bc = \frac{\sqrt{2}}{2}\cdot\frac{\sqrt{2}}{2} - \frac{\sqrt{2}}{2}\left(-\frac{\sqrt{2}}{2}\right) = \frac{2}{4} + \frac{2}{4} = 1$, *and*

$A^{-1} = \begin{bmatrix} \frac{\sqrt{2}}{2} & \frac{\sqrt{2}}{2} \\ \frac{-\sqrt{2}}{2} & \frac{\sqrt{2}}{2} \end{bmatrix}$.

d. Prove that the inverse matrix of a pure dilation with a factor of 4 is $\begin{bmatrix} \frac{1}{4} & 0 \\ 0 & \frac{1}{4} \end{bmatrix}$, which is the same as $\begin{bmatrix} \frac{d}{ad-bc} & \frac{-c}{ad-bc} \\ \frac{-b}{ad-bc} & \frac{d}{ad-bc} \end{bmatrix}$.

The matrix for a pure dilation with a factor of 4 *is* $A = \begin{bmatrix} 4 & 0 \\ 0 & 4 \end{bmatrix}$. *Det* $A = 16 - 0 = 16$, *and*

$$A^{-1} = \frac{1}{16}\begin{bmatrix} 4 & 0 \\ 0 & 4 \end{bmatrix} = \begin{bmatrix} \frac{1}{4} & 0 \\ 0 & \frac{1}{4} \end{bmatrix}.$$

e. Prove that the matrix used to undo a $\frac{\pi}{3}$ radians clockwise rotation and a dilation of a factor of 2 is $\frac{1}{2}\begin{bmatrix} \cos\left(\frac{\pi}{3}\right) & -\sin\left(\frac{\pi}{3}\right) \\ \sin\left(\frac{\pi}{3}\right) & \cos\left(\frac{\pi}{3}\right) \end{bmatrix}$, which is the same as $\begin{bmatrix} \frac{d}{ad-bc} & \frac{-c}{ad-bc} \\ \frac{-b}{ad-bc} & \frac{d}{ad-bc} \end{bmatrix}$.

The matrix for undoing the rotation of a $\frac{\pi}{3}$ *radians clockwise and dilating a factor of* 2 *is*

$$\frac{1}{2}\begin{bmatrix} \cos\left(\frac{\pi}{3}\right) & -\sin\left(\frac{\pi}{3}\right) \\ \sin\left(\frac{\pi}{3}\right) & \cos\left(\frac{\pi}{3}\right) \end{bmatrix} = \frac{1}{2}\begin{bmatrix} \frac{1}{2} & \frac{-\sqrt{3}}{2} \\ \frac{\sqrt{3}}{2} & \frac{1}{2} \end{bmatrix} = \begin{bmatrix} \frac{1}{4} & \frac{-\sqrt{3}}{4} \\ \frac{\sqrt{3}}{4} & \frac{1}{4} \end{bmatrix}.$$

The matrix for a $\frac{\pi}{3}$ *radians clockwise rotation and a dilation of a factor of* 2 *is*

$$A = 2\begin{bmatrix} \cos\left(\frac{-\pi}{3}\right) & -\sin\left(\frac{-\pi}{3}\right) \\ \sin\left(\frac{-\pi}{3}\right) & \cos\left(\frac{-\pi}{3}\right) \end{bmatrix} = 2\begin{bmatrix} \frac{1}{2} & \frac{\sqrt{3}}{2} \\ \frac{-\sqrt{3}}{2} & \frac{1}{2} \end{bmatrix} = \begin{bmatrix} 1 & \sqrt{3} \\ -\sqrt{3} & 1 \end{bmatrix}$$

, Det $A = 1 + 3 = 4$*, and*

$$A^{-1} = \frac{1}{4\begin{bmatrix} 1 & -\sqrt{3} \\ \sqrt{3} & 1 \end{bmatrix}} = \begin{bmatrix} \frac{1}{4} & \frac{-\sqrt{3}}{4} \\ \frac{\sqrt{3}}{4} & \frac{1}{4} \end{bmatrix}.$$

f. Prove that any matrix whose determinant is not 0 will have an inverse matrix to "undo" a transformation. For example, use the matrix $A = \begin{bmatrix} a & c \\ b & d \end{bmatrix}$ and the point $\begin{bmatrix} x \\ y \end{bmatrix}$.

$$A = \begin{bmatrix} a & c \\ b & d \end{bmatrix}, \quad \begin{bmatrix} a & c \\ b & d \end{bmatrix}\begin{bmatrix} x \\ y \end{bmatrix} = \begin{bmatrix} ax+cy \\ bx+dy \end{bmatrix} \quad A^{-1} = \begin{bmatrix} \frac{d}{ad-bc} & \frac{-c}{ad-bc} \\ \frac{-b}{ad-bc} & \frac{a}{ad-bc} \end{bmatrix}$$

$$\begin{bmatrix} \frac{d}{ad-bc} & \frac{-c}{ad-bc} \\ \frac{-b}{ad-bc} & \frac{a}{ad-bc} \end{bmatrix}\begin{bmatrix} ax+cy \\ bx+dy \end{bmatrix} = \begin{bmatrix} \frac{d(ax+cy)}{ad-bc} - \frac{c(bx+dy)}{ad-bc} \\ \frac{-b(ax+cy)}{ad-bc} + \frac{a(bx+dy)}{ad-bc} \end{bmatrix} = \begin{bmatrix} \frac{adx+cdy-bcx-cdy}{ad-bc} \\ \frac{-abx-bcy+abx+ady}{ad-bc} \end{bmatrix}$$

$$= \begin{bmatrix} \frac{(ad-bc)x}{ad-bc} \\ \frac{(ad-bc)y}{ad-bc} \end{bmatrix} = \begin{bmatrix} x \\ y \end{bmatrix}$$

9. Perform the transformation $\begin{bmatrix} 2 & 2 \\ 2 & 2 \end{bmatrix}$ on the unit square.

a. Can you find the inverse matrix that will "undo" the transformation? Explain your reasons arithmetically.

No. The determinant of the matrix is 0. Therefore, there is no inverse matrix that can be found to undo the transformation.

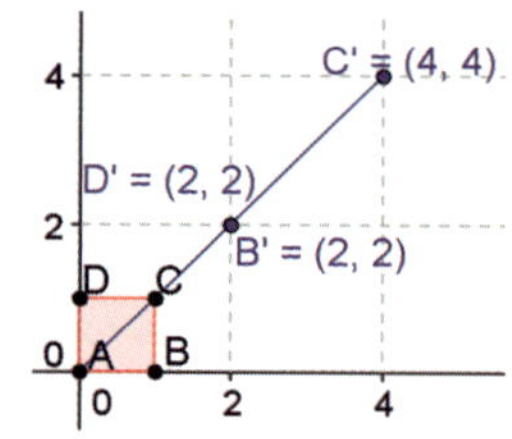

b. When all four vertices of the unit square are transformed and collapsed onto a straight line, what can be said about the inverse?

The determinant of the matrix is 0.

c. Find the equation of the line that all four vertices of the unit square collapsed onto.

$y = x$

d. Find the equation of the line that all four vertices of the unit square collapsed onto using the matrix $\begin{bmatrix} 1 & 3 \\ 2 & 6 \end{bmatrix}$.

$y = 2x$

e. A function has an inverse function if and only if it is a one-to-one function. By applying this concept, explain why we do not have an inverse matrix when the transformation is collapsed onto a straight line.

When doing transformations, we are mapping the four vertices to new coordinates; however, when we reverse this process, there should be a one-to-one property. However, we see that the point $\begin{pmatrix} 2 \\ 2 \end{pmatrix}$ will map onto two different points. Because there is no one-to-one property, this means there is no inverse matrix.

10. The determinants of the following matrices are 0. Describe what pattern you can find among them.

a. $\begin{bmatrix} 1 & 2 \\ 1 & 2 \end{bmatrix}$, $\begin{bmatrix} 1 & 1 \\ 2 & 2 \end{bmatrix}$, $\begin{bmatrix} 1 & 2 \\ 4 & 8 \end{bmatrix}$, and $\begin{bmatrix} 1 & -2 \\ 2 & -4 \end{bmatrix}$

If one column is the multiple of the other column, or one row is the multiple of the other row, then the determinant is 0, and there is no inverse matrix.

b. $\begin{bmatrix} 0 & 1 \\ 0 & 1 \end{bmatrix}$, $\begin{bmatrix} 1 & 0 \\ 1 & 0 \end{bmatrix}$, $\begin{bmatrix} 0 & 0 \\ 1 & 1 \end{bmatrix}$, $\begin{bmatrix} 1 & 1 \\ 0 & 0 \end{bmatrix}$, and $\begin{bmatrix} 0 & 0 \\ 0 & 0 \end{bmatrix}$

If either column or row is 0, then the determinant is 0, and there is no inverse matrix.

Name ______________________________ Date ______________

1. Consider the transformation on the plane given by the 2×2 matrix $\begin{pmatrix} 1 & k \\ 0 & k \end{pmatrix}$ for a fixed positive number $k > 1$.

   a. Draw a sketch of the image of the unit square under this transformation (the unit square has vertices$(0,0)$, $(1,0)$, $(0,1)$, $(1,1)$). Be sure to label all four vertices of the image figure.

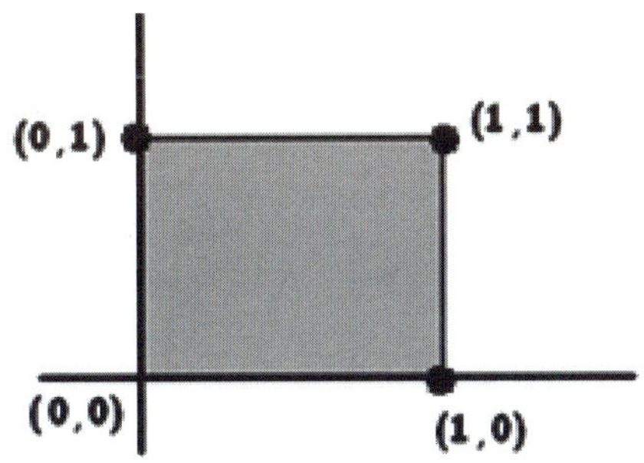

The Unit Square

b. What is the area of the image parallelogram?

c. Find the coordinates of a point $\begin{pmatrix} x \\ y \end{pmatrix}$ whose image under the transformation is $\begin{pmatrix} 2 \\ 3 \end{pmatrix}$.

d. The transformation $\begin{pmatrix} 1 & k \\ 0 & k \end{pmatrix}$ is applied once to the point $\begin{pmatrix} 1 \\ 1 \end{pmatrix}$, then once to the image point, then once to the image of the image point, and then once to the image of the image of the image point, and so on. What are the coordinates of a tenfold image of the point $\begin{pmatrix} 1 \\ 1 \end{pmatrix}$, that is, the image of the point after the transformation has been applied 10 times?

2. Consider the transformation given by $\begin{pmatrix} \cos(1) & -\sin(1) \\ \sin(1) & \cos(1) \end{pmatrix}$.

a. Describe the geometric effect of applying this transformation to a point $\begin{pmatrix} x \\ y \end{pmatrix}$ in the plane.

b. Describe the geometric effect of applying this transformation to a point $\begin{pmatrix} x \\ y \end{pmatrix}$ in the plane twice: once to the point and then once to its image.

c. Use part (b) to prove $\cos(2) = \cos^2(1) - \sin^2(1)$ and $\sin(2) = 2\sin(1)\cos(1)$.

3.

a. Explain the geometric representation of multiplying a complex number by $1 + i$.

b. Write $(1 + i)^{10}$ as a complex number of the form $a + bi$ for real numbers a and b.

c. Find a complex number $a + bi$, with a and b positive real numbers, such that $(a + bi)^3 = i$.

d. If z is a complex number, is there sure to exist, for any positive integer n, a complex number w such that $w^n = z$? Explain your answer.

e. If z is a complex number, is there sure to exist, for any negative integer n, a complex number w such that $w^n = z$? Explain your answer.

4. Let $P = \begin{pmatrix} 0 & 0 \\ 1 & 0 \end{pmatrix}$ and $O = \begin{pmatrix} 0 & 0 \\ 0 & 0 \end{pmatrix}$.

a. Give an example of a 2×2 matrix A, not with all entries equal to zero, such that $PA = O$.

b. Give an example of a 2×2 matrix B with $PB \neq O$.

c. Give an example of a 2×2 matrix C such that $CR = R$ for all 2×2 matrices R.

d. If a 2×2 matrix D has the property that $D + R = R$ for all 2×2 matrices R, must D be the zero matrix O? Explain.

e. Let $E = \begin{pmatrix} 2 & 4 \\ 3 & 6 \end{pmatrix}$. Is there a 2×2 matrix F so that $EF = \begin{pmatrix} 1 & 0 \\ 0 & 1 \end{pmatrix}$ and $FE = \begin{pmatrix} 1 & 0 \\ 0 & 1 \end{pmatrix}$? If so, find one. If not, explain why no such matrix F can exist.

5. In programming a computer video game, Mavis coded the changing location of a space rocket as follows: At a time t seconds between $t = 0$ seconds and $t = 2$ seconds, the location $\begin{pmatrix} x \\ y \end{pmatrix}$ of the rocket is given by

$$\begin{pmatrix} \cos\left(\frac{\pi}{2}t\right) & -\sin\left(\frac{\pi}{2}t\right) \\ \sin\left(\frac{\pi}{2}t\right) & \cos\left(\frac{\pi}{2}t\right) \end{pmatrix} \begin{pmatrix} -1 \\ -1 \end{pmatrix}.$$

At a time of t seconds between $t = 2$ seconds and $t = 4$ seconds, the location of the rocket is given by $\begin{pmatrix} 3 - t \\ 3 - t \end{pmatrix}$.

a. What is the location of the rocket at time $t = 0$? What is its location at time $t = 4$?

b. Petrich is worried that Mavis may have made a mistake and the location of the rocket is unclear at time $t = 2$ seconds. Explain why there is no inconsistency in the location of the rocket at this time.

c. What is the area of the region enclosed by the path of the rocket from time $t = 0$ to time $t = 4$?

d. Mavis later decided that the moving rocket should be shifted five places farther to the right. How should she adjust her formulations above to accomplish this translation?

A Progression Toward Mastery

Assessment Task Item		STEP 1 Missing or incorrect answer and little evidence of reasoning or application of mathematics to solve the problem.	STEP 2 Missing or incorrect answer but evidence of some reasoning or application of mathematics to solve the problem.	STEP 3 A correct answer with some evidence of reasoning or application of mathematics to solve the problem, OR an incorrect answer with substantial evidence of solid reasoning or application of mathematics to solve the problem.	STEP 4 A correct answer supported by substantial evidence of solid reasoning or application of mathematics to solve the problem.
1	a N-VM.C.11 N-VM.C.12	Student provides a solution that does not apply matrix multiplication or transformations to determine the coordinates of the resulting image. The sketch is missing.	Student computes two or more coordinates of the image incorrectly, and the sketch of the image is incomplete or poorly labeled, or the image is a parallelogram with no work shown and no vertices labeled.	Student computes coordinates of the image correctly, but the sketch of the image may be slightly inaccurate. Work to support the calculation of the image coordinates is limited. OR Student computes three out of four coordinates correctly, and the sketch accurately reflects the student's coordinates.	Student applies matrix multiplication to each coordinate of the unit square to get the image coordinates and draws a fairly accurate sketch of a parallelogram with vertices correctly labeled. Values for k vary, but the resulting image should look like a parallelogram, and the distance k in the vertical and horizontal direction should appear equal.
	b N-VM.C.12	Student does not compute the area of a parallelogram or his sketched figure correctly.	Student computes the area of his sketched figure correctly but does not use determinant of the 2×2 matrix in his calculation.	Student computes the area of his figure using the determinant of the 2×2 matrix, but the solution may contain minor errors.	Student computes the area of the parallelogram correctly using a determinant. Work shows understanding that the area of the image is the product of the area of the original figure and the absolute value of the determinant of the transformation matrix.

	c **N-VM.C.10** **N-VM.C.11**	Student does not provide a solution. OR Student provides work that is unrelated to the standards addressed in this problem.	Student computes an incorrect solution or setup of the original matrix equation. Limited evidence is evident that the student understands that the solution to the matrix equation finds the point in question. OR Student creates a correct matrix equation, and no additional work is given. OR Student creates the correct system of linear equations, and no additional work Is given.	Student creates a correct matrix equation to solve for the point and translates the equation to a system of linear equations. Work shown may be incomplete, and final answer may contain minor errors. OR Student has the correct solution, but the matrix equation or the system of equations is missing from the solution. Very little work is shown to provide evidence of student thinking.	Student creates a correct matrix equation to solve for the point. Student translates the equation to a system of linear equations and solves the system correctly. Work shown is organized in a manner that is easy to follow and uses proper mathematical notation.
	d **N-VM.C.11** **N-VM.C.12**	Student provides a solution that does not correctly apply the transformation one time. AND Student does not attempt a generalization for the tenfold image.	Student provides a solution that does not correctly apply the transformation more than one time. Student may attempt to generalize to the tenfold image, but the answer contains major conceptual errors.	Student provides a solution that includes evidence that the student understood the problem and observed patterns, but minor errors prevent a correct solution for the tenfold image. OR Student provides a solution that shows correct repeated application of the transformation at least three times, but the student is unable to extend the pattern to the tenfold image.	Student gives correct solution for the tenfold image. Student solution provides enough evidence and explanation to clearly illustrate how she observed and extended the pattern.
2	**a** **N-VM.C.11** **N-VM.C.12**	Student does not recognize the transformation as a rotation of the point about the origin.	Student identifies the transformation as a rotation but cannot correctly state the direction or the angle measure.	Student correctly identifies the transformation as a rotation about the origin, but the answer contains an error, such as the wrong direction or the wrong angle measurement.	Student correctly identifies the transformation as a counterclockwise rotation about the origin through an angle of 1 radian.

	b **N-VM.C.11** **N-VM.C.12**	Student does not identify the repeated transformation as a rotation.	Student identifies the transformation as a rotation, but the solution does not make it clear that the second rotation applies to the image of the original point. OR Student identifies the transformation as an additional rotation, but the answer contains two or more errors.	Student correctly identifies the repeated transformation as an additional rotation, but the answer contains no more than one error.	Student correctly identifies the repeated transformation as a rotation of the image of the point another 1 radian clockwise about the origin for a total of 2 radians.
	c **N-VM.C.8**	Student makes little or no attempt at multiplying the point (x, y) by either of the rotation matrices.	Student sets up and attempts the necessary matrix multiplications, but solution has too many major errors. OR Student provides too little work to make significant progress on the proof.	Student provides a solution that includes multiplication of (x, y) by the original rotation matrix twice and multiplication of (x, y) by the 2-radian rotation matrix. Student fails to equate the two answers to finish the proof. The solution may contain minor computation errors.	Student provides a solution that details multiplication by the original rotation matrix twice, compares that result to multiplication by the 2-radian rotation matrix, and equates the two answers to verify the identities. Student uses correct notation, and the solution illustrates his thinking clearly. The solution is free from minor errors.
3	**a** **N-CN.B.5**	Student makes little or no attempt to explain the geometric relationship of multiplying by $1 + i$.	Student attempts to explain the geometric relationship of multiplying by $1 + i$ but makes mistakes.	Student attempts to explain the geometric relationship of multiplying by $1 + i$ but mentions either the dilation or rotation, not both.	Student fully explains the geometric relationship of multiplying by $1 + i$ in terms of a dilation and a rotation.
	b **N-CN.B.4**	Student makes little or no attempt to find the modulus and argument.	Student attempts to find the modulus and argument, but solution has major errors that lead to an incorrect answer.	Student has the correct answer, but it may not be in proper form, or student makes minor computational errors in finding the modulus and argument.	Student writes the correct answer in the proper form and correctly solves for the modulus and argument of the expression, showing all steps.

	c N-CN.B.4 N-CN.B.5	Student makes little or no attempt to solve for a complex number.	Student attempts to find a complex number but lacks the proper steps in order to do so, resulting in an incorrect answer.	Student may find a correct answer but does not show any steps taken to solve the problem. OR Student has an answer that does not have a and b as positive real numbers.	Student correctly finds a complex number in the form $a + bi$, where a and b are positive real numbers, that satisfies the given equation and shows all steps such as finding the modulus and argument of i.
	d N-CN.B.4 N-CN.B.5	Student does not give any explanation as to whether a complex number, w, exists for the given equation and conditions and answers incorrectly.	Student answers incorrectly but gives an explanation that has somewhat valid points but is lacking proper information.	Student answers correctly but does not give an accurate written and algebraic explanation such as stating the modulus and argument of z and w for both zero and nonzero cases.	Student answers correctly and provides correct reasoning as to why w is sure to exist, including stating the modulus and argument of z and w if they are nonzero.
	e N-CN.B.4 N-CN.B.5	Student does not give any explanation as to whether a complex number, w, exists for the given equation and conditions and answers incorrectly.	Student answers incorrectly but gives an explanation that has somewhat valid points but is lacking proper information.	Student answers correctly but lacks proper reasoning to support the answer.	Student answers correctly and provides correct reasoning as to why w is sure to exist, including an algebraic solution.
4	a N-VM.C.8 N-VM.C.10	Student makes little to no attempt to find matrix.	Student sets up a matrix equation but does not use the correct matrices in order to solve the problem.	Student correctly sets up the matrix equation, but, due to errors in calculations, fails to find the correct matrix.	Student correctly sets up and solves the matrix equation leading to the correct matrix.
	b N-VM.C.8 N-VM.C.10	Student makes little to no attempt to find matrix.	Student sets up a matrix equation but does not use the correct matrices in order to solve the problem.	Student correctly sets up the matrix equation, but, due to errors in calculations, fails to find the correct matrix.	Student correctly sets up and solves the matrix equation leading to the correct matrix.
	c N-VM.C.8 N-VM.C.10	Student makes little to no attempt to find matrix.	Student sets up a matrix equation but does not use the identity matrix in order to solve the problem.	Student identifies the identity matrix as the answer but writes the matrix incorrectly.	Student identifies the identity matrix as the answer and writes it correctly.
	d N-VM.C.8 N-VM.C.10	Student makes little to no attempt to find matrix.	Student sets up a matrix equation but does not use the correct matrices in order to solve the problem.	Student correctly sets up the matrix equation, but, due to errors in calculations, fails to find the correct matrix.	Student correctly sets up and solves the matrix equation leading to the correct matrix.

	e **N-VM.C.8** **N-VM.C.10**	Student makes little to no attempt to find matrix.	Student sets up a matrix equation but does not use the correct matrices in order to answer the question.	Student correctly sets up one or both matrix equations, but, due to errors in calculations, fails to arrive at the correct answer.	Student correctly sets up and solves both matrix equations leading to the correct answer.
5	**a** **N-VM.C.10** **N-VM.C.11** **N-VM.C.12**	Student makes little to no attempt to solve for the location of the rocket at either time given.	Student sets up a matrix equation but does not use the correct matrices in order to solve the problem.	Student correctly sets up the matrix equation, but, due to errors in calculations, fails to reach a correct final answer for the location of the rocket at both times.	Student correctly solves for the location of the rocket at both times given, using the correct matrix equation.
	b **N-VM.C.10** **N-VM.C.11** **N-VM.C.12**	Student makes little to no attempt to find the location of the rocket at the given time for either set of instructions and gives no explanation.	Student sets up matrix equations to solve for the location of the rocket but fails to properly solve the equations and produce an accurate explanation.	Student correctly finds the location of the rocket for one set of instructions but fails to verify that the location of the rocket for the other set of instructions is consistent with the first.	Student correctly gives the location of the rocket for the given time for both sets of instructions and correctly makes the correlations between the two.
	c **N-VM.C.10** **N-VM.C.11** **N-VM.C.12**	Student makes little to no attempt to solve for the area.	Student attempts to find the area of the region enclosed by the path of the rocket but does not make the correct conclusion that it travels in a semicircle.	Student correctly finds that the path traversed is a semicircle but has minor errors in calculations that prevent the correct area from being found.	Student correctly finds the area of the enclosed path of the rocket including finding the radius of the traversed path.
	d **N-VM.C.10** **N-VM.C.11** **N-VM.C.12**	Student makes little to no attempt to adjust the matrix five places farther right.	Student sets up matrix/matrices for one or both sets of instructions but incorrectly translates the points 5 units to the right.	Student correctly sets up the shifted matrix for one set of instructions but fails to correctly set up the shifted matrices for both sets of instructions.	Student correctly sets up the matrices for both sets of instructions that results in a shift of the rocket five places to the right.

Name ______________________________ Date ______________

1. Consider the transformation on the plane given by the 2×2 matrix $\begin{pmatrix} 1 & k \\ 0 & k \end{pmatrix}$ for a fixed positive number $k > 1$.

   a. Draw a sketch of the image of the unit square under this transformation (the unit square has vertices $(0,0)$, $(1,0)$, $(0,1)$, and $(1,1)$). Be sure to label all four vertices of the image figure.

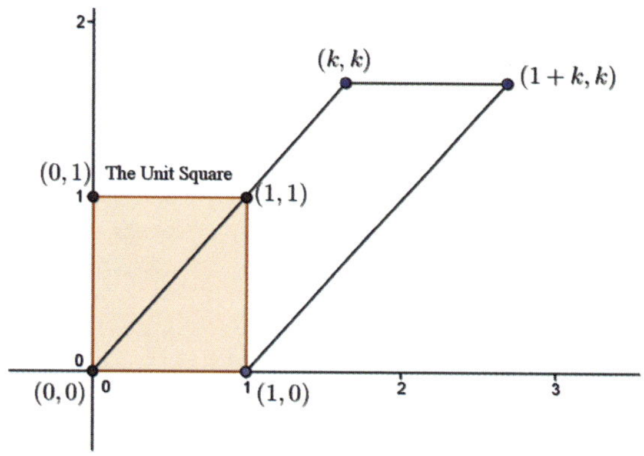

To find the coordinates of the image, multiply the vertices of the unit square by the matrix.

$$\begin{pmatrix} 1 & k \\ 0 & k \end{pmatrix}\begin{pmatrix} 0 \\ 0 \end{pmatrix} = \begin{pmatrix} 0 \\ 0 \end{pmatrix}$$

$$\begin{pmatrix} 1 & k \\ 0 & k \end{pmatrix}\begin{pmatrix} 0 \\ 1 \end{pmatrix} = \begin{pmatrix} k \\ k \end{pmatrix}$$

$$\begin{pmatrix} 1 & k \\ 0 & k \end{pmatrix}\begin{pmatrix} 1 \\ 1 \end{pmatrix} = \begin{pmatrix} 1 + k \\ k \end{pmatrix}$$

$$\begin{pmatrix} 1 & k \\ 0 & k \end{pmatrix}\begin{pmatrix} 1 \\ 0 \end{pmatrix} = \begin{pmatrix} 1 \\ 0 \end{pmatrix}$$

The image is a parallelogram with base = 1 and height = k.

b. What is the area of the image parallelogram?

To find the area of the image figure, multiply the area of the unit square by the absolute value of $\begin{bmatrix} 1 & k \\ 0 & k \end{bmatrix}$.

$$\begin{vmatrix} 1 & k \\ 0 & k \end{vmatrix} = (1 \times k) - (0 \times k) = k$$

$$\text{Area} = 1 \times |k| = k \text{ since } k > 0$$

c. Find the coordinates of a point $\begin{pmatrix} x \\ y \end{pmatrix}$ whose image under the transformation is $\begin{pmatrix} 2 \\ 3 \end{pmatrix}$.

Solve the equation to find the coordinates of $\begin{pmatrix} x \\ y \end{pmatrix}$.

$$\begin{pmatrix} 1 & k \\ 0 & k \end{pmatrix}\begin{pmatrix} x \\ y \end{pmatrix} = \begin{pmatrix} 2 \\ 3 \end{pmatrix}$$

Converting the matrix equation to a system of linear equations gives us

$$x + ky = 2$$
$$ky = 3.$$

Solve this system.

$$y = \frac{3}{k}$$
$$x + k\left(\frac{3}{k}\right) = 2$$
$$x + 3 = 2$$
$$x = -1$$

The point is $\begin{pmatrix} x \\ y \end{pmatrix} = \begin{pmatrix} -1 \\ \frac{3}{k} \end{pmatrix}$.

d. The transformation $\begin{pmatrix} 1 & k \\ 0 & k \end{pmatrix}$ is applied once to the point $\begin{pmatrix} 1 \\ 1 \end{pmatrix}$, then once to the image point, then once to the image of the image point, and then once to the image of the image of the image point, and so on. What are the coordinates of a tenfold image of the point $\begin{pmatrix} 1 \\ 1 \end{pmatrix}$, that is, the image of the point after the transformation has been applied 10 times?

Multiply to apply the transformation once: $\begin{pmatrix} 1 & k \\ 0 & k \end{pmatrix}\begin{pmatrix} 1 \\ 1 \end{pmatrix} = \begin{pmatrix} 1+k \\ k \end{pmatrix}$.

Multiply again by the 2×2 matrix: $\begin{pmatrix} 1 & k \\ 0 & k \end{pmatrix}\begin{pmatrix} 1+k \\ k \end{pmatrix} = \begin{pmatrix} 1+k+k^2 \\ k^2 \end{pmatrix}$

Multiply again by the 2×2 matrix: $\begin{pmatrix} 1 & k \\ 0 & k \end{pmatrix}\begin{pmatrix} 1+k+k^2 \\ k^2 \end{pmatrix} = \begin{pmatrix} 1+k+k^2+k^3 \\ k^3 \end{pmatrix}$.

By observing the patterns, we can see that the result of n multiplications is a 2×1 matrix whose top row is the previous row plus k^n and whose bottom row is k^n. The tenfold image would be $\begin{pmatrix} 1+k+k^2+k^3+\dots k^{10} \\ k^{10} \end{pmatrix}$.

2. Consider the transformation given by $\begin{pmatrix} \cos(1) & -\sin(1) \\ \sin(1) & \cos(1) \end{pmatrix}$.

a. Describe the geometric effect of applying this transformation to a point $\begin{pmatrix} x \\ y \end{pmatrix}$ in the plane.

This transformation will rotate the point $\begin{pmatrix} x \\ y \end{pmatrix}$ counterclockwise about the origin through an angle of 1 radian.

b. Describe the geometric effect of applying this transformation to a point $\begin{pmatrix} x \\ y \end{pmatrix}$ in the plane twice: once to the point and then once to its image.

This transformation will rotate the point $\begin{pmatrix} x \\ y \end{pmatrix}$ counterclockwise about the origin an additional 1 radian for a total rotation of 2 radians.

c. Use part (b) to prove $\cos(2) = \cos^2(1) - \sin^2(1)$ and $\sin(2) = 2\sin(1)\cos(1)$.

To prove this, multiply $\begin{pmatrix} x \\ y \end{pmatrix}$ by the transformation matrix:

$$\begin{pmatrix} \cos(1) & -\sin(1) \\ \sin(1) & \cos(1) \end{pmatrix}\begin{pmatrix} x \\ y \end{pmatrix} = \begin{pmatrix} x\cos(1) - y\sin(1) \\ x\sin(1) + y\cos(1) \end{pmatrix}$$

Then, multiply this answer by the transformation matrix:

$$\begin{pmatrix} \cos(1) & -\sin(1) \\ \sin(1) & \cos(1) \end{pmatrix}\begin{pmatrix} x\cos(1) - y\sin(1) \\ x\sin(1) + y\cos(1) \end{pmatrix}$$

Apply matrix multiplication:

$$\begin{pmatrix} \cos(1)(x\cos(1) - y\sin(1)) - \sin(1)(x\sin(1) + y\cos(1)) \\ \sin(1)(x\cos(1) - y\sin(1)) + \cos(1)(x\sin(1) + y\cos(1)) \end{pmatrix}$$

Distribute:

$$\begin{pmatrix} x(\cos(1))^2 - y\cos(1)\sin(1) - x\sin(1)^2 - y\sin(1)\cos(1) \\ x\sin(1)\cos(1) - y\sin(1)^2 + x\cos(1)\sin(1) + y(\cos(1))^2 \end{pmatrix}$$

Rearrange and factor:

$$\begin{pmatrix} x((\cos(1))^2 - \sin(1)^2) - y(2\sin(1)\cos(1)) \\ x(2\sin(1)\cos(1)) + y(\cos(1)^2 - \sin(1)^2) \end{pmatrix}$$

This matrix is equal to the matrix resulting from the 2-radian rotation.

$$\begin{pmatrix} \cos(2) & -\sin(2) \\ \sin(2) & \cos(2) \end{pmatrix}\begin{pmatrix} x \\ y \end{pmatrix} = \begin{pmatrix} x\cos(2) - y\sin(2) \\ x\sin(2) + y\cos(2) \end{pmatrix}$$

When you equate the answers and compare the coefficients of x and y, you can see that

$\cos(2) = \cos(1)^2 - \sin(1)^2$ and $\sin(2) = 2\sin(1)\cos(1)$.

The matrices are equal because they represent the same transformation.

$$\begin{pmatrix} x((\cos(1))^2 - \sin(1)^2) - y(2\sin(1)\cos(1)) \\ x(2\sin(1)\cos(1)) + y(\cos(1)^2 - \sin(1)^2) \end{pmatrix} = \begin{pmatrix} x\cos(2) - y\sin(2) \\ x\sin(2) + y\cos(2) \end{pmatrix}$$

3.

a. Explain the geometric representation of multiplying by $1 + i$.

$1 + i$ has argument $\frac{\pi}{4}$ and modulus $\sqrt{2}$, so geometrically this represents a dilation with a scale factor of $\sqrt{2}$ and a counterclockwise rotation of $\frac{\pi}{4}$ about the origin.

b. Write $(1 + i)^{10}$ as a complex number of the form $a + bi$ for real numbers a and b.

$1 + i$ has argument $\frac{\pi}{4}$ and modulus $\sqrt{2}$, and so $(1+i)^{10}$ has argument $10 \times \frac{\pi}{4} = \frac{\pi}{2} + 2\pi$ and modulus $(\sqrt{2})^{10} = 2^5 = 32$. Thus, $(1+i)^{10} = 32i$.

c. Find a complex number $a + bi$, with a and b positive real numbers, such that $(a + bi)^3 = i$.

i has argument $\frac{\pi}{2}$ and modulus 1. Thus, a complex number $a + bi$ of argument $\frac{\pi}{6}$ and modulus 1 will satisfy $(a+bi)^3 = i$. We have $a + bi = \frac{\sqrt{3}}{2} + i\frac{1}{2}$.

d. If z is a complex number, is there sure to exist, for any positive integer n, a complex number w such that $w^n = z$? Explain your answer.

Yes. If $z = 0$, then $w = 0$ works. If, on the other hand, z is not zero and has argument θ and modulus m, then let w be the complex number with argument $\frac{\theta}{n}$ and modulus $m^{\frac{1}{n}}$:

$$w = m^{\frac{1}{n}}\left(\cos\left(\frac{\theta}{n}\right) + i\sin\left(\frac{\theta}{n}\right)\right).$$

e. If z is a complex number, is there sure to exist, for any negative integer n, a complex number w such that $w^n = z$? Explain your answer.

If $z = 0$, then there is no such complex number w. If $z \neq 0$, then $\frac{1}{w}$, with w as given in part (c), satisfies $\left(\frac{1}{w}\right)^{-n} = z$, showing that the answer to the question is yes in this case.

4. Let $P = \begin{pmatrix} 0 & 0 \\ 1 & 0 \end{pmatrix}$ and $O = \begin{pmatrix} 0 & 0 \\ 0 & 0 \end{pmatrix}$.

a. Give an example of a 2×2 matrix A, not with all entries equal to zero, such that $PA = O$.

Notice that for any matrix $A = \begin{pmatrix} a & b \\ c & d \end{pmatrix}$, *we have* $PA = \begin{pmatrix} 0 & 0 \\ 1 & 0 \end{pmatrix}\begin{pmatrix} a & b \\ c & d \end{pmatrix}\begin{pmatrix} 0 & 0 \\ a & b \end{pmatrix}$.

If we choose $A = \begin{pmatrix} 0 & 0 \\ 1 & 1 \end{pmatrix}$, *for example, then* $PA = O$.

b. Give an example of a 2×2 matrix B with $PB \neq O$.

Following the discussion in part (a), we see that choosing $A = \begin{pmatrix} 1 & 1 \\ 0 & 0 \end{pmatrix}$ *gives* $PA = \begin{pmatrix} 0 & 0 \\ 1 & 1 \end{pmatrix}$, *which is different from* O.

c. Give an example of a 2×2 matrix C such that $CR = R$ for all 2×2 matrices R.

Choose $C = \begin{pmatrix} 1 & 0 \\ 0 & 1 \end{pmatrix}$. *The identity matrix has this property.*

d. If a 2×2 matrix D has the property that $D + R = R$ for all 2×2 matrices R, must D be the zero matrix O? Explain.

Write $D = \begin{pmatrix} a & b \\ c & d \end{pmatrix}$ *and* $R = \begin{pmatrix} x & y \\ z & w \end{pmatrix}$. *Then, for* $D + R = \begin{pmatrix} a+x & b+y \\ c+z & d+w \end{pmatrix}$ *to equal* $\begin{pmatrix} x & y \\ z & w \end{pmatrix}$ *no matter the values of* x, y, z, *and* w, *we need:*

$$a + x = x$$

$$b + y = y$$

$$c + z = z$$

$$d + w = w$$

to hold for all values x, y, z, *and* w. *Thus, we need* $a = 0$, $b = 0$, $c = 0$, *and* $d = 0$. *That is,* D *must indeed be the zero matrix.*

e. Let $E = \begin{pmatrix} 2 & 4 \\ 3 & 6 \end{pmatrix}$. Is there a 2×2 matrix F so that $EF = \begin{pmatrix} 1 & 0 \\ 0 & 1 \end{pmatrix}$ and $FE = \begin{pmatrix} 1 & 0 \\ 0 & 1 \end{pmatrix}$? If so, find one. If not, explain why no such matrix F can exist.

The determinant of E is $|2 \cdot 6 - 3 \cdot 4| = 0$, and so no inverse matrix like F can exist.

Alternatively:

Write $F = \begin{pmatrix} a & b \\ c & d \end{pmatrix}$. Then, $EF = \begin{pmatrix} 2a+4c & 2b+4d \\ 3a+6c & 3b+3d \end{pmatrix}$. For this to equal $\begin{pmatrix} 1 & 0 \\ 0 & 1 \end{pmatrix}$, we need, at the very least:

$2a + 4c = 1$

$3a + 6c = 0$

The first of these equations gives $a + 2c = \frac{1}{2}$ and the second $a + 2c = 0$. There is no solution to this system of equations, and so there can be no matrix F with the desired property.

5. In programming a computer video game, Mavis coded the changing location of a space rocket as follows: At a time t seconds between $t = 0$ seconds and $t = 2$ seconds, the location $\begin{pmatrix} x \\ y \end{pmatrix}$ of the rocket is given by:

$$\begin{pmatrix} \cos\left(\frac{\pi}{2}t\right) & -\sin\left(\frac{\pi}{2}t\right) \\ \sin\left(\frac{\pi}{2}t\right) & \cos\left(\frac{\pi}{2}t\right) \end{pmatrix} \begin{pmatrix} -1 \\ -1 \end{pmatrix}.$$

At a time of t seconds between $t = 2$ seconds and $t = 4$ seconds, the location of the rocket is given by $\begin{pmatrix} 3-t \\ 3-t \end{pmatrix}$.

a. What is the location of the rocket at time $t = 0$? What is its location at time $t = 4$?

At time $t = 0$, the location of the rocket is

$$\begin{pmatrix} \cos(0) & -\sin(0) \\ \sin(0) & \cos(0) \end{pmatrix} \begin{pmatrix} -1 \\ -1 \end{pmatrix} = \begin{pmatrix} 1 & 0 \\ 0 & 1 \end{pmatrix} \begin{pmatrix} -1 \\ -1 \end{pmatrix} = \begin{pmatrix} -1 \\ -1 \end{pmatrix}.$$

At time $t = 4$, the location of the rocket is

$$\begin{pmatrix} 3-4 \\ 3-4 \end{pmatrix} = \begin{pmatrix} -1 \\ -1 \end{pmatrix},$$

the same as at the start.

b. Petrich is worried that Mavis may have made a mistake and the location of the rocket is unclear at time $t = 2$ seconds. Explain why there is no inconsistency in the location of the rocket at this time.

According to the first set of instructions, the location of the rocket at time $t = 2$ is

$$\begin{pmatrix} \cos(\pi) & -\sin(\pi) \\ \sin(\pi) & \cos(\pi) \end{pmatrix}\begin{pmatrix} -1 \\ -1 \end{pmatrix} = \begin{pmatrix} -1 & 0 \\ 0 & -1 \end{pmatrix}\begin{pmatrix} -1 \\ -1 \end{pmatrix} = \begin{pmatrix} 1 \\ 1 \end{pmatrix}.$$

According to the second set of instructions, its location at this time is

$$\begin{pmatrix} 3-2 \\ 3-2 \end{pmatrix} = \begin{pmatrix} 1 \\ 1 \end{pmatrix}.$$

These are consistent.

c. What is the area of the region enclosed by the path of the rocket from time $t = 0$ to time $t = 4$?

The path traversed is a semicircle with a radius of $\sqrt{2}$. The area enclosed is $\frac{1}{2} \times 2\pi = \pi$ squared units.

d. Mavis later decided that the moving rocket should be shifted five places farther to the right. How should she adjust her formulations above to accomplish this translation?

Notice that:

$$\begin{pmatrix} \cos\left(\frac{\pi}{2}t\right) & -\sin\left(\frac{\pi}{2}t\right) \\ \sin\left(\frac{\pi}{2}t\right) & \cos\left(\frac{\pi}{2}t\right) \end{pmatrix}\begin{pmatrix} -1 \\ -1 \end{pmatrix} = \begin{pmatrix} -\cos\left(\frac{\pi}{2}t\right) + \sin\left(\frac{\pi}{2}t\right) \\ -\sin\left(\frac{\pi}{2}t\right) - \cos\left(\frac{\pi}{2}t\right) \end{pmatrix}.$$

To translate these points 5 units to the right, use

$$\begin{pmatrix} -\cos\left(\frac{\pi}{2}t\right) + \sin\left(\frac{\pi}{2}t\right) + 5 \\ -\sin\left(\frac{\pi}{2}t\right) - \cos\left(\frac{\pi}{2}t\right) \end{pmatrix} \text{ for } 0 \le t \le 2.$$

Also, use

$$\begin{pmatrix} 3-t+5 \\ 3-t \end{pmatrix} = \begin{pmatrix} 8-t \\ 3-t \end{pmatrix} \text{ for } 2 \le t \le 4.$$